FLORA ZAMBESIACA

Flora terrarum Zambesii aquis conjunctarum

VOLUME ONE

FLORA ZAMBESIACA

MOZAMBIQUE

FEDERATION OF RHODESIA AND NYASALAND

BECHUANALAND PROTECTORATE

VOLUME ONE

Edited by

A. W. EXELL and H. WILD

on behalf of the Editorial Board

Published on behalf of the Governments of
Portugal
The Federation of Rhodesia and Nyasaland
and the United Kingdom by the
Crown Agents for Oversea Governments and Administrations,
4, Millbank, London, S.W.1
1960–61

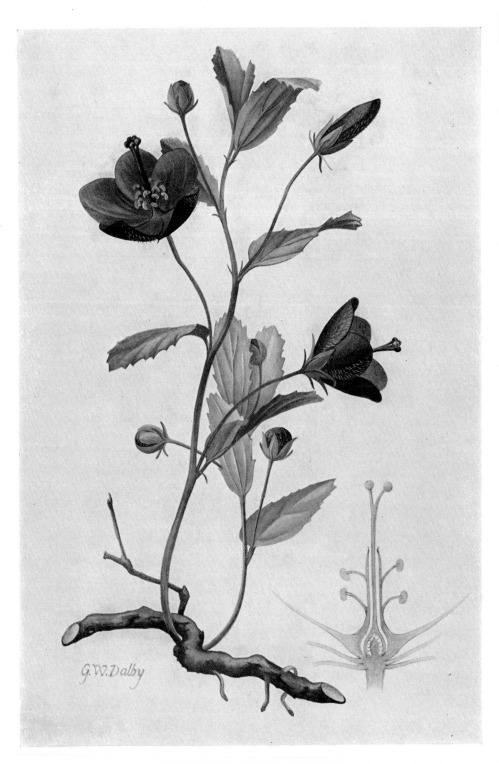

HIBISCUS RHODANTHUS

FLORA ZAMBESIACA

MOZAMBIQUE
FEDERATION OF RHODESIA AND NYASALAND
BECHUANALAND PROTECTORATE

VOLUME ONE: PART TWO

Edited by

A. W. EXELL and H. WILD

on behalf of the Editorial Board:

J. P. M. BRENAN
Royal Botanic Gardens, Kew

A. W. EXELL
British Museum (Natural History)

F. A. MENDONÇA
Junta de Investigações do Ultramar, Lisbon

H. WILD
Federal Department of Agriculture, Salisbury

Published on behalf of the Governments of
Portugal
The Federation of Rhodesia and Nyasaland
and the United Kingdom by the
Crown Agents for Oversea Governments and Administrations,
4, Millbank, London, S.W.1
April 14, 1961

© *Flora Zambesiaca Managing Committee, 1961*

Printed at the University Press Glasgow
by Robert MacLehose & Company Limited

CONTENTS

v

LIST OF FAMILIES INCLUDED IN VOL. I, PART 2

ANGIOSPERMAE

LIST OF NEW NAMES PUBLISHED IN THIS WORK

21. CARYOPHYLLACEAE

by H. Wild

(*Dianthus* by Sheila S. Hooper)

Annual or perennial herbs or shrublets. Leaves opposite, often arranged in false whorls; stipules present or absent. Inflorescence cymose, often loosely dichasial but occasionally secund or capitate, rarely flowers solitary. Flowers actinomorphic (at least in our genera), bisexual or unisexual, 5-merous or rarely 4-merous, perianth hypogynous or perigynous, often with an anthophore between the calyx and corolla. Sepals free or calyx gamosepalous, often persistent and frequently more or less scarious. Petals free, in the gamosepalous genera usually with well-differentiated lamina and claw and coronal scales often present, in the polysepalous genera less differentiated and entire to more or less deeply bifid, sometimes absent. Stamens 5 + 5 or fewer by reduction. Ovary superior, sessile or shortly stalked, 1-locular or incompletely or more rarely completely divided into 2–5 loculi; ovules 2–many with axile, central, free-central or basal placentation. Fruit capsular.

Saponaria officinalis L., the Soapwort of the British Isles, which is a native of Europe and Asia, has been recorded as a cultivated plant in Nyasaland. *Agrostemma githago* L., a native of southern and central Europe, with mauve flowers is commonly sold as a cut flower in S. Rhodesia. This is the Corncockle, an occasional cornfield weed in Britain.

Sepals free, hypogynous or on the rim of a perigynous tube :
  Leaves stipulate :
    Styles united at least below :
      Petals entire or slightly emarginate or dentate ; leaves never ovate, 1-nerved from
        the base :
        Style arms and capsule valves 5 ; sepals 9–13 - - - **1. Krauseola**
        Style arms and capsule valves 3 ; sepals 5 :
          Sepals strongly keeled, greenish with scarious margins - **2. Polycarpon**
          Sepals not keeled, entirely scarious, silvery or brown to purple
                              3. Polycarpaea
      Petals deeply bifid ; leaves ovate, 3–7 nerved from the base - **4. Drymaria**
      Styles free - - - - - - - - - **5. Spergula**
  Leaves without stipules :
    Styles usually 4–5 ; capsule ellipsoid or subcylindric - - **6. Cerastium**
    Styles usually 2–3 ; capsule subglobose or if ellipsoid then shorter than the persistent
      sepals - - - - - - - - - - **7. Stellaria**
Sepals united to form a tubular calyx :
  Styles 3 ; calyx with 10 principal veins ; calyx bracts lacking - - **8. Silene**
  Styles 2 ; calyx with numerous parallel often obscure veins, invested at the base by
    4–10 calyx bracts - - - - - - - - **9. Dianthus**

1. KRAUSEOLA Pax & K. Hoffm.

Krauseola Pax & K. Hoffm. in Engl. & Prantl, Nat. Pflanzenfam. ed. 2, **16c** : 295, 308 (1934).
    Pleiosepalum Moss in Journ. of Bot. **69** : 65 (1931) non Hand.-Mazz. (1922).

Annual or perennial herbs with a sparse indumentum of branched hairs ; stems erect or diffuse. Leaves opposite but usually arranged in false whorls, subsessile, obovate to narrowly oblanceolate, entire; stipules scarious. Flowers several–∞ in loose or compact cymes. Sepals 9–13, spirally arranged, imbricate. Petals 5–8, hyaline, much shorter than the sepals. Stamens 5–8, alternating with the petals. Ovary with numerous ovules ; style with 5 stigmatic arms. Capsule 5-valved. Seeds numerous.

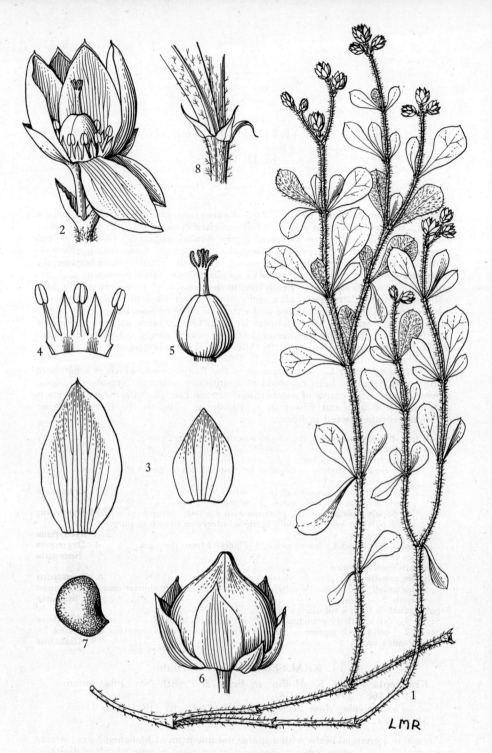

Tab. 59. KRAUSEOLA MOSAMBICINA. 1, part of plant (× ⅔) *Pedro* 50; 2, flower with some sepals bent outwards (× 6) *Mogg* 26941; 3, sepals (× 8) *Mogg* 26941; 4, arrangement of petals and stamens (× 8) *Mogg* 26941; 5, gynoecium (× 8) *Mogg* 26941; 6, capsule with persistent sepals (× 6) *Pedro* 50; 7, seed (× 16) *Pedro* 50; 8, node with stipules (× 4) *Mogg* 26941.

Krauseola mosambicina (Moss) Pax & K. Hoffm. in Engl. & Prantl, Nat. Planzenfam. ed. 2, **16c** : 308 (1934). TAB. **59**. Type : Mozambique, Marracuene, *Moss* 8026 (BM ; J, holotype).

    Pleiosepalum mosambicinum Moss in Journ. of Bot. **69** : 65, t. 596 (1931). Type as above.

Rather straggling annual or perennial herb up to c. 40 cm. tall, branching near the base, with the stems and leaves sparsely pilose, with branched hairs or with some hairs simple. Leaves 2–3 × 0·8–2·5 cm., pseudoverticillate, 4–6 in each false whorl, 2 often smaller than the rest, subsessile, oblanceolate to narrowly obovate, obtuse and mucronulate at the apex, cuneate at the base ; stipules up to 7 mm. long, scarious, deltoid. Flowers in 3–9-flowered terminal or axillary cymes ; pedicels up to 7 mm. long, sparsely pilose ; bracts c. 2 mm. long, oblong, with hyaline, ciliolate margins. Sepals 9–13, green with hyaline margins, 1·7–6·0 × 1·0–2·3 mm., oblong-ovate, acute, 5–7-nerved, the outer ones gradually smaller. Petals 5–6, 2–4 × 1 mm., membranous, lanceolate to ovate, acute. Stamens 5–6 ; filaments 2–3 mm. long, broad and membranous ; anthers c. 1 mm. long, broadly oblong. Ovary very broadly ovoid, shortly stipitate ; style c. 3 mm. long, stigmatic arms twisted. Capsule c. 7 × 6 mm., broadly ovoid, smooth, brown, surrounded by the persistent sepals. Seeds reddish-brown, 0·7 × 0·5 mm., reniform; testa minutely reticulate.

Mozambique. LM : Marracuene, fl. 6.xi.1945, *Pedro* 50 (K ; PRE).
Known only from the Lourenço Marques Province. In open scrub on coastal dunes.

2. POLYCARPON L.

Polycarpon L., Syst. Nat. ed. 10, **2** : 881 (1759).

Annual or perennial herbs. Leaves opposite, often in false whorls; stipules membranous. Flowers small in axillary or terminal cymes ; bracts membranous. Sepals 5, with a broad green keel and membranous margins. Petals up to 5, or occasionally absent, entire or emarginate, much shorter than the sepals. Stamens 3–5. Ovary 1-locular, multiovulate; style short with 3 stigmatic arms. Capsule 3-valved. Seeds many.

Polycarpon prostratum (Forsk.) Aschers. & Schweinf. apud Aschers. in Oest. Bot. Zeitschr. **39** : 128 (April, 1889) ; tom. cit. : 325 (1889).—Pax in Engl. & Prantl, Nat. Pflanzenfam. **3**, 1b : 87 (1889).—Exell & Mendonça, C.F.A. **1**, 1 : 110 (1937) ; op. cit. **1**, 2 : 368 (1951).—Milne-Redh. in Kew Bull. **1948** : 451 (1949). TAB. **60**. Type from Egypt.

    Pharnaceum depressum L., Mant. Pl. Alt. : 562 (1771). Type from India.
    Alsine prostrata Forsk., Fl. Aegypt.-Arab. : LXIV, 207 (1775). Type as for *Polycarpon prostratum*.
    Hapalosia loeflingii Wall. ex Wight & Arn., Prodr. Fl. Penins. Ind. Or. **1** : 358 (1834) *nom. illegit.*
    ? *Polycarpaea mozambica* Kunth & Bouché in Index Sem. Hort. Bot. Berol. **1848** : 15 (1848). Type : a plant cultivated in Berlin from seed collected by Peters in Mozambique (B, holotype †).
    Arversia depressa (L.) Klotzsch in Peters, Reise Mossamb. Bot. **1** : 140 (1861). Type as for *Pharnaceum depressum*.
    Polycarpon loeflingii Benth. in Benth. & Hook., Gen. Pl. **1** : 153 (1862) *nom. illegit.*— Oliv., F.T.A. **1** : 144 (1868).
    Polycarpon depressum (L.) Rohrb. in Mart., Fl. Bras. **14**, 2 : 257, t. 59 (1872) non Nutt. (1838).—Balle, F.C.B. **2** : 148 (1951).

Tufted annual herb with ascending, spreading branches up to c. 15 cm. long ; branches pubescent and sometimes densely so in the lower internodes, the upper with a line of hairs or all the internodes glabrescent or glabrous. Leaves 0·3–3 × 0·1–0·7 cm., opposite or subverticillate, often in unequal pairs, subsessile, oblanceolate, narrowly obovate, rarely elliptic or linear, obtuse or subacute at the apex, cuneate at the base, margins entire, pubescent with white hairs or glabrous ; stipules whitish, 1–2 mm. long, membranous, ovate to lanceolate. Inflorescences of terminal or axillary, many-flowered cymes ; peduncles up to c. 0·5 cm. long, with a line of hairs like the internodes or glabrous ; bracts whitish, 1–2 mm. long, membranous, lanceolate or ovate, acute ; central flower sessile, laterals on pedicels up to 6 mm. long. Sepals 5, 3–4 × 1·5 mm., lanceolate with a broad, green, some-

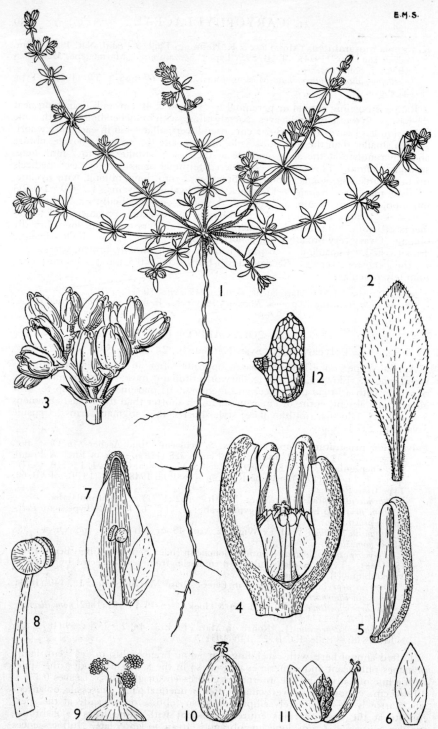

Tab. 60. POLYCARPON PROSTRATUM. 1, plant (×1) ; 2, lower surface of leaf (×4) ; 3, inflorescence (×4) ; 4, flower with one sepal removed (×14) ; 6, petal (×14) ; 7, sepal, petal and stamen (×14) ; 8, stamen (×30) ; 9, style and stigmatic branches (×30) ; 10, capsule (×14) ; 11, dehisced capsule showing seeds (×14) ; 12, seed (×80). From F.T.E.A.

what keeled median zone and whitish, hyaline margins, apex blunt and hooked. Petals 5, sometimes 2–3 or rarely absent, c. 1.5×0.75 mm., membranous and transparent, elliptic, acute. Stamens 3–5, alternating with the petals ; filaments whitish, c. 1 mm. long, flattened, tapering upwards ; anthers 0.2 mm. in diam., orbicular. Ovary ellipsoid, multiovulate ; style 0.25 mm. long with 3 spreading stigmatic arms c. 0.2 mm. long. Capsule ellipsoid, slightly shorter than the persistent sepals, with 3 membranous semi-transparent valves. Seeds many, 0.4–0.5×0.3 mm., subcylindric and slightly arcuate with a minute lateral hilum nearer one end than the other ; testa smooth, brown.

N. Rhodesia. B : Sesheke, fl., *Macaulay* 209 (K). E : Lundazi R., above dam, fl. 19.xi.1958, *Robson* 668 (BM ; K ; LISC ; SRGH). S : Mapanza, fl. 8.xi.1953, *Robinson* 363 (K). **S. Rhodesia.** N : Urungwe, Mgunje, fl. 16.xi.1953, *Wild* 4150 (BR ; K ; LISC ; S ; SRGH). W : Bulawayo, fl. ii.1923, *Borle* 363 (PRE). C : Salisbury, fl. 17.xi.1945, *Greatrex* in GHS 13972 A (SRGH). S : Fort Victoria, Umshandige Dam, fl. 10.x.1949, *Wild* 3054 (K ; SRGH). **Nyasaland.** S : Nsessi R., fl. xii.1887, *Scott* (K). **Mozambique.** T : Boroma, fl. vii.1891, *Menyhart* 1100a (K). MS : Sena, fl., *Peters* (B†). SS : Aldeia da Barragem, R. Limpopo, fl. 20.xi.1957, *Barbosa & Lemos* in *Barbosa* 8219 (K ; LMJ).

A pantropical species also occurring to a limited extent in the subtropics and commonest in the Western Hemisphere. In damp sandy places ; common, although rather inconspicuous.

The very involved synonymy and nomenclature of this species can be found more completely set out by Milne-Redhead (loc. cit.).

3. POLYCARPAEA Lam.

Polycarpaea Lam. in Journ. Hist. Nat. Par. **2** : 3, t. 25 (1792) *nom. conserv.*
Polia Lour., Fl. Cochinch. : 164 (1790).

Annual or perennial herbs, sometimes somewhat woody near the base ; stems ascending or spreading, sometimes much branched. Leaves opposite or apparently whorled, linear to oblanceolate or ovate; stipules scarious. Inflorescences of loose or compact terminal cymes ; bracts scarious. Perianth and stamens hypogynous or somewhat perigynous. Sepals 5, scarious, not keeled, silvery, brownish, purple or pinkish. Petals 5, shorter than the sepals, entire or emarginate. Stamens 5 or fewer by abortion; filaments free or slightly connate at the base, sometimes with a ring of five staminodes. Ovary 1-locular ; ovules few to ∞ ; style with 3 short stigmatic arms or capitate. Capsule 3-valved.

Ovary usually 3 (2–4)-ovulate ; a bushy usually much-branched annual ; branches
  spreading ; stipule-margins ciliate in the upper half   -   -   -   1. *eriantha*
Ovary usually 7–12 (5–13)-ovulate ; annual herb with unbranched striate stems or if
  branching then side-branches forming a narrow angle with main stem ; stipule-margins
  not ciliate -   -   -   -   -   -   -   -   -   -   -   2. *corymbosa*

1. **Polycarpaea eriantha** Hochst. ex A. Rich., Tent. Fl. Abyss. **1** : 303 (1848).—Exell & Mendonça, C.F.A. **1**, 1 : 111 (1937).—Garcia in Bol. Soc. Brot., Sér. 2, **20** : 34 (1946). —Balle in F.C.B. **2** : 144 (1951).—Milne-Redh. in Mem. N.Y.Bot. Gard. **8**, 3 : 221 (1953).—Turrill, F.T.E.A. Caryophyll. : 7 (1956). Type from Ethiopia.
  Polycarpaea corymbosa var. *effusa* Oliv., F.T.A. **1** : 145 (1868).—R.E.Fr., Wiss. Ergebn. Schwed. Rhod.-Kongo-Exped. **1** : 38 (1914). Syntypes from Nigeria, Uganda, Angola and Mozambique : Sena, *Kirk* (K).

Bushy annual herb up to 19 cm. tall, branching from the base ; internodes with a rather woolly indumentum of whitish hairs. Leaves opposite or whorled, sessile ; basal leaves 1–1.5×0.3–0.4 cm. (absent from older plants), spathulate; cauline leaves 5–18×0.5–1 mm., linear, acute and terminating in a hair-like, somewhat caducous appendage c. 1 mm. long, with a woolly whitish indumentum when young, glabrescent later, 1-nerved ; stipules whitish, 3–4 mm. long, scarious, lanceolate-acuminate, with one or more long hairs at the apex, margin ciliate in the upper half. Inflorescence of many-flowered, loose or somewhat compact, terminal, dichotomous cymes with a single pedicellate flower between the main branches ; bracts whitish, c. 3 mm. long, scarious, lanceolate-acuminate, bifid, margins ciliate, usually hirsute on the back ; pedicels 1.5–9 mm. long. Sepals 1.5–2×0.7 mm., scarious, silvery pink or whitish, lanceolate-acuminate, with somewhat curly hairs outside or glabrous. Petals c. 0.6 mm. long, elliptic, apex emarginate. Stamens 5 or more often reduced to 2, c. $\frac{1}{2}$ the length of the petals.

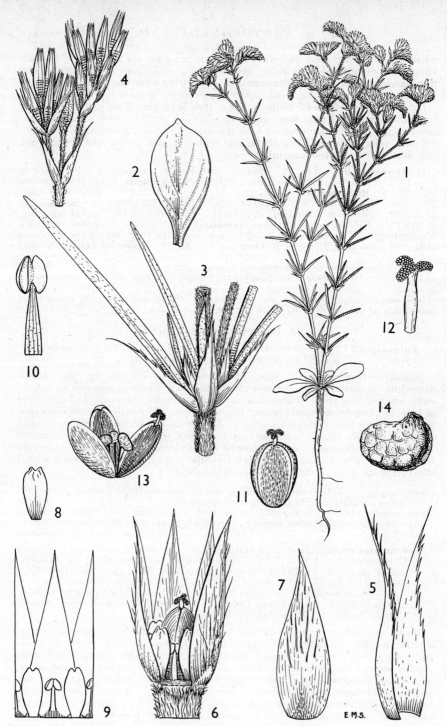

Tab. 61. POLYCARPAEA ERIANTHA VAR. ERIANTHA   1, plant (× ⅔); 2, upper surface of basal
leaf (× 3); 3, node showing leaves and stipules (× 8); 4, part of inflorescence (× 5); 5,
bract (× 20); 6, flower with one sepal removed (× 20); 7, sepal (× 20); 8, petal (× 20);
9, part of flower from inside showing arrangement of petals and stamens (diagram-
matic) (× 20); 10, stamen (× 60); 11, gynoecium (× 20); 12, style and stigmatic
lobes (× 120); 13, dehisced capsule (× 20); 14, seed (× 60).   From F.T.E.A.

Ovary ellipsoid, 3-ovulate, with a very short style c. 0·15 mm. long and 3 spreading stigmatic arms. Capsule c. 1 × 0·5 mm., ellipsoid ; valves brownish with paler margins. Seeds brown, 0·4 × 0·3 mm., subreniform ; testa lightly tessellated.

Var. **eriantha** TAB. **61**.
    Polycarpaea corymbosa var. *effusa* Oliv., F.T.A. **1** : 145 (1868) pro parte quoad specim. *Speke & Grant* et *Kirk*.

Sepals hairy.

N. Rhodesia. N : Abercorn, Lake Chila, fl. 15.iii.1955, *Richards* 4959 (K). S : Mapanza, fl. 11.v.1953, *Robinson* 216 (K). **S. Rhodesia.** N : Darwendale, fl. 20.iv.1948, *Rodin* 4322 (K ; PRE ; SRGH). C : Marandellas, Ziyambe, fl. iii.1955, *Davies* 956 (SRGH). **Nyasaland.** N : Nyika, Nymkowa, fl. ii.1903, *McClounie* 168 (K). C : Kasungu, fl. 26.viii.1946, *Brass* 17424 (K ; SRGH). S : Zomba, fl. 23.iv.1955, *Banda* 88 (K ; SRGH). **Mozambique.** N : Mutuali, fl. 28.ii.1953, *Gomes e Sousa* 4042 (COI ; K ; PRE). Z : between Naburi and Muliguge, fl. 6.x.1949, *Barbosa & Lemos* in *Barbosa* 4321 (K ; LMJ). T : Boroma, fl. 9.v.1891, *Menyharth* (K). MS : Sena, fl. iv.1860, *Kirk* (K).
    Widely distributed in tropical Africa. In open woodland and grassland, common on sandy soils and as a weed in cultivated ground and fallows.

Var. **effusa** (Oliv.) Turrill in Kew Bull. **1954** : 503 (1954). Lectotype from Nigeria.
    Polycarpaea corymbosa var. *effusa* Oliv., F.T.A. **1** : 145 (1868) pro parte quoad specim. Barter. et Welwitsch., emend. Pax in Engl., Bot. Jahrb. **17** : 590 (1893).— Garcia in Bol. Soc. Brot., Sér. 2, **20** : 33 (1946). Lectotype as above.
    Polycarpaea rhodesica Suesseng. in Proc. & Trans. Rhod. Sci. Ass. **43** : 83 (1951). Type : S. Rhodesia, Marandellas, *Dehn* 766 (M, holotype ; SRGH).

Sepals glabrous.

N. Rhodesia. B : Namushakende, 24 km. S. of Mongu, fl. 25.vi.1955, fl. *King* 3 (K). N : Abercorn, Inono valley, fl. 13.iv.1955, *Richards* 5418 (K). W : Ndola, fl. 21.vii.1954, *Fanshawe* 1387 (K ; SRGH). C : Kapiri Mposhi, fl. 22.i.1955, *Fanshawe* 1826 (K). **S. Rhodesia.** W : Wankie, fl. iv.1932, *Levy* 37 (PRE). C : Salisbury, fl. 15.iv.1947, *Greatrex* in GHS (K ; SRGH). E : Umtali, Dora R., fl. 12.vi.1948, *Fisher* 1587 (K ; SRGH). **Mozambique.** N : Mecaloja, Boronango, fl. 13.ix.1934, *Gomes e Sousa* 545 (COI ; LISC).
    Much the same range as var. *eriantha*. Ecology as for var. *eriantha*.

2. **Polycarpaea corymbosa** (L.) Lam., Tabl. Encycl. Méth., Bot. **2** : 129 (1797).—Sond. in Harv. & Sond., F.C. **1** : 133 (1860).—Oliv., F.T.A. **1** : 145 (1868) pro parte.— Bak. f. in Journ. Linn. Soc., Bot. **40** : 25 (1911).—R.E.Fr., Wiss. Ergebn. Schwed. Rhod.-Kongo-Exped. **1** : 38 (1914).—Eyles in Trans. Roy. Soc. S. Afr. **5** : 351 (1916).—Burtt Davy, F.P.F.T. **1** : 152 (1926).—Weim. in Bot. Notis. **1934** : 91 (1934).—Exell & Mendonça, C.F.A. **1**, 1 : 110 (1937).—Garcia in Bol. Soc. Brot., Sér. 2, **20**: 33 (1946).—Balle, F.C.B. **2** : 145 (1951).—Turrill, F.T.E.A. Caryophyll.: 8 (1956). Type from Ceylon.
    Achyranthes corymbosa L., Sp. Pl. **1** : 205 (1753). Type as above.
    Polycarpaea glabrifolia sensu Klotzsch in Peters, Reise Mossamb. Bot. **1** : 139 (1861).
    Polycarpaea corymbosa var. *parviflora* Oliv., F.T.A. **1** : 145 (1868).—Garcia in Bol. Soc. Brot., Sér. 2, **20** : 33 (1946). Type from Angola.

Annual erect herb up to 40 cm. tall, usually with strict stems, often unbranched, if branched, branches not very spreading, internodes with white curly hairs, glabrescent with age. Leaves 5–35 × 0·5–1 mm., opposite or more often subverticillate, sessile, linear, apex acute and with a terminal somewhat caducous hair-like bristle 1 mm. long, margins usually revolute, whitish woolly-hairy when young but soon glabrescent, 1-nerved ; stipules 2–9 mm. long, membranous, whitish or brownish, narrowly lanceolate, terminated by a single hair or the lower ones with 2–several hairs. Inflorescence of many-flowered, dichotomous, loose or compact, terminal cymes, often with a single pedicellate flower between the main branches ; bracts 2–6 mm. long, similar to the stipules but usually bifid, glabrous or somewhat hairy. Sepals silvery white, pink, purplish or a rich brown, 1·5–3·5 mm. long, membranous, lanceolate-acuminate, glabrous. Petals pinkish, 0·6–1·25 mm. long, ovate, rounded, slightly emarginate or erose at the apex. Stamens usually 5, 0·5–0·75 mm. long. Ovary ovoid or ellipsoid, 5–13-ovulate ; style 0·15–0·25 mm.

long, with 3 stigmatic arms. Capsule c. $1 \cdot 5 \times 1$ mm., ellipsoid ; valves shining and horny, brown with thin paler margins. Seeds pale brown, c. $0 \cdot 45 \times 0 \cdot 27$ mm., reniform.

Caprivi Strip. Fl. 1946, *Kanget* (PRE). **N. Rhodesia.** B : Senanga, fl. 4.viii.1952, *Codd* 7395 (K ; PRE). N : Abercorn, Lake Tanganyika, fl. 18.iii.1955, *Richards* 5011 (K). W : Luanshya, fl. 29.iii.1957, *Fanshawe* 3125 (K). C : 57 km. W. of Lusaka, fl. 5.v.1957, *Noak* 225 (K ; SRGH). S : 27 km. NE of Choma, fl. 11.vii.1930, *Hutchinson & Gillett* 3538 (K). **S. Rhodesia.** N : Trelawney, fl. 4.v.1943, *Jack* 204 (K; PRE ; SRGH). W : Nyamandhlovu, fl. 13.iv.1953, *Plowes* 1595 (K ; SRGH). C : Marandellas, fl. 20.iv.1948, *Corby* 103 (K ; SRGH). E : Odzi R., fl. 18.iv.1948, *Chase* 668 (K ; SRGH). S : Fort Victoria, fl. 17.iv.1946, *Greatrex* in GHS 14775 (SRGH). **Nyasaland.** N : Mzimba, Mbawa, fl. 19.vi.1952, *Jackson* 848 (K). C : Lilongwe, fl. 16.iv.1956, *Banda* 254 (BM). S : Zomba, Makwapala Exp. Sta., fl. 3.vii.1937, *Lawrence* 436 (K). **Mozambique.** N : Ribaué, fl. 12.ix.1931, *Gomes e Sousa* 730 (K ; SRGH). Z : Quelimane, Namagoa, fl. vii–ix, *Faulkner* 234 (COI ; K ; SRGH). T : between Lupata and Tete, fl. ii.1859, *Kirk* (K). MS : Chimoio, fl. 3.iii.1948, *Barbosa* 1075 (BM ; LISC; SRGH). SS : Inhambane, Cumbana, fl. 6.vii.1938, *Gomes e Sousa* 2158 (K). LM : Lourenço Marques, fl. iii.1893, *Quintas* 1 (COI ; K).

Widely distributed in the tropics of the Old and the New World. Ecology similar to that of *P. eriantha*. It also occurs as a weed of cultivation in waste places and fallows.

This species varies considerably in size, amount of branching, in the size and number of the flowers and in the compactness of the inflorescences.

4. DRYMARIA Willd.

Drymaria Willd. apud Schult. in L., Syst. Veg., ed. nov. **5** : XXXI, 406 (1819).

Dichotomously branching herbs. Leaves flat, opposite (in our species petiolate); stipules small, often caducous. Flowers small, solitary in the leaf-axils or in usually rather loose cymes at the ends of the branches. Sepals (4) 5, sometimes with a scarious margin. Petals (4) 5, deeply bifid, the lobes sometimes themselves lobed. Stamens 5 or fewer by abortion, $\pm$ perigynous, filaments shortly connate at the base. Ovary 1-locular, 2–∞-ovulate ; style with 3 stigmatic arms. Capsule 3-valved.

Drymaria cordata (L.) Willd. apud Schult. in L., Syst. Veg. ed. nov. **5** : 406 (1819).— Sond. in Harv. & Sond., F.C. **1** : 136 (1860).—Oliv., F.T.A. **1** : 143 (1868).—Bak. f. in Journ. Linn. Soc., Bot. **40** : 25 (1911).—Pax & K. Hoffm. in Engl. & Prantl, Nat. Pflanzenfam. ed. 2, **16c** : 307, fig. 29 (1934).—Exell & Mendonça, C.F.A. **1**, 1 : 109 (1937).—Balle, F.C.B. **2** : 139 (1951).—Milne-Redh. in Mem. N.Y.Bot. Gard. **8**, 3 : 221 (1953).—Turrill, F.T.E.A. Caryophyll.: 9, t. 4 (1956).—Mizushima in Journ. Jap. Bot. **32** : 69 (1957). TAB. **62** fig. A. Syntypes from Jamaica and Surinam. *Holosteum cordatum* L., Sp. Pl. **1** : 88 (1753). Syntypes as above.

Tender herb with stems straggling, procumbent or ascending, branching dichotomously, often rooting at the lower nodes, quadrangular, glabrous or papillose (especially above) ; leaves rather widely spaced. Leaves on glabrous petioles up to 1 cm. long ; lamina $1–3 \cdot 5 \times 0 \cdot 6–3 \cdot 0$ cm., ovate to very broadly ovate, apex acute or subacute and shortly apiculate or occasionally rounded or obtuse, cordate, truncate or abruptly cuneate at the base, 3–7-nerved at the base, glabrous ; stipules c. 1 mm. long, interpetiolar, deeply fringed or composed of several subulate segments. Flowers in axillary or terminal, rather loose dichotomous cymes 2–25 cm. long, peduncles slender and often elongate ; bracts 1–4 mm. long, narrowly lancolate, membranous ; pedicels $1 \cdot 5–5$ (12) mm. long, filiform, glabrous or papillose. Sepals green with whitish margins, narrowly lanceolate, acute, keel viscid-papillose, with an additional nerve on each side of the midrib. Petals white, deeply bilobed, somewhat shorter than the sepals, sometimes absent. Stamens 5 or reduced to 3 ; filaments c. 2 mm. long ; anthers c. $0 \cdot 5 \times 0 \cdot 4$ mm. Ovary ovoid-ellipsoid, usually 3-ovulate ; style c. $0 \cdot 25$ mm. long with 3 spreading stigmatic arms. Capsule c. $2 \times 1 \cdot 5$ mm. ellipsoid. Seeds 1–3, c. $1 \times 0 \cdot 75$ mm., somewhat flattened-reniform ; testa brown, minutely and bluntly tubercled.

N. Rhodesia. N : Abercorn, fl. 10.iii.1955, *Richards* 4872 (K). **S. Rhodesia.** E : Himalayas, Banti N., fl. 4.iii.1954, *Wild* 4518 (K ; LISC; SRGH). **Nyasaland.** N : Misuku Hills, fl. vii.1896, *Whyte* (K). S : Blantyre, fl. 1891, *Buchanan* 591 (BM)

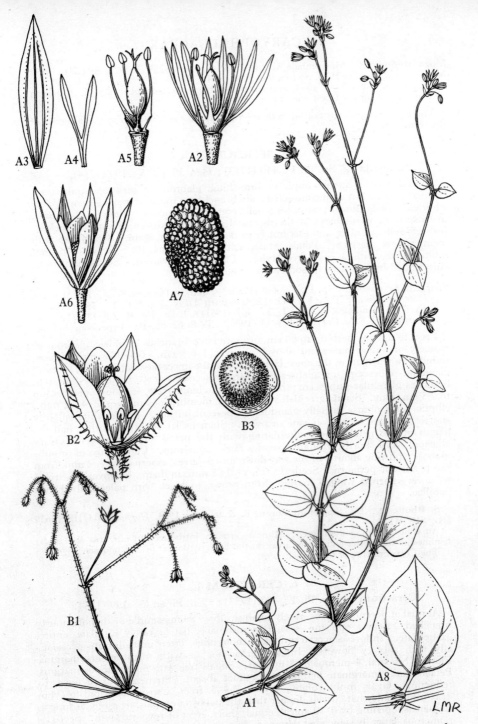

Tab. 62. A.—DRYMARIA CORDATA. A1, part of plant (×⅔) *Chase* 6049 ; A2, flower, sepals
and a petal removed (×8) *Chase* 6049 ; A3, sepal (×8) *Chase* 6049 ; A4, petal (×8)
Chase 6049 ; A5, gynoecium and stamens (×8) *Chase* 6049 ; A6, fruit with persistent
sepals (×6) *Swynnerton* 2046 ; A7, seed (×16) *Swynnerton* 2046 ; A8, leaf and node
(×2) *Chase* 6049. B.—SPERGULA ARVENSIS. B1, inflorescence with leaves (×⅔) ; B2,
flower, one sepal and two petals removed (×6) ; B3, seed (×16), all from *Verdcourt*
325.

Mozambique. Z: Morrumbala Mt., fl. 1.v.1943, *Torre* 5258 (LISC; SRGH). MS: Mt. Maruma, fl. 13.ix.1906, *Swynnerton* 2046 (BM; K).

A pantropical species. Usually in shady places in our areas of higher rainfall. Common on forest floors and at forest margins.

The glandular pedicels fall off with the ripe capsules and so help in the dispersal of the seed.

5. SPERGULA L.

Spergula L., Sp. Pl. **1**: 440 (1753); Gen. Pl. ed. 5: 199 (1754).

Annual herbs; stems simple or branching, glabrous or glandular-pubescent. Leaves opposite but with contracted leafy branches in the axils making them appear whorled, sessile, linear; stipules small, scarious, caducous. Flowers 5-merous, in loose terminal dichasia or the dichasial structure reduced in some branches; pedicels reflexed after anthesis but later becoming erect again. Sepals free with membranous margins. Petals white, entire. Stamens 10. Ovary 1-locular, multiovulate, with 3 or 5 free styles. Capsule 3–5-valved. Seeds numerous, surrounded by a narrow or somewhat winged margin.

Spergula arvensis L., Sp. Pl. **1**: 440 (1753).—Sond. in Harv. & Sond., F.C. **1**: 135 (1860).—Oliv., F.T.A. **1**: 143 (1868).—Burtt Davy, F.P.F.T. **1**: 151 (1926).—Exell & Mendonça, C.F.A. **1**, 2: 367 (1951).—Balle, F.C.B. **2**: 138 (1951).—Turrill, F.T.E.A. Caryophyll.: 11 (1956). TAB. **62** fig. B. Type from Europe.

Erect branching herb up to 60 cm. or more tall; branches cylindric, glabrous or sparsely glandular-pubescent above. Leaves 1–6·5 cm. ×0·5–0·75 mm., linear, rounded or subacute at the apex, channelled on the lower (abaxial) surface, sparsely glandular-pubescent or glabrescent. Inflorescences few- to many-flowered, branches glandular-pubescent; bracts c. 1 mm. long, ovate, scarious; pedicels up to 2·5 cm. long. Sepals greenish with a white membranous margin, c. 4 ×2 mm., elliptic, apex obtuse, usually glandular-pubescent outside. Petals white, slightly shorter than the sepals, obovate or ovate. Stamens 10, slightly less than half the length of the petals, those alternating with the petals slightly longer. Ovary ellipsoid, very shortly stipitate; ovules fairly numerous. Capsule ovoid or sub-globose, splitting about ⅔ of the way down into 5 valves, exserted for c. 2 mm. from the persistent perianth. Seeds 15–25, c. 0·85 mm. in diam., biconvex, encircled by a very narrow winged margin, from brown to black with paler club-shaped papillae.

S. Rhodesia. E: Stapleford Forest, fl. & fr. 2.viii.1944, *Hopkins* in GHS 12639 (K; SRGH).

An almost cosmopolitan weed. Almost certainly introduced in our area and so far confined to the wetter and cooler areas on the Eastern Border of S. Rhodesia. The Corn Spurrey of the British Isles.

6. CERASTIUM L.

Cerastium L., Sp. Pl. **1**: 437 (1753); Gen. Pl. ed. 5: 199 (1754).

Annual or perennial herbs, rarely small shrubs. Stems simple or dichotomously branched. Leaves opposite, sessile or subsessile, flat or rarely subulate, entire, exstipulate. Inflorescences terminal, dichotomous, often umbelliform, occasion-ally with solitary flowers; bracts herbaceous or with membranous margins. Flowers bisexual, 4-merous or 5-merous. Sepals free with membranous margins. Petals white, emarginate or bifid, sometimes absent. Stamens 5 + 5 or fewer by reduction. Ovary multiovulate; styles (3) 4–5, free. Capsule ellipsoid to cylin-dric, often somewhat curved, longer than the persistent sepals, or rarely as long as the sepals, opening by 6–10 apical, equal, short, erect or revolute teeth. Seeds ± numerous, laterally somewhat flattened, discoid or reniform, granular or verrucose.

Capsule equalling or only slightly exceeding the persistent sepals; petals always present; stamens 10 - - - - - - - - - - 1. *indicum*
Capsule almost twice the length of the persistent sepals; petals usually absent in the lower flowers; stamens usually 5 - - - - - - - 2. *glomeratum*

1. **Cerastium indicum** Wight & Arn., Prodr. Fl. Penins. Ind. Or. **1**: 43 (1834).—

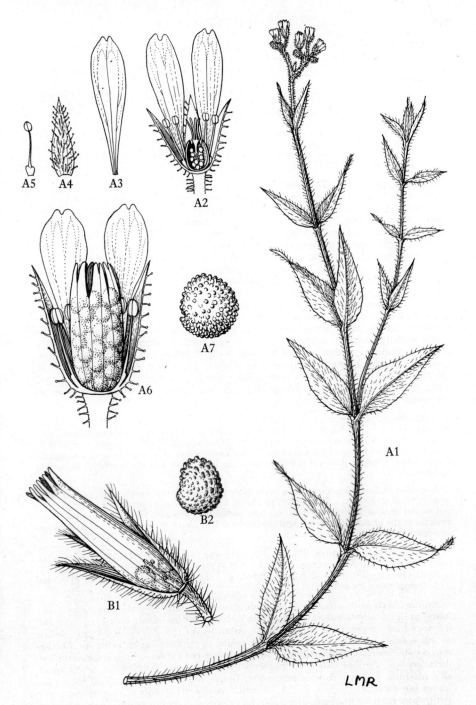

Tab. 63. A.—CERASTIUM INDICUM. A1, part of plant (× ⅔) *Brass* 16292 ; A2, longitudinal
section of flower (×4) *Brass* 16292 ; A3, petal (×4) *Brass* 16292 ; A4, sepal (×4)
Brass 16292 ; A5, stamen (×4) *Brass* 16292 ; A6, capsule and section of persistent
perianth (×6) *Jackson* 134 ; A7, seed (×16) *Jackson* 134. B.—CERASTIUM GLOMERATUM.
B1, capsule and persistent calyx, one sepal removed (×6) ; B2, seed (×20), all from
Boughey 119.

Möschl in Mem. Soc. Brot. **7**: 53 (1951).—Turrill, F.T.E.A. Caryophyll.: 19 (1956). TAB. **63** fig. A. Type from India (Nilgiri Hills).
  Arenaria africana Hook. f. in Journ. Linn. Soc., Bot. **7**: 184 (1864). Type from Cameroons.
  Cerastium africanum (Hook. f.) Oliv., F.T.A. **1**: 141 (1868).—Weim. in Svensk Bot. Tidskr. **27**: 415 (1933); in Bot. Notis. **1934**: 90 (1934).—Balle, F.C.B. **2**: 135 (1951).—Milne-Redh. in Mem. N.Y.Bot. Gard. **8**, 3: 221 (1953). Type as above.
  Cerastium africanum var. *ruwenzoriense* F. N. Williams in Journ. of Bot. **36**: 342 (1898). Type from Uganda.
  Cerastium indicum var. *ruwenzoriense* (F. N. Williams) Möschl, tom. cit.: 55 (1951).—Turrill, tom. cit.: 21 (1956). Type as above.

Perennial, pilose and glandular herb with spreading and ascending branches up to 60 (75) cm. long. Leaves opposite, sessile or the lower ones with a pilose petiole c. 1 mm. long; lamina 1–6·5 × 0·3–2·3 cm., narrowly lanceolate, lanceolate or oblong-elliptic, apex acute and mucronate to shortly acuminate, cuneate or rounded at the base with somewhat sparse silky hairs on both sides. Flowers 3–13 together in terminal dichotomous inflorescences which are usually rather contracted; lower bracts usually foliaceous, the remainder linear-lanceolate to linear and progressively reduced to c. 1 mm. long, pilose and glandular; pedicels up to 2 cm. long but usually less, densely glandular and pilose. Perianth 5-merous. Sepals green, 3–6 mm. long, lanceolate, apex acute or acuminate, glandular outside except towards the apex, inner sepals with whitish membranous margins. Petals white, 5–8 mm. long, narrowly obovate, emarginate at the apex. Stamens 10 with white filaments slightly wider at the base and up to 3 mm. long, those alternating with the petals slightly longer; anthers yellow, c. 0·3 mm. in diam., globular. Ovary ellipsoid, glabrous; styles 3–5, 0·8–1·5 mm. long, papillose towards the base within. Capsule straight, ellipsoid, equalling or slightly exceeding the persistent sepals, splitting at the apex into 6–10 blunt teeth. Seeds 10–20, rich brown, c. 1·5 × 1·2 mm., slightly compressed, oblong-ellipsoid, verrucose or bluntly papillose in concentric rings.

S. Rhodesia. E: Inyanga, fl. 30.i.1931, *Norlindh & Weimarck* 4753 (BM; K; LD). **Nyasaland.** S: Zomba, fl. 7.vi.1946, *Brass* 16292 (K; SRGH). **Mozambique.** Z: Guruè, Marrequélo Munhano, fl. 18.iv.1944, *Mendonça* 2099 (BM; LISC; SRGH).
  In Eastern Africa from Ethiopia to the Cape Province and also in the Belgian Congo and Cameroons. Outside Africa it is known from the Mascarene Islands, southern India and Celebes.

A species of forest margins, grasslands and roadsides in the wetter parts of our area, usually from 2–3000 m. It has been recorded as a weed of cultivation in East Africa but not in our area so far. The ripe fruits and pedicels of this species fall off together and, because of the glands on pedicels and persistent calyx, attach themselves to animals and so provide a means of seed dispersal.

2. **Cerastium glomeratum** Thuill., Fl. Env. Par. ed. 2: 226 (1799).—Aschers. & Graebn., Syn. Mitteleurop. Fl. **5**, 1: 674 (1918).—Möschl in Mem. Soc. Brot. 7: 85 (1951).—Turrill, F.T.E.A. Caryophyll.: 22 (1956). TAB. **63** fig. B. Type from France (near Paris).

Annual herb with straggling, ascending stems up to 45 cm. long, glandular-hairy at least above. Leaves opposite; basal leaves oblanceolate to obovate, narrowed into a petiole c. 4 mm. long; stem leaves 0·5–2·5 × 0·3–0·8 cm., sessile or subsessile, broadly ovate to elliptic-ovate or elliptic, apex acute or obtuse, mucronate, base cuneate, with long silky hairs on both sides. Flowers in terminal, dichotomous, cymose clusters; pedicels glandular-hairy, very short or occasionally reaching 5 mm.; bracts lanceolate, herbaceous, hairy, similar to the upper leaves but smaller. Sepals 5, c. 5 × 1 mm., narrowly lanceolate, very acute, green with white scarious margins, glandular-hairy outside. Petals white, often absent on the lower flowers or rarely all absent, ± equalling or somewhat shorter than the sepals, bifid c. ¼ of the way down, claw somewhat ciliate. Stamens 10 or often only 5, c. 2 mm. long. Ovary ovoid, glabrous; styles 5, c. 0·5 mm. long, free. Capsule up to twice the length of the calyx, subcylindric, shortly stalked, slightly curved, opening by 10 blunt, apical teeth. Seeds numerous, 0·5–0·75 mm. in diam., pale brown, with concentric rings of minute tubercles.

S. Rhodesia. C: Salisbury Distr., Umwindsi R., fl. & fr., 4.ix.1955, *Boughey* 119 (K; SRGH). E: Manchester Park, Vumba, fl. and fr. 27.viii.1958, *Whellan* 1560 (SRGH).

A cosmopolitan temperate, subtropical and tropical weed. Introduced in our area. The Sticky Mouse-eared Chickweed of the British Isles.

7. STELLARIA L.

Stellaria L., Sp. Pl. **1** : 421 (1753) ; Gen. Pl. ed 5 : 193 (1754).

Annual or perennial herbs usually with slender diffuse stems. Leaves opposite, sessile or petiolate, simple, entire, flat or very rarely subulate, exstipulate. Inflorescences of terminal dichasial cymes, rarely flowers solitary. Sepals 4–5, free. Petals 4–5, white, deeply bilobed, or occasionally absent. Stamens 10 (5 + 5) or fewer by reduction. Ovary 1-locular; styles (2) 3, free. Capsule usually more or less rounded, opening at the apex by 3 or 6 (rarely 2 or 4) valves. Seeds numerous or occasionally 1–4, roundish-reniform.

Sepals 5 ; styles 3 :
 Internodes with glandular hairs all round or the lower sometimes glabrescent (indigenous species) - - - - - - - - - - - 1. *mannii*
 Internodes with a longitudinal line of hairs (introduced weeds) :
  Sepals 4·5–5 mm. long ; petals usually present ; stamens 3–7 - 2. *media*
  Sepals 2–3·5 mm. long ; petals absent or vestigial ; stamens 1–3 (5) - 3. *pallida*
Sepals 4 ; styles 2 - - - - - - - - - - - 4. *sennii*

1. **Stellaria mannii** Hook. f. in Journ. Linn. Soc., Bot. **7** : 183 (1864).—Oliv., F.T.A. **1** : 141 (1868).—Milne-Redh. in Mem. N.Y. Bot. Gard. **8**, 3 : 221 (1953).—Turrill, F.T.E.A. Caryophyll. : 24 (1956). Type from the Cameroons.

Weak, procumbent or ascending herb, sometimes rooting at the nodes ; stems with scattered hairs in the upper internodes, often glabrous in the lower internodes or with hairs retained in the stem furrows. Leaves all petiolate ; lamina up to 4·5 × 3 cm., ovate, apex acute and apiculate or shortly acuminate, rounded or slightly cuneate at the base, with scattered and often glandular hairs on both sides, midrib conspicuous, lateral nerves obscure but anastomosing close to the margins and producing a submarginal nerve, petiole up to c. 2 (3) cm. long, slightly amplexicaul and connate at the base, glandular-hirsute. Inflorescences lax, terminal, densely glandular-pubescent ; bracts c. 3·5 × 1 mm., lanceolate, glandular-pubescent ; pedicels short at first but elongating to c. 8 mm. long. Sepals 5, 4·5–6 × 1·5–2 mm., lanceolate to ovate-lanceolate, acute or shortly acuminate, glandular outside. Petals 5, white, somewhat rhombic, bifid c. ⅓ of the way down, from slightly shorter to 1½ times the length of the sepals. Stamens usually 10 but sometimes fewer by reduction, filaments c. 3 mm. long, swollen at the base. Ovary ovoid, c. 3-ovulate ; styles 3, c. 1·5 mm. long, slender. Capsule globose, slightly shorter than the persistent sepals, opening at the apex by 6 valves. Seeds (often only one developing to maturity) brown, c. 3 mm. in diam., discoid ; testa lightly striate.

S. Rhodesia. E: Melsetter, Heathfield, fl. 27.iv.1947, *Wild* 1960 (K; SRGH). **Nyasaland.** S: Cholo Mt., fl. & fr. 20.ix.1946, *Brass* 17662 (K; SRGH). **Mozambique.** Z: Massingire, fl. 6.viii.1942, *Torre* 4523 (LISC; SRGH). MS: Mavita, fl. 16.iv.1949, *Pedro & Pedrógão* 6628 (LMJ; SRGH).

Also in Ethiopia, Uganda, Kenya, Tanganyika, Cameroons, S. Tomé, Fernando Po, Belgian Congo and Madagascar. Evergreen forest floors.

2. **Stellaria media** (L.) Vill., Hist. Pl. Dauph. **3** : 615 (1789).—Sond. in Harv. & Sond., F.C. **1** : 130 (1860).—Oliv., F.T.A. **1** : 141 (1868).—Burtt Davy, F.P.F.T. **1** : 151 (1926).—Exell & Mendonça, C.F.A. **1**, 1 : 109 (1937).—Clapham in Clapham, Tutin & Warburg, Fl. Brit. Is. : 306 (1952).—Turrill, F.T.E.A. Caryophyll. : 24 (1956). Type from Europe.
 Alsine media L., Sp. Pl. **1** : 272 (1753). Type as above.

Annual herb with diffuse leafy stems ; stems with a single line of hairs down each internode. Lower leaves on petioles of c. 1·5 cm. ; lamina c. 1 × 1 cm., ovate to very broadly ovate, apex mucronate, base broadly cuneate, upper leaves usually somewhat larger, ± sessile, ovate or broadly elliptic, apex acute or shortly acuminate, glabrous or ± ciliate near the base. Flowers in terminal, often leafy dichasia ; pedicels c. 1 cm. long, slender, usually with a line of hairs. Sepals 5, 3·5–5 mm. long, ovate-lanceolate, acute or obtuse with a narrow membranous margin, usually glandular-hairy outside. Petals white, somewhat shorter than the sepals, bifid to near the base, sometimes absent. Stamens 3–10, c. ½ the length of the sepals, with red-violet anthers. Ovary globose ; styles 3, 1 mm. long, slender, spreading. Capsule ovoid, opening from the apex by 6 teeth or valves, exceeding the persistent sepals in length only very slightly, pedicels usually turning downwards in fruit. Seeds fairly numerous, reddish-brown, 0·9–1·3 mm. in diam., somewhat discoid, minutely tuberculate.

S. Rhodesia. N : Nyamanetsi, Horseshoe Block, N. Umvukwes Range, fl. iii.1960, *Drummond* 6824 (K ; LISC ; PRE ; SRGH). W : Bulawayo, fl. & fr. vii.1955, *Miller* 2928 (SRGH). C : Salisbury, fl. & fr. ii.1919, *Eyles* 1513 (BM ; PRE ; SRGH). E : Inyanga, fr. & fr. 16.vi.1957, *Goodier & Phipps* 112 (K ; SRGH).
A cosmopolitan weed.

Rather rare as an introduction in our area and probably a weed of gardens rather than agricultural land so far. The Chickweed of Britain.

3. **Stellaria pallida** (Dumort.) Piré in Bull. Soc. Bot. Belg. **2** : 49 (1863). Type from Belgium.
   Alsine pallida Dumort., Florul. Belg. : 109 (1827). Type as above.

Annual much-branched prostrate herb very similar to *S. media* but the leaves are usually all petiolate and less than 7 mm. long, the sepals are 2–3·5 mm. long, the petals are absent or vestigial, the stamens are 1–3 and the seeds 0·6–1·0 mm. in diam.

N. Rhodesia. N : Abercorn, Chilongowelo Gardens, fl. & fr. 20.iv.1952, *Richards* 1496 (K).
Indigenous in Gt. Britain, Channel Is. and Eastern, Central and Southern Europe, where it usually grows on sandy soils. An introduced weed in our area.
The Lesser Chickweed of Britain.

4. **Stellaria sennii** Chiov. in Atti Reale Accad. Ital. Mem. **11**, 2 : 20 (1940).—Turrill, F.T.E.A. Caryophyll. : 26, t. 11 (1956). TAB. **64**. Type from Ethiopia.

Tender herb with procumbent or weakly ascending stems up to c. 40 cm. long ; internodes glabrous or with 1–2 lines of hairs. Leaves all petiolate ; lamina 0·3–1·8 × 0·3–1·4 cm., ovate to broadly ovate, apex acute, shortly apiculate, base ± cordate, glabrous or with a few scattered hairs, midrib and main veins fairly conspicuous, the latter anastomosing near the margin and producing a submarginal nerve, petiole up to 1·7 cm. long, pilose near the base, glabrescent above. Flowers solitary and axillary towards the ends of the branches ; pedicels up to 1·8 cm. long, slender, with a few sparse hairs or glabrous. Sepals 4, 3–4 mm. long, narrowly lanceolate, apex acute or shortly acuminate, with a few long hairs outside and at the margins, margins scarious. Petals absent or represented by vestigial teeth between the stamens. Stamens 4, filaments swollen at the base, c. 1 mm. long. Ovary ellipsoid, (2) 4-ovulate ; styles 2, 0·75 mm. long, slender. Capsule ellipsoid, somewhat shorter than the persistent calyx, opening at the apex by 4 valves. Seeds 1–4, dark brown, c. 1·75 × 1·25 mm., oblong-ellipsoid, somewhat flattened ; testa verrucose-papillose.

Mozambique. ?N or Z : Mts. E. of Lake Nyasa, fl., *Johnson* (K).
Also in Ethiopia, Uganda, Kenya, Tanganyika and the Belgian Congo. A species of forests and forest margins at altitudes of 1500–3300 m.

Johnson's specimen may possibly be from mountains E. of Lake Nyasa in Tanganyika as he collected in this area also, but even if this specimen is not from our area the species may be expected to occur on the mountains of Mozambique and Nyasaland.

8. SILENE L.

Silene L., Sp. Pl. **1** : 416 (1753) ; Gen. Pl. ed. 5 : 193 (1754).

Herbs or shrublets. Leaves opposite, sometimes slightly connate at the base, exstipulate. Flowers in cymose, paniculate, spicate or aggregate-capitulate inflorescences, or rarely single, unisexual or bisexual. Calyx tubular or dilated,

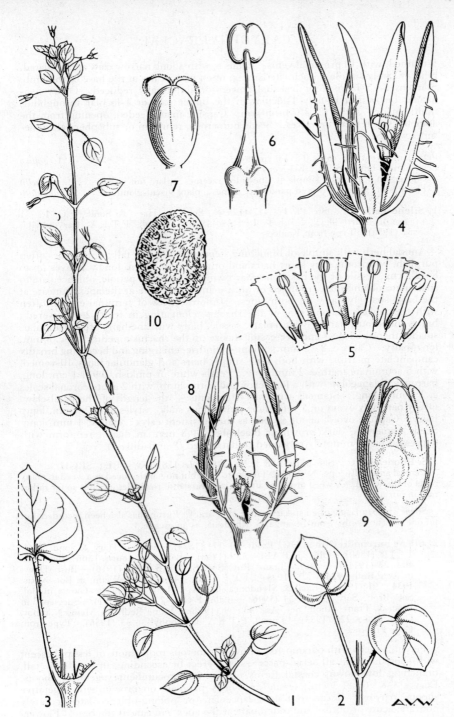

Tab. 64. STELLARIA SENNII. 1, part of plant (×1) ; 2, pair of larger leaves (×1) ; 3, node showing indumentum of petiole (×2) ; 4, flower (×12) ; 5, flower opened to show vestigial petals and stamens (upper part of sepals and gynoecium removed) (×12) ; 6, stamen viewed from the back (×36) ; 7, gynoecium (×12) ; 8, persistent calyx enclosing capsule (×8) ; 9, dehiscing capsule (×8) ; 10, seed (×18). From F.T.E.A.

5-toothed, with 10 principal veins. Petals 5, with a long narrow claw and a dilated, bifid or rarely simple or laciniate lamina, often with scales at the base of the limb. Stamens 5 + 5 in male or bisexual flowers or abnormally reduced. Ovary 3–5-locular below but usually 1-locular in the upper part or 1-locular throughout, multiovulate ; styles (2) 3 (5), filiform. Capsule many-seeded, opening from the apex by 3 or 6 teeth or valves. Seeds numerous, reniform or subspherical, sometimes winged.

Petal-lamina entire ; calyx 0·8–1·0 cm. long - - - - - - 1. *gallica*
Petal-lamina bifid ; calyx 1·1–3·5 cm. long :
  Flowers in apparently simple ± one-sided racemes ; plant not viscid   2. *burchellii*
  Flowers in lax few-flowered paniculate cymes ; plant viscid-glandular   3. *undulata*

1. **Silene gallica** L., Sp. Pl. **1** : 417 (1753).—Sond. in Harv. & Sond., F.C. **1** : 31 (1860).—Burtt Davy, F.P.F.T. **1** : 149 (1926).—Turrill, F.T.E.A. Caryophyll. : 31 (1956). Type from France.

Annual herb with simple or branching stems up to 40 cm. tall ; internodes often purplish, with some very short hairs and others up to 2 mm. long. Leaves up to 8 × 2·5 cm., sessile or the basal ones narrowing into a short petiole, oblanceolate to spathulate-oblanceolate, rounded or obtuse and mucronate at the apex, cuneate at the base, sparsely hairy on both sides. Inflorescences of terminal cymes, often with the flowers on one side of the rhachis, elongating in fruit, 2–9-flowered ; bracts foliaceous, narrowly lanceolate, sessile, hairy with the hairs often glandular, smaller than the leaves and decreasing in size up the rhachis ; pedicels 0–c. 8 mm. long, hairy. Calyx c. 8 × 2 mm., oblong-cylindric, enlarging and becoming broadly campanulate in fruit, with both long simple hairs and glandular hairs, 10-veined, with 5 acuminate teeth c. 2 mm. long. Petals white or pinkish, 0·9–1·1 cm. long, narrowly oblanceolate with a long claw and entire limb, with 2 acute coronal scales c. 1·5 mm. long. Stamens with hairy filaments ± the length of the petal-claw. Ovary narrowly ovoid on a very short puberulous stalk ; styles 3, c. 2·5 mm. long, hairy. Capsule ovoid, about as long as the persistent calyx ; stalk c. 1 mm. long. Seeds numerous, dark brown or blackish, c. 0·8 mm. in diam., reniform with depressed faces, minutely tubercled to flat-plated (" armadillo ").

S. Rhodesia. C : Salisbury, fl. & fr. ix.1919, *Eyles* 1787 (K ; PRE ; SRGH).
Central and S. Europe, N. Africa, Turkey to Iran but now introduced as a weed in many parts of the world. A weed of wheat and maize, grassland, roadsides etc. but so far rather rare in our area.

Silene armeria L. a garden species native of S. and C. Europe has also been recorded, but on one occasion only, from Inyanga, S. Rhodesia, as an escape.

2. **Silene burchellii** Otth in DC., Prodr. **1** : 374 (1824).—Sond. in Harv. & Sond., F.C. **1** : 128 (1860).—Oliv., F.T.A. **1** : 139 (1868).—Bak. f. in Journ. Linn. Soc., Bot. **40** : 25 (1911).—Eyles in Trans. Roy. Soc. S. Afr. **5** : 351 (1916).—Burtt Davy, in Kew Bull. **1924** : 228 (1924) ; F.P.F.T. **1** : 149 (1926).—Weim. in Bot. Notis. **1934** : 91 (1934).—Exell & Mendonça, C.F.A. **1**, 1 : 113 (1937).—Garcia in Bol. Soc. Brot., Sér. 2, **20** : 34 (1946).—Balle, F.C.B. **2** : 150 (1951).—Suesseng. in Proc. & Trans. Rhod. Sci. Ass. **43** : 9 (1951).—Milne-Redh. in Mem. N.Y.Bot. Gard. **8**, 3 : 220 (1953).—Turrill, F.T.E.A. Caryophyll. : 33 (1956). Type from Cape Province.

Perennial herb with parsnip-like or ovoid-tuberous roots, more or less pubescent with short hairs on all aerial parts ; stems erect or ascending, up to 70 cm. tall, branching low down, coastal forms often with procumbent vegetative shoots. Leaves 1–8 × 0·1–1·3 cm., sessile or shortly petiolate on procumbent vegetative shoots, narrowly linear to oblong-lanceolate or oblong-oblanceolate or rarely obovate, acute or obtuse and mucronate at the apex, cuneate at the base. Flowers in simple, lax, one-sided, 2–7-flowered racemes at the ends of the branches ; peduncle up to 20 cm. long, usually rather elongate ; bracts up to 8 mm. long but often less, subulate to linear-lanceolate ; pedicels very short or up to c. 1·5 cm. long on the lowest flowers. Calyx 1·1–2·5 (3·5) cm. long, tubular-clavate with 10 darker nerves ; teeth 2–3 mm. long, lanceolate to ovate, obtuse or acute to shortly acuminate. Petals reddish-brown, purple, pinkish or white, 1–3 (–5) mm. long, often minutely puberulous ; claw about the same length as the calyx ; limb about

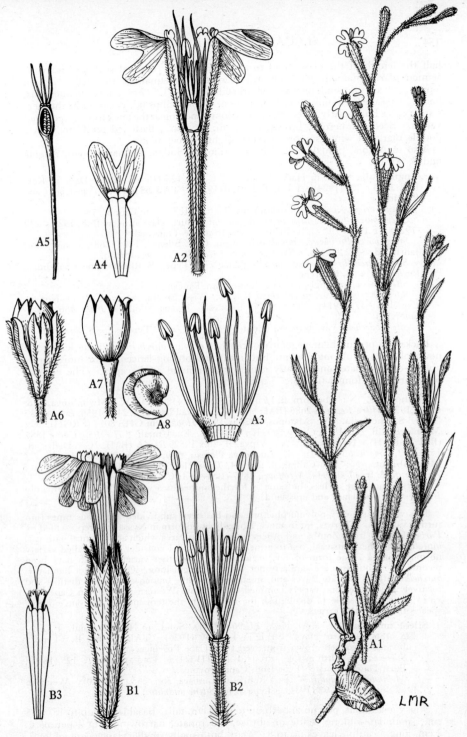

Tab. 65. A.—SILENE BURCHELLII VAR. ANGUSTIFOLIA. A1, part of plant (×⅔) *Robinson*
1922 ; A2, flower, part of calyx removed (×3) *Robinson* 1922 ; A3, stamens (×3)
Robinson 1922 ; A4, petal (×3) *Robinson* 1922 ; A5, gynoecium, ovary wall removed
to show ovules, and gynophore (×3) *Robinson* 1922 ; A6, capsule with persistent calyx
(×2) *Fanshawe* 2428 ; A7, capsule and gynophore (×2) *Fanshawe* 2428 ; A8, seed
(×8) *Fanshawe* 2428. B.—SILENE UNDULATA. B1, flower (×⅘) ; B2, flower, calyx
and petals removed (×⅘) ; B3, petal (×⅘), all from *Wild* 3051.

half the length of the claw, bifid to about half way ; coronal scales whitish, ±
semicircular. Stamens with filiform filaments ± equalling the petal claw in
length. Ovary ellipsoid, on a very short stalk ; styles 3, 5–7 mm. long, filamentous,
hairy. Capsule on a stalk up to 6 mm. long, oblong-ellipsoid, enclosed in the persistent calyx but slightly exserted, c. 1 cm. long, opening at the apex by 6 ± recurved
valves. Seeds numerous, brown, c. 2 mm. in diam., flattened-reniform with a
double marginal wing, smooth or very slightly tuberculate.

Also in Arabia, Somaliland, Eritrea, Ethiopia, Sudan, Belgian Congo, Angola
and S. Africa.

Var. **angustifolia** Sond. in Harv. & Sond., F.C. **1** : 128 (1860).—Burtt Davy in Kew
    Bull. **1924** : 228 (1924) ; F.P.F.T. **1** : 150 (1926). TAB. **65** fig. A. Type from Cape
    Province.
    Silene cernua sensu Bartl. in Linnaea, **7** : 623 (1832) non Thunb. (1800).
    Silene burchellii var. *latifolia* Sond., loc. cit.—Burtt Davy in Kew Bull. **1924** : 229
    (1924) ; F.P.F.T. **1** : 150 (1926). Type from the Transvaal.
    Silene burchellii var. *cernua* Rohrb., Mon. Gatt. Silene : 121 (1868).—Engl., Bot.
    Jahrb. **48** : 382 (1912). Type as for *S. burchellii* var. *angustifolia*.
    Silene burchellii var. *maschonica* Engl., loc. cit. Type : S. Rhodesia, near Salisbury,
    Engler 3061a (B, holotype †).
    Silene meruensis Engl., loc. cit. Type from Tanganyika.
    Silene burchellii var. *macrorrhiza* R.E.Fr., Wiss. Ergebn. Schwed. Rhod.-Kongo-
    Exped. **1:** 37 (1914). Type : N. Rhodesia, Bwana Mkubwa, *Fries* 480 (UPS,
    holotype).
    Silene burchellii var. *meruensis* (Engl.) Weim., loc. cit. Type as for *S. meruensis*.

Lacks the procumbent vegetative branches of var. *burchellii* and the leaves are
never obovate to oblong-obovate but linear and oblong-lanceolate or more rarely
oblong-oblanceolate, and mostly acute or acuminate at the apex. The coronal
scales are often longer (up to c. 3 mm. long).

N. Rhodesia. N : Abercorn, fl. 14.ii.1955, *Richards* 4488 (K). W : Mwinilunga, fl. &
fr. 9.x.1937, *Milne-Redhead* 2685 (BM ; K). E : Nyika Plateau, fl. 24.ix.1956, *Benson* 137
(BM). **S. Rhodesia.** W : Matopos, fl. 20.iv.1941, *Hopkins* in GHS 8014 (SRGH). C :
Marandellas, fl. 23.x.1948, *Corby* 160 (K ; SRGH). E : Umtali, fl. 7.i.1950, *Chase* 1887
(BM ; SRGH). **Nyasaland.** N : Vipya, fl. 26.x.1952, *Rees* 6 (BM). C : Dedza, fl.
15.x.1937, *Longfield* 36 (BM). S : Luchenya Plateau, fl. 9.vii.1946, *Brass* 16758 (K).
Mozambique. N : Vila Cabral, fl. iii.1934, *Torre* 21 (COI ; LISC). LM : Maputo,
Zitunde, fl. & fr. 17.xi.1944, *Mendonça* 2897 (BM ; LISC).

Distribution more or less the same as for the species as a whole. Woodland (mainly
Brachystegia woodland) and grassland.

This variety is a very polymorphic one and has been subdivided at various times into
further varieties. There is, for instance, a broad-leaved form in Nyasaland, represented by
Purves 164 (K) from Zomba and others from the same area which are identical with var.
latifolia from the Transvaal, but intermediates exist and I consider this so-called variety
no more than a shade form of var. *angustifolia*. The other varieties which have been
recorded from our area are similarly not worth distinguishing. On the other hand var.
burchellii with its obovate leaves and prostrate vegetative branches is a most distinct and
relatively uniform coastal variety confined to the South Western and Eastern Cape Province; so it is advisable to distinguish the material in the remainder of the range of this
species from the type variety.

3. **Silene undulata** Ait., Hort. Kew. **2** : 96 (1789).—Sond., in Harv. & Sond., F.C. **1** :
    125 (1860).—Burtt Davy, F.P.F.T. **1** : 149 (1926). TAB. **65** fig. B. Type a
    cultivated specimen grown from seed from Cape Province.
    Silene capensis Otth in DC., Prodr. **1** : 379 (1824).—Sond., tom. cit. : 125 (1860).
    —Burtt Davy, loc. cit. Type from Cape Province.
    Melandrium undulatum (Ait.) Rohrb. in Linnaea, **36** : 245 (1869–70).—Weim. in
    Bot. Notis. **1934** : 92 (1934). Type as for *Silene undulata*.

Sticky, glandular, perennial herb up to c. 60 cm. tall. Basal leaves up to 15 × 2·5
cm., spathulate-oblong, acute or obtuse, mucronate, narrowing into a petiole c.
2 cm. long ; cauline leaves up to 8 × 2 cm. but usually smaller, lanceolate, oblong-
lanceolate or narrowly elliptic, apex acute, cuneate and sessile or subsessile at the
base, margin sometimes undulate. Flowers in terminal, lax, rather few-flowered
paniculate cymes ; bracts similar to the upper leaves, but progressively smaller
upwards, lanceolate-acuminate ; pedicels 0·6–2 cm. long, viscid-glandular, often
in triads with the middle one shortest. Calyx 2·5–3·5 × c. 0·5 cm., cylindric, but

dilated in fruit, 10-ribbed, viscid-glandular, teeth lanceolate-subulate, c. 5 mm. long. Petals white or pinkish, lamina c. 0·9 × 0·4 cm., spreading, bifid, with a linear claw somewhat longer than the calyx and lobes denticulate or entire ; coronal scales dentate, c. 1·5 mm. long. Stamens on slender biseriate filaments, the longer c. 3·5 cm. long, the shorter c. 1·7 cm. long. Ovary narrowly oblong-ovoid, 1-locular, on a puberulous stalk up to 7·5 mm. long ; styles 3, c. 1·5 cm. long, slender, papillate along one side. Capsule 1·2–1·8 × 0·8 cm., horny, oblong-ovoid, opening by recurved valves at the apex, on a stout stalk ⅓–½ the length of the capsule. Seeds many, almost black, c. 1·2 × 1 mm., reniform with flattened sides, minutely and concentrically tuberculate.

S. Rhodesia. C : Salisbury, fl. 15.xi.1926, *Eyles* 4550 (K ; SRGH). E : Inyanga, fl. 3.ii.1952, *Chase* 5503 (BM ; SRGH). S : Victoria, Morgenster Mission, fl. 4.x.1949, *Wild* 3051 (K ; SRGH).

Also in Cape Province, Natal, Orange Free State and the Transvaal. In open grassland or woodland usually in higher rainfall areas of the eastern border of S. Rhodesia or in localities like Zimbabwe with a locally high rainfall.

This species has a 1-locular ovary and so was placed by Rohrbach (loc. cit.) in the genus *Melandrium*. The view followed here, however, is that of Chowdhuri and Davis (in Not. Roy. Bot. Gard. Edinb. **22**, 3 : 221 (1957)) who conclude that *Melandrium* is a very artificial genus and should be distributed between various parts of *Silene*.

9. DIANTHUS L.
by
Sheila S. Hooper

DIANTHUS L., Sp. Pl. **1** : 409 (1753) ; Gen. Pl. ed. 5 : 191 (1754).

Annual, biennial or generally short-lived perennial, occasionally suffruticose herbs, generally with a woody tap root. Leaves exstipulate, usually linear or linear-lanceolate. Inflorescence cymose. Flowers solitary, fasciculate or clustered into heads. Calyx tubular, generally striate-nervose, 5-lobed, invested at the base by 1–12 pairs of calyx bracts. Internode between calyx and corolla (anthophore) variable. Gynophore short. Petals 5, claw of petal with two central longitudinal ridges (sometimes very inconspicuous). Stamens 10, united at the base into a short ring. Ovary ovoid, oblong or cylindric, placentation free-central ; stigmas 2, free to base. Capsule with 4 equal teeth.

Petal-lamina obovate, entire, crenulate or shallowly dentate :
  Leaves 1·5–4 mm. wide, sheath 3–7 mm. long ; upper pairs of calyx bracts
    elliptic, acute or acuminate ; petals pink or pale purple - - 1. *angolensis*
  Leaves 4–8 mm. wide, sheath 7·5–12 mm. long ; upper pairs of calyx bracts
    broadly elliptic, subobtuse and mucronate ; petals white (sometimes pink when
    fading) - - - - - - - - - - - - 2. *excelsus*
Petal-lamina elliptic or obovate, fimbriate or laciniate :
  Petal-lamina deeply pinnately-fimbriate, fimbriae up to 12 mm. long ; flowering shoots
    branched from near the base, branches leafy, single-flowered ; leaves with 3 equal
    prominent nerves running from base to apex - - - 3. *chimanimaniensis*
  Petal-lamina palmately-fimbriate or laciniate, fimbriae or lobes not more than 8 mm.
    long ; leaves with more than 3 nerves, at least at base - - - - 4. *zeyheri*

1. **Dianthus angolensis** Hiern ex F. N. Williams in Journ. of Bot. **24** : 301 (1886).—Exell & Mendonça, C.F.A. **1** : 112 (1937).—Hooper in Hook., Ic. Pl. **37** : t. 3603 (1959). Type from Angola (Benguela).

Slender erect or semi-decumbent glabrous herb with a narrowly turbinate, woody tap-root and short, branched, woody stem. Flowering shoots 40–110 cm. high, ± virgate, often branched near base, leafy below, terminating in a branched inflorescence. Leaves towards base of shoot 4·5–7·5 cm. × 1·5–4·0 mm., linear or linear-lanceolate ; sheath 3–6 mm. long ; upper leaves smaller, grading into bracts. Inflorescence axis elongated, branched, bearing 3–28 flowers ; pedicels slender, erect, bracteolate, generally shorter than the corresponding internode on the axis, some or all the flowers somewhat drooping. Calyx bracts 6–10, investing lower half of calyx, uppermost elliptic or narrowly elliptic, acute or acuminate. Calyx 1·5–2·4 cm. × 2·5–5·0 mm., narrowly elliptic, lobes c. 4 mm. long, lanceolate. Petal-lamina pink or pale purple, 3–6 × c. 3 mm., obovate, rounded, entire, crenulate or shallowly dentate.

N. Rhodesia. W : S. of Mwinilunga, between Mukimina and Kauku, fl. & fr. 24.viii.1930, *Milne-Redhead* 952 (K).

Otherwise confined to Angola, chiefly on the central plateau. In *Brachystegia* woodland at 1200–2000 m.

2. **Dianthus excelsus** Hooper in Hook., Ic. Pl. **37** : t. 3604 (1959). TAB. **66**. Type from Tanganyika (Ufipa).

    Dianthus angolensis subsp. *orientalis* Turrill in Kew Bull. **1954** : 49 (1954) ; F.T.E.A., Caryophyll. : 34, t. 14 (1956). Type as for *D. excelsus*.

Stiffly erect herb with a short branched woody base. Flowering shoots robust, virgate, 60–120 cm. high, occasionally branched near base, leafy below, terminating in a generally freely branched inflorescence. Leaves towards the base of shoot 6·5–12·5 cm. × 5·0–9·0 mm., linear; sheath 7·5–12·0 mm. long. Inflorescence bearing up to 30 flowers; axis elongated; primary branches erect or slightly spreading, generally branched, only the uppermost shorter than the corresponding internode on the axis; flowers erect. Calyx bracts 8 or 10, closely imbricate, covering the calyx nearly to the base of the calyx-lobes, uppermost broadly elliptic, subobtuse, mucronate, margin fringed with short, white hairs. Calyx cylindric or elliptic 1·2–1·8 cm. × 2·5–4·0 mm., lobes c. 6 mm. long. Petal-lamina pure white, turning pink with age, 5·0–8·0 × 3·5–5·0 mm., obovate or rounded, crenulate or shallowly dentate.

N. Rhodesia. N : near Senga Hill, fl. 1.iv.1957, *Richards* 8955 (K); 64 km. SE. of Kasama, fl. 6.iv.1932, *St. Clair Thompson* 1265 (K).

Also on the adjacent high ground in SW. Tanganyika and the Belgian Congo. In *Brachystegia-Isoberlinia* woodland and sometimes in dambos (seasonal swamps), at 1200–1800 m.

    D. excelsus is closely related to *D. angolensis* but may be distinguished by the more robust erect branched flowering shoots, wider leaves with longer sheaths, subobtuse and mucronate upper calyx bracts and generally white flowers.

3. **Dianthus chimanimaniensis** Hooper in Kew Bull. **1958** : 318 (1958) ; in Hook., Ic. Pl. **37** : t. 3605 (1959). Type : Mozambique, Chimanimani Mts., Musapa Gap, *Phipps* 838 (K, holotype; SRGH).

Caespitose, glaucous herb. Flowering shoots branched from near the base, leafy below, branches bearing solitary terminal flowers. Leaves rather short, 2 × 2·5 cm. × 1·5–2 mm., about half as long as the subtended internode; lamina thin, conspicuously and evenly 3-nerved from base to apex; sheath very short, minutely hispid. Calyx bracts 4, lower pair markedly smaller than upper pair, upper pair lanceolate-elliptic, c. 1·0 cm. × 4 mm. with an acute and minutely apiculate apex. Calyx c. 3·5 cm. × 4 mm., with lobes c. 7 mm. long. Anthophore c. 3 mm. long. Petal-lamina white or pale pink, obovate-elliptic, deeply pinnately fimbriate, fimbriae up to 1·2 cm. long ; stigmas exserted c. 1·0 cm. beyond petals.

Mozambique. MS : Chimanimani Mts., Musapa Gap, fl. 20.xii.1957, *Phipps* 838 (K ; SRGH).

Known only from the type collection. The type was collected on quartzite slopes above the level of scattered *Brachystegia tamarindoides* in submontane scrub. Readily distinguished from the other Central and South African species of this genus by the deeply fimbriate petal-laminae.

4. **Dianthus zeyheri** Sond. in Harv. & Sond., F.C. **1** : 124 (1860).—Burtt Davy in Kew Bull. **1922** : 222 (1922). Type from the Transvaal.

    D. mecistocalyx F. N. Williams in Journ. of Bot. **27** : 199 (1889). Type from the Transvaal.

Glaucous herb with a woody stem-base. Flowering shoots robust, erect, often virgate, 20–75 cm. high, leafy below with a solitary flower or branched terminal inflorescence. Leaves 2·0–9·0 cm. × 2–12 mm., lanceolate-elliptic, generally equalling or exceeding the corresponding internode in length, variable in width. Calyx bracts 4, 6 or 8, covering about ⅓ of the calyx-length. Calyx 2·8–4·5 cm. × 4–5 mm., cylindric. Petal-lamina 8–24 × 5–12 mm., obovate, dentate or lacero-fimbriate, white or pink.

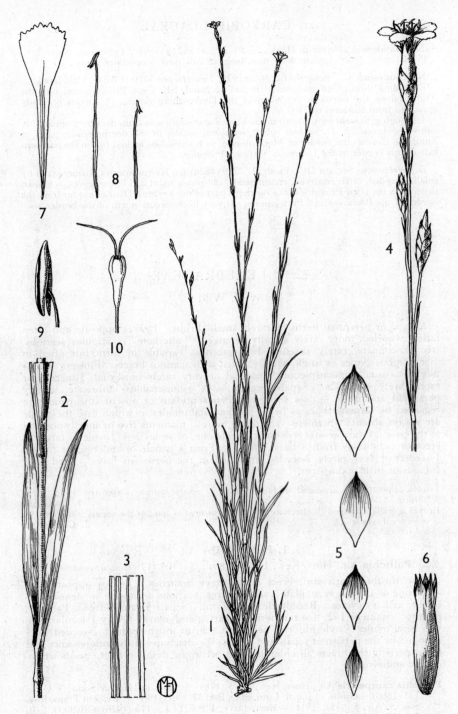

Tab. 66. DIANTHUS EXCELSUS. 1, plant (× ⅓); 2, part of lower stem with leaves (× 1); 3, portion of leaf to show venation (× 2); 4, inflorescence (× 1); 5, series of bracts (× 2); 6, calyx (× 2); 7, petal (× 2); 8, stamens (× 2); 9, anther (× 10); 10, gynoecium and gynophore (× 2). From F.T.E.A.

Subsp. **natalensis** Hooper in Hook., Ic. Pl. **37**: t. 3622 (1959). Type from Natal.
Petal-lamina finely dentate, 8–17 mm. long, usually pink, sometimes white.

N. Rhodesia. C: Broken Hill, fl. xi.1928, *Van Hoepen* 1370 (PRE). Cultivated.
Extending throughout the coastal region of Natal, SE. Cape Province and into the
Transvaal on the northern extension of the Drakensberg range. A species of sandy
grassland from sea-level to 2300 m.

Although at present only known in the Flora Zambesiaca area from the above apparently
cultivated specimen, *D. zeyheri* subsp. *natalensis* occurs in the northernmost parts of
Zululand, close to the border of Mozambique, so it should be looked for on the northern
side of this border in the Lourenço Marques Province.

D. micropetalus Ser. (in DC., Prodr. **1**: 359 (1824)), a Karroo species characterized by
small, rounded, often recurved, cream petals, caespitose habit and linear leaves, is known
to occur in the Cape Province in the northern part of the Vryburg District not far from the
border of the Bechuanaland Protectorate. It very likely occurs north of the border also.

22. ILLECEBRACEAE

by H. Wild

Annual or perennial herbs or rarely small shrubs. Leaves opposite and often
falsely whorled, more rarely spirally arranged (" alternate "); stipules scarious,
free or connate, rarely absent. Inflorescences variable in form but often in
dichotomous cymes or in glomerules, often with scarious bracts. Flowers actinomorphic, usually inconspicuous, bisexual or very rarely unisexual, 5-merous or
more rarely 4-merous. Sepals free or calyx gamosepalous, imbricate, usually
persistent and more or less scarious. Petals present or absent (the petals are
regarded by some authors as being petaloid staminodes in which case the petals
are always absent). Stamens 1–5 (rarely more); filaments free or shortly connate
at the base. Ovary sessile, 1-locular; style one, or styles 2–3; ovule 1, rarely 2,
erect or pendulous from a basal funicle. Fruit a utricle or indehiscent nutlet,
1-seeded or very rarely 2-seeded, included in the persistent calyx. Seed with
copious or little endosperm.

Leaves opposite but arranged in false whorls; petals minute; stamens 1–2; style
  filiform - - - - - - - - - - - - **1. Pollichia**
Leaves spirally arranged (" alternate "); petals almost as long as the sepals; stamens 5;
  stigma subsessile - - - - - - - - - - - **2. Corrigiola**

1. POLLICHIA Ait.

Pollichia Ait., Hort. Kew. **1**: 5 (1789); **3**: 505 (1789) *nom. conserv.*

Low shrublet with stiff, erect or suberect branches. Leaves opposite but
appearing as though verticillate; stipules free, scarious. Flowers in dense, subsessile, axillary cymes. Receptacle cupuliform. Sepals 5, small, free. Petals 5,
minute. Stamens 1–2, like the perianth-parts perigynous. Ovary 1-locular with
two basal ovules; style filiform, persistent. Fruit indehiscent, 1–2-seeded, surrounded by the persistent sepals and bracts; the rhachides of the inflorescence and
lower parts of the bracts all enlarging and becoming fleshy in fruit. Seeds with a
curved embryo.

Pollichia campestris Ait., Hort. Kew. **1**: 5 (1789).—Sond. in Harv. & Sond., F.C. **1**:
  133 (1860).—Gibbs, in Journ. Linn. Soc., Bot. **37**: 465 (1906).—Eyles in Trans. Roy.
  Soc. S. Afr. **5**: 351 (1916).—Burtt Davy, F.P.F.T. **1**: 173 (1926).—Balle, F.C.B.
  2: 137 (1951).—Suesseng. in Proc. & Trans. Rhod. Sci. Ass. **43**: 83 (1951).—O. B.
  Mill. in Journ. S. Afr. Bot. **18**: 13 (1952).—Turrill, F.T.E.A. Caryophyll.: 13, t. 5
  (1956). TAB. **67**. Type a cultivated specimen grown from seed from the Cape
  Province.

Small shrublet, often much branched, up to c. 60 cm. tall; branches terete with
a white woolly indumentum at least when young. Leaves grey-green, 0·5–3·2 ×

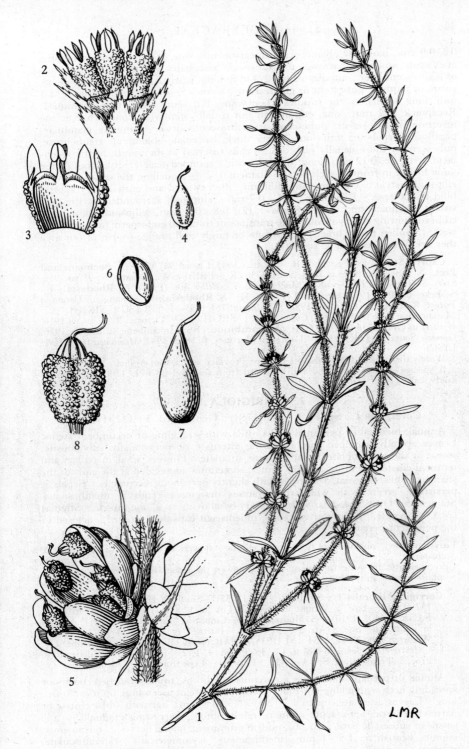

Tab. 67. POLLICHIA CAMPESTRIS. 1, part of plant (×⅔) ; 2, part of inflorescence (×8) ; 3, calyx opened to show petals and stamen (×12) ; 4, gynoecium (×12) ; 5, infructescence (×4) ; 6, seed (×12) ; 7, fruit (×12) ; 8, fruit with persistent calyx (×12), all from *Corby* 84.

0·5–0·9 cm., narrowly oblanceolate, linear-oblanceolate or very narrowly elliptic, apex acute, apiculate, narrowed to a subsessile base, hairy with white crinkled hairs or glabrescent later ; stipules white, 2–4 mm. long, scarious, lanceolate-acuminate, entire or toothed, pubescent or glabrescent. Cymes 5–∞–flowered ; bracts c. 1·5 mm. long, similar to the stipules when young but acute rather than acuminate. Receptacle c. 1 mm. long, cupuliform but usually somewhat constricted at the insertion of the sepals, pubescent or glabrescent outside, minutely glandular-papillose. Sepals 0·5 mm. long, rather thick, lanceolate-oblong, apex subobtuse, pubescent or more usually glabrous. Petals inserted at the mouth of the receptacle, white, c. 0·12 mm. long, deltoid, with a rounded scale (? disk-lobe) of the same height in front of each petal. Stamens 1–2, ± equalling the sepals. Ovary ellipsoid ; style c. 0·6 mm. long, filiform, often twisted and curled, very shortly bifid at the apex. Capsule c. 1·25 × 0·7 mm., ellipsoid, surrounded by the persistent receptacle and calyx. Seeds 1 (2), 0·8 × 0·5 mm., ellipsoid, the curved embryo showing clearly through the transparent testa and endosperm (at least when wet). Swollen fleshy bracts waxy-white or rarely dull orange (north of our area they are often recorded as scarlet).

Caprivi Strip. E. of Kwando R., fl. & fr. x.1945, *Curson* 981 (PRE). **Bechuanaland Prot.** N : Sepopa, fl. 25.ix.1954, *Story* 4736 (K ; PRE). SW : Ghanzi, fr. 25.vii.1955, *Story* 5040 (K ; PRE). SE : Mochudi, fl. i., *Miller* 406 (PRE). **N. Rhodesia.** B : Sesheke, fl. 20.xii.1952, *Angus* 982 (FHO ; K). **S. Rhodesia.** W : Shangani, Gwampa Forest Reserve, fl. & fr. vii.1955, *Goldsmith* 161/1955 (K ; LISC ; SRGH). C : Marandellas, fl. & fr. 3.iv.1948, *Corby* 84 (K ; SRGH). S : Victoria, fl. & fr. vii.1916, *Walters* in GHS 2336 (K ; SRGH). **Mozambique.** SS : Inhambane, fl. & fr. x.1936, *Gomes e Sousa* 1865 (K). LM : Lourenço Marques, fl. 18.xii.1942, *Mendonça* 1605 (BM ; LISC).

Widely distributed from Ethiopia to Cape Province and also in Arabia.

A species of open woodland and bushland from sea-level up to c. 1700 m., often on sandy soils. Fruit edible.

2. CORRIGIOLA L.

Corrigiola L., Sp. Pl. **1** : 271 (1753) ; Gen. Pl. ed. 5 : 132 (1754).

Annual, biennial or perennial herbs often with straggling or decumbent stems. Leaves spirally arranged but appearing alternate or occasionally subopposite, sessile or shortly petiolate, linear to ovate, stipulate. Flowers small in axillary and terminal clusters or in lax axillary cymes, sometimes aggregated at the ends of the stem-branches, bisexual, 5-merous, with slightly perygnous receptacles. Sepals 5, persistent, green, with white membranous margins. Petals 5, membranous. Stamens 5. Ovary 1-locular with a solitary basal ovule on a long funicle. Stigmas (2) 3, subsessile. Fruit a crustaceous, indehiscent capsule, 1-seeded, enclosed in the persistent perianth.

Leaves narrowly oblanceolate to narrowly elliptic ; flowers in dense sessile or subsessile
clusters - - - - - - - - - - 1. *litoralis*
Leaves elliptic to ovate or oblong-ovate ; flowers in rather lax pedunculate cymes ;
inflorescences up to 13 cm. long - - - - - - - 2. *drymarioides*

1. **Corrigiola litoralis** L., Sp. Pl. **1** : 271 (1753).—Sond. in Harv. & Sond., F.C. **1** : 132 (1860).—Bak. & Wright in Dyer, F.T.A. **6**, 1 : 12 (1909).—Cooke in Dyer, F.C. **5**, 1 : 401 (1910).—S. Moore in Journ. Linn. Soc., Bot. **40** : 181 (1911).—Eyles in Trans. Roy. Soc. S. Afr. **5** : 351 (1916).—Burtt Davy, F.P.F.T. **1** : 173 (1926).—Weim. in Bot. Notis. **1934** : 91 (1934). TAB. **68** fig. B. Type from Europe.
　Corrigiola capensis Willd. in L., Sp. Pl. ed. 4, **1**, 2 : 1507 (1798).—Suesseng. in Proc. & Trans. Rhod. Sci. Ass. **43** : 83 (1951). Type from Cape Province.

Annual or perhaps sometimes a perennial, glabrous, much-branched, bushy or spreading herb with a long tap-root ; stems prostrate or ascending, up to c. 30 cm. long. Leaves alternate, up to 1·7 (2·5) × 0·3 (0·5) cm., narrowly oblanceolate or narrowly elliptic, apex acute or more rarely obtuse, narrowing gradually to the sessile or subsessile base, glaucous, midrib prominent, lateral nerves inconspicuous ; stipules whitish, c. 1·5 × 1 mm., membranous, asymmetrically ovate-auriculate, margins subentire or irregularly crenate. Flowers in compact, axillary or terminal, sessile or very shortly pedunculate clusters on leafy branches ; pedicels very short and scarcely visible or sometimes up to 1 mm. long ; bracts up to 1 mm. long,

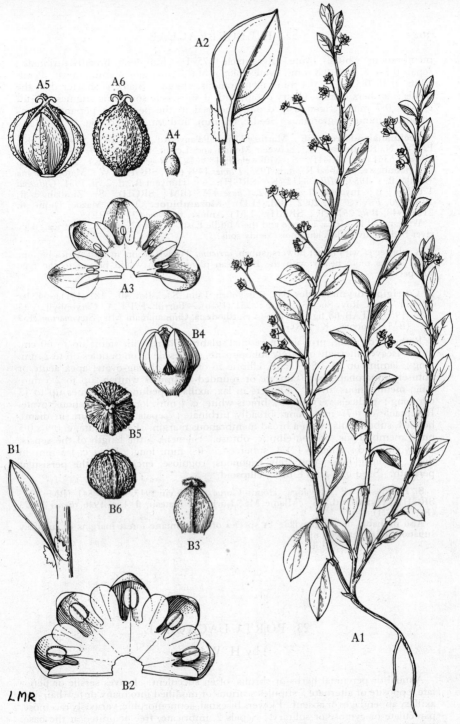

LMR

Tab. 68. A.—CORRIGIOLA DRYMARIOIDES. A1, plant (×⅔) *Jackson* 1960 ; A2, leaf and
stipules (×3) *Jackson* 1960 ; A3, dissected perianth showing sepals, petals and stamens (×8) *Jackson* 1960 ; A4, gynoecium (×8) *Jackson* 1960 ; A5, fruit with persistent calyx split at the base by expansion of the fruit (×8) *Wild* 3595 ; A6, fruit
(×8) *Wild* 3595. B.—CORRIGIOLA LITORALIS. B1, leaf and stipules (×4) ; B2,
dissected flower showing sepals, petals and stamens (×24) ; B3, gynoecium (×24) ; B4,
fruit with persistent calyx (×12) ; B5, fruit from above (×2) ; B6, fruit, side view
(×12), all from *Wild* 2579.

membranous, whitish, ovate. Receptacle 0·75–1·35 mm. long, broadly turbinate ;
sepals 0·5–0·8 × 0·3–0·8 mm., ovate-oblong, obtuse, margins membranous. Petals
white, 0·38–0·55 × 0·3–0·4 mm., membranous, oblong. Stamens shorter than the
petals ; anthers ovate. Ovary ovoid, with 3 subsessile spreading stigmas. Fruit
c. 0·9 × 0·7 mm., trigonously ovoid, enclosed in the persistent perianth, horn-
coloured, minutely granular. Seed c. 0·7 mm. in diam., subglobose.

Bechuanaland Prot. SE : Martin's Drift—Palapye Rd., fl. 6.ix.1954, *Story* 4613 (K ;
PRE). **N. Rhodesia.** B : between Mongu and Lealui, fl. 12.xi.1959, *Drummond &
Cookson* 6361 (K ; SRGH). S : Mukwela, fl. 18.vi.1920, *Rogers* 26059 (K). **S. Rhodesia.**
N : Sebungwe, Zambezi R., fl. ix.1955, *Davies* 1491 (K ; SRGH). W : Matobo, Besna
Kobila, fl. vi.1957, *Miller* 4415 (K ; SRGH). C : Hunyani R., fl. 2.viii.1931, *Gilliland*
1 (BM). E : Inyanga, fl. 24.i.1952, *Chase* 4355 (BM ; SRGH). S : Zimbabwe, fl.
19.x.1930, *F.N. & W.* 2092 (BM ; LD). **Mozambique.** T : R. Mazoe, Dique, fl.
21.ix.1948, *Wild* 2579 (K ; SRGH). LM : Anluke, fl., *Junod* 337 (PRE).
Widespread in Africa, Europe and the Middle East. Also in N. and S. America. Dry
river beds, alluvial flats and moist sandy soils.

A subspecies with larger flowers (subsp. *africana* Turrill) occurs in East Africa but our
plant agrees more nearly with the European plant and so would belong to the type
subspecies.

2. **Corrigiola drymarioides** Bak. f. in Journ. Linn. Soc., Bot. **40** : 181 (1911).—Eyles
in Trans. Roy. Soc. S. Afr. **5** : 351 (1916).—Turrill, F.T.E.A. Caryophyll. : 13
(1956). TAB. **68** fig. A. Type : S. Rhodesia, Chimanimani Mts., *Swynnerton* 2159
(BM, holotype ; K ; SRGH).

Rather straggling, probably perennial, glabrous herb with stems up to 60 cm.
long. Leaves alternate or rarely subopposite, subsessile or on petioles up to 2 mm.
long ; lamina up to 2·5 × 1 cm., elliptic to ovate or oblong-ovate, apex acute or
obtusely mucronate, base cuneate or rounded ; stipules whitish, up to 2·5 mm.
long, membranous, ovate. Flowers in lax, axillary, pedunculate cymes up to 13
cm. long ; pedicels very short ; bracts white, c. 1 mm. long, membranous, ovate.
Receptacle c. 0·75 mm. long, broadly turbinate ; sepals 1·25–2 mm. in diam.,
hooded, suborbicular with a broad membranous margin. Petals white, c. 0·9 × 0·5
mm., membranous, oblong-elliptic, obtuse. Stamens ½ the length of the sepals.
Ovary ellipsoid ; stigmas (2) 3, suberect, c. 0·3 mm. long. Fruit c. 1·5 mm. in
diam., ovoid-globose, obscurely trigonous, rugulose, enclosed in the persistent
perianth. Seed c. 1·2 × 0·8 mm., ellipsoid.

S. Rhodesia. E : Inyanga, Hondi Gorge, fl. 20.viii.1947, *Chase* 541 (BM ; K ;
SRGH). **Nyasaland.** S : Mlanje Mt., Luchenya Plateau, fl. 1.vii.1946, *Brass* 16573
(K ; SRGH).
Also probably in Tanganyika. A species of submontane forest margins or in rocky
situations at about 2000 m.

23. PORTULACACEAE

by H. Wild

Annual or perennial herbs or shrubs, often succulent. Leaves sessile or petiolate, opposite or alternate ; stipules scarious or modified into many or few hair-like
axillary appendages or absent. Flowers bisexual, actinomorphic, variously racemose,
paniculate or cymose or solitary. Sepals 2, imbricate, free or united at the base.
Petals 4–6 (in all African genera), imbricate, free or connate up to half-way or more,
often fugacious. Stamens as many as the petals or more numerous, free or adnate
to the petals. Ovary superior or half inferior, 1-locular or partially divided into
several loculi near the base ; placentation basal ; ovules 1–∞ ; style simple or
variously divided. Fruit a capsule dehiscing by longitudinal valves or circumscissile, very rarely an indehiscent nutlet.

The Spekboom (*Portulacaria afra* Jacq.) is a native of the Cape Province and the Transvaal. It does not occur naturally in our area but is occasionally cultivated as a browse for cattle in S. Rhodesia and Mozambique.

Capsule pyxidiform, circumscissile - - - - - - - **1. Portulaca**
Capsule dehiscing longitudinally into 3–4 valves :
  Stipules membranous silvery and scarious, hiding the minute leaves
                                                    2. Anacampseros
  Stipules absent ; leaves obvious - - - - - - - **3. Talinum**

1. PORTULACA L.

Portulaca L., Sp. Pl. **1** : 445 (1753) ; Gen. Pl. ed. 5 : 204 (1754).

Annual, biennial or perennial herbs; branches erect or often prostrate. Leaves shortly petiolate or sessile, opposite or alternate, fleshy, cylindric or plane, glabrous (at least in our species); stipules divided into numerous hairs or more rarely membranous or absent. Flowers sessile, solitary or a few together at the ends of the branches and surrounded by a group of 2–several apical leaves. Sepals 2, unequal, the larger slightly enfolding the margins of the smaller, connate into a tube below, which is adnate to the base of the capsule. Petals 4–5 (6), fugacious, free or ± united, marcescent and enfolding the ripe capsule. Stamens 4–c. 100, inserted on the corolla or at its base. Ovary semi-inferior, 1-locular, at least in the upper part ; placentation free central ; ovules many or very rarely 1 or several ; style simple with 2-several stigmas. Capsule dehiscing transversely. Seeds 1-∞, usually reniform, smooth or variously tuberculate or granulate.

Stamens 7 or more :
  Leaves obovate-spathulate ; stipular hairs very few, inconspicuous and caducous, c.
    1 mm. long - - - - - - - - - - - - - 1. *oleracea*
  Leaves linear-cylindric, lanceolate-acuminate, elliptic or rarely ovate ; stipular hairs
    numerous and more than 3 mm. long, or if few then leaves cylindric :
    Leaves linear, cylindric :
      Corolla yellow, orange or occasionally pink-tinged ; stipular hairs few, incon-
        spicuous, up to 3 mm. long - - - - - - - - 2. *foliosa*
      Corolla bright carmine ; stipular hairs more numerous, up to 7 mm. long
                                      3. *kermesina*
    Leaves flattened, lanceolate-acuminate or elliptic to ovate :
      Stipular hairs dilated towards the base ; a small stout perennial herb up to 10 cm.
        tall - - - - - - - - - - - - - - - 4. *collina*
      Stipular hairs not dilated ; a prostrate annual herb - - - 5. *quadrifida*
Stamens 4 :
  Flowers pink ; stipular hairs absent - - - - - - 6. *rhodesiana*
  Flowers purplish or white ; stipular hairs fairly numerous, c. 2 mm. long
                                        7. *hereroensis*

1. **Portulaca oleracea** L., Sp. Pl. **1** : 445 (1753).—Sond. in Harv. & Sond., F.C. **2** : 381 (1862).—Oliv., F.T.A. **1** : 148 (1868).—R.E.Fr., Wiss. Ergebn. Schwed. Rhod.-Kongo-Exped. **1** : 37 (1914).—Burtt Davy, F.P.F.T. **1** : 165 (1926).—Poellnitz in Fedde, Repert. **37** : 258 (1934).—Exell & Mendonça, C.F.A. **1**, 1 : 115 (1937).—Hauman, F.C.B. **2** : 126 (1951). Type locality doubtful ; specimen probably cultivated.
    Portulaca oleracea var. *sylvestris* DC., Prodr. **3** : 353 (1828). Type from Europe.
    Portulaca oleracea subsp. *sylvestris* (DC.) Thell., Fl. Advent. Montpell.: 222 (1912).—Brenan in Mem. N.Y.Bot. Gard. **8**, 3 : 221 (1953). Type as above.

Annual glabrous rather fleshy herb with numerous spreading or prostrate branches up to 30 cm. long and up to 5 mm. in diam. Leaves alternate and often somewhat crowded towards the ends of the branches, sessile or shortly stalked, up to 3·0 × 1·2 cm., obovate-spathulate, apex rounded or truncate; stipular hairs very few, inconspicuous, caducous, c. 1 mm. long. Flowers terminal, solitary or up to 5, surrounded by a cluster of subverticillate leaves ; bracts membranous, c. 3 mm. long, ovate-acuminate. Sepals 2–4 mm. long, united below into a tube c. 2 mm. long, free portions fleshy, oblong-ovate, keeled or slightly winged. Petals yellow, 4–8 mm. long, united at the base, obovate-oblong to obovate, sometimes emarginate. Stamens 7–12. Ovary ovoid ; style short, with 3–6 subulate lobes. Capsule obovoid to ovoid, enveloped in the marcescent corolla, transversely dehiscent across the middle. Seeds many, c. 0·5 mm. in diam., dull black, reniform, bluntly verrucose-granulate.

Bechuanaland Prot. N : Ngamiland, fl. & fr. xii.1930, *Curson* 164 (PRE). **N. Rhodesia,** N : Lake Bangweulu, fl. 27.x.1911, *Fries* 813 (UPS). C : Chilanga, Quien Sabe, fl. & fr. 17.ix.1929, *Sandwith* 168 (K). S : 110 km. upstream from Kariba Gorge, fl. & fr. 17.iii.1957, *Scudder* 51 (SRGH). **S. Rhodesia.** W : Nyamandhlovu, fl. & fr. 20.ii.1956, *Plowes* 1933 (SRGH). C : Salisbury, fl. & fr. 27.iii.1941, *Hopkins* in GHS 7996 (SRGH). E : Chipinga, fl. & fr. 22.i.1957, *Phipps* 107 (SRGH). S : Victoria District, *Monro* 2231 (BM). **Nyasaland.** N : Nyungwe R., fl. & fr. 20.ix.1930, *Migeod* 943 (BM). S : Lower Mwanza R., fl. 4.x.1946, *Brass* 17953 (K ; SRGH). **Mozambique.** SS : between Vilanculos and Macovane, fr., *D'Orey* 19 (LISC).

A cosmopolitan weed of temperate and tropical regions. A weed of cultivation, railway tracks and roadsides. Usually a garden weed in our area.

Cultivated in some European countries as a vegetable. The Common Purslane of Britain.

2. **Portulaca foliosa** Ker-Gawl. in Bot. Reg. **10** : t. 793 (1824).—Oliv., F.T.A., **1** : 148 (1868).—Poellnitz in Fedde, Repert. **37** : 304 (1934).—Exell & Mendonça, C.F.A. **1,** 1 : 114 (1937).—Hauman, F.C.B. **2** : 124 (1951). TAB. **69** fig. D. Type from Ghana.

Rather robust annual or perennial herb ; branches prostrate or suberect, often woody at the base, up to 35 cm. long and 5 mm. in diam. Leaves alternate, up to 2·5 × 0·3 cm., linear, subcylindric, glaucous, crowded towards the ends of the branches, sessile or almost sessile, apex acute or obtuse, sometimes purplish towards the apex, longer than the internodes ; stipular hairs few, not conspicuous, up to 3 mm. long. Flowers in terminal clusters of 2–6 or rarely solitary, surrounded by a terminal cluster of subverticillate leaves and dense tufts of axillary hairs ; bracts whitish, often with purple tips, membranous, ovate, long-acuminate. Sepals c. 3 mm. long, ovate-acuminate, often pinkish. Petals yellow or orange (or sometimes pink-tinged?), almost free, up to 8 mm. long, lanceolate-ovate, apex emarginate or retuse. Stamens 8–13. Ovary ovoid ; style simple with 4 recurved stigmas. Capsule c. 5 mm. long, ovoid, enclosed in the marcescent corolla, transversely dehiscent below the middle, tipped by the persistent style base. Seeds numerous, c. 0·5 mm. in diam., subreniform, black to dark grey, covered with concentric rings of minute impressed stelliform tubercles.

N. Rhodesia. B : Nangweshi, Zambezi R., fl. & fr. 26.vii.1952, *Codd* 7198 (BM ; K ; PRE ; SRGH). S : Mumbwa, fl. & fr., *Macaulay* 1059 (K). **Mozambique.** T : Boroma, Msusa, fl. & fr. 26.vii.1950, *Chase* 2703 (BM ; SRGH). MS : Chiramba, fl. & fr. 13.iv.1860, *Kirk* (K).

Also from Senegal southwards to Angola. In dry places, confined more or less to the Zambezi valley in our area.

3. **Portulaca kermesina** N.E.Br. in Kew Bull. **1909** : 91 (1909).—Burtt Davy, F.P.F.T. **1** : 165 (1926).—Poellnitz in Fedde, Repert. **37** : 305 (1934).—Hauman, F.C.B. **2** : 123 (1951). Type : Bechuanaland Prot., Kwebe Hills, *Lugard* 88 (K, holotype). *?Portulaca crocodilorum* Poellnitz in Fedde, Repert. **50** : 62 (1941). Type : Bechuanaland Prot., Crocodile R., *Klingberg* (B, holotype †).

Erect herb very similar to *P. foliosa* but stems always erect or suberect, the stipular hairs usually more persistent, more numerous and up to 7 mm. long. Petals bright carmine in colour and the stamen number varying from 10–25.

Bechuanaland Prot. N : 12 km. S. of Nata R., fl. & fr. 23.iv.1957, *Drummond & Seagrief* 5198 (K ; SRGH). SW : 80 km. N. of Kang, fl. 18.ii.1960, *Wild* 5072 (SRGH). SE : Lobatsi, fl. 15.xi.1948, *Hillary & Robertson* 547 (PRE). **S. Rhodesia.** W : Matobo South, fl. & fr. 16.iii.1957, *Garley* 101 (K ; SRGH). S : Triangle Sugar Estates, fl. & fr., *Wild* 3713 (SRGH). **Mozambique.** SS : Chibuto, fl. 15.viii.1944, *Torre* 6763 (BM ; LISC). LM : Lourenço Marques, fl. 14.iii.1920, *Borle* 375 (PRE).

Also in Eritrea, Kenya, Tanganyika, Belgian Congo, SW. Africa, Cape Province and the Transvaal. On dry sandy soils in the hotter and drier parts of our area, either in open woodland or at the margins of pans and seasonal swamps.

Although it has not been possible to see the type of *Portulaca crocodilorum* it is probably not distinct from *P. kermesina* as it appears from the description to differ only in having blunter petals.

4. **Portulaca collina** Dinter in Fedde, Repert. **19** : 139 (1923).—Poellnitz in Fedde, Repert. **37** : 309 (1934).   Type from SW. Africa.

Perennial branching herb up to 10 cm. tall ; roots rather swollen and parsnip-like ; stems often reddish, c. 2·5 mm. in diam.   Leaves opposite, sessile, fleshy, up to 7 × 2 mm., lanceolate-acuminate, margins somewhat revolute, ± flattened on both sides; stipular hairs whitish, narrowing gradually from a dilated base up to 2 mm. wide to a hair-like apex, somewhat longer than both the leaves and the internodes.   Flowers in terminal clusters of 3–4, sessile, surrounded by four apical leaves and numerous hairs scarcely dilated at the base and 4–7 mm. long.   Sepals c. 4 × 3 mm., broadly ovate.   Petals 4, yellow, almost free, c. 4 mm. long, sub-spathulate.   Stamens 8, 4 mm. long.   Ovary ovoid ; style 4 mm. long with 3 stigma-lobes.   Capsule c. 3 mm. long, ovoid-globose, surrounded by the marces-cent corolla, dehiscing horizontally across the middle.   Seeds numerous, c. 0·45 mm. in diam., dark brown, subreniform, somewhat laterally compressed, with concentric lines of minute tubercles.

S. Rhodesia.   W : Bulawayo, st., *Rogers* 13563 (SRGH).
Also in SW. Africa and the Transvaal.   Ecology not sufficiently well known but prob-ably a species of stony dry ground.

5. **Portulaca quadrifida** L., Syst. Nat. ed. 12, **2** : 328 et Mant. Pl. : 73 (1767).—Sond. in Harv. & Sond., F.C. **2** : 382 (1862).—Oliv., F.T.A. **1** : 149 (1868).—Burtt Davy, F.P.F.T. **1** : 165 (1926).—Poellnitz in Fedde, Repert. **37** : 275 (1934).—Exell & Mendonça, C.F.A. **1**, 1 : 115 (1937).—Hauman in F.C.B. **2** : 125 (1951).   TAB. **69** fig. C.   Type a cultivated specimen grown from seed sent from Egypt.
  Portulaca foliosa sensu Bak. f. in Journ. Linn. Soc., Bot. **40** : 25 (1911).

Prostrate annual herb with a somewhat swollen tap-root ; stems often reddish and sometimes rooting at the nodes, sometimes reaching 25 cm. long and c. 1 cm. in diam.   Leaves opposite ; lamina fleshy, up to 10 × 4 mm. but often much less and frequently somewhat shrunk in dried specimens, elliptic, lanceolate, elliptic-oblong or rarely cordate-ovate, apex acute or obtuse, both surfaces more or less flattened ; petiole c. 1 mm. long ; stipular hairs whitish, numerous, 3–5 mm. long.   Flowers 1–4 at the ends of the branches, surrounded by 4 leaves often somewhat larger than the cauline ones, and by numerous hairs c. 4 mm. long.   Sepals 2–4 × 1·5–3·5 mm., ± triangular, obtuse, united at the base.   Petals 4 (5), yellow or orange or very rarely pink or purplish, almost free, c. 5 × 3 mm., elliptic to ovate.   Stamens 8–12.   Ovary conical-ovoid ; style c. 5 mm. long, thickened at the apex and with 4 spreading stigmas.   Capsule conical-ovoid, dehiscing horizontally c. ⅓ of the way up.   Seeds many, c. 1 mm. in diam., greyish, reniform, verrucose with blunt tubercles.

S. Rhodesia.   N : Kariba, fr. vi.1960, *Goldsmith* 78/60 (K ; LISC ; PRE ; SRGH). E : Sabi Valley, Muhenye, fl. & fr. 24.i.1957, *Phipps* 149 (SRGH).   S : Tuli Circle, st. 4.v.1959, *Drummond* 6077 (SRGH).   **Nyasaland.**   N : Nyika, fl. vi.1896, *White* (K).   S : Palombe Plain, fl. 29.vii.1956, *Newman & Whitmore* 303 (BM ; SRGH).   **Mozambique.** N : Mozambique I., fl. 31.x.1942, *Mendonça* 1169 (BM ; LISC).   Z : Vicente, R. Zambeze, fl. 3.x.1887, *Scott* (K).   MS : Vila Machado, fl. & fr. vii.1947, *Pimento* in GHS 17232 (SRGH).   SS : Bilene, Macia, fl. 9.xii.1940, *Torre* 2272 (LISC).   LM : Delagoa Bay, fl. 31.xii.1897, *Schlechter* 11960 (BM ; K).
Cosmopolitan as a weed in the tropics and subtropics.   Apparently confined to the hotter, lower altitude parts to the east of our area, most of our recorded specimens being weeds from cultivated land.

6. **Portulaca rhodesiana** R. A. Dyer & E. A. Bruce in Fl. Pl. Afr. **27** : t. 1069 (1949). TAB. **69** fig. A.   Type : S. Rhodesia, 67 km. E. of Salisbury, *Eyles* 8821 (K ; PRE, holotype ; SRGH).

Ephemeral annual herb with a small tuberous root ; branches reddish, decum-bent, up to 7 cm. long and 1 mm. in diam.   Leaves opposite, succulent and bladder-like when fresh, up to 4 × 3 mm. and 1·75 mm. thick, suborbicular, apex rounded, base cuneate, upper surface red or a mottled olive-green, often whitish or " frosted "-translucent below.   Flowers terminal, solitary and sessile.   Sepals 2 mm. long, ovate, rounded or subacute at the apex, colourless and translucent. Petals bright pink, c. 2·3 × 1·5 mm., obovate, obtuse at the apex, united below for 0·2–0·3 mm.   Stamens 4, as long as or slightly longer than the petals and alter-

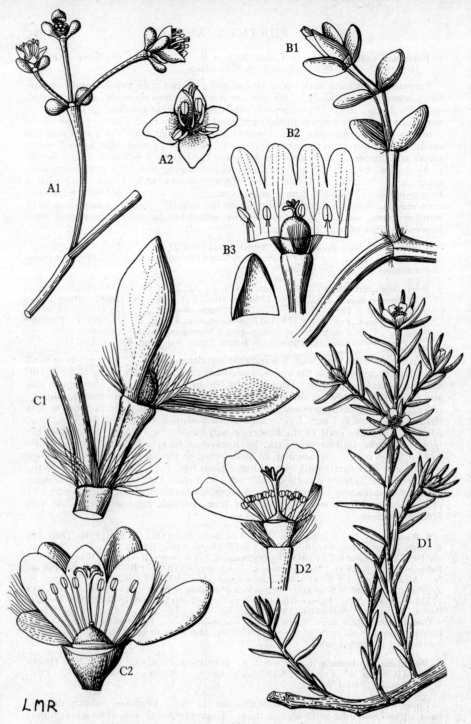

LMR

Tab. 69. A.—PORTULACA RHODESIANA. A1, flowering branchlet (×4) *Wild* 3777 ; A2,
flower (×8) *Wild* 3777. B.—PORTULACA HEREROENSIS. B1, flowering branchlet (×4)
Wild 2489 ; B2, flower with perianth opened out to show gynoecium (×8) *Wild*
2489 ; B3, sepal (×8) *Wild* 2489. C.—PORTULACA QUADRIFIDA. C1, flowering
branchlet (×4) *Schlechter* 11960 ; C2, flower with one sepal, one petal and 4 stamens
removed (×4) *Schlechter* 11960. D.—PORTULACA FOLIOSA. D1, flowering branchlet
(×⅔) *Codd* 7198 ; D2, flower with two petals and one sepal removed (×4) *Codd* 7198.

nating with them ; filaments twisted at the apex, mauve-tinged ; anthers golden. Ovary subglobose ; style c. 1·3 mm. long with 3 stigmas. Capsule c. 1·4 × 1·6 mm., depressed-ovoid, dehiscing horizontally near the middle. Seeds 7–16, blackish, c. 0·5 mm. in diam., suborbicular-reniform, minutely rugulose-granulate (not concentrically).

S. Rhodesia. W : Matopos, fl. & fr. ii.1957, *Garley* 134 (SRGH). C : Chindamora Reserve, Ngomakurira, fl. & fr. 25.iii.1952, *Wild* 3777 (K ; SRGH).
Known at present only from S. Rhodesia. A pioneer in the hollows of bare, granite outcrops, developing as an ephemeral in the course of the rainy season.

7. **Portulaca hereroensis** Schinz in Mém. Herb. Boiss. **20** : 18 (1900).—Poellnitz in Fedde, Repert. **37** : 308 (1934).—Exell & Mendonça, C.F.A. **1**, 1 : 116 (1937). TAB. **69** fig. B. Type from SW. Africa.
Sedopsis hereroensis (Schinz) Poellnitz in Bol. Soc. Brot., Sér. 2, **15** : 152 (1941). Type as above.

Delicate annual herb with spreading or prostrate often reddish branches up to c. 7 cm. long and 1 mm. in diam. Leaves opposite, fleshy, 2–4 × 1–2 mm., elliptic to broadly elliptic or broadly ovate, apex obtuse; stipular hairs fairly numerous, c. 2 mm. long. Flowers solitary or 2–3 together at the ends of the branches, surrounded by a cluster of apical leaves and hairs equalling the length of the leaves. Sepals up to 3 mm. long, ovate, acute at the apex, pale and rather thin-textured. Petals 4, purplish or white, 1–1·5 mm. long, united to about half-way. Stamens 4, alternating with the petals. Ovary ovoid ; style c. 0·5 mm. long, with 4 stigmas. Capsule c. 2 mm. long, ovoid-globose, dehiscing horizontally below the middle. Seeds many, grey, c. 0·5 mm. in diam., reniform, somewhat laterally compressed, with concentric lines of minute blunt tubercles.

Bechuanaland Prot. N : between Leshumo and Thneamansfly, fl. & fr. 21.ii.1877, *Holub* (K). SE : Metsimaklaba near Gaberones, fl. & fr. 14.iii.1940, *van Son* in Herb. Transv. Mus. 28985 (BM ; PRE ; SRGH). **S. Rhodesia.** N : Kariba, fl. & fr. vi.1960, *Goldsmith* 77/60 (K ; SRGH). W : Matobo, fl. & fr. 16.iii.1957, *Garley* 152 (SRGH). C : Hartley, Poole, fl. & fr. 22.iv.1951, *Hornby* 3276 (SRGH). E : Sabi Valley, Muhenye, fl. & fr. 23.i.1957, *Phipps* 140 (K ; SRGH). S : Sabi R., West Bank, fl. & fr. 2.ii.1948, *Wild* 2489 (BM ; K ; SRGH). **Nyasaland.** N : Nymkowa, fl. & fr. ii–iii.1903, *McClounie* 175 (K). **Mozambique.** T : Tete, ii.1859, *Kirk* (K). LM : Magude, Chobela, fl. & fr. 12.ii.1953, *Myre & Balsinhas* 1453 (K ; LM).
Also in SW. Africa and the Transvaal. In shallow soil on rocky outcrops and often a pioneer on bare alluvial soils.

This rather inconspicuous species has long been confused with *P. quadrifida* but its 4 stamens render it quite distinct. It was placed in the genus *Sedopsis* (Engl.) Exell & Mendonça by von Poellnitz (loc. cit.) but the view taken here is that of Dyer & Bruce (Fl. Pl. Afr.: t. 1069 (1949)) that the genus *Sedopsis* cannot be separated satisfactorily from *Portulaca* since, among other reasons, the more recently described *P. rhodesiana* is intermediate in several characters between the two genera. *P. hereroensis* was described as having a single stigma and in young flowers this often appears to be so. In mature flowers the 4 (? or sometimes 3) stigmas separate out.

2. ANACAMPSEROS Boehm.

Anacampseros Boehm. in Ludw., Defin. Gen. Pl. : 309 (1760) *nom. conserv.*

Small perennial herbs or dwarf undershrubs. Leaves rather fleshy, often minute. Stipules membranous, often scarious and silvery, imbricate and hiding the leaves or forming hair-like subscarious fascicles in the axils of the leaves. Flowers sessile at the ends of the branches or in elongated racemes ; bracts membranous, entire or lobed with setaceous segments. Sepals 2, shortly united at the base. Petals 5. Stamens 5–∞, united with the petals at the base. Ovary superior, 1-locular, multiovulate ; style filiform, trifid at the apex or entire. Capsule oblong-ovoid or conical, 3–4-valved, sometimes with the valves longitudinally divided. Seeds numerous, winged or with a minute blunt protuberance.

Anacampseros rhodesica N.E.Br. in Kew Bull. **1914** : 132 (1914).—Poellnitz in Engl., Bot. Jahrb. **65** : 411 (1933). TAB. **70**. Syntypes from S. Rhodesia : Matopos, *McDonald* (K) and *Dowsett* (K) ; Salisbury, *Mundy* (K).
Anacampseros bremekampii Poellnitz in Fedde, Repert. **28** : 29 (1930) quoad

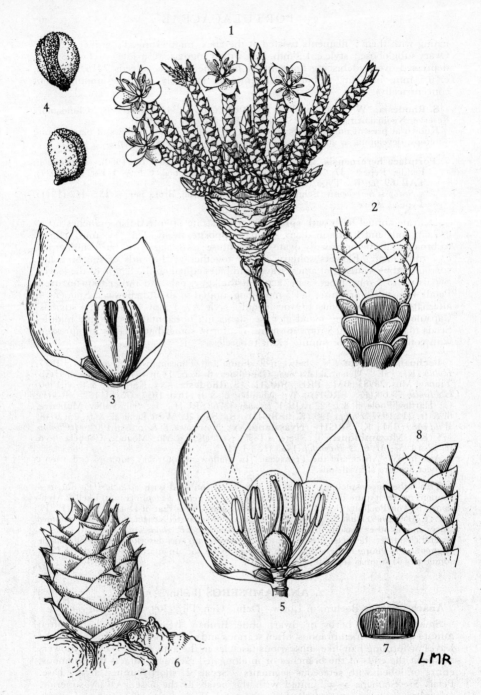

Tab. 70. ANACAMPSEROS RHODESICA.   1, whole plant (×1) *Dowsett* s.n. and photo by *Leach*;
2, stem, lower stipules removed to shew leaves (×6) *Dowsett* s.n. ; 3, capsule and sepals
(×8) *Dowsett* s.n. ; 4, seed (×14) *Dowsett* s.n. ; 5, flower, 2 petals and 3 stamens
removed (×6) *Pienaar* in GHS 34819 ; 6, dry season stem (×6) *McDonald* s.n. ; 7,
leaf (×8) *Dowsett* s.n. ; 8, young rainy-season stem (×6) *Dowsett* s.n.

specim. Eyles.; in Engl., Bot. Jahrb. **65** : 415 (1933).—Pole Evans in Bot. Surv. S. Afr. Mem. **21** : 30 (1948).

Densely caespitose glabrous branching perennial up to c. 3 cm. tall ; branches erect, usually simple, c. 3 mm. in diam. Leaves sessile, green, fleshy, 1 × 1·5–2 mm., oblate, very broadly acute at the apex, entire; stipules submembranous, scaly, silvery, overlapping and hiding the stem and leaves, c. 2·5 × 2 mm., ovate-orbicular, rounded at the apex (particularly low down on the stems), or cuspidate, or very shortly split, or the cusp split into two more or less divergent horns less than 1 mm. long, clasping the stem and erect, or somewhat spreading, or the cusps recurved. Flowers pale pink, apricot or white, terminal, solitary, hidden by the bracts or visible if grown under humid conditions ; bracts 4–7 mm. long, ovate-lanceolate, acute, imbricate, similar in appearance to the stipules. Sepals 3 × 2 mm., broadly ovate, subacute or obtuse. Petals 2·5–6 × 1·5–4 mm., broadly elliptic, obtuse. Stamens 6–8; filaments shortly connate at the base, c. 2 mm. long. Ovary ovoid ; style 1·5 mm. long ; stigma simple. Capsule c. 3 mm. long, 4-valved, greenish, oblong-ovoid. Seeds c. 0·6 × 0·5 mm., ellipsoid with a minute blunt horn or outgrowth, pale brown with an interior transparent integument covered over most of its surface by a paler outer integument.

Bechuanaland Prot. N : Tati Concession, Tantabane, st. 17.iv.1931, *Pole Evans* 3239 (PRE). **S. Rhodesia.** W : Bulalima-Mangwe, Marula, fl. 7.xi.1951, *Plowes* 1307 (SRGH). C : Marandellas, st. 29.vii.1951, *Pienaar* in GHS 33760 (SRGH), flowered in cultivation 21.xi.1951 (*Pienaar* in GHS 34819 (SRGH)). S : Zimbabwe, st. 1920, *Wilman* (PRE).

Not definitely known outside our area but may occur in the Transvaal and SW. Africa. On bare rocky outcrops growing in crevices and shallow pockets of soil with such species as *Actiniopteris australis* (L.f.) Link and *Myrothamnus flabellifolius* Welw.

The flowers are ephemeral and open only in the afternoon. Von Poellnitz has at various times named material of this species *A. rhodesica*, *A. bremekampii* (whose type comes from the Transvaal), or *A. fissa* Poellnitz (Fedde, Repert. **27** : 132 (1930)), whose type is a cultivated specimen originally from the Transvaal. Material from our area named *A. bremekampii* has more closely appressed stipules than typical *A. rhodesica*, and *A. fissa* has recurved shortly divided stipules. All these forms can be seen in the various syntypes of *A. rhodesica* and their differences are probably due to different conditions of humidity and the relative age of the stems. Young rapidly growing stems in the wet season tend to be of the " *A. bremekampii* " type and old stems in the dry season tend to be of the " *A. fissa* " type. I have not seen the type of *A. fissa* but it is very likely synonymous with *A. rhodesica*. The type of *A. bremekampii* has stipules with brown markings and may possibly represent a distinct species.

A. rhodesica is used by natives for beer making and infusions of the plant have been used by them in the treatment of blackwater fever. Its native names in S. Rhodesia are Quilika (Sindebele) and Tirika (Chishona, Kalanga). There is a " Quilika Act " (Ordinance No. 10 of 1916) in existence in S. Rhodesia prohibiting the sale, possession, or use of any plant material or extract designated by the names Quilika or Tirika because of the violently intoxicating effects on human beings of such extracts. Unfortunately it is clear from the debate held in the Legislature when the Act was being passed that two different plants were concerned and the principal one was not an *Anacampseros* sp. but an Arum-like vlei (swamp) plant—this may possibly be the common araceous plant *Zantedeschia melanoleuca* var. *tropicalis* N.E.Br.—although it is clear that the intention was almost certainly to include *Anacampseros rhodesica* also. Apparently no chemical investigations were carried out in support of the evidence produced and no botanical names were used to designate the plants concerned, so the Act is a very unsatisfactory one from both the legal and botanical points of view. It is very desirable that chemical investigations should be carried out on *Anacampseros rhodesica* to see what its properties really are.

3. TALINUM Adans.

Talinum Adans., Fam. Pl. **2** : 245, 609 (1763).

Perennial herbs or shrubs. Leaves alternate or rarely subopposite or rosulate, somewhat succulent; stipules absent. Flowers in terminal or axillary cymes, racemes or panicles, rarely solitary. Sepals 2, opposite. Petals 5 (at least in our species), free or joined at the base. Stamens 5–∞. Ovary superior, 1-locular, multiovulate ; style filiform, with 3 stigma lobes. Capsule chartaceous, 1-locular, many-seeded, opening by 3 valves (6-valved in one species outside our area). Seeds reniform or lenticular, often black or greyish and somewhat compressed ; testa smooth or variously tuberculate, pitted or ridged, with a distinct hilum; embryo annular.

Flowers yellow or golden-yellow, axillary, solitary or more rarely peduncles 2-flowered:
  Seeds with prominent concentric ridges; typical leaves lanceolate to elliptic-lanceolate or ovate, rarely almost linear - - - - - - - - 1. *caffrum*
  Seeds without concentric ridges:
    Seeds smooth, testa with narrowly oblong cells in concentric rings sometimes accompanied by very minute pits; leaves typically linear, very rarely lanceolate to oblanceolate - - - - - - - - - - 2. *crispatulatum*
    Seeds concentrically verruculose; leaves oblong to broadly obovate or ovate, rarely lanceolate-elliptic - - - - - - - - - 3. *arnotii*
Flowers purplish, rose-pink to almost white, in an elongated terminal racemose panicle or raceme - - - - - - - - - - 4. *portulacifolium*

1. **Talinum caffrum** (Thunb.) Eckl. & Zeyh., Enum. Pl. Afr. Austr. Extratrop. **2**: 282 (1836).—Sond. in Harv. & Sond., F.C. **2**: 385 (1862).—Burtt Davy, F.P.F.T., **1**: 166 (1926).—Poellnitz in Fedde, Repert. **35**: 12 (1934). TAB. **71** fig. C. Type from Cape Province.
  Portulaca caffra Thunb., Prodr. Pl. Cap. **2**: [85] (1800). Type as above.

Perennial glabrous herb c. 40 cm. tall from a thickened turnip-like root c. 12 cm. long; stems suberect or erect. Leaf-lamina somewhat fleshy, 1·5–6 × 0·2–1·5 cm., very variable in shape, typically lanceolate to elliptic-lanceolate or ovate, rarely almost linear, apex subacute, mucronate, cuneate at the base, margin sometimes revolute, petiole up to 4 mm. long. Inflorescences axillary, 1– or rarely 2–3-flowered; peduncle c. 2 cm. long, thickened towards the apex, with a pair of subulate bracteoles c. 1·5 mm. long in the upper half. Sepals 0·5–1·0 cm. long, ovate, acuminate at the apex. Petals yellow or golden, about twice the length of the sepals, ovate, rounded at the apex. Stamens numerous; filaments filamentous, c. 4 mm. long. Ovary subglobose; style 2–5 mm. long; stigmas 3, 1–2 mm. long, spreading. Capsule subglobose. Seeds c. 20, shining black with a white hilum, 1·5–2 mm. in diam., reniform-globose, with well defined, elevated, concentric ridges.

Bechuanaland Prot. SE: Kanye, fl. & fr. 13.xi.1948, *Hillary & Robertson* 506 (PRE). **S. Rhodesia.** W: Bulawayo, Hillside, fl. & fr. iii.1944, *Martineau* 280 (SRGH). C: Salisbury Kopje, fl. & fr. 1.xi.1945, *Wild* 321 (K; SRGH). **Nyasaland.** S: Ncheu Distr., Bilila, fl. & fr. 1.ii.1959, *Robson* 1410 (BM; K; LISC; SRGH). **Mozambique.** LM: Sabié, Moamba, fl. & fr. 28.xi.1944, *Mendonça* 3094 (BM; LISC).
Also in Kenya, Tanganyika, Angola, SW. Africa, Cape Province, Orange Free State, Natal and the Transvaal and possibly in Ethiopia and the Sudan. In open woodland or grassland up to altitudes of c. 1550 m.
Von Poellnitz (loc. cit.) describes the seeds of this species as being rather indistinctly concentrically ridged. All the Thunberg material of that species from the Cape, however, has strongly ridged seeds, as is the case with material from our area.

2. **Talinum crispulatum** Dinter in Fedde, Repert. **23**: 369 (1927).—Poellnitz in Fedde, Repert. **35**: 18 (1934). TAB. **71** fig. A. Type from SW. Africa.

Very similar to *T. caffrum* but leaves more typically linear to narrowly linear-lanceolate and only very rarely lanceolate to oblanceolate and the margins revolute and often finely crisped. Petals yellow; seeds shining black and smooth but without concentric ridges. The smooth testa is actually made up of narrowly oblong cells arranged in concentric rings and radiating from the hilum, the outer wall of these cells is very slightly raised in the middle and they are sometimes furnished with very minute pits at the ends of the cells. These details can be seen clearly at a magnification of 25 × or more.

Bechuanaland Prot. N: Kwebe Hills, fl. & fr. xii.1897, *Lugard* 60 (K). SW: 12 km. S. of Kang, fl. & fr. 18.ii.1960, *Wild* 5010 (K; PRE; SRGH). **N. Rhodesia.** E: Forest Reserve W. of Nyimba, Great East Road, fl. & fr. 12.xii.1958, *Robson* 929 (BM; K; LISC; SRGH). S: Nega Nega Siding, fl. & fr. x.1909, *Rogers* 8423 (SRGH). **S. Rhodesia.** N: Urungwe, Rifa R., fl. & fr. 24.ii.1953, *Wild* 4072 (K; SRGH). W: Wankie, fl. & fr. iv.1954, *Levy* 1133 (PRE; SRGH). E: Umtali, fl. & fr. 5.xi.1955, *Chase* 5905 (BM; SRGH). S: Birchenough Bridge, fl. & fr. i.1938, *Obermeyer* 2474 (PRE; SRGH). **Nyasaland.** S: Lake Nyasa, Boadzulu I., fl. & fr. 14.iii.1955, *E.M. & W.* 889 (BM; LISC; SRGH). **Mozambique.** N: Montepuez, fl. & fr. 17.x.1942, *Mendonça* 929 (LISC). T: Lupata Gorge, fl. & fr. 31.xii.1859, *Kirk* (K). MS: Cheringoma, fl. &

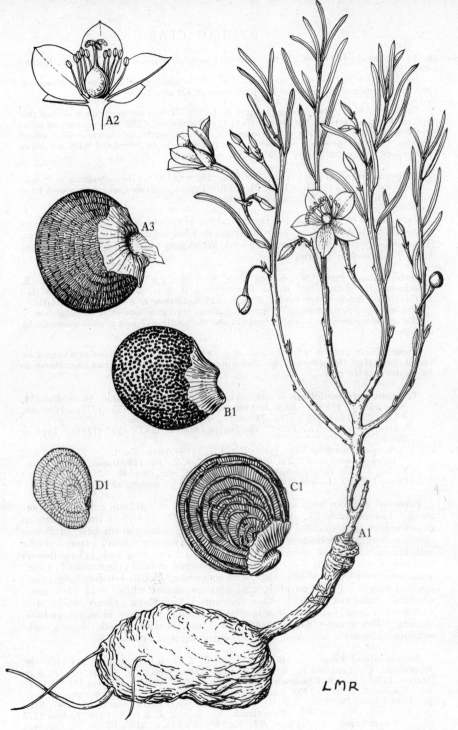

Tab. 71. A.—TALINUM CRISPATULATUM. A1, whole plant (× ⅔) *Lang* in Herb. Transv. Mus.
32199; A2, longitudinal section of flower (× ⁴⁄₃) *Trapnell 523*; A3, seed (× 16) *Trapnell 523*. B.—TALINUM ARNOTII. B1, seed (× 16) *Obermeyer 2431*. C.—TALINUM
CAFFRUM. C1, seed (× 16) *Mendonça 3094*. D.—TALINUM PORTULACIFOLIUM. D1,
seed (× 16) *Obermeyer* in Herb. Transv. Mus. 37448.

fr. 18.ii.1948, *Andrada* 1060 (LISC). LM: Lourenço Marques, fr. 15.xi.1942, *Hornby* 4603 (PRE).

Also in SW. Africa and the Transvaal. With similar ecology and habit to *T. caffrum* but apparently confined to the hotter and drier parts of our area below 1000 m.

The isotype of *T. transvaalense* Poellnitz in the SRGH does not have ripe seeds but this species is almost certainly not distinct from *T. crispatulatum*. The very minute pits usual in the seed of this species are lacking in some specimens, particularly towards the north of its area, but their presence or absence does not seem to be correlated with any other differences, so this can hardly be of specific significance.

3. **Talinum arnotii** Hook. f., Curt. Bot. Mag. **102**: t.6220 (1876).—Poellnitz in Fedde, Repert. **35**: 13 (1934). TAB. **71** fig. B. Type a cultivated specimen grown from seed from the Cape Province (Griqualand).

Very similar to *T. caffrum* but the typical leaves are oblong to broadly obovate or ovate and up to 4 × 2·75 cm., only rarely are they lanceolate-elliptic and c. 4 × 1 cm. The petals are golden-yellow and the seeds are shining black and noticeably concentrically verruculose but not ridged.

Bechuanaland Prot. N: Kwebe Hills, fl. & fr. 22.ii.1898, *Lugard* 188 (K). **S. Rhodesia.** E: Sabi valley, Rupisi Hot Springs, fl. & fr. 28.i.1948, *Wild* 2309 (K; SRGH). S: Birchenough Bridge, fl. & fr. i.1938, *Obermeyer* 2431 (PRE; SRGH).

Also in Cape Province and SW. Africa. Ecology similar to that of *T. crispatulatum*.

The leaves are eaten as a salad by natives and the turnip-like root is often eaten out by animals.

It seems likely that the type specimen of this species was not preserved as it cannot be traced in the Kew Herbarium. The illustration in the Botanical Magazine must therefore be regarded as the type.

4. **Talinum portulacifolium** (Forsk.) Aschers. ex Schweinf. in Bull. Herb. Boiss. **4**, App. 2: 172 (1896).—Exell & Mendonça, C.F.A. **1**, 1: 113 (1937).—Hauman, F.C.B. **2**: 119 (1951). TAB. **71** fig. D. Type from Arabia.

    Orygia portulacifolia Forsk., Fl. Aegypt.-Arab.: CXIV, 103 (1775). Type as above.

    Portulaca cuneifolia Vahl, Symb. Bot. **1**: 33 (1790) *nom. illegit.*

    Talinum cuneifolium Willd. in L., Sp. Pl. ed. 4, **2**: 864 (1800) *nom. illegit.*—Oliv., F.T.A. **1**: 150 (1868).—Bak. f. in Journ. Linn. Soc., Bot. **40**: 25 (1911).—Burtt Davy, F.P.F.T. **1**: 166 (1926).—Poellnitz in Fedde, Repert. **35**: 14 (1934).

Perennial glabrous herb with annual branches c. 1 m. tall from a thickened rootstock. Leaves subfleshy, almost sessile, 2–9 (14·5) × 1–3 (5·5) cm., obovate to oblanceolate, apex rounded or obtuse, mucronate, cuneate at the base. Inflorescence a terminal, elongated, racemose panicle or raceme ; lower bracts leaf-like but diminishing in size upwards ; peduncles c. 1·5 cm. long with 1–3 (6) flowers per peduncle ; pedicels 0·7–1·5 cm. long, recurved in fruit ; bracteoles 1–4 mm. long, membranous, ovate-lanceolate, apex acuminate. Sepals 4–6 mm. long, ovate, apex apiculate. Petals purplish, rose-pink or almost white, 9–12 × 5–6 mm., obovate. Stamens c. 25, with yellow filaments 3–4 mm. long. Ovary ovoid ; style c. 1·5 mm. long ; stigmas 3, spreading. Capsule 5–7 mm. in diam., globose, shining yellow-green. Seeds 30–40, c. 1·2 mm. in diam., black, shining, subreniform, obscurely patterned with concentric rings of oblong cells.

Bechuanaland Prot. N: Botletle valley, fl. & fr. ii.1897, *Lugard* 214 (K). **N. Rhodesia.** E: Luangwa R., fl. & fr. 11.i.1930, *Bush* 4 (K). S: Mazabuka, fl. & fr. *Martin* (FHO ; K). **S. Rhodesia.** N: Chirundu, fl. 20.iii.1952, *Whellan* 642 (SRGH). W: Wankie, fl. & fr. 5.i.1935, *Leach in Eyles* 8281 (SRGH). E: Hot Springs, fl. & fr. 26.i.1949, *Chase* 1160 (BM ; SRGH). S: Triangle Sugar Estate, fl. & fr. 15.i.1947, *Bates in GHS* 15653 (K ; SRGH). **Nyasaland.** S: Zomba, fl. & fr. 5.i.1956, *Jackson* 1773 (K). **Mozambique.** T: Tete, fl. & fr. 1.ii.1860, *Kirk* (K). MS: Báruè, Vila Gouveia, fl. & fr. 1.vii.1941, *Torre* 2964 (BM ; LISC). LM: Maputo, Catuane, fl. & fr. 26.x.1940, *Torre* 1922 (BM ; LISC).

Widely distributed through Africa from the Cape Province to Ethiopia, also in Arabia and India.

A species of open woodland in the hotter and drier parts of our area. The leaves are eaten as spinach.

24. ELATINACEAE
by H. Wild

Herbs or low shrubs. Leaves opposite or verticillate, simple, entire or toothed ; stipules present, paired. Flowers small, actinomorphic, bisexual, axillary, solitary or in small cymes, fascicles or verticillate clusters. Sepals 3–5, imbricate. Petals 3–5(6), imbricate, persistent. Stamens as many, or twice as many as the sepals; anthers 2-thecous, opening by longitudinal slits. Ovary superior, 3–5-locular, multiovulate ; styles 2–5, free. Fruit a septicidal capsule. Seeds without endosperm, straight or curved ; embryo with short cotyledons.

Flowers 4–5-merous ; sepals acute or acuminate with a well-marked midrib ; stamens
5 or more - - - - - - - - - - - - 1. **Bergia**
Flowers 2–3-merous ; sepals obtuse without a well-marked midrib ; stamens 3 **2. Elatine**

1. BERGIA L.
Bergia L., Mant. Pl. Alt. : 152, 241 (1771).

Prostrate or erect annual or perennial herbs or undershrubs, often pubescent, sometimes glandular-pubescent. Leaves opposite, usually sessile or subsessile, obovate, oblanceolate, lanceolate, elliptic or linear, entire or serrate ; stipules usually denticulate or puberulous at the margin. Flowers axillary, sessile or pedicelled, solitary or fascicled or in small cymes, sometimes crowded and verticillate. Sepals 4–5 (6), free, lanceolate, elliptic or oblong, usually keeled and with hyaline margins, imbricate. Petals 4–5 (6), free, as long as or slightly shorter than the sepals, oblong, oblanceolate, or linear-oblong. Stamens 5–12 (usually 5 or 10), those opposite the petals shorter or absent, those alternating with the petals often with filaments broadened at the base ; anther-thecae dehiscing by slits. Ovary 5 (rarely 4)-locular, with many axile ovules in each loculus ; carpels almost free ; styles free, one per carpel. Fruit a septicidal capsule with the valves separating from the central column from above and so releasing the numerous seeds. Seeds subcylindric with rounded ends or 3-angled, often shining, brown or almost black and with the cells of the testa often longitudinally tessellated ; embryo straight or curved.

Annual herbs :
Styles elongated, as long as or almost as long as the ovary   -   2. *mossambicensis*
Styles very short, much shorter than the ovary :
Flowers subsessile, in dense verticillate clusters - - - - 1. *ammannioides*
Flowers single or occasionally paired in the upper axils ; pedicels up to 1·8 cm. long
3. *polyantha*
Perennial herbs with woody stems and/or a strong woody taproot :
Leaves membranous, glandular on both sides, and all other vegetative parts viscid-glandular - - - - - - - - - - - 4. *glutinosa*
Leaves coriaceous, eglandular, sometimes a few gland-tipped hairs on stems, stipules, bracts or sepals :
Leaves lanceolate or ovate-lanceolate ; sepals acuminate, up to 2·5 mm. long
5. *prostrata*
Leaves linear to linear-lanceolate ; sepals with a subulate apex, 3–4 mm. long
6. *decumbens*

1. **Bergia ammannioides** Heyne ex Roth, Nov. Pl. Sp. : 219 (1821).—Roxb., [Hort. Beng. : 34 (1814) *nom. nud.*] Fl. Ind. ed. 2, **2** : 456 (1832) (" ammanoiodes ").—Oliv., F.T.A. **1** : 152 (1868), (" ammanoides ").—Engl., Pflanzenw. Afr. **3**, 2 : 523, t. 237 fig. K–R (1921).—Keay, F.W.T.A. ed. 2, **1**, 1 : 128 (1954). TAB. **72** fig. A. Type from India.

Small erect or decumbent annual, branched or with simple often pinkish stems, pilose with the hairs often capitate, or sometimes quite glabrous, occasionally reaching 30 cm. tall but usually a good deal less. Leaves opposite, sessile or with a very short petiole in our material not exceeding 2 mm. in length ; leaf-lamina

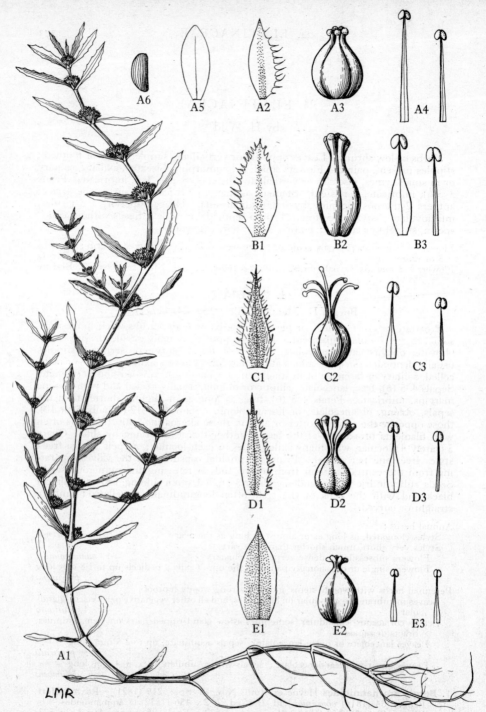

LMR

Tab. 72. A.—BERGIA AMMANNIOIDES. A1, plant (× ⅔); A2, sepal (×20); A3, gynoecium (×20); A4, stamens (×20); A5, petal (×20), all from *Wild* 3336a. A6, seed
(×25). B.—BERGIA PROSTRATA. B1, sepal (×20); B2, gynoecium (×14); B3,
stamens (×20), all from *Codd* 7184. C.—BERGIA MOSSAMBICENSIS. C1, sepal (×14);
C2, gynoecium (×14); C3, stamens (×14), all from *Gazaland Expedition* s.n.
D.—BERGIA DECUMBENS. D1, sepal (×7); D2, gynoecium (×7); D3, stamens (×7),
all from *F.N. & W.* 2113. E.—BERGIA POLYANTHA. E1, sepal (×7); E2, gynoecium
(×7); E3, stamens (×7), all from *Robinson* 738.

0·6–5 × 1·5–2 cm., oblanceolate or oblong-elliptic, acute or blunt at the apex, cuneate and narrowing gradually into the petiole, margin distantly serrulate or almost entire, teeth often pink, with filamentous hyaline hairs on both sides or glabrescent ; stipules 2–3 mm. long, lanceolate to subulate, membranous with ciliolate or serrulate margins.   Flowers white or pinkish, in dense verticillate clusters, subsessile or on slender pubescent pedicels c. 1 mm. long ; bracts c. 1 mm. long, linear or linear-lanceolate, membranous, ciliolate. Sepals 1–1·3 × 0·4–0·5 mm., lanceolate, acute or acuminate, somewhat keeled with the keel pilose or ciliolate, margins membranous and hyaline, ciliolate or entire or often only one margin ciliolate the entire margin being that covered by the adjacent imbricate sepal.   Petals equalling or almost equalling the sepals in length, slightly narrower, lanceolate or oblanceolate, blunt at the apex.   Stamens 5–12, with the alternate ones slightly longer and with filaments usually slightly broader towards the base, the longer c. 1 mm., the shorter c. 0·7 mm. long.   Ovary c. 1 mm. in diam., globose, longitudinally 5-ribbed, 5-locular, glabrous, many-ovuled ; styles 5, very short, c. 0·2 mm. long, swollen and subcapitate at the stigmatic apices.   Fruit a crustaceous capsule, 1–2 mm. in diam.   Seeds c. 0·4 × 0·22 mm., subcylindric with rounded ends, dark brown, shining, very minutely tessellated longitudinally.

N. Rhodesia. B : 13 km. N. of Nangweshi, fl. 23.vii.1952, *Codd* 7177 (K ; PRE). S : Kasungula-Katombora, fl. 25.viii.1947, *Greenway & Brenan* 7977 (K). **S. Rhodesia.** N : Sebungwe, fl. ix.1955, *Davies* 1517 (K ; SRGH). E : Lower Sabi, east bank, fl. & fr. 28.i.1948, *Wild* 2319 (K ; SRGH). S : Sabi-Lundi Junction, Chitsa's Kraal, fl. 4.vi.1950, *Wild* 3336a (K ; SRGH).

Widespread in Africa, Asia and Australia. Usually found in damp or swampy situations.

This species varies greatly in size and degree of pubescence. *Wild* 2319, for instance, represents a completely glabrous form not matched, it appears, by any other African gatherings to date. Glabrous forms like this, however, are fairly common in Asia and *Haussknecht* (K) from South Persia and *Stocks* 1129 (K) from Baluchistan are almost identical with our plant. This apparent isolation of representatives of a particular form or variety may be due either to the fact that this is a small and inconspicuous species often missed by collectors, or, alternatively, because the very small seeds are carried long distances by migratory wading birds.

2. **Bergia mossambicensis** Wild in Bol. Soc. Brot., Sér. 2, **31** : 93 (1957). TAB. **72** fig. C. Type : Mozambique, Sul do Save, Guíjà District, R. Limpopo, *Gazaland Exped.* (PRE ; SRGH, holotype).

Erect herb up to 20 cm. tall and somewhat branched ; stems brownish, pilose but becoming glabrous later. Leaves opposite, sessile or subsessile, 0·7–2·5 × 0·25–0·8 cm., rather membranous, narrowly elliptic, narrowly oblanceolate or oblanceolate, acuminate or acute at the apex, cuneate at the base, margin serrulate with the teeth gland-tipped ; stipules 2–3 mm. long, subulate, pilose. Flowers in axillary 5–9-flowered fascicles or cymes, on very slender, pilose pedicels up to 6 mm. long ; bracts up to 2 mm. long, subulate, single-nerved, with ciliolate-pubescent margins. Sepals 1·5–2 × 0·5 mm., lanceolate, subulate at the apex, pilose on the back with a green keel and purplish margins. Petals probably white, slightly shorter than the sepals, c. 1·5 × 0·75 mm., subacute or obtuse. Stamens 10 or occasionally 12, filaments broadened at the base and narrowing gradually to a subulate apex, those opposite the sepals c. 1·5 mm. long and those opposite the petals c. 1·3 mm. long ; anthers versatile, broadly oblong and c. 0·3 mm. long. Ovary ovoid, glabrous, 5-locular, longitudinally 5-ribbed, loculi many-ovuled ; styles 5, slender, recurved, as long or almost as long as the ovary. Capsule c. 1·5 mm. in diam., splitting into 5 valves. Seeds brown, 0·4–0·5 × 0·15–0·2 mm., very numerous, narrowly oblong-cylindric, minutely longitudinally striate.

Mozambique. SS : Guíjà District, R. Limpopo, fl. & fr. vii.1915, *Gazaland Exped.* (PRE ; SRGH).

Recorded as being found along the banks of the R. Limpopo.

Known only from the type collection.

3. **Bergia polyantha** Sond. in Linnaea, **23** : 16 (1850).—Sond. in Harv. & Sond., F.C. **1** : 116 (1860).—Oliv., F.T.A. **1** : 153 (1868).—Burtt Davy, F.P.F.T. **1** : 147 (1926).— Exell & Mendonça, C.F.A. **1**, 1 : 118 (1937). TAB. **72** fig. E. Type from Orange Free State (Vaal River).

Small annual with prostrate or decumbent, 4-sided, glabrous, often purplish stems 5–8 cm. long, often forming a rosette. Leaves opposite, sessile or subsessile, 0·5–1·7 × 0·2–0·7 cm., narrowly elliptic, elliptic or oblanceolate, acute or subacute at the apex, cuneate at the base, margin distantly denticulate or entire, glabrous and glaucous on both sides ; stipules 1–2 mm. long, linear-lanceolate, with scarious-denticulate margins. Flowers solitary or rarely two together in the leaf-axils, on glabrous slender pedicels up to 1·8 cm. long, crowded towards the ends of the branches ; bracts c. 2 mm. long, subulate. Sepals 3–5 × 2–3 mm., ovate, acute or acuminate, exposed margins purplish, margins overlapped by other sepals hyaline and membranous. Petals white or rose, 2·5–3 × 1·5–2 mm., elliptic, blunt or sub-acute at the apex. Stamens 10, subequal; filaments c. 2 mm. long, somewhat broadened at the base, tapering to the subulate apex ; anthers c. 0·3 mm. long, ovate. Ovary ovoid, longitudinally 5-ribbed; styles very short, less than 0·5 mm. long, recurved; stigmas capitate. Capsule c. 2·5 mm. in diam., splitting into 5 valves. Seeds brown, c. 0·4 × 0·3 mm., very numerous, broadly oblong-cylindric, minutely longitudinally striate.

Bechuanaland Prot. N : Okovango Swamps, Karo I., fl. 28.ix.1954, *Story* 4773 (K ; PRE). **N. Rhodesia.** S : Mapanza, fl. & fr. 9.v.1954, *Robinson* 738 (K ; SRGH). **S. Rhodesia.** S : Gwanda, Machachuta, fl. & fr. v.1955, *Davies* 1281 (SRGH).
Also in the Orange Free State and Angola. An inconspicuous herb of damp places.

4. **Bergia glutinosa** Dinter & Schulze-Menz in Notizbl. Bot. Gart. Berl. **15** : 453 (1941).
   Type from SW. Africa.

Very viscid glandular-pubescent perennial branching herb with spreading or ascending stems up to 30 cm. long and a long woody tap-root. Leaves opposite, sessile or subsessile, up to 4·5 × 0·8 cm., lanceolate to elliptic-lanceolate, apex acute, cuneate at the base, margin finely and irregularly serrate; stipules up to 4 mm. long, lanceolate to linear-lanceolate, glandular. Flowers in dense, axillary, sub-sessile, 2–5-flowered cymes ; pedicels up to 6 mm. long, but mostly shorter, glandular. Sepals 4–5 mm. long, lanceolate, apex long-acuminate, margin hyaline and often pinkish, somewhat keeled, glandular outside. Petals 2·5–3 mm. long, oblong-elliptic, mucronulate. Stamens 10, the outer 5 slightly longer than the inner and with filaments slightly more dilated towards the base. Ovary ovoid, longitudinally 5-ribbed, many-ovulate; styles spreading, c. 1 mm. long ; stigmas capitate. Ripe capsules and seeds not seen.

Bechuanaland Prot. N : Nata R., 11 km. from mouth near Madsiara Drift, fl. 23.iv.1957, *Drummond & Seagrief* 5186 (K ; SRGH). **S. Rhodesia.** S : Beitbridge Distr., Chiturupadzi Store, fl. 12.v.1958, *Drummond* 5762 (K ; SRGH).
Also in SW. Africa. Sandy river banks and seasonal swamps.

5. **Bergia prostrata** Schinz in Mém. Herb. Boiss. **20** : 22 (1900).—Engl., Pflanzenw. Afr. **3**, 2 : 524 (1921). TAB. **72** fig. B. Type from SW. Africa (Hereroland).
   Bergia pallide-rosea Gilg in Warb., Kunene-Samb.-Exped. Baum. : 308 (1903).— Exell & Mendonça, C.F.A. **1**, 1 : 118 (1937). Type from Angola (Huila).

Perennial herb or small shrub with prostrate branches ; branchlets with a pubes-cence of hyaline hairs, some of them gland-tipped, becoming glabrous with age ; bark thin and flaking. Leaves crowded on short fertile suberect branchlets, opposite, subsessile, 0·4–1·6 × 0·15–0·7 cm., ovate-lanceolate to oblong, apex acute or subacute, rounded or subcuneate at the base, denticulate or almost entire, glabrous on both sides or puberulous below, ciliolate at the margins ; stipules 2–3 mm. long, linear-lanceolate, glandular-puberulous on back and margins. Flowers in the leaf-axils, white or pink, solitary or in 2–3-flowered fascicles or in few-flowered cymes, often crowded towards the ends of the short fertile branchlets giving the appearance of a dense, verticillate spike or raceme; pedicels c. 1 mm. long, puberulous ; bracts c. 2 mm. long, linear-lanceolate, membranous, one-nerved, glandular-ciliolate at the margins. Sepals 5, 1·8–2·5 × 0·8–1 mm., oblong or lanceolate, apiculate at the apex, the apiculus usually crowned with a small gland, with membranous margins, sparsely puberulous on the dorsal keel, ciliolate at the margins with a proportion of the hairs gland-tipped. Petals slightly shorter than the sepals, obovate-oblong, blunt at the apex. Stamens 10, with the five larger 1·3–1·5 mm. long opposite the petals and with broadened filaments ; the five shorter c. 0·8–1·0 mm. long. Ovary c. 1 mm. in diam., 5-locular, longitudinally 5-ribbed, ovoid or globose, glabrous loculi many-ovuled, ; styles 5, c. 0·5 mm. long, slightly spreading, swollen at the stigmatic apices. Capsule splitting into 5 valves. Seeds c. 0·4 ×0·15 mm., subcylindric, almost black, longitudinally tessellated.

Caprivi Strip. E. of Kwando R., fl. x.1945, *Curson* 1025 (PRE). **Bechuanaland Prot.** N : Ngamiland, Botletle Valley, fl. ii.1897, *Lugard* 203 (K). **N. Rhodesia.** B : Nangweshi, fl. 27.vii.1952, *Codd* 7184 (BM ; K ; PRE ; SRGH). S : Mazabuka, fl. 17.xi.1957, *Robinson* 2498 (K ; SRGH).

Also in Angola, SW. Africa, the northern areas of the Cape Province and in the Orange Free State. Sandy river banks or sandy soils in dry, deciduous *Acacia* or *Colophospermum-Sclerocarya-Combretum-Spirostachys* mixed woodland.

6. **Bergia decumbens** Planch. ex Harv., Thes. Cap. **1** : 15, t. 24 (1859).—Harv. in Harv. & Sond., F.C. **1** : 116 (1860).—Oliv., F.T.A. **1** : 153 (1868).—Gibbs in Journ. Linn. Soc., Bot. **37** : 430 (1906).—Eyles in Trans. Roy. Soc. S. Afr. **5** : 421 (1916).— Engl., Pflanzenw. Afr. **3**, 2 : 524 (1921).—Burtt Davy, F.P.F.T. **1** : 147 (1926).— Garcia in Bol. Soc. Brot., Sér. 2, **20** : 34 (1946). TAB. **72** fig. D. Type from the Transvaal.

A perennial with spreading or decumbent stems from a woody rootstock ; branchlets pubescent with several-celled hyaline hairs or soon glabrescent. Leaves opposite, sessile or subsessile, 1–4 ×0·15–0·8 cm., linear to linear-lanceolate, acute or acuminate at the apex, gradually narrowed to the base, margin sharply serrate, glabrous on both sides or with a pubescence of several-celled hyaline hairs ; stipules 3–5 mm. long, subulate, scarious-margined, glandular-denticulate, attenuate to a sharp point. Flowers white with ovary showing pinkish or purple, in few- or many-flowered verticillate clusters towards the ends of the branches; pedicels 1–4 mm. long, with a hyaline pubescence ; bracts c. 1·5 mm. long, setaceous, sparsely glandular-ciliolate. Sepals 5, 3–4 ×0·75–1·25 mm., oblong-lanceolate or lanceolate, subulate-acuminate and curving outwards somewhat at the apex, dorsally keeled, membranous and ciliolate at the margins, or the latter often entire when overlapped by an adjacent sepal, sometimes pubescent on the back. Petals 5, as long as the sepals, oblanceolate. Stamens 10, those opposite the petals shorter, 2–2·3 mm. long, filaments not broadened; those alternating with the petals 2·2–2·5 mm. long, filaments noticeably widened towards the base. Ovary c. 1 ×1·3 mm., 5-locular, 5-ribbed longitudinally, ovoid, pinkish or purplish, very minutely and darkly pustulate; ovules many ; styles 5, c. 1·5 mm. long ; stigmatic apices slightly capitate. Capsule splitting when ripe into 5 valves. Seeds c. 0·55 ×0·15 mm., subcylindric, blunt at both ends, minutely tessellated, almost black.

Bechuanaland Prot. SE : Mochudi, fl. i–iv.1914, *Harbor* in Herb. Rogers 6582 (K ; PRE). **S. Rhodesia.** W : Bulawayo, fl. x.1902, *Eyles* 1092 (BM ; SRGH). C : Gwelo, fl. x.1919, *Eyles* 1829 (K ; PRE ; SRGH). E : Umtali, fl. 18.xi.1955, *Chase* 5860 (K ; SRGH). S : Fort Victoria, fl. 19.x.1930, *F.N. & W.* 2113 (BM ; K ; LD ; PRE ; SRGH). **Mozambique.** Z : Morrumbala, Massingire, fl. 26.x.1945, *Pedro* 464 (K ; LMJ ; PRE). MS : near Sena, fl. xi.1859, *Kirk* (K). LM : Catembe, fl. 5.xii.1897, *Schlechter* 11607 (BM ; COI ; K).

Also in the Transvaal. Usually in sandy soil near rivers.

Some of the Mozambique plants collected near the coast, including *Schlechter* 11607, are much more pubescent than the majority. The youngest stems of the type are, however, sparsely pubescent and a fairly continuous series seems to exist between the almost glabrous and the rather densely pubescent forms.

2. ELATINE L.

Elatine L., Sp. Pl. **1** : 367 (1753) ; Gen. Pl. ed. 5 : 172 (1754).

Small glabrous herbs, aquatic or found in damp situations. Leaves opposite or more rarely verticillate ; stipules present. Flowers minute, solitary in the leaf axils. Sepals 2–4, shortly connate at the base, obtuse. Petals 2–4, free, white or pink. Stamens as many as the petals or twice the number. Ovary globose, 3–4-locular ; styles minute, 3–4, spreading. Fruit globose, somewhat depressed at the apex, wall membranous ; seeds numerous, straight or curved; testa with a scalariform reticulation.

Flowers sessile or very obscurely pedicellate   -   -   -   -   -   -   1. *triandra*
Flowers on slender pedicels up to 5 mm. long   -   -   -   -   -   2. ? *ambigua*

1. **Elatine triandra** Schkuhr, Bot. Handb. ed. 2, **1** : 345, t. 1096 (1808).—Niedenzu in Engl. & Prantl, Pflanzenfam. ed. 2, **21** : 276, fig. 120k–l (1925).—Keay, F.W.T.A. ed 2, **1**, 1 : 128 (1954). Type from Germany.

A tender glabrous creeping annual herb, often much branched and rooting at the nodes ; branches up to 15 cm. long. Leaves opposite, very shortly petiolate, 3–15 × 1–4 mm., ovate-oblong or linear-lanceolate, rounded or emarginate at the apex, cuneate at the base, entire, penninerved with 2–4 pairs of nerves ; stipules minute, deciduous, ovate-triangular, margin undulate or dentate. Flowers sessile or obscurely pedicellate in the leaf-axils. Sepals (2) 3, c. 0·5 mm. long, triangular, apex obtuse. Petals (2) 3, white or pink, 1–1·5 mm. long, ovate, apex obtuse, Stamens 3, slightly longer than the sepals. Ovary globose, with 3 minute spreading styles. Capsule c. 1·5 mm. in diam., subglobose. Seeds c. 0·5 mm. long, yellowish-brown, oblong-cylindric, slightly curved.

Also in Europe, N. Africa, Senegal and America. Pools and mud banks.

S. Rhodesia. C : Rusape, fl. & fr., *Hislop* Z. 143 (K).

2. **Elatine ? ambigua** Wight in Hook., Bot. Misc. **2** : 103, Suppl. t. 5 (1831).—Niedenzu in Engl. & Prantl, Pflanzenfam. ed. 2, **21** : 276 (1925). Type from India.

A tender, creeping herb very similar to *E. triandra* but the petioles sometimes reach 2 mm. long and the slender pedicels occasionally reach 5 mm. long in fruit. In floral structure the two species are remarkably alike.

Caprivi Strip. Pool close to Linyanti R., Kazungula and Kasane, fl. & fr. 6.vii.1930, *Stephens* 56 (BM).
Also in Asia, Polynesia and eastern Europe. A submerged aquatic plant or occasionally on mudbanks.

Only one gathering of this plant is known from our area and so far it has not been recorded elsewhere in Africa. It is very desirable that fresh or spirit material should be examined to confirm or disprove this tentative determination. In authentic material of *E. ambigua* the pedicels do not exceed 2 mm. in length. Miss Stephens's plant is not localized with certainty in the Caprivi Strip and may have been collected in Bechuanaland Prot. or even N. Rhodesia.

25. GUTTIFERAE
(incl. *HYPERICACEAE*)

By N. K. B. Robson

Trees, shrubs, woody climbers or perennial or annual herbs ; juice resinous. Leaves opposite or rarely subopposite to alternate or whorled, simple, usually entire, exstipulate, containing various glandular secretions. Flowers actinomorphic, bisexual, dioecious or polygamous. Sepals (2) 4–5 (6 or more), quincuncial or decussate. Petals 4–5 (6 or more), free, usually convolute (sometimes imbricate or decussate, rarely absent), alternating with the sepals. Androecium basically of two whorls of stamen fascicles, the outer (antisepalous) one often sterile (" fasciclodes ") or absent, filaments variously united or free (when the androecium may appear polyandrous) ; antipetalous fascicles variously united or free, very rarely each one reduced to a single stamen. Ovary superior, 1–5-locular (rarely to 12-locular), placentation usually axile, sometimes ± parietal, loculi 1–∞-ovulate ; styles free, ± united or absent ; stigmas equal in number to the loculi. Fruit a septicidal (rarely loculicidal) capsule or a berry or drupe. Seeds sometimes winged, carinate or arillate, without endosperm.

Although *Hypericum* and some allied genera are frequently treated as a separate family, *Hypericaceae*, a general view shows that they differ from the rest of the *Guttiferae* no more than certain subfamilies within the latter group do from one another.

Styles 1–5, ± elongate; flowers always bisexual; fasciclodes free or absent:
  Stamen-fascicles free, or stamens apparently free:
    Styles (2) 3–5, free or ± united; fruit with 5 or more small seeds; leaves usually
      with ± densely reticulate venation:
      Petals glabrous; fasciclodes absent; fruit capsular or rarely baccate
                                      1. Hypericum
      Petals hirsute within; 5 free fasciclodes present; fruit baccate or drupaceous:
        Fruit a 5-seeded berry -   -   -   -   - **2. Psorospermum**
        Fruit a drupe with 5 coherent pyrenes -   -   - **3. Harungana**
    Style single with peltate stigma; fruit with 1 large seed; leaves with numerous
      parallel lateral nerves -   -   -   -   - **4. Calophyllum**
  Stamen-fascicles united to form a tube round the ovary; fasciclodes united to form a
    cupule outside the stamen-tube -   -   -   -   **5. Symphonia**
  Styles absent or almost so, stigmas sessile; flowers polygamous or dioecious, rarely
    bisexual -   -   -   -   -   -   -   -   - **6. Garcinia**

1. HYPERICUM L.

Hypericum L., Sp. Pl. **2** : 783 (1753); Gen. Pl. ed. 5 : 341 (1754).

Trees, shrubs, or herbs. Leaves opposite, rarely whorled, sessile or shortly petiolate, entire (very rarely glandular-ciliate), with translucent glandular dots or linear canals and frequently with dark submarginal glandular dots. Flowers terminal, solitary or in a cymose inflorescence, sometimes forming a corymb or panicle, bisexual. Sepals usually 5. Petals usually 5, usually yellow, often red-tinged especially in bud, asymmetrical, often with dark or translucent glandular dots or streaks, persistent after flowering (in tropical African species), swollen nectariferous tissue usually absent. Androecium usually of 5 fascicles of stamens, single or merged together in various ways, few to many stamens in each fascicle, with the filaments free for most of their length; fasciclodes nearly always absent. Ovary with (2) 3–5 axile or parietal placentas; styles (2) 3–5, free or variously united. Fruit a septicidal capsule, rarely indehiscent. Seeds minute, cylindric to ovoid, with a reticulate, pitted or papillose testa.

A large, almost cosmopolitan genus (300–400 species) whose tropical members occur mostly at high altitudes. The species in Africa south of the Sahara belong to four sections, three of which are represented in our area.

Styles 5, ± united; trees or shrubs (Sect. *Campylosporus*):
  Flowers solitary at branch ends -   -   -   -   -   -   3. *revolutum*
  Flowers in corymbose cymes:
    Leaves with tertiary veins forming a conspicuous reticulation, and with a trans-
      lucent glandular dot within each reticulation -   -   - 2. *roeperanum**
    Leaves without visible tertiary veins or glandular dots; linear longitudinal glands
      embedded below the upper surface - -   -   -   - 1. *quartinianum*
Styles (2) 3–5, free; shrublets or herbs:
  Stems terete or 2-lined above; dark glands present; translucent glands usually con-
    spicuous; petals lemon- to golden-yellow; placentation axile; stigmas ±
    capitate (Sect.*Humifusoideum*):
    Sepals acute, subequal; styles 3 (4); leaves sessile or almost so; shrubs or erect
      herbs:
      Petals unspotted (except round the margin) or shortly streaked with faint reddish or
        orange glands; stems usually branched, more rarely erect and unbranched,
        eglandular; slender or bushy shrub, or erect woody herb   4. *conjungens*
      Petals spotted or streaked with blackish glands; stems usually erect and un-
        branched from the base, spotted with dark glands; perennial tufted herb
                             5. *aethiopicum* subsp. *sonderi*
    Sepals obtuse or rounded, unequal; styles (3) 4–5; leaves ± petiolate;
      decumbent or prostrate herbs:

---

* *H. chinense* L., an E. Asia species which is cultivated in gardens, has been confused with *H. roeperanum*, but differs in having deciduous petals and stamen-fascicles, oblong to oblanceolate leaves with rounded apices, and a low shrubby habit. Dark glands are completely absent from this plant.

Fruit capsular; styles 3–4; pedicels usually erect in fruit   -   - 6. *wilmsii*
Fruit baccate; styles (4) 5; pedicels usually reflexed in fruit   7. *peplidifolium*
Stems quadrangular; dark glands completely absent; pale glands usually small; petals apricot- to orange-yellow; placentation parietal; stigmas ± peltate (Sect. *Brathys* subsect. *Spachium*):
Stems ± erect; at least upper leaves ± lanceolate, acute; flowers usually in regular dichasia - - - - - - - - - - - 8. *lalandii*
Stems decumbent or prostrate, creeping, rarely almost erect; all leaves ± oblong-ovate or oblong-obovate, rounded; flowers usually in monochasia and often apparently axillary:
Styles (2) 3 (4), c. 1 mm. long; montane species   -   -   - 9. *scioanum*
Styles (4) 5, c. 0·5 mm. long; plateau species   -   -   - 10. *oligandrum*

1. **Hypericum quartinianum** A. Rich., Tent. Fl. Abyss. **1** : 97 (1847).—Oliv., F.T.A. **1** : 156 (1868) pro parte excl. syn. Schimp.—R. Good in Journ. of Bot. **65** : 332, t. 382 fig. 10–11 (1927).—Milne-Redh., F.T.E.A. Hyperic.: 3 (1953). Type from Ethiopia.

Hypericum ulugurense Engl., Bot. Jahrb. **28** : 434 (1900). Type from Tanganyika.

Shrub 0·5–4·5 m. high. Young stems 4-angled, soon becoming woody and terete. Leaves sessile; lamina 35–90 × 5–27 mm., ovate-lanceolate to oblong-elliptic or oblanceolate, acute or subacute at the apex, broadly cordate or sub-cordate and clasping at the base, with 2–3 pairs of nerves (often inconspicuous) running very obliquely from near the base and anastomosing with a few less oblique secondaries in the apical half (the anastomoses rarely extending nearer the base), without noticeable tertiary veins and reticulations but with pale longitudinal glands, mostly streaks or lines, often inconspicuous, embedded near the upper surface, and dark marginal dots. Flowers in few- to many-flowered terminal corymbose cymes. Sepals lanceolate (more rarely ± ovate), acute, with dark marginal dots. Petals, 1·9–4·1 cm. long, c. 3–4 times as long as the sepals, with dark marginal dots. Androecium of 5 stamen-fascicles, each with c. 30 stamens. Ovary 5-locular; styles 5, united ± to the apex, 8–13 mm. long. Fruit capsular, 5-valved.

N. Rhodesia. N : Fwambo Area, Fisa R., fl. 3.ix.1956, *Richards* 6072 (BM ; K). **Nyasaland.** N : Nyika Plateau, Nacheri, fl. ix.1902, *McClounie* 143 (K). **Mozambique.** N : Lago, Maniamba, R. Lualeze, fl. & fr. 10.x.1942, *Mendonça* 720 (BM ; LISC).

In N. Rhodesia, Nyasaland, and eastern tropical Africa. Rocky places, gullies and river banks in upland and submontane grassland or deciduous woodland, 1120–2250 m.

Most specimens from our area and adjacent areas of Tanganyika have the large flowers and broad leaves with cordate bases that are typical of *H. ulugurense*. However, other specimens from S. Tanganyika have oblanceolate leaves with narrow bases, and in general the variation in leaf shape and flower size is not very well correlated. Hence *H. ulugurense* cannot be recognized as a separate species (cf. Milne-Redhead, loc. cit.).

2. **Hypericum roeperanum** Schimp. ex A. Rich., Tent. Fl. Abyss. **1** : 96 (1847).— R. Good in Journ. of Bot. **65** : 331, t. 582 fig. 6 (1927).—Staner in Bull. Jard. Bot. Brux. **13** : 76 (1934).—Exell & Mendonça, C.F.A. **1**, 1 : 119 (1937).—Bredell in Bothalia, **3** : 582 (1939).—Suesseng. & Merxm. in Proc. & Trans. Rhod. Sci. Ass. **43** : 88 (1951).—Milne-Redh., F.T.E.A. Hyperic.: 3 (1953). Type from Ethiopia.

Hypericum lanceolatum sensu Bak. f. in Journ. Linn. Soc., Bot. **40** : 26 (1911) pro parte quoad specim. *Swynnerton* 681 & 681a pro parte.

Hypericum quartinianum sensu Steedman, Trees, etc. S. Rhod. : 50 (1933).

Shrub or small tree 0·6–5 m. high. Young stems terete, older ones becoming woody, sometimes ± flattened near the inflorescence. Leaves sessile; lamina (25) 30–80 (115) × (6) 10–25 (30) mm., lanceolate or oblong-elliptic, acute or subacute at the apex, narrowed to a clasping base, with several pairs of secondary veins and a reticulation of tertiary veins visible on both surfaces, with translucent glandular dots (or dashes in the uppermost leaves) between the veins, and dark marginal dots. Flowers in few- to many-flowered terminal corymbose cymes. Sepals ovate (more rarely ± lanceolate), acute or obtuse, with marginal (sessile or fringing) dark glands and submarginal dots or streaks. Petals (1·2) 2–2·5 (3·5) cm. long, about 5 times as long as the sepals, bright yellow, usually with a varying number of dark marginal dots (often few). Androecium of 5 stamen fascicles each with c. 45 stamens. Ovary 5-locular; styles 5, united ± to the apex, 9–11 mm. long. Fruit capsular, 5-valved.

N. Rhodesia. W: Mwinilunga Distr., by Matonchi R., fl. & fr. 21.x.1937, *Milne-Redhead* 2880 (K; PRE). **S. Rhodesia.** C: Makoni, fl. & fr. vi.1917, *Eyles* 795 (BM; K; SRGH). E: Inyanga, fl. & fr. 2.viii.1950, *Wild* 3506 (K; PRE; SRGH). **Mozambique.** MS: Manica, Macequece, Serra da Vumba, fl. & fr. 22.xi.1943, *Torre* 6208 (BM; LISC).

In the Transvaal, Mozambique, the Rhodesias, eastern tropical Africa, Ethiopia, Sudan, Belgian Congo and Angola (Huila). Evergreen forest and bushland, moist bamboo thickets and grassland in upland and submontane regions, often by rivers or streams. Normally found between 1500 m. and 2900 m. but occurs at lower altitudes in areas of high rainfall.

H. schimperi Hochst. from Ethiopia and *H. riparium* A. Chev. from W. Africa may not be specifically distinct.

3. **Hypericum revolutum** Vahl, Symb. Bot. **1**: 66 (1790).—Christensen in Dansk Bot. Arkiv. **4**, 3: 39 (1922).—N. Robson in Kew Bull. **14**, 2: 251 (1960). Type from " Arabia felix."

Hypericum kalmii Forsk., Fl. Aegypt.-Arab.: CXVIII (1775) *nom. nud.* Type as above.

Hypericum lanceolatum Lam., Encycl. Méth. Bot. **4**: 145 (1797).—Oliv., F.T.A. **1**: 156 (1868).—Bak. f. in Journ. Linn. Soc., Bot. **40**: 26 (1911).—Eyles in Trans. Roy. Soc. S. Afr. **5**: 420 (1916).—R. Good in Journ. of Bot. **65**: 330, t. 582 fig. 1 (1927).—Staner in Bull. Jard. Bot. Brux. **13**: 74 (1934).—Robyns, Fl. Parc Nat. Alb. **1**: 620, t. 62 (1948).—Milne-Redh., F.T.E.A. Hyperic.: 4 (1953); in Mem. N.Y. Bot. Gard. **8**, 3: 221 (1953).—Keay & Milne-Redh., F.W.T.A. ed. 2, **1**, 1: 287, t. 109 (1954). Type from Réunion.

Hypericum leucoptychodes Steud. ex A. Rich., Tent. Fl. Abyss. **1**: 96 (1847).—R. Good in Journ. of Bot. **65**: 330, t. 582 fig. 3–5 (1927).—Norlindh in Bot. Notis. **1934**: 100 (1934).—Bredell in Bothalia, **3**: 580 (1939).—Phillips in Fl. Pl. S. Afr. **20**: t. 787 (1940).—Pardy in Rhod. Agr. Journ. **53**: 514, cum tt. (1956). Type from Ethiopia.

Much-branched evergreen shrub or tree (1) 2–6 (12) m. high, with scaly bark. Young stems 4-angled, soon becoming woody and terete. Leaves sessile; lamina (11) 20–30 (37) × 2·5–6 (10) mm., lanceolate or oblong-lanceolate, acute at the apex, narrowing to a clasping base, with pinnate or ± reticulate venation, translucent glands (mostly longitudinal streaks or interrupted lines) and small marginal glandular dots (translucent or sometimes dark). Flowers solitary, terminal. Bracts (when present) leaflike, with dark marginal dots. Sepals ovate, somewhat concave, with dark submarginal dots, and sometimes also dark fringing glands. Petals yellow, (1·5) 2·5–3 cm. long, about 3 times as long as the sepals, often tinged with orange, with a few dark marginal dots or eglandular. Androecium of 5 stamen-fascicles, each with c. 30 stamens. Ovary 5-locular; styles 5 with only the lower parts united or almost free, 4–7 mm. long. Fruit capsular, 5-valved.

N. Rhodesia. E: Nyika Plateau, fl. & fr. 27.xi.1955, *Lees* 95 (K). **S. Rhodesia.** E: Inyanga, fl. & fr. 28.x.1946, *Wild* 1549 (K; SRGH). **Nyasaland.** N: Nyika Plateau, fl. xi.1903, *Henderson* (BM). C: Dedza Mt., fl. & fr. 10.ix.1929, *Burtt Davy* 21555 (FHO). S: Luchenya Plateau, Mlanje Mt., fl. & fr. 25.vi.1946, *Brass* 16428 (BM; K; PRE; SRGH). **Mozambique.** MS: Serra da Gorongoza, fl. & fr. 10.x.1944, *Mendonça* 2403 (BM; LISC).

In south-western Arabia and the mountains of East Africa from Eritrea and Ethiopia to Cape Prov.; also in the Cameroons and Fernando Po; and in Madagascar, the Comoro Is. and Réunion. Submontane evergreen forest and bushland, and stream-sides in upland and submontane grassland, 1800–2900 m. (900–3360 m. in East Africa). *H. revolutum* occurs most frequently in the mist belt of montane vegetation. A specimen labelled " Broken Hill " (*van Hoepen* 1231 (PRE)) is probably wrongly localised.

4. **Hypericum conjungens** N. Robson in Kew Bull. **1958**: 397 (1959). Type from Tanganyika (Njombe).

Hypericum sp. A.—Milne-Redh., F.T.E.A. Hyperic.: 12 (1953).

Hypericum sp. B.—Milne-Redh., loc. cit.

Hypericum conjunctum N. Robson in Kew Bull. **1957**: 437 (1958) non Kimura (1938). Type as above.

Slender or bushy shrub or perennial herb 0·3–1·5 m. high (or higher?). Stems erect or somewhat ascending, usually branched above, terete, usually with cortex exfoliating, eglandular. Leaves sessile; lamina (15) 20–30 × 6–12 (17) mm., oblong-ovate to broadly ovate, rounded or retuse at the apex, rounded to cordate-amplexicaul at the base, with ± undulate margins, reticulate venation, numerous prominent yellowish translucent glandular dots, and intramarginal dark glandular dots. Flowers numerous, in rather dense corymbose cymes. Sepals lanceolate or linear, acute or acuminate, subequal, with longitudinal translucent glandular streaks or lines and marginal dark glandular dots. Petals primrose-yellow, red-tinged or red-veined, with dark glandular marginal dots, and sometimes with short orange or reddish streaks dispersed over the lamina. Stamens c. 80, irregularly arranged or in 3 indistinct groups, with filaments ± united at the base ; anthers with a dark gland at the end of the connective. Ovary 3-locular ; styles 3, free, 3–4·5 mm. long. Fruit capsular, 3-valved, erect.

N. **Rhodesia.** E : Nyika Plateau, 4 km. W. of Rest House, fl. 22.x.1958, *Robson* 269 (BM ; K ; LISC ; SRGH).

In south-western Kenya, south-western Tanganyika, south-eastern Belgian Congo and eastern N. Rhodesia. Open grassland, grassy valleys or forest margins, 1800–2550 m.

In many respects *H. conjungens* is intermediate between *H. kiboense* Oliv. (from E. Africa and Angola) and *H. aethiopicum* Thunb. (from S. Africa and Angola). It differs from the former species, however, by its larger, sessile or amplexicaul leaves and from the latter by its habit and by the absence of dark glands from the petal-lamina and the stem. The N. Rhodesian locality is not far from the Nyasaland border, so *H. conjungens* may well occur in that country also.

5. **Hypericum aethiopicum** Thunb., Prodr. Pl. Cap. **2** : 138 (1800).—Sond. in Harv. & Sond., F.C. **1** : 117 (1860). Type from S. Africa (Cape Province).

Perennial herb. Stems erect, or decumbent at the base, tufted, unbranched, arising from an underground crown, (8) 10–45 (60) cm. long (usually 20 cm. or less in our area), or sometimes ± flattened or 2-lined above, eglandular or spotted with dark glands. Leaves sessile, rarely shortly petiolate ; lamina 5–25 × 3–15 mm., variable in shape and size, ovate, oblong or rotund, obtuse to rounded (rarely ± acute) at the apex, rounded to cordate-amplexicaul at the base, with margins often ± revolute, with numerous translucent glandular dots which vary in size, the larger ones prominent and yellowish, and dark glandular dots round the margins and sometimes elsewhere. Flowers in lax or compact terminal few- to many-flowered cymes, rarely solitary ; pedicels short (rarely exceeding 5 mm.), erect in fruit. Sepals lanceolate, acuminate or attenuate, subequal, with translucent glandular dots or short streaks, and dark glandular dots both marginal and dispersed over the lamina. Petals (8) 10–13 (15) mm. long, 2–4 times as long as the sepals, primrose-yellow, usually red-tinged, with dark glandular marginal dots and dots or short streaks dispersed over the lamina (very rarely with only marginal dots). Stamens c. 50–70, irregularly arranged or in 3 or 4 indistinct groups, with filaments ± united at the base ; anthers usually with a dark or orange gland at the end of the connective. Ovary 3 (4)-locular ; styles 3 (4), free, 3–6·5 mm. long. Fruit capsular, 3–4-valved, erect.

Upland regions from S. Rhodesia and Mozambique to Cape Province and in Angola (Huila).

Subsp. **sonderi** (Bredell) N. Robson in Kew Bull. **1957** : 440 (1958). TAB. **73** fig. A. Lectotype from Natal.

Hypericum aethiopicum var. *glaucescens* Sond., tom. cit. : 118 (1860).—Burtt Davy, F.P.F.T. **1** : 251 (1926).—Syntypes from the Transvaal (Apies R. and Magalisberg). *Hypericum aethiopicum* sensu Bak. f. in Journ. Linn. Soc., Bot. **40** : 26 (1911).— Eyles in Trans. Roy. Soc. S. Afr. **5** : 420 (1916).—Norlindh in Bot. Notis. **1934** : 101 (1934).—Exell & Mendonça, C.F.A. **1**, 1 : 120 (1937). *Hypericum sonderi* Bredell in Bothalia, **3** : 578 (1939).—Verdoorn in Fl. Pl. S. Afr. **23** : t. 897 (1943). Lectotype from Natal.

Stems spotted with dark glands, rarely completely eglandular. Sepals and bracts with sessile or sometimes slightly protruding marginal dark glands.

S. **Rhodesia.** E : Chirinda, fl. 22.x.1947, *Sturgeon* in GHS 18203 (K ; SRGH). **Mozambique.** MS : Gogoi, R. Lucite, fl. 21.vii.1949, *Pedro & Pedrógão* 7562 (SRGH).

In north-eastern Cape Province, Orange Free State, Basutoland, Natal, Swaziland, Transvaal, S. Rhodesia, Mozambique and Angola (Huila). Open grassland, less frequently on bare or cultivated areas or in vleis (seasonal swamps) ; from 750 m. to c. 1850 m. in our area.

The typical subspecies differs from subsp. *sonderi* in having eglandular stems and ± glandular-fimbriate sepals ; it is confined to Cape Province. The Rhodesian and Mozambique specimens, which are smaller and tend to be more branched than most of the S.

African specimens, appear to belong to two distinct populations. Those from Chipinga, Melsetter and Mozambique have ovate to oblong, sessile leaves, sometimes longer than the internodes, while the specimens from the Umtali and Inyanga districts have smaller, elliptic or rotund leaves, which may be shortly petiolate and are usually much shorter than the internodes. The distinctness is rather blurred by the Mozambique specimens which, although nearer the Melsetter type than the other, nevertheless show some intermediate characters. The variation may be clinal, and the two populations may only appear to be distinct owing to lack of material from intermediate regions.

6. **Hypericum wilmsii** R. Keller in Bull. Herb. Boiss., Sér. 2, **8**: 179 in clav. (1908).—Bredell in Bothalia, **3**: 579 (1939).—N. Robson in Kew Bull. **1957**: 440 (1958). TAB. **73** fig. B. Type from the Transvaal.

    Hypericum rupestre sensu Perrier in Arch. Bot. Bull. Mens. **1**: 9 (1927) excl. syn., non Jaubert et Spach nec Bojer mss.

    Hypericum nigropunctatum Norlindh in Bot. Notis. **1934**: 103, fig. 8 (1934). Type: S. Rhodesia, Inyanga, *F.N. & W.* 3634 (BM; LD, holotype; PRE; SRGH).

    Hypericum bojeranum sensu Perrier, Fl. Madag. Hypéric.: 4, t. 1 fig. 14–17 (1951), excl. syn., non *H. bojeranum* Perrier in Notul. Syst. **13**: 269 (1949).

Perennial herb. Stems 10–20 (30) cm. long, ± procumbent, numerous, ± branched, arising from a persistent tap root, terete, or sometimes ± flattened or 2-lined above, ± densely spotted with dark glands or eglandular. Leaves shortly petiolate, rarely almost sessile; lamina 3–15 × 2–8 mm., variable in shape and size, obovate or ovate to oblong or almost orbicular, rounded at the apex, cuneate to rounded at the base, with numerous translucent glandular dots which vary in size, the larger ones prominent and yellowish, and with dark glandular dots round the margins. Flowers in terminal few-flowered leafy cymes or solitary; pedicels 4–10 mm. long, usually erect in fruit (± reflexed in some S. Rhodesian specimens). Sepals oblong, or the outer ones obovate, rounded at the apex, unequal; with translucent and dark dots, the dark ones usually confined to the margin, but occasionally occurring elsewhere. Petals 5–9 mm. long, 1½–2 times as long as the sepals, primrose to bright yellow, occasionally red-tinged, with dark glandular dots confined to the margin (but with dark streaks on the lamina in two S. Rhodesian gatherings). Stamens c. 25–30, irregularly arranged or in 3 or 4 indistinct groups, with filaments united at the base; anthers with a dark gland at the end of the connective. Ovary 3–4-locular; styles 3–4, free, 1·5–2·5 mm. long. Fruit capsular, 3- to 4-valved, erect (rarely ± reflexed).

S. Rhodesia. E: Inyanga, fl. & fr. 8.xii.1930, *F.N. & W.* 3634 (BM; LD; PRE; SRGH).

In Cape Province, Basutoland, Transvaal, S. Rhodesia and Madagascar (Centre). Among rocks and in grassy places in high-rainfall areas from 1200 m. to c. 2000 m. in our area.

    H. wilmsii differs from the S. African *H. natalense* Wood & Evans by its decumbent habit, its more slender and usually shorter stems, and its shortly petiolate leaves. Some S. Rhodesian specimens approach *H. peplidifolium* in having ± reflexed fruits, while the S. African representatives usually have eglandular stems like those of *H. natalense* and *H. aethiopicum* subsp. *aethiopicum*. The Madagascar specimens are less variable than those on the mainland. They tend to have broader, more nearly rotund or orbicular leaves, and the ovary is always 4-merous. The specimen to which Bojer gave the name of *H. rupestre* is actually *H. japonicum* Thunb. Perrier's name, *H. bojeranum*, is therefore a synonym for that species.

7. **Hypericum peplidifolium** A. Rich., Tent. Fl. Abyss. **1**: 95 (1847).—Oliv., F.T.A. **1**: 155 (1868).—Staner in Bull. Jard. Bot. Brux. **13**: 68 (1934).—Norlindh in Bot. Notis. **1934**: 101 (1934).—Exell & Mendonça, C.F.A. **1**, 2: 370 (1951).—Milne-Redh., F.T.E.A. Hyperic.: 9 (1953).—Keay & Milne-Redh., F.W.T.A. ed. 2, **1**, 1: 287 (1954).—N. Robson in Kew Bull. **1957**: 443 (1958). TAB. **73** fig. C. Type from Ethiopia.

    Hypericum peplidifolium var. *robustum* Bak. f. in Trans. Linn. Soc., Bot., Ser. 2, **4**: 6 (1894). Type: Nyasaland, Mt. Mlanje, *Whyte* 143 (BM, holotype; K).

Perennial herb. Stems prostrate to procumbent or ascending, rarely ± erect, tufted, usually ± slender, ± branched, arising from an underground crown or sometimes from adventitious buds on the horizontal roots, (5) 10–60 (90) cm. long (usually less than 30 cm. in our area), terete or sometimes slightly 2-lined above, spotted with dark glands (or eglandular outside our area). Leaves shortly petiolate,

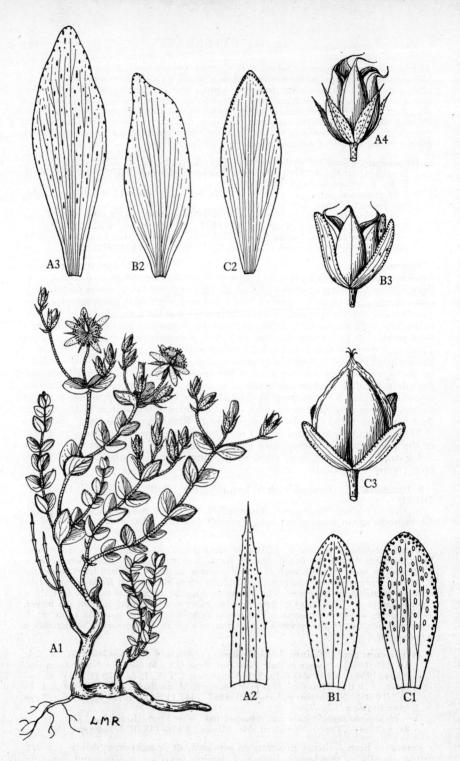

Tab. 73. A.—HYPERICUM AETHIOPICUM SUBSP. SONDERI. A1, part of plant (×⅔); A2, sepal (×5); A3, petal (×5); A4, fruit (×2), all from *Williams* 105 except fruit, *Fries* 3635. B.—HYPERICUM WILMSII. B1, sepal (×5); B2, petal (×5); B3, fruit (×2), all from *E.M. & W.* 124. C.—HYPERICUM PEPLIDIFOLIUM. C1, sepal (×5); C2, petal (×5); C3, fruit (×2), all from *Whyte* s.n.

rarely subsessile ; lamina 3–26 × 2–17 mm., variable in shape and size, ovate, or elliptic to oblong or linear-oblong or obovate, rounded at the apex, cuneate or rounded at the base, with numerous translucent glandular dots sometimes varying in size and ± prominent and yellowish, and dark glandular dots round the margin and occasionally elsewhere. Flowers terminal in few-flowered cymes or solitary, often apparently axillary ; pedicels 5–40 mm. long, ± reflexed in fruit. Sepals elliptic obovate or oblong, obtuse, unequal or very unequal, the inner ones usually much narrower, with a varying number of translucent glandular dots and dark glandular dots which are usually submarginal only but may also occur elsewhere. Petals (5) 7–8 (14) mm. long, 1½–2 times as long as the sepals, primrose to bright yellow, often red-tinged, with dark glandular dots usually confined to the margin (very rarely also dispersed over part of the lamina). Stamens c. 20–40 (60), irregularly arranged or in 3, 4 or 5 indistinct groups, with filaments ± united at the base ; anthers with a dark gland at the end of the connective. Ovary 4–5-locular ; styles 5, less frequently 4, free, 1–4 mm. long. Fruit fleshy, usually subglobose or broadly ovoid, indehiscent.

N. Rhodesia. N : Lake Chila, fl. 29.xii.1954, *Richards* 3786 (K). W : Mwinilunga Distr., SE. of Dobeka Bridge, fl. & fr. 17.xi.1937, *Milne-Redhead* 3283 (K). E : Nyika Plateau, 4 km. W. of Rest House, fr. 22.x.1958, *Robson* 267 (BM ; K ; LISC ; SRGH). **S. Rhodesia.** E : Inyanga, Trias Hill, fl. & fr. xii.1919, *Philomena* in Herb. Eyles 5181 (K ; SRGH). **Nyasaland.** N : Vipya, Chikangawa, fl. & fr. 22.i.1956, *Chapman* 358 (BM). C : Dedza Mt., fl. 10.ix.1929, *Burtt Davy* 21577 (K). S : Shire Highlands, Ndirandi, fl. & fr. xii.1893, *Scott Elliot* 8483 (BM ; K). **Mozambique.** Z : Gúruè, Pico Namuli, fl. & fr. 29.ix.1944, *Mendonça* 2257 (BM ; LISC).
In highlands from Ethiopia and the Sudan (Imatong Mts.) south to S. Rhodesia, N. Rhodesia and Angola (Benguela) ; also in the Cameroons highlands and Fernando Po. Pastures, open woodlands, grassland, disturbed ground, pathsides and streamsides in high rainfall areas from c. 1100–2000 m. in our area, but reaching 3600 m. in East Africa.

H. peplidifolium attains its southernmost limits in S. Rhodesia, where the plants tend to differ in some respects from those of the E. African mountains. All the specimens examined had glandular dots on the stems, unlike practically all the plants seen from other parts of its range. Furthermore, some specimens from Nyasaland and S. Rhodesia were more robust and less prostrate than usual. However, it does not seem possible to recognize satisfactory infraspecific taxa in *H. peplidifolium*.
The specimens from Gurué (Zambesia) resemble *H. natalense* Wood & Evans from SE. Africa in having stout erect stems and almost sessile leaves ; but they also have spotted stems, 5 carpels and succulent fruits on reflexed pedicels—all of which are characters typical of *H. peplidifolium* in adjacent areas. A specimen from Rungwe, SW. Tanganyika (*Geilinger* 2180, sp. C. in F.T.E.A.), which is similar in habit to the Gúruè ones, approaches *H. natalense* even more closely in having eglandular stems and completely sessile leaves, but the fruit characters are those of *H. peplidifolium*.

8. **Hypericum lalandii** Choisy in DC., Prodr. **1** : 550 (1824).—Gibbs in Journ. Linn. Soc., Bot. **37** : 430 (1906).—Bak. f. in Journ. Linn. Soc., Bot. **40** : 26 (1911).—Eyles in Trans. Roy. Soc. S. Afr. **5** : 420 (1916).—Staner in Bull. Jard. Bot. Brux. **13** : 70 (1934).—Norlindh in Bot. Notis. **1934** : 102 (1934).—Gomes e Sousa in Bol. Soc. Estud. Col. Moçamb. **26** : 42 (1935).—Exell & Mendonça, C.F.A. **1, 1** : 120 (1937). —Bredell in Bothalia, **3** : 575 (1939).—Suesseng. & Merxm. in Proc. & Trans. Rhod. Sci. Ass. **43** : 88 (1951).—Perrier, Fl. Madag., Hypéric.: 14, t. 1 fig. 7–8 (1951).— Milne-Redh. F.T.E.A., Hyperic.: 7 (1953). Type from Cape of Good Hope.
Hypericum baumii Engl. & Gilg in Warb., Kunene-Samb.-Exped. Baum : 306 (1903). Type from Angola.
Hypericum lalandii var. *valderamosum* Suesseng. & Merxm., loc. cit. Type : S. Rhodesia, Marandellas, *Dehn* 26a (M, holotype).

Perennial herb (? rarely annual). Stems erect, 8–70 cm. long, single or slightly tufted, sometimes shortly decumbent below, slender, quadrangular. Leaves sessile ; lamina 7–23 × 1–7 mm., that of the middle and upper ones usually lanceolate or narrowly elliptic, acute or obtuse at the apex, clasping at the base, that of the basal leaves and those on sterile shoots relatively shorter and broader, with minute inconspicuous translucent glandular dots. Flowers up to c. 50 (often much fewer), in a loose dichasial cyme or solitary. Sepals ± equal, lanceolate, acute, with longitudinal translucent veins. Petals 6–8 (10) mm. long, 1–2 times as long as the sepals, apricot- or orange-yellow, sometimes marked with red, eglandular. Stamens 40–60, irregularly arranged. Ovary 1-locular ; styles (2) 3–4, free. Fruit capsular, (2) 3–4-valved.

N. Rhodesia. N : Abercorn Distr., L. Chila, fl. & fr. 20.vi.1956, *Robinson* 1696 (K ; SRGH). W : Mwinilunga Distr., by Luao R., fl. & fr. 1.xi.1937, *Milne-Redhead* 3048 (BM ; K ; PRE). C : Broken Hill, Molungushi R., fl. & fr. i.1906, *Allen* 468 (K ; SRGH). S : Choma, fl. & fr. 26.iii.1955, *Robinson* 1174 (K ; SRGH). **S. Rhodesia.** N : Miami, fl. vii.1926, *Rand* 187 (BM). W : Matobo Distr., Besna Kobila Farm, fl. & fr. i.1954, *Miller* 2089 (K ; PRE ; SRGH). C : Salisbury, fl. & fr. 15.iii.1921, *Godman* 101 (BM). E : Inyanga, fl. & fr. 25.xi.1930, *F.N. & W.* 3222 (BM ; PRE ; SRGH). S : Nuanetsi, Rhino Hotel, fl. & fr. xii.1955, *Davies* 1745 (SRGH). **Nyasaland.** N : Vipya Mts., fr. 12.vi.1947, *Benson* 1304 (BM). C : Dedza Mt., fl. & fr. 22.x.1956, *Banda* 282 (BM ; COI ; K ; LISC ; SRGH). S : Cholo, fl. 13.vi.1950, *Wiehe* N/582 (SRGH). **Mozambique.** N : Metonia, Vila Cabral, fl. & fr. 30.x.1934, *Torre* 558 (COI ; LISC). MS : Lower Umswirizwi, fr. 24.xi.1906, *Swynnerton* 1748 (BM ; K). SS : between Cuguno and Inhambane, fl. & fr. 26.x.1935, *Lea* 116 (PRE.) LM : Delagoa Bay, Rikatla, fl. & fr. 1908, *Junod* 2919 (BM; G).

From Sudan (Equatoria) south to Cape Province and west to SW. Africa (Waterberg Plateau), Angola and Nigeria (Bauchi Plateau) ; Madagascar ; also in Bhutan, Khasia and SW. Yunnan. Marshes and wet places in grassland from c. 1200 m. upwards, in areas where the rainfall exceeds c. 75 cm. per annum.

H. lalandii is variable in size, habit, leaf-shape and flower-size, but it does not appear possible to recognize any of the named segregates, at least in our area. In general the plants which flower in the dry season are shorter, more tufted and more branched than those which grow in the wet season. *H. baumii* and *H. lalandii* var. *valderamosum* were described from small plants which might be termed " formae ".

9. **Hypericum scioanum** Chiov. in Ann. Bot. Rom. **9** : 317 (1911).—Gillett & Milne-Redh. in Kew Bull. **1950** : 343 (1951).—Milne-Redh., F.T.E.A. Hyperic. : 12, t. 2 (1953). Type from Ethiopia (Shoa).
   Hypericum stolzii Briq. in Ann. Conserv. Jard. Bot. Genève, **20** : 391 (1919).— Milne-Redh. in Kew Bull. **1948** : 455 (1949) pro parte. Type from Tanganyika.
   Hypericum thoralfi T.C.E. Fr. in Notizbl. Bot. Gart. Berl. **8** : 566 (1923). Type from Kenya.
   Hypericum afropalustre Lebrun & Taton in Bull. Jard. Bot. Brux. **18** : 279 (1947). —Robyns, Fl. Parc Nat. Alb. **1** : 619 (1948). Type from the Belgian Congo.

Perennial herb. Stems prostrate to procumbent or ascending, up to 30 cm. long but often considerably less, sometimes much branched and forming tufts or mats, rooting at the lower nodes, slender, quadrangular. Leaves ± sessile ; lamina 3–9 × 2–5·5 mm., rounded at the apex, broadly cuneate to rounded or subamplexi-caul at the base, with plane margin and minute inconspicuous translucent glandular dots. Flowers solitary, terminal, or in few-flowered monochasial cymes and then appearing axillary. Sepals subequal or unequal, oblong to deltoid or lanceolate, obtuse or acute, with translucent glandular longitudinal lines and submarginal dots. Petals 2–5 (7) mm. long, 1–2 times as long as the sepals, apricot-yellow becoming orange-yellow or orange, occasionally marked with red, eglandular. Stamens 15–27, irregularly arranged or in three indistinct groups. Ovary 1-locular ; styles (2) 3 (4), 0·7–1·5 mm. long, free. Fruit capsular, (2) 3 (4)-valved.

Nyasaland. N : Nyika Plateau, Lake Kaulime, fl. 23.x.1958, *Robson* 291 (BM ; K ; LISC ; SRGH).
From Ethiopia, Kenya and Uganda to Tanganyika, eastern Belgian Congo, northern Nyasaland and northern N. Rhodesia. Damp places, usually in acid peat ; at 2200–2300 m. in our area, (1350) 2190–3360 m. in E. Africa.

H. scioanum differs from *H. lalandii* mainly by its prostrate to ascending habit with a tendency to monochasial development of the inflorescence. In these respects it resembles *H. peplidifolium*, from which it can easily be distinguished by the square stem, the complete absence of dark glandular dots, the apricot to orange-yellow petals and the broadly capitate or peltate styles—all characters typical of Sect. *Brathys* subsect. *Spachium*.

I also observed this species near the Government Rest House on the Nyika Plateau, just within the N. Rhodesian boundary, but did not collect a specimen.

10. **Hypericum oligandrum** Milne-Redh. in Kew Bull. **1948** : 454 (1949). Type : N. Rhodesia, Mwinilunga Distr., *Milne-Redhead* 3399 (BM ; K, holotype).

Closely allied to *H. scioanum* and showing similar variations in habit, but differing in the following characters :
Stems up to 25 cm. long, sometimes becoming almost erect. Leaves (2) 3–12 × (1·5) 2–6 mm., usually with an undulate margin. Petals 2–3·5 mm. long. Stamens c. 12. Styles 4–5, 0·3–0·5 mm. long. Fruit capsular, 4–5-valved.

N. Rhodesia. B : Lungwebungu R., fl. 4.x.1957, *West* 3503 (K ; SRGH). N : Shiwa Ngandu, Lake Young, fl. & fr. 19.i.1959, *Richards* 10739 (K). W : Mwinilunga Distr., Lunga R. just below Madjanyama R., fl. & fr. 25.xi.1937, *Milne-Redhead* 3399 (BM ; K).

Known at present only from N. Rhodesia. In water or mud at lake margins and on river banks, 1200–1400 m.

H. oligandrum is a species of the N. Rhodesian plateau in contrast to the montane *H. scioanum*, and can be distinguished from the latter species by the length (and usually also by the number) of its styles.

2. PSOROSPERMUM Spach

Psorospermum Spach in Ann. Sci. Nat., Bot., Sér. 2, **5** : 157 (1836).

Trees, shrubs or shrublets. Leaves opposite, less frequently subopposite or alternate, petiolate, entire, often furnished with opaque glandular dots and stellate indumentum. Inflorescence a terminal panicle, usually cymose. Flowers bisexual. Sepals 5, with longitudinal linear glands. Petals 5, villous within, furnished with longitudinal glandular lines and swollen nectariferous tissue at the base. Androecium of 5 fascicles of stamens, with few to many stamens in each fascicle and filaments united for most of their length ; fasciclodes 5, fleshy, scale-like, alternating with the fascicles. Ovary 5-locular, with 1 (2)-ovulate loculi and ascending ovules ; styles 5, free. Fruit a berry. Seeds large, with a fleshy glandular-punctate testa.

Note. In *P. chevalieri* Hochr., from Upper Oubangui, all floral parts vary from 5 to 6.

This genus differs from *Vismia* Vand. in having usually one ovule per loculus instead of many and seeds with a fleshy testa. Apart from continental Africa it occurs only in Madagascar, where it is represented by 21 endemic species. No species of *Vismia* is known from our region, but *V. orientalis* Engl. may be present in northern Mozambique.

Leaves coriaceous, broadly cuneate to rounded or cordate at the base, sessile or with petioles rarely over 2 mm. long, pallid below with conspicuous reticulation ; leaves and stems glabrous to densely rusty-tomentose ; tree or shrub - 1. *febrifugum*
Leaves usually chartaceous, narrowly cuneate or attenuate at the base, with petioles usually 3–8 mm. long, green below, the reticulation usually inconspicuous or absent ; leaves and stems usually glabrous, rarely densely rusty-tomentose :
Tree or shrub (1–5 m.), much branched ; petioles (3) 4–8 mm. long ; leaves usually sparsely rusty-tomentose below, at least on the midrib, with dark glands usually confined to near the apex and margins - - - - - - - 2. *baumii*
Shrublet or perennial herb (15–40 cm.), with stems little branched or branching from the base ; petioles (2) 3–4 (6) mm. long ; leaves usually completely glabrous, with dark glands scattered ± densely over the whole under surface. - 3. *mechowii*

1. **Psorospermum febrifugum** Spach in Ann. Sci. Nat., Bot., Sér. 2, **5** : 163 (1836).— Oliv., F.T.A. **1** : 158 (1868).—Ficalho, Pl. Ut. Afr. Port. : 94 (1884).—Sim, For. Fl. Port. E. Afr. : 14 (1909).—Bak. f. in Journ. Linn. Soc., Bot. **40** : 26 (1911).— R.E.Fr., Wiss. Ergebn. Schwed. Rhod.-Kongo-Exped. **1** : 151 (1914).—Eyles in Trans. Roy. Soc. S. Afr. **5** : 420 (1916) pro parte excl. specim. *Rogers* 7259.—Nor-lindh in Bot. Notis. **1934** : 104 (1934).—Gomes e Sousa in Bol. Soc. Estud. Col. Moçamb. **26** : 42 (1935).—Exell & Mendonça, C.F.A. **1**, 1 : 122 (1937) excl. syn. *P. baumii* Engl.—Suesseng. & Merxm. in Proc. & Trans. Rhod. Sci. Ass. **43** : 89 (1951).—Milne-Redh., F.T.E.A. Hyperic. : 17 t. 4 (1953).—Pardy in Rhod. Agr. Journ. **52** : 235, cum tt. (1955). TAB. **74**. Type from Angola.

Psorospermum ferrugineum Hook. f. in Hook., Niger Fl. : 241 (1849). Type from Sierra Leone.

Psorospermum febrifugum var. *albidum* Oliv., tom. cit. : 159 (1868). Syntypes from Angola and Mozambique : R. Zambezi below Tete, *Kirk* (K).

Psorospermum febrifugum var. *glabrum* Oliv., loc. cit. Type from Nigeria.

Psorospermum albidum (Oliv.) Engl., Bot. Jahrb. **17** : 83 (1893).—R. E. Fries, loc. cit. (1914). Same types as *P. febrifugum* var. *albidum*.

Psorospermum campestre Engl., tom. cit. : 84 (1893). Type from Angola.

Psorospermum stuhlmannii Engl., Pflanzenw. Ost-Afr. C : 274 (1895). Type from Tanganyika.

Psorospermum stuhlmannii var. *cuneifolium* Engl., loc. cit. Syntypes from Tanganyika and Mozambique : Quelimane, *Stuhlmann* 663 (B).

Psorospermum febrifugum var. *ferrugineum* (Hook. f.) Keay & Milne-Redh. in Kew Bull. **1953** : 290 (1953). Same type as *P. ferrugineum*.

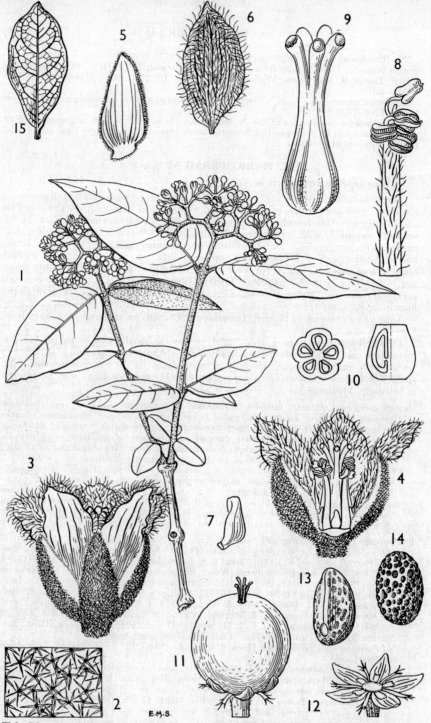

Tab. 74. PSOROSPERMUM FEBRIFUGUM. 1, flowering branch ($\times\frac{2}{3}$); 2, lower surface of leaf ($\times 14$); 3, flower ($\times 8$); 4, flower with a sepal and two petals removed ($\times 8$); 5, sepal from within ($\times 8$); 6, petal from within ($\times 8$); 7, fasciclode ($\times 24$); 8, stamen-fascicle ($\times 16$); 9, pistil ($\times 16$); 10, ovary, longitudinal and transverse sections ($\times 16$); 11, berry ($\times 4$); 12, calyx, fasciclodes and stamen-fascicles after removal of berry ($\times 4$); 13, seed ($\times 4$); 14, seed after drying ($\times 4$); 15, leaf showing lower surface ($\times\frac{2}{3}$). From F.T.E.A.

Shrublet to shrub or small tree (0·3) 1–6 m. high, much branched, with bark peeling or flaking, often corky. Stems terete (sometimes flattened or ± quadrangular above), glabrous to densely rusty-tomentose. Leaves coriaceous, deciduous, usually strictly opposite, subsessile or shortly petiolate; lamina up to 11 × 8 cm. but often much smaller, varying from subrotund to ovate, obovate, or elliptic, acute, obtuse or rounded at the apex, subcordate to rounded or broadly cuneate at the base, glabrous or glabrescent above, varying from glabrous to densely rusty-tomentose below but always white or pallid with conspicuous reticulate venation, the tertiary veins often prominent, with dark glandular dots usually ± confined to the apex and margins; petiole up to 2 (4) mm. long, glabrous or rusty. Inflorescence terminating main or lateral shoots, cymose, appearing paniculate, variable in size and in degree of indumentum, usually pedunculate, branches ± quadrangular. Sepals 3–4 mm. long, elliptic to lanceolate, rusty-tomentose or glabrescent outside. Petals up to 6 mm. long, elliptic to obovate, acute, white or yellowish-white, streaked with dark purple-red. Fasciclodes glabrous, bfid to truncate or apiculate. Stamens 5–6 per fascicle; filaments villous. Berry bright red, c. 10 mm. in diam. Seeds c. 5 mm. long.

N. Rhodesia. N: Abercorn Distr., fl. & fr. 1.xi.1952, *Robertson* 196 (K; PRE; SRGH). W: Mwinilunga Distr., 0·4 km. S. of Matonchi Farm, fl. 6.x.1937, *Milne-Redhead* 2610 (BM; K; PRE). C: 9·6 km. E. of Lusaka, fl. 4.x.1955, *King* 171 (K). E: Fort Jameson, fl., *Gilges* 3 (SRGH). S: Mazabuka Distr., Choma, fr. 7.iii.1952, *White* 2220 (FHO; K). **S. Rhodesia.** N: Mtoko Distr., Vombosi R., fl. & fr. 16.xii.1953, *Phelps* 85 (K; PRE; SRGH). C: Salisbury Distr., Domboshawa, fr. 16.ii.1947, *Wild* 1653 (K; SRGH). E: Inyanga, fl. 29.x.1930, *F.N. & W.* 2369 (BM; K; PRE). S: Buhera, fl. xi.1953, *Davies* 624 (PRE; SRGH). **Nyasaland.** N: Katowo Agricultural House, Mwenemibisuka, fl. 11.xi.1952, *Chapman* 43 (FHO; K). C: Dedza Distr., fr. 19.iii.1955, *E.M. & W.* 1059 (BM; SRGH). S: Neno Hills, fl. 7.xi.1937, *Lawrence* 466 (K). **Mozambique.** N: Mutuali, Malema Rd., fr. 25.ii.1954, *Sousa* 4217 (K; PRE). Z: Mocuba, fl. & fr. 12.xii.1942, *Torre* 4795 (BM; LISC). MS: Chimoio, Gondola, fl. & fr. 7.xi.1941, *Torre* 3795 (BM; LISC). SS: Bilene, Macia, fl. & fr. 10.i.1943, *Torre* 4770 (BM; LISC).

From Angola, S. Rhodesia and Mozambique northward to Sierra Leone, Oubangui-Chari and the Sudan. A plant of open deciduous woods, scrub and wooded grasslands, 15–1950 m.

P. febrifugum is very variable, but it does not appear possible to subdivide it satisfactorily. Extreme types with densely tomentose or completely glabrous leaves are very distinct; but variation in the degree of indumentum is more or less continuous, and does not appear to be correlated with other variable characters such as leaf shape. An atypical specimen from Chirinda (*Wild* 2122) is a shrub 0·6 m. high branching from the base. It has very acute leaves which tend to be alternate or subopposite above, and the whole plant is densely rusty-tomentose. These aberrations are probably due to coppicing or burning and the latter may be the cause of the flowering of plants in the grasslands of the Nyika Plateau when only 0·3 m. high. The leaves of some N. Rhodesian specimens are greenish beneath and cuneate at the base. These may be hybrids with *P. baumii*.

2. **Psorospermum baumii** Engl., Bot. Jahrb. **55** : 383 (1919). Type from Angola (Bié).
    Psorospermum albidum sensu Engl. & Gilg in Warb., Kunene-Samb.-Exped. Baum : 306 (1903).
    Psorospermum tenuifolium sensu R.E.Fr., Wiss. Ergebn. Schwed. Rhod.-Kongo-Exped., **1** : 151 (1914).
    Psorospermum febrifugum sensu Eyles in Trans. Roy. Soc. S. Afr. **5** : 420 (1916) pro parte quoad specim. *Rogers* 7259.—Exell & Mendonça, C.F.A. **1**, 1 : 123 (1937) pro parte.
    Vismia corymbosa sensu Eyles, loc. cit.

Shrub or small tree 1–5 m. high, much branched, with bark often peeling and exuding resin. Stems quadrangular at first, eventually terete, glabrous or ± densely rusty-tomentose when young but later always glabrous. Leaves membranous, strictly opposite, petiolate; lamina 2·5–7·5 × 1·5–4 cm., elliptic or oblong to obovate, rounded to obtuse or slightly acuminate at the apex, cuneate or attenuate at the base (never rounded), glabrescent or with a persistent sparse rusty indumentum on the midrib below and rarely also on the lamina, glossy, dark to light green above, concolorous or paler green below (never whitish), with the tertiary venation not prominent and usually inconspicuous below, with dark glandular dots usually ± confined to the apex and margins; petiole 3–8 mm. long, glabrous or rusty-tomentose. Inflorescence terminating main or lateral shoots, cymose, appearing paniculate, few- to many-flowered, pedunculate, with branches quadrangular, ± rusty-tomentose. Sepals (2·5) 3–4 mm. long, elliptic to lanceolate, rusty-tomentose outside, usually glabrescent. Petals up to 6 mm. long, elliptic to obovate, white or yellowish-white streaked with dark purple-red. Fasciclodes glabrous, spathulate, truncate or apiculate. Stamens 5–6 per fascicle ; filaments glabrous or sparsely villous. Berry crimson, c. 6 mm. in diam. Seeds c. 3 mm. long.

N. Rhodesia. B : 16 km. N. of Senanga, fl. 31.vii.1952, *Codd* 7294 (BM ; K ; PRE ; SRGH). N : Mporokoso, Pansa R. east of boma, fl. & fr. 6.x.1949, *Bullock* 1148 (BM ; K). W : Mwinilunga Distr., near source of Matonchi R., fl. 7.x.1937, *Milne-Redhead* 2615 (BM ; K ; PRE). S : Livingstone, fl. i.1910, *Rogers* 7245 (K).
Occurs only in N. Rhodesia, Angola (Bié) and the Belgian Congo (Katanga). In riverine forest edges and open woodland, or in bush on Kalahari Sand.

P. baumii appears to be related to *P. tenuifolium* Hook. f., a species of hygrophilous forests in W. Africa, Cameroons and the Belgian Congo ; but in *P. baumii* the leaves are generally smaller and less distinctly acuminate, the pedicels stouter and the sepals larger than in *P. tenuifolium*, which also has a completely glabrous inflorescence.
P. baumii differs from *P. febrifugum* essentially in the shape of the leaves, the distinctness of the reticulation beneath, the shape of the leaf-base and the length of the petiole.

3. **Psorospermum mechowii** Engl., Bot. Jahrb. **55** : 386 (1919).—Exell & Mendonça, C.F.A. **1**, 1 : 122 (1937). Type from Angola (Malange).
  Psorospermum hundtii Exell & Mendonça in Journ. of Bot. **74** : 133 (1936) ; C.F.A. **1**, 1 : 123 (1937). Type from Angola (Benguela).

Shrublet or perennial herb 15–40 cm. high. Stems branched or almost un-branched, ascending or erect, caespitose, arising from a polycephalous rootstock, quadrangular at first (sometimes ± flattened), eventually terete, glabrous except when very young. Leaves chartaceous or ± coriaceous, opposite or sometimes sub-opposite, petiolate; lamina (2·5) 3–7·5 × 0·6–3·6 cm., broadly elliptic to oblong or oblanceolate, acute or acuminate at the apex, cuneate or attenuate at the base (never rounded), completely glabrous except when very young, dark or olive green and ± glossy above, paler green below (never whitish), with tertiary venation usually prominent, and with dark glandular dots ± densely scattered over the lower surface as well as at the apex and margin ; petiole (2) 3–4 (6) mm. long, glabrous. Inflores-cence terminating main or lateral shoots, cymose, appearing paniculate, few- to many-flowered, pedunculate, with branches quadrangular, glabrescent. Sepals 2–4 mm. long, elliptic to lanceolate, glabrescent. Petals up to 6 mm. long, elliptic to oblanceolate, white, sometimes tinged with pink or green, streaked with dark purple-red. Fasciclodes glabrous, spathulate, apiculate. Stamens c. 8 per fascicle ; filaments glabrous. Berry 6–10 mm. in diam., dark red or black. Seeds c. 3 mm. long.

N. Rhodesia. W : Mwinilunga Distr., slope E. of Matonchi Farm, fl. 6.x.1937, *Milne-Redhead* 2594 (BM ; K ; PRE).
In N. Rhodesia, Angola and the Belgian Congo (Katanga). Open forests, bush and grassland, often on Kalahari Sand.

P. mechowii can be distinguished from both *P. baumii* and *P. tenuifolium* by its habit. It resembles *P. baumii* in its flower size and stout pedicels, but the stem, leaves and mature inflorescence are nearly always completely glabrous, as in *P. tenuifolium*. However, one specimen from Angola (*Young* 1307) has a densely rusty indumentum over the whole plant.
P. hundtii Exell & Mendonça appears to be synonymous with *P. mechowii*. The type specimen does not have the shortly unguiculate petals attributed to it, and the other differentiating character—opacity of the leaves—will vary according to how the specimen is dried.

3. HARUNGANA Lam.

Harungana Lam., Tabl. Encycl. Méth. Bot. : t. 645 (1797).
  Haronga Thou., Gen. Nov. Madag. : 15 (1806) *nom. illegit.*

Trees or shrubs. Leaves opposite, petiolate, entire, with opaque glandular dots and stellate or dendroid indumentum. Inflorescence a terminal corymbose panicle. Flowers bisexual. Sepals 5, glanduliferous. Petals 5, ± villous within, glanduliferous, sometimes with swollen nectariferous tissue at the base. Androecium of 5 fascicles

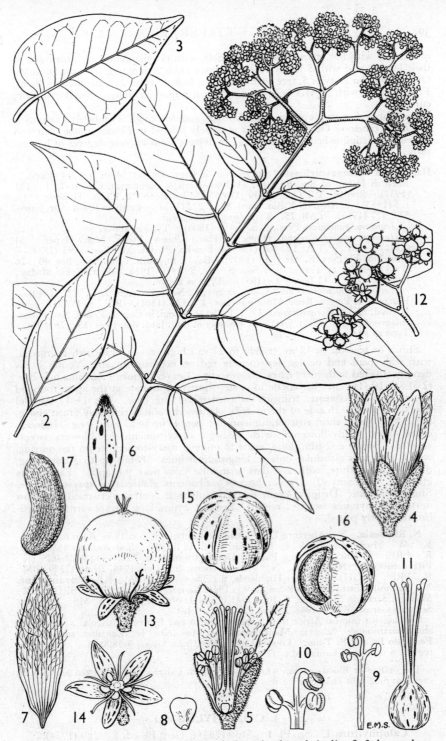

Tab. 75. HARUNGANA MADAGASCARIENSIS. 1, flowering branch ($\times \frac{1}{2}$); 2, 3, leaves showing variation in shape ($\times \frac{1}{2}$); 4, flower ($\times 8$); 5, flower with a sepal and two petals removed ($\times 8$); 6, sepal from within ($\times 12$); 7, petal from within ($\times 12$); 8, fasciclode ($\times 12$); 9, stamen-fascicle ($\times 12$); 10, stamens ($\times 24$); 11, pistil ($\times 12$); 12, part of infructescence ($\times 1$); 13, drupe ($\times 8$); 14, calyx, staminodes and stamen-fascicles after removal of drupe ($\times 8$); 15, coherent pyrenes ($\times 8$); 16, the same cut open to show seed ($\times 8$); 17, seed ($\times 16$). From F.T.E.A.

of stamens with few stamens in each fascicle, with the filaments fused for most of their length, and with 5 fleshy scale-like fasciclodes alternating with the fascicles. Ovary 5-locular ; styles 5, free or united at the base ; ovules 2 (3) per loculus, basal. Fruit a drupe with 5 pyrenes cohering to form a spherical mass. Seeds cylindric, curved.

Probably a monotypic genus, since *Haronga scandens* Engl. is probably a synonym of *Vismia rubescens* Oliv. *Harungana* is closely related to *Psorospermum*, but can be readily distinguished by its fruit. The flowers show varying degrees of hetero-styly.

Harungana madagascariensis Lam. ex Poir. in Lam., Encycl. Méth. **6** : 314 (1806).— Exell in Journ. of Bot. **68** : 181 (1930).—Exell & Mendonça, C.F.A. **1**, 1 : 121 (1937).—Brenan, T.T.C.L. : 249 (1949).—Milne-Redh., F.T.E.A. Hyperic. : 19, t. 5 (1953) ; in Mem. N.Y.Bot. Gard. **8**, 3 : 222 (1953).—Pardy in Rhod. Agr. Journ. **53** : 427 (1956). TAB. **75**. Type from Madagascar.

Arungana paniculata Pers., Syn. **2** : 91 (1806). Type as above.

Haronga madagascariensis (Lam. ex Poir.) Choisy, Prodr. Mon. Hypér. : 34 (1821).—Oliv., F.T.A. **1** : 160 (1868).—Ficalho, Pl. Ut. Afr. Port. : 95 (1884).—Sim, For. Fl. Port. E. Afr. : 14 (1909).—Bak. f. in Journ. Linn. Soc., Bot. **40** : 26 (1911).—Eyles in Trans. Roy. Soc. S. Afr. **5** : 420 (1916).—Perrier, Fl. Madag., Hypéric. : 12, t. 11 figs. 8–13 (1951). Type as above.

Haronga paniculata (Pers.) Lodd. ex Steud., Nom. Bot., ed. 2, **1** : 722 (1840).—Klotzsch in Peters, Reise Mossamb. Bot. **1** : 122 (1861).—R.E.Fr., Wiss. Ergebn. Schwed. Rhod.-Kongo-Exped. **1** : 151 (1914).—Engl. in Engl. & Prantl, Nat. Pflanz-enfam. ed. 2, **21** : 188, t. 76 (1925).—Staner in Bull. Jard. Bot. Brux. **13** : 78 (1934). Type as above.

Shrub or tree up to 12 m. (rarely to 27 m.) high, much branched, evergreen, with scaly bark and orange sap turning red on exposure to air. Young stems densely covered with rusty hairs, glabrescent. Leaves petiolate ; lamina 6·5–20 × (2·5) 3·5–10 (14) cm., lanceolate to ovate, shortly acuminate at the apex, rounded (rarely broadly cuneate, truncate or cordate) at the base, with 14–17 parallel lateral veins on each side of the midrib, glabrescent and dark glossy green above, pallid below with short rusty indumentum ; petiole up to 27 mm. long. Inflores-cence large, many-flowered, with pedicels rusty-tomentose. Flowers sweet-scented. Sepals c. 2 mm. long, ovate to oblong, rusty-tomentose on the outside, with a few dark glandular dots or longitudinal lines. Petals up to 3 mm. long, ovate-elliptic, white, with 2–4 dark glandular dots near the apex. Fasciclodes glabrous. Stamens (2) 3–4 per fascicle ; filaments glabrous or sparsely ciliate. Stigmas capitate. Drupe c. 4 mm. in diam., spherical ; pericarp crustaceous, yellow or orange ; pyrenes each 0–2-seeded. Seeds c. 2 mm. long ; testa varnished, red-brown, faintly pitted.

N. Rhodesia. N : Abercorn Distr., road to Kambole, fl. 6.vi.1936, *Burtt* 6169 (BM ; K). W : Mwinilunga Distr., by Matonchi R., fl. 12.ii.1938, *Milne-Redhead* 4550 (BM ; K ; PRE). **S. Rhodesia.** E : Pungwe Gorge, Inyanga, fr. 7.ix.1954, *Wild* 4607 (K ; PRE ; SRGH). **Nyasaland.** N : Nchena-chena, fr. 21.viii.1946, *Brass* 17379 (BM ; K ; PRE ; SRGH). S : Shire Highlands, fl., *Adamson* 313 (BM; K). **Mozambique.** N : Nampula Distr., near Molócué, fr. 9.vi.1949, *Gerstner* 7146 (K ; PRE ; SRGH). Z : Milange, Serra do Mumbine, fl. 10.xi.1941, *Torre* 3399 (BM ; LISC). MS : Chimoio, Serra de Garuso, fl. 5.iii.1948, *Garcia* 528 (BM ; LISC).

Throughout tropical Africa from Senegal, Sudan and Kenya to Angola, S. Rhodesia, and Mozambique ; also in Madagascar, the Mascarene Is., Zanzibar and Pemba, Fernando Po and S. Tomé. Lowland and upland rain forest, 0–1800 m. ; confined to regions with annual rainfall of c. 113 cm. or more.

The orange or blood-red sap which exudes from a slash or from broken leaves is an excellent field spot-character for this species.

4. CALOPHYLLUM L.

Calophyllum L., Sp. Pl. **1** : 513 (1753) ; Gen. Pl. ed. 5 : 229 (1754).

Trees or rarely shrubs secreting a milky or yellow or clear latex. Leaves opposite, almost always petiolate, entire, often coriaceous, with lateral nerves numerous, slender, close together and parallel (usually nearly at right angles to the midrib), alternating with ± translucent glandular canals. Flowers terminal or axillary, in

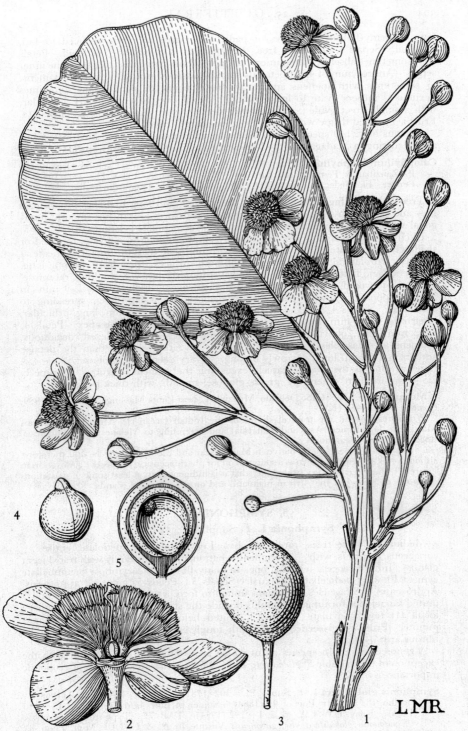

Tab. 76. CALOPHYLLUM INOPHYLLUM. 1, flowering branch (×⅔); 2, vertical section of flower (×2); 3, fruit (×⅔); 4, seed (×⅔); 5, vertical section of fruit with seed removed (×⅔). All from *Faulkner* 762.

few- to many-flowered racemes or paniculate cymes or rarely reduced to 1–3, bisexual. Sepals 4, decussate, free, the inner pair sometimes ± petaloid. Petals 4–8 (sometimes absent), white, imbricate, not always distinguishable from the inner sepals. Androecium of 4 antipetalous stamen-fascicles each of numerous stamens, or sometimes with stamens apparently free ; filaments slender and sometimes flexuous ; anthers ovate to linear-oblong ; fasciclodes absent. Ovary 1-locular, with a single erect ovule ; style simple, slender, often flexuous ; stigma peltate. Fruit a 1-seeded drupe with crustaceous pericarp. Seeds large.

A genus of c. 140 species, most abundant in tropical Asia and Australasia, but also occurring in Madagascar, the Mascarene Is., East Africa and tropical America.

Calophyllum inophyllum L., Sp. Pl. **1** : 513 (1753).—Williams, Useful & Ornamental Pl. Zanzibar & Pemba : 163 cum tab. (1949).—Brenan, T.T.C.L. : 241 (1948).— Perrier, Fl. Madagasc., Guttif. : 6 (1951). TAB. **76**. Type from India.

Tree 15–30 m. high, usually with a short trunk and long branches ; bark pale grey and fawn with shallow elliptic longitudinal fissures ; branches smooth, tetragonal when young, eventually cylindric. Leaves petiolate ; lamina (8) 10 × (4·5) 5–10 cm., coriaceous, concolorous, broadly elliptic-oblong to obovate, rounded or shallowly emarginate at the apex, often with undulate margin, broadly cuneate at the base, with lateral nerves prominent on both surfaces ; petiole 10–15 mm. long, broadened and flattened towards the apex. Inflorescences 7–8 cm. long, racemose, lax, 3–12-flowered, in the axils of the upper leaves. Flower-buds 7–9 mm. in diam., globose. Flowers pedicellate ; pedicels 1·5–4 cm. long, spreading to ascending. Sepals 4, reflexed, deciduous, outer ones c. 7–8 mm. long, orbicular, inner ones c. 10 mm. long, obovate, ± petaloid, rounded at the apex. Petals 4, 9–12 mm. long, obovoid, narrower than the inner sepals, reflexed, deciduous. Stamens in 4 fascicles, ∞ , orange, equal to or rather shorter than the petals ; anthers c. 1·5 mm. long, narrowly oblong. Ovary deep pink, globose ; style c. 4 times as long as the ovary, scarcely exceeding the stamens, flexuous. Drupe c. 2·5–4 cm. in diam. when ripe, green, globose, smooth, with thick pericarp.

Mozambique. N : Cabo Delgado, Muazimi I., near Ponta Massinge, fl. & fr. 4.v.1959, *Gomes e Sousa* 4458 (K).

Widespread round the tropical shores of the Indian Ocean (E. Africa, Madagascar, Mascarene Is., tropical Asia and Malaysia) and extending to Melanesia and Polynesia. Rocky and sandy sea-shores.

C. inophyllum appears to be native in Madagascar and the Mascarene Is. and therefore, although it is sometimes grown as a source of oil for pharmaceutical purposes (" Alexandrian Laurel "), there seems every probability that it is indigenous in at least some of its stations along the east coast of the African mainland and on the off-shore islands.

5. SYMPHONIA L. f.

Symphonia L. f., Suppl. Pl. : 49 (1781).

Medium or large trees, rarely shrubs. Leaves opposite, petiolate, entire, ± coriaceous, glabrous, with ± prominent venation, opaque or rarely with translucent glands. Inflorescence a terminal one- to many-flowered, corymbose or umbellate cyme. Flowers pedicellate, bisexual. Sepals 5. Petals 5, incurved at anthesis. Androecium of 5 fascicles of stamens united to form a tube round the ovary and 5 united fasciclodes forming an annulus outside the stamen tube. Ovary 5-locular ; loculi (1) few–∞ -ovulate ; styles 5, united below, spreading above ; stigmas minute. Fruit a 1–3-seeded berry with tough epidermis. Seeds large, with a fibrous aril.

A genus of about 16 species all of which are confined to Madagascar with the exception of the variable *S. globulifera* L. f. This species yields a resin of economic importance.

Symphonia globulifera L. f., Suppl. Pl. : 302 (1781).—Oliv., F.T.A. **1** : 163 (1868).— Ficalho, Pl. Ut. Afr. Port. : 95 (1884).—Staner in Bull. Jard. Bot. Brux. **13** : 143 (1934).—Keay, F.W.T.A. ed. 2, **1** : 292 (1954). TAB. **77**. Type from Surinam.
Symphonia globulifera var. *gabonensis* Vesque in A. & C. DC., Mon. Phan. **8** ; 231 (1893). Type from Gaboon.
Symphonia gabonensis (Vesque) Pierre in Bull. Soc. Linn. Par. **2** : 1228 (1896).— Exell & Mendonça, C.F.A. **1**, 1 : 130 (1937).—Brenan, T.T.C.L. : 244 (1949). Type as above.

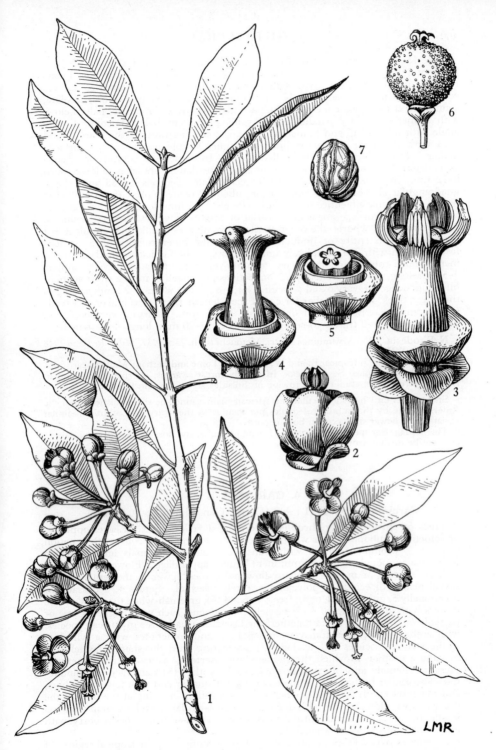

Tab. 77. SYMPHONIA GLOBULIFERA. 1, flowering branch (×⅔); 2, flower (×2); 3, flower, petals removed (×4); 4, flower, perianth and most of staminal tube removed (×4); 5, as 4, but showing the ovary in transverse section (×4); 6, fruit (×⅘); 7, seed (×⅘), all from *Milne-Redhead* 2948 except fruit and seed, *Holmes* 1249.

Tree 15–30 m. (rarely to 36 m.) high, with straight bole and a crown of short horizontal branches ; bark smooth, dark or reddish-brown. Branches longi- tudinally grooved and slightly flattened when young, eventually cylindric. Leaves coriaceous, petiolate ; lamina 5–12 × 1·3–4 cm., oblong or elliptic to oblanceolate, acuminate at the apex, cuneate at the base, dark glossy green above, paler below, with numerous closely parallel lateral veins prominent below forming a wide angle with the midrib and branching and anastomosing towards the margin, and with dark glandular streaks or lines sometimes visible on the lower surface ; petiole 5–15 (20) mm. long, channelled above. Inflorescences corymbose or sub- umbellate many-flowered cymes, terminating short lateral branches. Flower-buds 6–12 mm. in diam., globose. Flowers pedicellate; pedicels 12–25 mm. long, quadrangular above. Sepals unequal or subequal, ovate to orbicular, rounded at the apex, with margin entire or ciliolate. Petals 7–15 mm. long, 5–6 times as long as the sepals, carnose, ± orbicular, scarlet or bright- or yellowish-red, deciduous. Androecium of 2 whorls, the outer sterile, forming a ± pentagonal cupule 1·5–4 mm. high, with entire or undulate margins ; the inner comprising 5 stamen- fascicles each with 3–4 completely united stamens, the fascicles united to form a membranous tube round the ovary, free above, spreading at anthesis ; anthers 2–5 mm. long, linear, parallel, with connectives prolonged to form a short acute or bifid appendage. Ovary ovoid ; loculi (1) 2–4-ovulate ; styles 1–2 times as long as the ovary, thick, fused together for c. $\frac{2}{3}$ of their length, then spreading or recurved ; stigmas minute. Berry ovoid or globose, 15–35 mm. in diam., verrucose, red turning brown, 1–2 (rarely 3)-seeded. Seeds c. 15–20 mm. long, ± flattened.

N. Rhodesia. W : Mwinilunga Distr., Matonchi R., fl. 25.x.1937, *Milne-Redhead* 2948 (BM ; K ; PRE).

Also widespread in tropical E. and W. Africa, S. Tomé and perhaps Madagascar ; also in Central and tropical S. America. In muddy forest swamps, on river banks and in fringing forest ; sometimes also growing on hillsides above the rain forest.

Staner (loc. cit.) has shown that the African plant cannot be distinguished from the American one by the number of ovules per loculus in the ovary. In addition, similar variations in flower size occur in both regions.

The Madagascar species *S. urophylla* (Decne.) Vesque, and possibly *S. oligantha* Bak. f., may not be distinct from *S. globulifera*.

6. GARCINIA L.

Garcinia L., Sp. Pl. **1** : 443 (1753) ; Gen. Pl. ed. 5 : 202 (1754).

Trees or shrubs, rarely shrublets, secreting a yellow latex. Leaves opposite, or sometimes subopposite or whorled, petiolate, entire, coriaceous, with venation usually ± prominent, often with translucent glandular canals and brownish resin canals ; petiole with a ± prominent ligulate appendage. Flowers terminal or axillary, solitary or in few- to many-flowered cymes, fewer in the female or bi- sexual plants, dioecious or polygamous, rarely bisexual. Sepals 4, decussate (or occasionally 5, quincuncial, or 3), free. Petals 4 (5), greenish-white to yellow. Male flowers with androecium of 4 (5) fascicles of stamens, each with filaments free or partially or completely fused together, the fascicles usually free in African species, with anthers sometimes transversely septate ; and sometimes with a whorl of sterile stamen-fascicles (" fasciclodes ") alternating with the stamen-fascicles or forming a cushion in which the stamens are inserted ; ovary-vestige sometimes present. Female and bisexual flowers usually with 4 (5) stamen- or staminode- fascicles, similar to those of the male flowers but smaller and with fewer members, and sometimes with fasciclodes, free or fused together in a ring at the base of the ovary ; ovary globular, 2–5 (12)-locular ; loculi 1-ovulate ; styles absent ; stigma sessile, broad, 2–5-lobed or entire, sticky. Fruit a 1–4-seeded ± fleshy berry, with tough epidermis. Seeds large, arillate.

A genus of over 200 species, usually said to be confined to the tropical regions of the E. Hemisphere. However, the genus *Rheedia* L. from Madagascar and tropical America does not appear to be satisfactorily separable from *Garcinia*, and should probably be merged with it in any future monographic treatment (see N. Robson in Bol. Soc. Brot., Sér. 2, **32** : 171 (1958)).

Stamen-filaments incompletely fused or free, anthers ovoid ; fasciclodes present ; ligule
usually prominent :
Stamen- and staminode-fascicles 5 (rarely 4), filaments united for c. ¾ of their length ;
    inflorescence terminal (Subgen. *Xanthochymus*) - - - - 1. *volkensii*
Stamen- and staminode-fascicles 4, filaments united for ⅔ of their length or less, or
    stamens and staminodes free ; inflorescence axillary (Subgen. *Rheediopsis*) :
Androecium of 4 stamen- or staminode-fascicles alternating with 4 fasciclodes ;
    leaves in pairs :
Leaves ± oblong, rounded to cordate at the base ; main lateral veins curving up-
    wards, with shorter laterals between them ; pedicels and sepals crimson
                                                            2. *smeathmannii*
Leaves lanceolate or oblong-lanceolate, ± cuneate at the base ; lateral veins
    straight, ± equally prominent ; pedicels and sepals pale green   3. *mlanjiensis*
Androecium of free stamens or staminodes, inserted in a fleshy cushion or annulus
    formed by the fused fasciclodes :
Fruit smooth ; leaves usually in whorls of 3 :
Leaf-margin plane or slightly undulate, entire or crenate, not or only slightly
    incrassate ; petals usually small (3–7 mm. long) ; pedicels variable in length
                                                            4. *livingstonei*
Leaf-margin strongly undulate, crenate, markedly incrassate ; petals large (7–10
    mm. long) ; pedicels usually short (9 mm. long or less)   -   5. *pachyclada*
Fruit verrucose ; leaves in opposite pairs   -   -   -   -   10. *sp. A*
Stamen-filaments completely fused, anthers oblong, curved ; fasciclodes absent ; ligule
inconspicuous (Subgen. *Garcinia*) :
Stamen-fascicles 4, antipetalous ; leaves caudate or obtusely acuminate to rounded at
    the apex ; young stems terete or flattened, with 2 pairs of decurrent raised lines :
Leaves with a long cauda and with lateral veins forming an angle of 60°–80° with the
    midrib ; anthers transversely septate   -   -   -   -   6. *punctata*
Leaves shortly acuminate to rounded at the apex and with lateral veins forming an
    angle of 30°–60° with the midrib ; anthers not transversely septate :
Plant a tree or shrub ; leaves always opposite, oblong or elliptic (rarely oblanceolate)
    and with the apex often ± acuminate   -   -   -   -   7. *huillensis*
Plant a rhizomatous shrublet ; leaves often subopposite, usually narrowly oblanceo-
    late or narrowly elliptic and with the apex acute or rounded   -   8. *buchneri*
Stamen-fascicles 2, antisepalous ; leaves shortly and acutely acuminate at the apex ;
    young stems narrowly 4-winged   -   -   -   -   -   9. *acutifolia*

1. **Garcinia volkensii** Engl., Pflanzenw. Ost-Afr. **C** : 275 (1895).—Brenan, T.T.C.L.:
    243 (1949). Type from Tanganyika (Kilimanjaro).
        Garcinia usambarensis Engl., Bot. Jahrb. **40** : 561 (1908). Type from Tanganyika
    (E. Usambaras).
        Garcinia albersii Engl., Bot. Jahrb. **40** : 562 (1908). Type from Tanganyika (W.
    Usambaras).
        Garcinia bangweolensis R.E.Fr., Wiss. Ergebn. Schwed. Rhod.-Kongo-Exped. **1** :
    152, t. 11 (1914). Type : N. Rhodesia, L. Bangweulu, N. of Kasomo, *Fries* 708
    (UPS, holotype).
        Garcinia sp.—R.E.Fr., tom. cit. : 153 (1914). (*Fries* 1235).
        Garcinia bullata Staner in Bull. Jard. Bot. Brux. **13** : 157 (1934).—Exell &
    Mendonça, C.F.A. **1, 1** : 130 (1937). Type : Mozambique, Zambezia, Morrumbala,
    Luja 438 (BR, holotype).

Much-branched glabrous evergreen tree or shrub, (2) 4–20 m. high ; branches
often forming a wide angle with the stem, stiff, grooved or winged, usually ±
flattened at first, eventually ± quadrangular or cylindric (triangular when leaves
are whorled) ; bark grey-brown. Leaves opposite (rarely in whorls of 3), ±
coriaceous, petiolate ; lamina (2) 4–11 × (1) 1·5–6 cm., very variable in size, shape
and texture, lanceolate or oblanceolate to broadly ovate or obovate, acute to rounded-
apiculate at the apex, cuneate to rounded at the base, dark bluish- or yellowish-
green above, paler below, flat or ± bullate, with main lateral veins prominent on
both sides, tertiary venation usually ± prominent above, margins sometimes
incrassate, and with branched translucent glandular canals visible in the young
leaves at least, and longitudinal opaque canals visible below ; petiole 3–18 mm.
long and grooved above, transversely wrinkled ; ligule prominent. Inflorescence
terminal, cymose, loosely branched or ± condensed ; branches quadrangular or ±
flattened ; peduncle up to 4 cm. or absent ; bracts scale-like, carinate. Flowers
dioecious, numerous to single, pedicellate, rarely almost sessile, globose or campanulate. Sepals 5, ± unequal, coriaceous, carinate, orbicular to triangular-ovate,
with the margin often ± papillose. Petals 5, 4–9 mm. long, c. 4–5 times as long as

the sepals, ± carnose, orbicular to broadly obovate, cream to greenish-white, some-times tinged with pink, with yellowish glandular canals radiating from the base. Male flowers with 5 spongy fasciclodes uniting in the centre of the flower, alternat-ing with 5 fascicles each of 5–9 stamens with cream filaments united for most of their length and bearing brown or red anthers. Female flowers with small fasci-clodes alternating with staminode-fascicles ; ovary 2–4-locular, pale green or yellowish ; stigma 5-lobed, peltate. Berry globose, ovoid, or 2–4 lobed, 1–3 cm. in diam., dark green turning yellow, 1–4-seeded. Seeds ovoid, 1–2 (3) cm. long.

N. Rhodesia. N : Lake Bangweulu, Kasomo, ♀ fl. & fr. 21.ix.1911, *Fries* 708 (UPS). W : Chingola, ♂ fl. 14.x.1955, *Fanshawe* 2525 (K). C : Kafue R., fr. 2.iv.1955, *E.M. & W.* 1397 (BM ; SRGH). **Nyasaland.** N : Panda Peaks, ♂ fl ix.1902, *McClounie* 151 (K). **Mozambique.** Z : Serra de Gúruè, Marrequélo, ♂ fl. 20.ix.1944, *Mendonça* 2137 (BM ; LISC).

In Kenya, Tanganyika, N. Rhodesia, Nyasaland and Mozambique. A closely allied species occurs in Natal and the Transvaal. An understorey tree of evergreen rain forest, mist forest and fringing forest. From 1300–2000 m. in our area, but extending from 10 m. (on Mafia I.) to c. 2500 m. in E. Africa.

The leaves of this species vary in thickness and size according to whether it is growing in an open or a shaded locality. It has not been found possible to recognize any subspecies or varieties because the variations in size and shape of the leaves and inflorescence, though great, do not seem to be correlated.

G. natalensis Schlechter (*G. gerrardii* Harv. *nom. illegit.* ; *G. transvaalensis* Burtt Davy) is very closely allied to *G. volkensii*, but the outer floral whorls are 4-merous in this species, whereas they are 5-merous in all the E. African material examined except in one specimen from Tanganyika. The difference may be only of subspecific value.

G. bullata Staner differs from the other specimens only in the degree of bullation of the leaves, and hence is considered to be conspecific with *G. volkensii*. It was collected in Mozambique, not Angola as stated by Staner (l.c.).

2. **Garcinia smeathmannii** (Planch. & Triana) Oliv., F.T.A. **1** : 168 (1868).—Vesque in A. and C. DC., Mon. Phan. **8** : 334 (1893).—Keay, F.W.T.A. ed. 2, **1** : 295 (1954). TAB. **78**. Type from Sierra Leone.

Rheedia smeathmannii Planch. & Triana, Mém. Guttif. : 157 (1867). Type as above.

Garcinia polyantha Oliv., tom. cit. : 166 (1868).—Staner in Bull. Jard. Bot. Brux. **13** : 124.—Exell & Mendonça, C.F.A. **1**, 1 : 127 (1937). Syntypes from Nigeria and Fernando Po.

Garcinia chevalieri Engl. ex R.E.Fr., Wiss. Ergebn. Schwed. Rhod.–Kongo-Exped. **1** : 151 (1914). Type : N. Rhodesia, Bwana-Mkubwa, *Fries* 435 (UPS, holotype).

Garcinia mbulwe Engl., Bot. Jahrb. **55** : 389 (1919). Type from Tanganyika.

Garcinia stolzii Engl., tom. cit. : 391 (1919). Type from Tanganyika.

Much-branched glabrous evergreen tree or shrub (2·4) 4·5–13·5 m. high in our area (up to 21 m. in W. Africa), erect or ± trailing; branches yellowish-green, transversely wrinkled and longitudinally grooved, ± flattened when young, eventually cylindric. Leaves opposite, coriaceous, petiolate ; lamina 12–23 (30) × 4–9·5 (12) cm., oblong or ovate-oblong, obtuse to rounded-apiculate at the apex, broadly cuneate to slightly cordate at the base (rarely more narrowly cuneate), bluish green above, paler below, with prominent venation on both sides, the 16–25 main laterals curving upwards towards the leaf margin, interspersed with shorter secondary laterals and tertiary reticulation, but with no visible secretory system ; petiole 8–c. 20 mm. long, channelled above, longitudinally grooved ; ligule promi-nent. Flowers dioecious (? or polygamous), in fascicles of 5–30 in the axils of the older leaves and on old wood ; pedicels (10) 15–45 mm. long, crimson. Sepals 4, decussate, unequal, ovate to orbicular, obtuse, crimson. Petals 4, 4–8 (10) mm. long, 1–1·5 times as long as the inner sepals, ± carnose, obovate, cream-white, sometimes pink-tinged, with longitudinal translucent glandular lines. Male flowers with 4 spongy fasciclodes uniting in the centre of the flower, alternating with 4 fascicles each of 6–10 or more stamens (rarely less), with filaments connate for up to ⅔ their length. Female (and bisexual?) flowers with small denticulate fasciclodes alternating with fascicles of 1–4 staminodes (or stamens ?) ; ovary globose to ovoid, 2 (3–4)-locular, surmounted by a 2 (3–4)-lobed fleshy stigma. Berry 1–2·5 cm. in diam., globose (or 2–4-lobed, at least when dry), purplish-green turning yellow, 1–4-seeded. Seeds c. 1 cm. long, plano-ovoid.

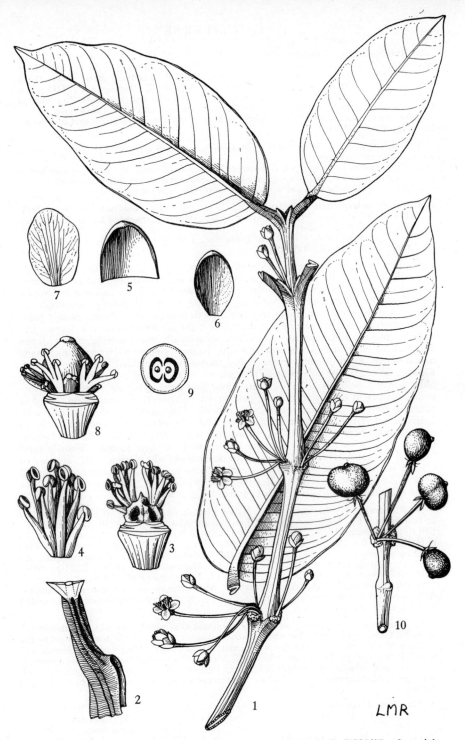

Tab. 78.—GARCINIA SMEATHMANNII. 1, flowering branch (×⅔) *Duff* 221/37 ; 2, petiole showing ligule (×2½) *Fanshawe* 1917 ; 3, male flower, perianth and two stamen-fascicles removed (×3) *Duff* 221/37 ; 4, stamen-fascicle (×6) *Duff* 221/37 ; 5–9, female flower (×3), all *Fanshawe* 1464 (5, outer sepal ; 6, inner sepal ; 7, petal ; 8, flower, perianth removed ; 9, ovary in trans. sect.) ; 10, section of branch with immature fruits (×⅔) *Fanshawe* 1917.

N. Rhodesia. B : Balovale, fl. vii.1933, *Trapnell* 1219 (K). N : Lunzua R., W. of Abercorn, ♀ fl. 21.vii.1933, *Michelmore* 489 (K). W : Old Ndola, ♂ fl. 31.viii.1937, *Duff* 220/37 (FHO ; K). **Nyasaland.** C : Nchisi Mt., st. 3.ix.1929, *Burtt Davy* 21211 (FHO). S : Mt. Mlanje, Tuchila R., c. 900 m., fl. 11.ix.1957, *Chapman* 427 (FHO ; K).

From French and Portuguese Guinea to Br. Cameroons and Fernando Po ; also in Gaboon, Angola, Belgian Congo, N. Rhodesia, Nyasaland and SW. Tanganyika.

A species of rain forest and river banks, which in our area is usually found in riverine forest from 900–1500 m.

Although the leaves, flowers and pedicels of this species show considerable variation, the leaf-venation is distinctive.

G. smeathmannii has 4 sepals and therefore belongs to *Garcinia*, not *Rheedia*, whether or not the latter is maintained as a separate genus. Since it does not differ essentially in any way from *G. polyantha* Oliv., it is necessary to replace that well-known name.

3. **Garcinia mlanjiensis** Dunkley [in N. C. L. : 46 (1936) *nom. nud.*] in Kew Bull. **1937** : 467 (1938). Type : Nyasaland, Mt. Mlanje, Luchenya Plateau, *Burtt Davy* 22045 (FHO ; K, holotype).

    Garcinia sp.—Bak. f. in Journ. Linn. Soc., Bot. **40** : 26 (1911).—Eyles in Trans. Roy. Soc. S. Afr. **5** : 421 (1916).

Shrub to large robust tree, 3–15 m. high (or more?), glabrous. Branches longitudinally grooved, flattened or quadrangular when young, eventually cylindric. Leaves opposite, coriaceous, petiolate ; lamina 6–16 × 3–6 cm., elliptic to lanceolate-elliptic or elliptic-oblong, obtuse to acute or rarely shortly acuminate or apiculate at the apex, cuneate at the base, bluish to sage green above, usually paler below, with prominent venation on both sides, the 30–60 main laterals leaving the midrib almost at right angles, ± straight, occasionally anastomosing, and with the longitudinal glandular canals usually visible below ; petiole 6–16 mm. long, channelled above, longitudinally grooved and transversely wrinkled; ligule prominent. Flowers dioecious (or polygamous?), single or in fascicles of 2–7 in the axils of the older leaves ; pedicels 7–15 (27) mm. long, pale green. Sepals 4, decussate, unequal, oblong to orbicular, obtuse, pale green. Petals 4, 4–6 mm. long, c. 1·5–2 times as long as the inner sepals, ± orbicular, cream-white or yellowish, with longitudinal translucent glandular lines. Male flowers with 4 spongy fasciclodes uniting in the centre of the flower, alternating with 4 fascicles each of 8–10 stamens (rarely less), with filaments connate for up to ⅔ of their length. Female (and bisexual?) flowers with small denticulate fasciclodes alternating with fascicles of (2) 3–5 staminodes (or stamens?) ; ovary globose to ovoid, 2-locular, surmounted by a ± 2-lobed fleshy stigma. Berries 1–2·5 cm. in diam., turning orange-yellow when mature, 1–2-seeded. Seeds c. 1 cm. long, plano-ovoid.

N. Rhodesia. E : Lundazi Distr., Nyika Plateau, st. 8.v.1952, *White* 2796 (FHO) **S. Rhodesia.** E : Melsetter Distr., Chimanimani Mts., ♀ fl. 9.x.1950, *Wild* 3552 (K ; SRGH). **Nyasaland.** N : Mugesse Forest, Misuku Hills, ♂ buds, ix.1953, *Chapman* 150 (FHO ; K). C : Nchisi Mt., st. 1929, *Burtt Davy* 21244 (FHO). S : Mlanje Mt., Luchenya Plateau, ♂ buds 24.ix.1929, *Burtt Davy* 22045 (FHO ; K). **Mozambique.** Z : Serra de Gúruè, R. Namituè, ♂ fl. 20.ix.1944, *Mendonça* 2155 (BM ; LISC). MS : Planalto da Serra da Gorongoza, ♂ fl. 29.ix.1943, *Torre* 5987 (BM ; LISC).

In N. Rhodesia, S. Rhodesia, Nyasaland, Mozambique and Tanganyika (Uluguru, Morogoro and Bukoba districts). An understorey tree of evergreen rain forest and wooded ravines, 1350–2100 m.

    G. mlanjiensis is very closely allied to *G. ovalifolia* Oliv. sensu lato, which is distributed from Fr. Guinea to Uganda, Belgian Congo and Angola. The latter is distinguished by the number of stamens (3 in each fascicle in male fls. and only 4 single ones in female and bisexual fls.), and by the shorter pedicels (2–5 mm.) and smaller flowers. *G. mlanjiensis* is also very near *Rheedia aphanophlebia* (Bak. f.) Perrier, a plant from Madagascar in which the stamen-filaments are free to the base and the calyx and corolla are linked by transitional forms.

4. **Garcinia livingstonei** T. Anders. in Journ. Linn. Soc., Bot. **9** : 263 (1866).—Oliv., F.T.A. **1** : 165 (1868).—Gibbs in Journ. Linn. Soc., Bot. **37** : 430 (1906).—Sim, For. Pl. Port. E. Afr. : 15, t. 4 (1909).—Bak. f. in Journ. Linn. Soc., Bot. **40** : 26 (1911).—Eyles in Trans. Roy. Soc. S. Afr. **5** : 421 (1916).—O. B. Mill., B.P.C.L.: 41 (1948) ; in Journ. S. Afr. Bot. **18** : 58 (1852).—Brenan, T.T.C.L. : 241 (1949).—Keay, F.W.T.A. ed. 2, **1** : 294 (1954).—Pardy in Rhod. Agr. Journ. **53** : 958, cum tab. (1956). Type : specim. cult. in hort. bot. Calcutt. ex Mozambique, " In rupibus schistosis prope flumen Zambesi ", *Kirk* (CAL, holotype).

Garcinia angolensis Vesque in A. & C. DC., Mon. Phan. **8** : 335 (1893).—R.E.Fr., Wiss. Ergebn. Schwed. Rhod.-Kongo-Exped. **1** : 151 (1914).—Burtt Davy, F.P.F.T. **1** : 252 (1926).—Staner in Bull. Jard. Bot. Brux. **13** : 120 (1934).—Exell & Mendonça, C.F.A. **1, 1** : 127 (1937). Type from Angola.
Garcinia baikieana Vesque, tom. cit. : 336 (1893). Type from Nigeria.
Garcinia ferrandii Chiov. in Stefan. & Paoli, [Miss. Somal. Ital. Merid. 228 (1916) *nom. nud.*] Result. Sci. Miss. Stefan.-Paoli Somal. Ital. **1** : 26 (1916). Type from Somalia.

Glabrous shrub or tree, (3) 4·5–12 (18) m. high, ± pyramidal when young, later bushy; branches striate when young, later smooth, cylindric or somewhat triangular, spreading at first, becoming virgate. Leaves in whorls of 3 (rarely of 4 or opposite), coriaceous, petiolate ; lamina 4–14 × 1·5–11 cm., very variable, lanceolate or oblanceolate to oblong or obovate (or rarely ± orbicular), emarginate or rounded to acute or apiculate at the apex, cuneate to rounded at the base, sometimes with shallowly crenate margin, dark to pale green above, pale green and ± glaucous below, with prominent reticulate venation on both sides, the main laterals variable in number, not always distinct from the secondary laterals, and with no visible secretory system; petiole 4–8 mm. long, channelled above, transversely wrinkled; ligule prominent. Flowers polygamous, in fascicles of 5–15 or more in the axils of the older leaves and on the old wood ; pedicels 4–20 (35) mm. long, varying in thickness. Sepals unequal, 4 in 2 opposite and decussate pairs, or 3, oblong to orbicular, cucullate. Petals 3–7 (9) mm. long, 1–1½ times as long as the inner sepals, usually 5, but sometimes up to 8 (when they are linked with the sepals by transitional forms), obovate to orbicular, greenish-white to cream or pale yellow, with ± translucent, orange or reddish longitudinal glandular lines. Male flowers with numerous apparently free stamens inserted in a fleshy cushion formed by the united fasciclodes. Bisexual (and female?) flowers with fewer stamens (or staminodes?) inserted in a fleshy fasciclodal ring below the ovary ; ovary globose, 2 (3)-locular, surmounted by a bilobed fleshy stigma. Berry 1–2·5 (3) cm. in diam., obovoid to globose, orange-yellow to reddish, 1–2 (3)-seeded. Seeds cylindric or ± planoovoid, 1·5–2 cm. long.

Caprivi Strip. N : Ngamiland, E. of the Kwando R., st. x.1945, *Curson* 936 (PRE). **Bechuanaland Prot.** N : Maun Camp, Thamalakan R., st. 29.vi.1937, *Erens* 324 (K ; PRE). **N. Rhodesia.** B : Shangombo, near Mashi R., ♂ fl. 15.viii.1952, *Codd* 7556 (BM ; K ; PRE ; SRGH). N : Mpika Distr., Luangwa R., fr. 3.x.1933, *Michelmore* 630 (K). W : Kitwe, ♂ fl 15.ix.1959, *Fanshawe* 5211 (K). C : Chingombe, fr. 26.ix.1957, *Fanshawe* 37 (K). E : Petauke Distr., Luangwa R., ♂ fl. 5.ix.1947, *Brenan & Greenway* 7808 (FHO ; K). S : Livingstone, N. bank of Zambezi R., ♀ or ☿ fl. 15.xi.1911, *Rogers* 7463 (K). **S.Rhodesia.** N : Mkota Res., Mazoe R., ♂ and ♀ or ☿ fl. 1.x.1948, *Wild* 2689 (K ; SRGH). W : Sebungwe, Zambezi valley, ♂ fl. ix.1955, *Davies* 1458 (K ; SRGH). E : Umtali Distr., Umvumvumvu R., st. 21.iii.1947, *Chase* 311 (K ; SRGH). S : Fort Victoria Distr., Tokwe R., ♀ or ☿ fl. 10.x.1951, *McGregor* 60/51 (SRGH). **Nyasaland.** S : Chiromo, st. 29.x.1929, *Burtt Davy* 22284 (FHO). **Mozambique.** N : Nampula, estrada de Murrupula, fr. 3.xi.1942, *Mendonça* 1209 (BM ; LISC). Z : Mocuba Distr., Namagoa, ♂ and ♀ or ☿ fl. 24.ix.1949, *Faulkner* K 470 (COI ; K ; SRGH). T : R. Luenha, ♂ fl. 27.ix.1948, *Wild* 2647 (K ; SRGH). MS : Manica, Matarara do Lucite, R. Lucite, fr. 10.x.1953, *Pedro* 4275 (K ; PRE). SS : Chibuto, Limpopo, fr. 12.xii.1940, *Torre* 2370 (BM; LISC). LM : Maputo, Bela Vista, st. 22.xi.1940, *Torre* 2153 (BM).

From Natal (Zululand), Swaziland and Transvaal to Kenya, Uganda and Somalia ; Zanzibar ; westward to Bechuanaland Prot., SW. Africa (Caprivi Strip), Angola, N. Rhodesia and Belgian Congo (Katanga). Also from Fr. Guinea to Nigeria. Practically confined to riverine forest and river banks. Sea level to c. 1050 m. (up to 1650 m. in E. Africa).

G. livingstonei is extremely variable in leaf-shape, pedicel-length and flower-size, but its virgate habit is characteristic. In general *G. livingstonei* sensu stricto has minute papillae on the lower epidermis, while in the other " species " (*G. ferrandii*, *G. angolensis* and *G. baikieana*) these are almost absent. However, intermediate conditions occur, so this cannot be used as a diagnostic character. The Tanganyikan species, *G. pallidinervia* Engl. and *G. bussei* Engl. should probably be included in *G. livingstonei*.

5. **Garcinia pachyclada** N. Robson in Bol. Soc. Brot., Sér. 2, **32** : 170 (1958). Type : N. Rhodesia, Abercorn Distr., Chambezi R., *Michelmore* 511 (K, holotype).

Glabrous shrubby tree 2·7–4·5 m. high, with the habit of *Uapaca kirkiana*; branches thick, with a rough greyish bark, transversely plicate on the young shoots.   Leaves in whorls of 3, coriaceous, shortly petiolate ; lamina 6–17 × 3·5–7·8 cm., lanceolate or oblanceolate to oblong-elliptic, emarginate or rounded to acute or apiculate at the apex, cuneate to rounded at the base, with margin incrassate, shallowly crenate and strongly undulate, and colour and venation as in *G. livingstonei*, but with no visible secretory system ; petiole 2–4 mm. long, channelled above, transversely wrinkled ; ligule prominent. Male flowers 15–20 mm. in diam., single or in few-flowered fascicles, axillary, on the old wood ; pedicels short (up to 9 mm. long), brown.   Sepals usually 4, reddish-brown, the outer pair ± orbicular, cucullate, equal, the inner oblong to ovate, obtuse or rounded, unequal and transitional to the corolla. Petals 5, (6) 7–10 mm. long, imbricate, orbicular or broadly obovate, white or pale cream-yellow, with translucent and reddish longitudinal glandular lines.   Stamens numerous (c. 100–120), apparently free, inserted in a fleshy cushion formed by the united fasciclodes. Ovary vestige absent. Female flowers with fewer free staminodes inserted in a lobed fleshy cushion. Ovary 3–4 mm. long, globose or broadly ovoid, 2–3-locular, with a fleshy 2–3-lobed stigma. Berry (immature) up to 1·3 cm. long, ovoid-globose, smooth, 1–3-seeded.

N. Rhodesia. N : Abercorn Distr., Kambole Escarpment, ♂ fl. 23.viii.1956, *Richards* 5923 (K).
Widespread in the plateau woodland between Abercorn and Lake Bangweulu on sandy soil.

G. pachyclada differs from *G. livingstonei* by its *Uapaca*-like habit (ascending branches), its thicker rough branches, its leaves with thickened, undulate margins, and its large flowers on short pedicels.   It is a plant of plateau woodland rather than riverine forest. The flowers and leaves are very similar to those of some Madagascar species of *Rheedia*.

6. **Garcinia punctata** Oliv., F.T.A. **1** : 167 (1868).—Staner in Bull. Jard. Bot. Brux. **13** : 138 (1934).—Exell & Mendonça, C.F.A. **1, 1** : 128 (1937).—Keay, F.W.T.A. ed 2, **1** : 295 (1954). Syntypes from Angola and Gaboon.

Evergreen shrub or tree, 5–9 m. high (up to 20 m. in Belgian Congo), erect or trailing, much branched, glabrous; branches spreading, slender, ± quadrangular at first, later cylindric. Leaves opposite, petiolate ; lamina 6–10 × 2–5 cm., oblong or elliptic, obtuse, with a long narrow obtuse cauda at the apex, cuneate to rounded at the base, bright green, concolorous, with numerous lateral veins prominent on both surfaces and forming a wide angle (60°–80°) with the midrib, and with dark ± longitudinal glandular lines visible on the lower surface, and translucent glandular dots and interrupted lines between, and parallel to, the lateral veins; petiole 4–6 mm. long, channelled above ; ligule inconspicuous.   Inflorescence terminal or axillary ; flowers c. 10 mm. in diam., dioecious, in single or in few-flowered, pedunculate or ± sessile, biparous cymes or fascicles ; pedicels 2–5 mm. long, pale green. Sepals 4, pale green, the outer pair shorter, triangular-ovate to orbicular, cucullate, the inner pair ovate to orbicular, obtuse to rounded at the apex. Petals 4, 4–5 mm. long, greenish-yellow or greenish-white, elliptic or obovate, ± reflexed, sometimes ± emarginate.   Male flowers with 4 stamen-fascicles each of 5–7 stamens with filaments completely united and anthers sessile, oblong, curved, transversely septate ; ovary-vestige orange, rugose, c. 1–2 mm. in diam. Female flowers with a yellowish, globular or obconical, 4-locular ovary surmounted by a broad peltate sticky orange-yellow ± 4-lobed stigma. Berry 15–18 mm. in diam., globular, yellow when ripe, 2–3-seeded. Seeds c. 9 mm. long, ± cylindric.

N. Rhodesia. W : Mwinilunga Distr., Luao R., ♂ fl. 23.i.1938, *Milne-Redhead* 4293 (BM ; K ; PRE).
In N. Rhodesia, Angola, Belgian Congo, Gaboon, Fr. Cameroons, Br. Cameroons and Nigeria.

A West African rain-forest tree which grows in the Mwinilunga Distr. along river banks and in *Cryptosepalum* woodland.

7. **Garcinia huillensis** Welw. ex Oliv., F.T.A. **1** : 167 (1868).—Staner in Bull. Jard. Bot. Brux. **13** : 134 (1934) pro parte excl. syn. *G. buchneri* Engl. et *G. sapinii* De Wild.—Exell & Mendonça, C.F.A. **1**, 1 : 128 (1937).—Brenan, T.T.C.L. : 242 (1949).—Milne-Redh. in Mem. N.Y.Bot. Gard. **8**, 3 : 222 (1953). Type from Angola (Huila).
   Garcinia buchananii Bak. in Kew Bull. **1894** : 354 (1894).—R.E.Fr., Wiss. Ergebn. Schwed. Rhod.-Kongo-Exped. **1** : 153 (1914).—Staner, tom. cit. : 133

(1934).—Norlindh in Bot. Notis. **1934** : 104 (1934).—Brenan, loc. cit. (1949).—Suesseng. & Merxm. in Proc. & Trans. Rhod. Sci. Ass. **43** : 88 (1951). Type : Nyasaland, *Buchanan* 183 (K, holotype).

Garcinia henriquesii Engl., Bot. Jahrb. **40** : 571 (1908).—R.E.Fr., loc. cit. Type from Angola (Malange).

Garcinia gossweileri Engl., loc. cit.—Exell & Mendonça, C.F.A. **1**, 1 : 129 (1937). Type from Angola (Malange).

Evergreen shrub or tree, 1·4–15 m. high, erect, much branched, glabrous; branches dense or ± spreading, angular or grooved. Leaves opposite, coriaceous, petiolate ; lamina 5–11 (13·5) × (2) 3–5 (7·5) cm., oblong or elliptic to oblanceolate, obtuse (rarely acute or rounded) at the apex, often with a rounded mucro or a short acumen, cuneate at the base, deep green above, paler below, purple-tinged when young, with numerous branched and anastomosing lateral veins prominent on both surfaces and forming an angle of 30°–60° with the midrib, and dark ± longitudinal glandular lines visible on the lower surface, but without any visible translucent glands ; petiole 4–6 mm. long, channelled above, sometimes red ; ligule inconspicuous. Inflorescence axillary, rarely terminal ; flowers polygamous, 10–15 mm. in diam., single or in few-flowered, sessile or shortly pedunculate, biparous cymes or fascicles ; pedicels 2–5 mm. long, pale green, ± quadrangular. Sepals 4, pale or yellowish green, ± unequal, the outer pair orbicular or semi-orbicular, sometimes very small, the inner pair larger, orbicular or broadly obovate, rounded at the apex. Petals 4, 5–8 mm. long, 2–3 times as long as the inner sepals, greenish to yellow, obovate, spreading in flower, rounded at the apex. Male flowers with 4 stamen-fascicles, each with 7–8 stamens with filaments completely united and anthers sessile, oblong, curved, not septate ; ovary-vestige orange, c. 2 mm. in diam. Female flowers with a pale green, globular or flask-shaped, 4-locular ovary, surmounted by a broad peltate sticky orange-yellow ± 4-lobed stigma. Bisexual flowers like female ones, but with 4 slender stamens, or 4 stamen-fascicles each of 2 stamens. Berry up to c. 20–25 mm. in diam., globular, yellow or orange when ripe, 1–3-seeded. Seeds curved, ± cylindric, c. 10 mm. long.

N. Rhodesia. N : Abercorn Distr., Kambole Escarpment, ♂ fl. 24.viii.1956, *Richards* 5950 (K). W : Mwinilunga Distr., Matonchi R., ♀ fl. & fr. 25.x.1937, *Milne-Redhead* 2952 (BM ; K ; PRE). C : Broken Hill, ♀ fl. 10.ix.1947, *Brenan* 7849 (FHO ; K). **S. Rhodesia.** C : Marandellas Distr., Cave Farm, fr. 5.iv.1950, *Wild* 3265 (K ; SRGH). E : Umtali, Odzani R., ♀ fl. 1915, *Teague* 303 (K). S : Buhera Distr., ♀ fl. & fr. xi.1953, *Davies* 619 (PRE ; SRGH). **Nyasaland.** N : Karonga Distr., Misuku Hills, fr. ii.1953, *Chapman* 64 (FHO). C : Kota Kota Distr., Chia area, ♀ fl. 5.x.1946, *Brass* 17543 (BM ; K ; PRE ; SRGH). S : Shire Highlands, Zambesia, ♀ fl. & fr., *Buchanan* 36 (K). **Mozambique.** N : Malema, Mutuali, R. Nalume, ♀ fl. & fr. 28.ix.1944, *Mendonça* 2298 (BM ; LISC). Z : Lugela, Mocuba Distr., Namagoa, ♀ fl. & fr. ix.–x., *Faulkner* K 84 (COI ; K). MS : Barue, Vila Gouveia, Serra Chôa, ♀ fl. & fr. 17.ix.1942, *Mendonça* 266 (BM ; LISC).

In Mozambique, Nyasaland, N. and S. Rhodesia, Angola, Belgian Congo, eastern tropical Africa and the Sudan (Equatoria). Riverine forest, woodland margins, open woodland, stony hillsides, often on heavy soils. In S. Rhodesia frequently occurring on granite. From 480–1560 m. in our area, but up to 1800 m. in Uganda.

G. buchananii cannot be separated from *G. huillensis* as the latter is very variable in leaf-shape and both plants have four sepals (cf. Staner, loc. cit. and Milne-Redhead, loc. cit.).

8. **Garcinia buchneri** Engl., Bot. Jahrb. **55** : 395 (1919).—Exell & Mendonça, C.F.A. **1**, 1 : 129 (1937). Syntypes from Angola (Malange and Huila).

Garcinia sapinii De Wild. in Ann. Soc. Sci. Brux. **41** : 378 (1922). Type from the Belgian Congo (Kasai).

Garcinia edulis Exell in Journ. of Bot. **65**, Suppl. Polypet. : 28 (1927).—Exell & Mendonça, loc. cit. Type from Angola (Bié).

Closely related to *G. huillensis*, but can be distinguished by the following characters :

Rhizomatous shrublet, 15–30 cm. high. Rhizome branched, red-brown, striate, with opposite or subopposite scale-leaves. Stems erect, flattened, with longitudinal raised lines. Leaves opposite or subopposite, sessile or shortly petiolate ; lamina 4·7–8 × (0·6) 1·5–3·5 cm., narrowly oblanceolate to obovate, acute to obtuse or rounded at the apex, sometimes with an apiculus, cuneate at the base ; petiole up to 2 mm. long. Inflorescence terminal, sometimes also axillary. Flowers up to 20 mm. in diam. Petals up to 11 mm. long, the margin sometimes denticulate. Male flowers with 4 fascicles each of 4–7 stamens. Fruit bright red (or orange?).

N. Rhodesia. W : Mwinilunga Distr., near source of Matonchi R., ♂ fl. 7.x.1937, *Milne-Redhead* 2626 (K). C : Kapiri Mposhi, ♂ buds, 6.viii.1957, *Fanshawe* 3430 (K).

In Angola, Belgian Congo (Kasai, Upper Katanga) and N. Rhodesia. Dry areas of scrub or open woodland, from 1700 to 2000 m. in Angola and N. Rhodesia but occurring at 350 m. in the Belgian Congo (Kasai).

9. **Garcinia acutifolia** N. Robson in Bol. Soc. Brot., Sér. 2, **34** : 95, t. 1 (1960) Type : Mozambique, Niassa, 2·3 km. from Muaguide, *Balsinhas* 61 (BM, holotype ; LMJ).

Shrub or small tree, 2–3 m. high, glabrous ; branches slender, narrowly 4-winged. Leaves opposite, petiolate ; lamina 6–9 × c. 2·5–5 cm., chartaceous, bright green, concolorous, ovate to lanceolate or elliptic, shortly and usually very acutely acuminate at the apex, cuneate to rounded at the base, with c. 12–20 main lateral nerves (sometimes alternating with weaker ones) ± prominent on both surfaces, forming a wide angle (60°–80°) with the midrib and anastomosing loosely towards the margin, and with translucent streaks and interrupted lines ± parallel to the lateral veins, but without visible dark longitudinal glandular lines ; petiole 4–6 mm. long, channelled above, yellow-green to orange ; ligule inconspicuous. Inflorescences terminal or axillary ; male flowers c. 5–6 mm. in diam., in few- or many-flowered pedunculate regularly biparous cymes ; pedicels 1·5–4 mm. long, pale green. Sepals 4, pale green, the outer pair shorter, broadly ovate to suborbicular, cucullate, the inner pair broadly obovate to orbicular, rounded at the apex. Petals 4, 3–3·5 mm. long, yellow, oblong or oblanceolate, rounded at the apex. Male flowers with 2 antisepalous stamen-fascicles, each of 3 stamens, with filaments completely united and anthers sessile, oblong or elliptic, curved or straight, not septate ; ovary-vestige c. 1 mm. in diam., flattened-obconic, with 4-lobed stigma. Female flowers and fruit as yet unknown.

Mozambique. N : between Muaguide and the Quissanga-Macomia crossroads, 2·3 km. from Muaguide, fl. 5.xi.1953, *Balsinhas* 61 (BM ; LMJ).

Also in north-eastern Tanganyika (Kisarawe Distr.). In evergreen forest, c. 50–500 m.

G. acutifolia differs from all the other species of *Garcinia* in our area in having 2 antisepalous stamen-fascicles, and from all species of Sect. *Tagmanthera* Pierre by its regularly quadrangular young shoots. The interrupted glandular lines of the leaves also serve to distinguish it from all other species of the genus in our area except *G. punctata*.

10. **Garcinia sp.** A.

Tree 9 m. high, glabrous; branches striate, grooved when young, becoming ± cylindric, spreading at an angle of 60°–80°. Leaves in pairs, coriaceous, petiolate ; lamina (5) 6–8·7 × 2·5–3·2 cm., oblanceolate to oblong or narrowly ovate, rounded or minutely apiculate at the apex, cuneate at the base, with margin shallowly crenate, dark to pale green above, pale greyish-green below, glossy, with reticulate venation prominent on both sides, the main laterals 20–30, not always distinct from the subsidiary laterals between them, and dark vertical secretory canals visible through the lower surface of the young leaves only ; petiole 6–8 mm. long, channelled above, transversely wrinkled ; ligule prominent. Female flowers single or in fascicles of 2–3 from axillary buds on the old wood ; pedicels (in fruit) 12–17 mm. long, green, striate. Perianth not complete on the available material. Staminodes numerous, apparently free, inserted in a fleshy lobed fasciclodal ring below the ovary. Berry globose to obovoid, c. 1·5–2 cm. in diam., orange, markedly verrucose, 1–2-seeded, with a peltate 2–3-lobed fleshy stigma terminating a short stylar attenuation. Seeds cylindric or plano-ovoid, c. 1·5 cm. long.

N. Rhodesia. N : Kawambwa, fr., 10.xi.1957, *Fanshawe* 3875 (K).
In riverine forest.

Known so far only from two gatherings, the above and one from the Lomami Distr. of Katanga (*Schmitz* 6686 (BR)). The only other African species of *Garcinia* with verrucose fruits are *G. elliotii* Engl.(Sect. *Rheediopsis*) and two species in Sect.*Xanthochymus*, all from W. Africa, whereas our plant belongs to Sect. *Tetracentrum* along with *G. livingstonei* ; but this character is more frequent in species of *Rheedia* L. (Madagascar and tropical America), a genus which does not appear to be satisfactorily distinguishable from *Garcinia*. (See N. Robson, Bol. Soc. Brot., Sér. 2, **32** : 172 (1958).)

26. THEACEAE
(TERNSTROEMIACEAE)
By N. K. B. Robson

Trees or shrubs. Leaves alternate or spiral, simple, usually coriaceous, exstipulate. Flowers axillary, solitary or sometimes paired or in fascicles, rarely cymose or in elongated racemes, actinomorphic, bisexual or rarely unisexual, often with one or more pairs of bracteoles below the calyx. Sepals usually 5, imbricate, free or shortly connate. Petals usually 5, imbricate, hypogynous, free or shortly connate. Stamens ∞ or rarely definite, hypogynous, free or fasciculate or connate at the base to form a tube, sometimes adnate to the base of the petals ; anthers versatile or basifixed, often with an apiculate connective-prolongation, dehiscing longitudinally or rarely at first by apical pores ; disk absent. Ovary superior (rarely apparently semi-inferior), syncarpous, sessile, (1) 2–5 (7)-locular ; styles free or united ; ovules 2 or more (rarely 1) in each loculus, axile, sometimes pendulous. Fruit a loculicidal capsule (often leaving a persistent central column) or a dry or rarely fleshy berry or drupe, indehiscent or rarely irregularly dehiscent at the apex. Seeds 1–∞ ; endosperm abundant, scanty or absent ; embryo straight or curved, sometimes folded or spirally twisted.

A family of about 20 genera and over 200 species, occurring mostly in E. Asia and America.

Ternstroemia polypetala Melch. and *Melchiora schliebenii* (Melch.) Kobuski both occur in Tanganyika and may eventually be found in the north of Nyasaland or N. Rhodesia. *Camellia sinensis* (L.) Kuntze, the Tea plant, is cultivated on a considerable scale at the foot of Mt. Mlanje in Nyasaland and also along the Eastern Border of S. Rhodesia.

Ficalhoa appears to be related to *Eurya* (*Ternstroemieae, Adinandrinae*), and it is therefore appropriate to include it in the *Theaceae*, although it has some characters which may prove to be sufficiently distinct to warrant treating it as a separate family.

FICALHOA
Ficalhoa Hiern in Journ. of Bot. **36** : 329 (1898).

Tree with terete branches and alternate serrulate leaves. Flowers bisexual, in solitary or paired axillary few- or many-flowered dichasia. Bracteoles 2 or 0. Sepals 5*, quincuncial, connate in the lower third, persistent. Petals 5*, quincuncial, alternating with the sepals, connate in the lower third to form a suburceolate corolla, eventually deciduous. Androecium pentadelphous*, with fascicles each of 3 stamens, opposite the sepals, 1-seriate ; anthers 2-thecous, basifixed, dehiscing by oblique apical pores later becoming slits ; filaments erect, adnate to near the top of the corolla tube. Ovary subhemispherical, obtusely 5*-lobed, 5*-locular, each loculus with numerous minute ovules on a central spongy placenta ; styles 5*, free almost to the base or connate for up to $\frac{2}{5}$ of their length, diverging. Fruit a 5*-valved loculicidal capsule, hemispherical, woody, many-seeded. Seeds small, irregularly compressed-ovoid, with a loosely reticulate testa and scanty endosperm ; embryo straight, terete.

A monotypic genus, placed originally in the *Ericaceae* but apparently most closely allied to *Eurya* and related genera in the *Theaceae*. From these genera, however, it differs in having a loculicidal capsule, seeds with an almost straight embryo which are borne on a spongy placenta, and in secreting latex. Cymose inflorescences are very rare in *Theaceae*, but they occur in species of *Eurya*.

Ficalhoa laurifolia Hiern in Journ. of Bot. **36** : 329, t. 390 (1898). TAB. **79**. Type from Angola (Huila).

* Occasional isomerous 6–merous flowers occur.

LMR

Tab. 79. FICALHOA LAURIFOLIA. 1, branch with flowers (×⅔) ; 2, flower (×8) ; 3, corolla and stamens (×8) ; 4, ovary (×8), all *Angus* 841 ; 5, dehisced capsule, showing numerous sterile seeds remaining on the five swollen placentae (×8) *Osmaston* 2060 ; 6, seed (×24) *Davies* 297.

Tree (2) 6–24 (36) m. high, much branched, evergreen, with rough fissured brown bark and copious white latex. Branches terete (or longitudinally ridged when dry), rather sparsely pilose at first, eventually glabrous. Leaves petiolate ; lamina (5) 7–12·4 × (1·8) 2–4·3 cm., oblong or lanceolate, acuminate at the apex, cuneate to rounded at the base, with a bluntly serrulate margin, subcoriaceous, dark glossy green above, paler and sometimes glaucous below, wholly glabrous or sparsely pilose below or on both sides, with slender pinnate nerves and frequently inconspicuous reticulate venation ; petiole 3–9 mm. long, pilose or glabrescent. Flowers small, in solitary or paired dichasia, on the second-year shoots ; primary peduncle 2–6 mm. long, pubescent ; pedicels short or very short, pubescent ; bracteoles half as long as the calyx, ovate or oblong, rounded. Sepals c. 1 mm. long, rounded, sparsely pubescent. Petals yellowish to white or greenish, 3 mm. long, oblong, rounded, glabrous. Stamens glabrous ; anthers c. 0·25 mm. long, elliptic-obovoid, glabrous ; filaments c. 1 mm. long. Ovary densely appressed-pilose ; styles pilose at the base, otherwise glabrous ; stigmas papillose. Capsule c. 3 mm. in diam., 1–1½ times as long as the calyx, with widely spreading valves. Seeds 1 mm. long, ovoid-oblong, slightly curved; testa loose, whitish-fawn, semi-translucent, reticulate.

N. Rhodesia. N : Mpika, fr.4.ii.1955, *Fanshawe* 1978 (K ; SRGH). W : Mwini-lunga, fl. 8.ix.1955, *Holmes* 1175 (K). **Nyasaland.** N : Vipya, above Luwawa Dam towards Kawendama, fl. 27.xii.1955, *Chapman* 262 (K).

In Angola (Huila), N. Rhodesia, Nyasaland, Tanganyika, Uganda, Ruanda-Urundi and the Belgian Congo. Evergreen rain-forest, riverine forest and streamsides, 1350–1800 m. (to 2400 m. in Uganda).

F. laurifolia is easily recognized by the axillary dichasia of flowers with 5 fascicles each of 3 stamens. The old capsules usually contain a large number of small, sterile seeds.

27. DIPTEROCARPACEAE

By P. Duvigneaud

Trees or shrubs. Leaves alternate, petiolate, entire, penninerved, in African species always with an extra-floral nectary at the base of the midrib on the upper face of the lamina ; stipules caducous. Flowers actinomorphic, bisexual. Sepals 5, accrescent. Petals 5, contorted to the right, free or slightly connate at the base. Stamens 5, 10, 15 or numerous (always numerous in African species). Ovary superior, (1) 3-locular with axile or parietal placentation ; carpels (1) 3, 2-ovulate. Fruit 1-seeded, indehiscent, surrounded by the wings (always 5 in the African species) of the accrescent calyx.

In Africa the family is represented by the subfam. *Monotoideae*, characterized by versatile anthers and the absence of secretory canals in the pith. Family distributed throughout the tropics ; 19 genera and about 380 species.

Tall trees with straight trunks with buttresses ; branchlets densely longitudinally chan-nelled ; leaves acuminate ; stamens and ovary inserted on a very distinct andro-gynophore ; anthers not appendiculate at the apex; ovary 3-locular at the base, 1-locular at the apex - - - - - - - - - 1. **Marquesia**
Trees or bushes with irregular trunks without buttresses ; branchlets ± smooth ; leaves very rarely acuminate ; stamens and ovary inserted on the scarcely produced receptacle ; anthers frequently produced at the apex into a triangular or ovate appendage ; ovary completely 3-locular - - - - - - - - 2. **Monotes**

1. MARQUESIA Gilg

Marquesia Gilg in Engl., Bot. Jahrb. **40** : 485 (1908).

Tall trees with buttresses. Leaves evergreen, acuminate ; midrib and lateral nerves slightly depressed above and prominent beneath ; reticulation very dense, isodiametrical ; indumentum composed only of simple isolated short pachyder-matous hairs ; minute spherical glands generally dispersed on nerves and reticulation. Flowers small, in large terminal panicles, with very hairy sepals and glabrescent petals; receptacle produced into a broad, elevated and conspicuous androgynophore. Stamens numerous; anthers short, not produced at the apex into a sterile appendage. Ovary hairy, 1-locular at the apex, with parietal placentation; ovules 6. Fruit ovoid-conical, with a parchment-like pericarp, surrounded by the 5 equal wings of the accrescent calyx.

Genus endemic in tropical Africa, where it is represented by 3–4 species.

Lower surface of the leaves glabrous except for a few sparse hairs on the principal nerves; petiole under 1 cm. long; basal gland transversely reniform or bilobed; fruits orange to brown - - - - - - - - - - 1. *acuminata*
Lower surface of the leaves with a short dense indumentum, and a not very prominent reticulation; petiole over 1 cm. long; basal gland elongated, linear or fusiform; fruits greenish-yellow - - - - - - - - - 2. *macroura*

1. **Marquesia acuminata** (Gilg) R.E.Fr. in Engl., Bot. Jahrb. **51**: 351 (1914). Type from Angola (Malange).

Monotes acuminata Gilg in Engl., Bot. Jahrb. **28**: 136 (1899). Type as above.

Tree 15–28 m. tall, with buttresses at the base of the trunk; branchlets sparsely puberulous, soon becoming glabrous. Leaf-lamina 7–10 × 3–5 cm., ovate to oblong-lanceolate, acuminate at the apex, rounded to slightly cordate at the base; upper surface finely reticulate, shining, glabrous; lower surface green, shining, glabrescent with only a very few small simple hairs scattered on the nerves; midrib slightly depressed above and prominent beneath; lateral nerves in 8–10 pairs, anastomosing well before reaching the margin of the leaf; reticulation prominent and very conspicuous; petiole under 1 cm. long. Flowers in large terminal panicles; pedicels 2–4 mm. long. Sepals 2 mm. long, greyish-tomentellous. Petals 7–8 mm. long, glabrous. Fruit 0·5 cm. in diam., ovoid-conical, subsericeous; wings 2–3 × 0·3–0·6 cm., orange to brownish, narrowly oblong, with sparse short subappressed hairs outside and inside.

N. Rhodesia. N: Kawambwa, fl. & fr. 22.viii.1957, *Fanshawe* 3501 (K). W: Mwinilunga Distr., Matonchi Farm, fr. 12.xi.1931, *Paterson* (K); Mwinilunga Distr., near Luao R., 16.ii.1938, *Milne-Redhead* 4611 (K).

Also in the Belgian Congo (Western Katanga) and in *Brachystegia* or mixed woodland and on loamy Kalahari Sand in the Lunda region of Angola. It reappears much more to the east, in the high-rainfall area of Kawambwa-Serenje.

2. **Marquesia macroura** Gilg in Engl., Bot. Jahrb. **40**: 485 (1908).—R.E.Fr. in Engl., Bot. Jahrb. **51**: 351 (1914). TAB. **80**. Type from Angola.

Monotes sapinii De Wild., Pl. Bequaert. **4**: 180 (1927). Syntypes from the Belgian Congo.

Tree up to 20 m. high; branchlets puberulous becoming glabrous. Leaf-lamina 6–9 × 2·5–4·5 cm., ovate to elliptic-lanceolate, acuminate at the apex, rounded to slightly cordate at the base; upper surface finely reticulate, shining, laxly puberulous with minute very short straight isolated hairs; lower surface whitish with very short dense straight simple hairs on the nerves and reticulation hiding the glabrous interreticular areoles; midrib slightly depressed above and prominent beneath; lateral nerves in 8–10 pairs, anastomosing long before reaching the margin of the leaf; reticulation not very prominent but very conspicuous; petiole over 1 cm. long. Inflorescences 2–9 cm. long, axillary, distinctly pedunculate, multiflorous, greyish-tomentose with long simple hairs, massing towards the end of the branches in large terminal white panicles; pedicels 2–3 mm. long. Sepals 2 mm. long, densely greyish-cottony-tomentose. Petals 8 mm. long, minutely and very laxly puberulous. Fruit 0·5 cm. in diam., ovoid-conical, shortly and laxly sericeous; wings 2–3·5 × 0·3–0·5 cm., greenish-yellow, narrowly oblong or subspathulate, with sparse simple very short hairs outside and inside.

N. Rhodesia. N: Abercorn, Kalambo Falls road, fl. 30.vii.1949, *Greenway & Hoyle* 8350 (K); Abercorn, Mpulungu, 17.ix.1950, *Bullock* 3330 (K); Chinsali, fl. & fr. 25.ix.1938, *Greenway* 5772 (K). W: Solwezi Distr., Mutanda Bridge, fl. 4.vii.1930, *Milne-Redhead* 671 (K); Ndola, fr. 28.ix.1947, *Brenan* 7975 (FHO; K); Mwinilunga Distr., fr. 28.x.1937, *Milne-Redhead* 2994 (K).

Also in Angola, Belgian Congo and very localized in Tanganyika. Very common and often dominant in the western part of its area where, in nearly pure stands or associated

Tab. 80. MARQUESIA MACROURA. 1, flowering branch ($\times\frac{2}{3}$); 2, flower ($\times 5$); 3, fruiting branch ($\times\frac{2}{3}$); 4, anther and part of filament ($\times 12$); 5, fruit ($\times 2$) all from *Carrisso & Mendonça* 283.

with *Marquesia acuminata* or with *Cryptosepalum pseudotaxus*, it forms dry evergreen forests on sandy or loamy soils. In the eastern part of its area, in the *Brachystegia* woodlands region, it becomes more and more localized on higher-rainfall rocky hills or plateaux, often associated with *Brachystegia microphylla*.

Timber used for building.

2. MONOTES A.DC.

Monotes A.DC. in DC., Prodr. **16**, 2 : 623 (1868).

Shurbs to medium-sized trees, without buttresses. Leaves very rarely acuminate, with a more or less rounded extra-floral nectary on the upper surface at the base of the midrib, and sometimes at the base of the lower lateral nerves ; midrib and lateral nerves depressed above and prominent beneath ; reticulation very dense, isodiametrical, scarcely to strongly prominent beneath, in the latter case turning the interreticular areoles into deep cavities ; indumentum extremely variable ; hairs simple, penicillate, or stellate, straight, curved, curled, or coiled, from very short to long, with punctiform spherical glands, sometimes making the lamina viscid, abundant or dispersed on the two surfaces of the leaf. Flowers small, in axillary clusters, with very hairy sepals and petals ; receptacle produced into a very short broad androgynophore. Stamens numerous ; anthers sohrt, very often produced at the apex into a triangular or ovate appendage. Ovary hairy, completely 3-locular with 2 ovules in each loculus. Fruit subglobose, with a hard thick pericarp and with 1 cavity containing 1 seed, surrounded by the 5 subequal wings of the accrescent calyx ; wings with minute penicillate hairs outside and inside.

Tropical Africa and Madagascar. Reported from the Tertiary Flora of Europe.

Upper surface of the leaf pubescent, ± scaberulous ; hairs straight, stiff or nearly so, usually exceeding 0·5 mm. in length, visible with a lens, usually developing on cushion-like emergences of the lamina ; reticulation of the under surface of the leaf prominent and conspicuous, covered with similar hairs ; upper surface finely reticulate or tuberculate-reticulate :

Leaves oblong to oblong-obovate or elliptic, at least 1½ times longer than broad, with a tendency to become obtuse, acute or even acuminate at the apex ; base subcuneate to rounded or slightly cordate ; petiole thick and relatively short (0·5–2 cm. long) :

Inflorescences many-flowered and rather dense, composed of subsessile clusters ; reticulation of the lower surface of the leaf very prominent, generally forming deep cavities ; interreticular areoles covered with stellate hairs ; anthers not produced into an appendage :

Leaves generally acute or acuminate at the apex ; reticulation of the lower surface of the leaf hidden or almost hidden by a woolly, white or rufous tomentum ; flowers in small remote clusters - - - - - - - 1. *dasyanthus*

Leaves generally obtuse to slightly emarginate at the apex, sometimes acute ; reticulation of the lower surface of the leaf clearly visible, despite the short fulvous tomentum ; flowers in clusters forming large terminal panicles

2. *katangensis*

Inflorescences few-flowered, on long slender axillary peduncles ; reticulation of the lower surface of the leaf not very prominent ; interreticular areoles glabrous ; anthers produced into an ovate appendage - - - - 3. *redheadii*

Leaves broadly elliptic to suborbicular, nearly as broad as long ; rounded to very deeply emarginate at the apex ; base truncate to deeply cordate :

Inflorescences few-flowered but rather dense, of small axillary pedunculate to subsessile clusters ; leaves never very deeply emarginate ; upper surface finely reticulate ; reticulation of the lower surface moderately prominent ; anthers produced into an ovate or triangular appendage :

Leaves very large, 13–23 × 11–17 cm., emarginate at the apex, cordate at the base, discolorous, floccose-tomentose on the lower surface ; petiole very thick and rather long (2–3·5 cm.); fruits 15–23 mm. in diam. (excluding wings)

4. *magnificus*

Leaves smaller (in our region), 6–12 × 4–10 cm., truncate to slightly emarginate at the apex, truncate to slightly cordate at the base, discolorous or concolorous, not floccose-tomentose on the lower surface ; fruits up to c. 15 mm. in diam. (excluding wings); petiole slender and relatively short (1–1·5 cm. long)

5. *adenophyllus*

Inflorescences many-flowered, flowers in clusters forming large terminal panicles ; leaves very deeply emarginate at the apex ; upper surface minutely tuberculate ;

reticulation of the lower surface very prominent, forming deep cavities; anthers
not produced into appendages - - - - - - 6. *autennei*
Upper surface of the leaf glabrous (except sometimes on the nerves), puberulous or
tomentellous, hairs very short, not exceeding 0·5 mm. in length and imperceptible
to the touch, only visible with a microscope, curled, simple, penicillate, or stellate;
reticulation of the under surface of the leaf either not prominent (*M. glaber*) or prominent, with all kinds of indumentum:
Branchlets glabrous; leaves glabrous on both surfaces (petioles sometimes slightly
hairy), always small and concolorous; anthers produced at the apex into a mucro
7. *glaber*
Branchlets hairy; leaves hairy, at least on the nerves of the lower surface; anthers
produced into a triangular or ovate appendage:
Leaves concolorous; interreticular areoles of the under surface glabrous; nerves,
veins and reticulation of the lower surface with minute short curled hairs
8. *africanus*
Leaves ± discolorous; interreticular areoles of the under surface covered with
minute coiled or stellate hairs:
Leaves broadly elliptic to suborbicular, up to 20 × 13 cm., finely reticulate and
glabrous above, sometimes shining - - - - - 9. *discolor*
Leaves elliptic to oblong, smaller, generally not exceeding 11 × 6 cm., puberulous
or tomentellous above:
Indumentum of the lower surface of the leaves soon becoming yellow, composed of minute, coiled, simple hairs which hide the stellate hairs of the
interreticular areoles; nerves and veins soon glabrescent, red-brown
10. *angolensis*
Indumentum of the lower surface of the leaves whitish or greyish, composed of
rather long curled simple hairs on the nerves, veins and reticulation, and
stellate hairs on the areoles:
Upper surface of the leaves smooth, with a dense persistent tomentellum;
lower surface tomentose - - - - - - 11. *elegans*
Upper surface of the leaves minutely reticulate, with an evanescent very lax
tomentellum; lower surface tomentellous - - - 12. *engleri*

1. **Monotes dasyanthus** Gilg in Warb., Kunene-Samb.-Exped. Baum: 307 (1903);
   in Engl., Bot. Jahrb. **41**: 288 (1908).—De Wild., Pl. Bequaert. **4**, 2: 170 (1927).—
   Hutch. in Kew Bull. **1931**: 253 (1931).—Bancroft in Bol. Soc. Brot., Sér. 2, **13**:
   372 (1939).—Duvign. in Lejeunea, **13**: 60 (1949).—Exell & Mendonça, C.F.A. **1**,
   2: 371 (1951). Type from Angola (Bié).

Tree 4–10 m. high; branchlets yellowish-brown or blackish-velutinous. Leaf-
lamina 7–12 × 3·5–6·5 cm., elliptic or obovate, apex obtuse to acute or acuminate
and mucronate, rounded to slightly cordate at the base; upper surface scaberulous,
with straight or nearly straight stiff simple and isolated pachydermatous hairs
mixed with minute stellate stiff hairs; lower surface discolorous, whitish- or
fulvous-cottony-tomentose or floccose with long curled leptodermatous (thin-
walled) hairs on the nerves and reticulation and with the interreticular areoles
densely covered with minute stellate hairs; midrib and lateral nerves slightly
depressed above and very prominent and strong beneath; lateral nerves in 10–16
pairs, nearly straight, incurving at the apex, reaching the margin of the leaf.
Inflorescences of subsessile condensed clusters forming multiflorous terminal
panicles 3–6 cm. long, greyish- or brownish-velutinous-tomentose; pedicels
2–3 mm. long. Sepals 3 mm. long, densely sericeous-tomentose. Petals 9 mm.
long, densely sericeous-tomentose. Stamens with anthers not produced at the
apex. Fruit 8–10 mm. in diam., subglobose, densely sericeous-tomentose, conical
at the apex; wings 3–4 × 1·0–1·5 cm., reddish-purple, generally broadly obovate
to spathulate.

N. Rhodesia. B: Balovale, fl. & fr. 18.iv.1954, *Gilges* 333 (K); Mankoya, fr. vi.1933,
Trapnell 1274 (K; NDO). W: Mwinilunga Distr., fr. 22.viii.1930, *Milne-Redhead*
941 (K).
Also in Angola and the Belgian Congo (Western Katanga). Very common on Kalahari
Sand, at the edges of forests or seasonal swamps.

2. **Monotes katangensis** (De Wild.) De Wild. in Ann. Mus. Cong. Bot., Sér. 4, **2**: 111
   (1913); Pl. Bequaert. **4**, 2: 177 (1927).—Bancroft in Bol. Soc. Brot., Sér. 2, **13**:
   372 (1939).—Duvign. in Lejeunea, **13**: 60, t. 14 fig. C (1949). TAB. **81** fig. A.
   Type from the Belgian Congo (Katanga).
   Vatica africana var. *glomerata* Oliv., F.T.A. **1**: 173 (1868). Type: N. Rhodesia,
   Batoka Hills, *Kirk* (K, holotype).

Vatica katangensis De Wild., Études Fl. Katanga, **1**: 92 (1903). Type as for *M. katangensis*.

Monotes glandulosissimus Hutch. in Kew Bull. **1931**: 246 (1931). Type: N. Rhodesia, near Lusaka, *Hutchinson & Gillett* 3595 (K, holotype).

Monotes obliquinervis Hutch., tom cit.: 247 (1931). Type: N. Rhodesia, Mpu-lungu, *Hutchinson & Gillett* 3963 (K, holotype).

Tree up to 13–14 m. high; branchlets tomentellous. Leaf-lamina 6–9 × 3·5–6 cm., elliptic to oblong or obovate-oblong, obtuse to slightly emarginate and some-times acute or apiculate at the apex, rounded to slightly cordate at the base; upper surface minutely tuberculate-subreticulate, slightly scaberulous with straight or nearly straight generally simple pachydermatous hairs developed on minute, white tubercles and sometimes mixed with smaller penicillate ones; lower surface fulvous-pubescent with straight or curled meso-pachydermatous (medium-thick-walled) hairs on the nerves, veins and reticulation, and with interreticular areoles covered with minute stellate hairs; midrib very thick and prominent beneath; lateral nerves in 10–12 pairs (with a weak tendency to the formation of short sub-sidiaries), nearly straight, some of them furcate near the apex, nerves and bifurca-tions reaching the margin of the leaf; reticulation very prominent and forming deep cavities; petiole thick, 1–2 cm. long. Inflorescences of subsessile condensed multi-florous clusters forming large terminal panicles, up to 15 cm. long, fulvous- or rufous-tomentose; pedicels 1–3 mm. long. Sepals 3–4 mm. long, densely sericeous-tomentose. Petals 8–10 mm. long, densely sericeous-tomentose. Stamens with anthers not produced at the apex. Fruits densely crowded (see TAB. **81** fig. A1) at the ends of the branches, 7–9 mm. in diam., subglobose, reticulate, tomen-tose, conical at the apex; wings 2·5–4 × 1–1·5 cm., reddish-purple, generally narrowly obovate-oblong to spathulate.

N. Rhodesia. N: Mpulungu, fl. & fr. 20.vii.1930, *Hutchinson & Gillett* 3963 (K); Chilongowelo, fl. 2.iv.1952, *Richards* 1404 (K). W: Chingola, fl. 17.iv.1954, *Fanshawe* 1098 (K; NDO); Ndola, fl. 30.iv.1937, *Duff* 216 (NDO); Bwana Mkubwa, 16.i.1911, *Fries* 326 (UPS). C: Lusaka, fl. 22.iv.1932, *Duff* 387/32 (FHO; NDO). E: between Minga and Sewa, fr. 2.viii.1929, *Burtt Davy* 950 (FHO). S: between Choma and Pemba, fl. & fr. 17.viii.1929, *Burtt Davy* 703 (FHO; K; NDO). **Mozambique.** T: Marávia, fr. 12.viii.1941, *Torre* 3265 (BM); Macanga, Furancungo, fr. 25.viii.1941, *Torre* 3334 (BM).

Also in the Belgian Congo (Katanga) and Tanganyika. One of the commonest trees in *Brachystegia* woodlands on all kinds of soils in N. Rhodesia and SE. Katanga.

Vernacular name: Chimpampa or Kimpampa throughout the whole distribution area.

3. **Monotes redheadii** Duvign. in Bol. Soc. Brot., Sér. 2, **33**: 101 (1959). Type: N. Rhodesia, Mwinilunga, *Milne-Redhead* 4608 (K, holotype).

Tree 4–7 m. high; branchlets densely ferruginous- or blackish-velutinous-tomentose. Leaf-lamina 6–9 × 2–3·5 cm., narrowly elliptic to oblong, obtuse to truncate at the apex, rounded to slightly cordate at the base; upper surface hairy with long straight simple isolated pachydermatous hairs; lower surface buff with long straight simple leptodermatous hairs on the nerves and reticulation and with the interreticular areoles quite glabrous; midrib thick and very prominent beneath; lateral nerves in 10–14 pairs (with a strong tendency to the formation of short subsidiaries), nearly straight, incurved at the apex without reaching the margin of the leaf; reticulation not very prominent but very conspicuous; petiole 0·7–1 cm. long, thick. Inflorescences 1–1·5 cm. long, of short few-flowered clusters solitary at the end of 2–3 cm. long axillary peduncles, blackish- or fulvous-tomentose; pedicels 5–7 mm. long. Sepals 2 mm. long, greyish-cottony-tomentose. Petals 7 mm. long, densely fulvous-sericeous-tomentose. Stamens with anthers produced into a short ovate appendage. Fruits unknown.

N. Rhodesia. W: Mwinilunga Distr., fl. 16.ii.1938, *Milne-Redhead* 4608 (K). Also in Angola. Edges of dry evergreen forest on Kalahari Sand.

Possibly a hybrid between *M. africanus* Gilg and *M. dasyanthus* Gilg. Vernacular name: Kasanya (chiLunda).

4. **Monotes magnificus** Gilg in Engl., Bot. Jahrb. **28**: 135 (1899); op. cit. **41**: 290 (1908).—De Wild., Pl. Bequaert. **4**, 2: 180 (1927).—Bancroft in Bol. Soc. Brot., Sér. 2, **13**: 377 (1939).—Duvign. in Lejeunea, **13**: 53 (1949). Type from Tanga-nyika (Uhehe).

Shrub to small tree up to 8 m. high; branchlets pubescent, becoming glabrous. Leaf-lamina very large, 13–23 × 11–17 cm., suborbicular, emarginate at the apex, cordate at the base; upper surface finely reticulate, substrigose, with straight single (rarely geminate) subappressed pachydermatous hairs (becoming sparse with age); lower surface discolorous, greyish- or brownish-floccose-tomentose, with relatively long dense curled lepto-mesodermatous (thin- to medium-walled) hairs on the nerves and veins, and with the reticulations and interreticular areoles covered with minute stellate hairs; extra leaf-glands in the axils of the lateral nerves; lateral nerves in 10–11 pairs (with a weak tendency to formation of short subsidiaries), slightly depressed above and very prominent and very thick beneath, slightly curving towards the apex, many of them producing 1–3 bifurcations on the side nearest the base of the leaf before reaching the margin, nerves and bifurcations anastomosing on the thickened margin; veins prominent and conspicuous; reticulation partially hidden by hairs; petiole 2–3·5 cm. long, thick. Inflorescences axillary, few-flowered, subsessile, often condensed in subterminal clusters, densely rufous-tomentose; pedicels 3–6 mm. long. Sepals 7 mm. long, rufous-sericeous-tomentose. Petals 11–12 mm. long, rufous-sericeous-tomentose. Stamens with anthers produced into a short triangular apiculus. Fruit 15–23 mm. in diam., spherical, slightly depressed at the apex, brownish, subsericeous-pubescent; wings 6–7 × 2·5 cm., yellow or reddish, broadly oblanceolate.

N. Rhodesia. N: Abercorn, fr. 22.iv.1936, *Burtt* 5917 (BM; K). Kawambwa, fl. 29.x.1952, *Angus* 666 (BRLU; FHO; K); North of Fort Rosebery, fl. 10.x.1947, *Brenan & Greenway* 8086 (K). **Nyasaland.** N: North Nyasa Distr., fl. viii.1940, *Lewis* 122 (FHO).
Also in Tanganyika and the Belgian Congo.

5. **Monotes adenophyllus** Gilg in Engl., Pflanzenw. Ost-Afr. **C**: 275 (1895); in Engl., Bot. Jahrb. **28**: 135 (1899). Type from Tanganyika (Ugogo).

Related to *M. magnificus* Gilg but leaves without additional glands in the axils of the lateral nerves on the upper face, and not floccose-tomentose on the lower surface, with a general tendency in the Zambezian region to be smaller (6–12 × 4–10 cm.); petiole thin and relatively short, 1–1·5 cm. long. Flowers and fruits also generally smaller.

This is a most variable species with many local races, the definite taxonomic status of which can be decided only after much more collecting. In the contact zone with the Guinean region, principally in the Belgian Congo, the leaves are large as in typical *M. adenophyllus*, but transitional forms occur: in more southern localities they become smaller. The difficulty of naming specimens from our area comes from the fact that most of the types of different subspecies of *M. adenophyllus* are large-leaved Belgian Congo specimens. The 3 subspecies cited below occur in our region, the typical subspecies being confined to Tanganyika.

Leaves concolorous; interreticular areoles on the lower surface glabrous, or with very few
    minute stellate hairs:
    Hairs on the two surfaces stiff, pachydermatous, straight or hooked - subsp. *delevoyi*
    Hairs on the two surfaces soft, lepto-mesodermatous, curled or coiled subsp. *subfloccosus*
Leaves discolorous; interreticular areoles on the lower face densely covered with small
    stellate hairs -   -   -   -   -   -   -   -   - subsp. *homblei*

Subsp. **delevoyi** (De Wild.) Duvign. in Bol. Soc. Brot., Sér. 2, **33**: 101 (1959). Type from the Belgian Congo (Katanga).
    Monotes delevoyi De Wild., Pl. Bequaert. **4**, 2. 171 (1927). Type as above.
    Monotes magnificus var. *paucipilosus* Duvign. in Lejeunea, **13**: 55, t. 13 figs. b–c (1949). Type from the Belgian Congo (Katanga).
    Monotes pwetoensis Robyns ex Duvign., loc. cit. *in synon.*
    Monotes magnificus var. *glabrescens* Duvign., loc. cit. *nom. provis.* Type as for *M. delevoyi.*

Small tree up to 4 m. high; branchlets brownish or blackish, tomentellous. Leaf-lamina 7·5–12 × 5·5–8·5 cm., broadly ovate to suborbicular, rounded to slightly emarginate at the apex, truncate to cordate at the base, concolorous; upper surface finely reticulate, ± bullate towards the margins, slightly scaberulous with short stiff hooked pachydermatous hairs borne singly or geminate or penicillate on small white cushions and with scattered minute yellow glands; lower surface with scattered minute yellow glands and with short stiff hooked pachydermatous simple or penicillate hairs on the nerves and reticulations and with the interreticular areoles generally glabrous, sometimes with scattered small meso-dermatous stellate hairs; midrib and lateral nerves slightly depressed above and usually prominent beneath; lateral nerves in 9–13 pairs (with a tendency to the formation of short subsidiaries), slightly curving towards the apex, many of them producing 1–3 bifurcations before reaching the margin, nerves and bifurcations anastomosing before reaching the thickened margin; veins and reticulation not hidden by hairs. Inflorescences 2–5 cm. long, axillary, brown-tomentellous, lax, on a 1·5–2 cm. long slender peduncle; pedicels 2 mm. long. Sepals 2·5 mm. long, rufous-cottony-tomentose. Petals densely sericeous-tomentose. Stamens with anthers produced at the apex into a long, oblong or triangular lobe. Fruit 1·2–1·4 cm. in diam., subglobose, subsericeous-tomentellous, mucronate at the apex; wings 2·5–3 × 1–2·5 cm., yellow, brownish or purplish, generally broadly elliptic to obovate, sometimes oblong and narrower.

N. Rhodesia. N: Mpulungu, near Abercorn, fr. 8.iii.1952, *Richards* 1406 (K); Abercorn Distr., fr. 22.v.1936, *Burtt* 5919 (BM; FHO; K).

Also in the Belgian Congo (Katanga).

Subsp. **subfloccosus** Duvign. in Bol. Soc. Brot., Sér. 2, **33**: 102 (1959). Type: N. Rhodesia, Northern Prov., Fort Rosebery-Luwingu road, fr. 19.v.1931, *Stevenson* 247/31 (FHO, holotype; NDO).

Differing from the type subspecies and from subsp. *delevoyi* by the greyish subfloccose aspect of the lower surface of the leaf, and by the nerves and reticulation being covered with minute soft white coiled leptodermatous hairs; interreticular areoles glabrous; upper surface with soft curled leptodermatous penicillate hairs. Fruits 1·5 cm. or more in diam., subspherical or ovoid (generally larger than in the other subspp.); wings attaining 8 × 3·5 cm.

N. Rhodesia. N: Fort Rosebery-Luwingu road, fr. 19.v.1931, *Stevenson* 247/31 (FHO; NDO).

Also in the Belgian Congo. Woodland and in wooded swampy depressions on rocky or lateritic soils.

In our area this subspecies is only known from two specimens collected by Stevenson in the region north of Fort Rosebery; they have been named *Monotes stevensonii* by Burtt Davy (in sched.). It most commonly occurs with larger leaves, in the adjacent Luapula-Mweru depression in the Belgian Congo.

Subsp. **homblei** (De Wild.) Duvign. in Bol. Soc. Brot., Sér. 2, **33**: 102 (1959). Type from the Belgian Congo.

Monotes homblei De Wild. in Bull. Jard. Bot. Brux. **5**: 55 (1915). Type as above.
Vatica homblei De Wild., loc. cit. Type as above.
Monotes magnificus var. *homblei* (De Wild.) Duvign. in Lejeunea, **13**: 55, t. 13 fig. a (1949). Type as above.

Small tree up to 8 m. high; branchlets brownish, tomentellous. Leaf-lamina 6–10 × 4–9 cm., broadly elliptic-oblong to orbicular, truncate to slightly emarginate at the apex, truncate to cordate at the base; upper surface finely reticulate, ± bullate towards the margin, slightly scaberulous with very short pachydermatous hairs which are generally penicillately arranged on small cushions; lower surface discolorous, greyish or whitish, with short pachydermatous penicillate hairs on the nerves mixed with long curled simple lepto-mesodermatous hairs on the nerves and reticulation; interreticular areoles densely covered with small leptodermatous stellate hairs; nerves, veins and reticulation as in subsp. *delevoyi*. Inflorescences 1–3 cm. long, axillary, brown-tomentellous, condensed in clusters on a short peduncle 0·5–1 mm. long. Flowers as in subsp. *delevoyi*. Fruit as in subsp. *delevoyi*, usually with short broad purplish or deep purple wings.

N. Rhodesia. B: Balovale, fr. vi. 1952, *Gilges* 90 (K; SRGH). W: Mwinilunga, Matonchi Farm, fl. 2.i.1938, *Milne-Redhead* 3923 (BM; K); Mwinilunga, fr. vii.1955, *Edmonds* 11/55 (BRLU; NDO); Solwezi, fr. 6.vii.1930, *Milne-Redhead* 677 (K).

Also in the Belgian Congo (Katanga) and Angola.

This subspecies is very common in E. Katanga (Belgian Congo), but with generally larger leaves. It probably hybridizes with *M. africanus* and I would refer to such a hybrid the sterile specimen collected by R. E. Fries at Kalambo (*R. E. Fries* 1371) which possesses the small narrow leaves of *M. africanus* with the indumentum of *M. adenophyllus* subsp. *delevoyi* (cited under *M. adenophyllus* Gilg by R. E. Fries, Wiss. Ergebn. Schwed. Rhod.-Kongo-Exped. **1**: 154 (1914)

6. **Monotes autennei** Duvign. in Bull. Soc. Roy. Bot. Belg. **90**: 183 (1958). Type from the Belgian Congo.

Tree up to 17 m. high; branchlets shortly puberulous. Leaf-lamina 12–18 × 9–12 cm., leathery, broadly elliptic, mostly obcordate at the apex, broadly cordate at the base and with extra leaf-glands in the axils of the lateral nerves; upper surface slightly scaberulous with short pachydermatous straight penicillate hairs on small white cushions; lower surface greyish- or fulvous-pubescent (sometimes glabrescent when old), with relatively long curved pachydermatous hairs mixed with tufted penicillate ones, and with the reticulation and the interreticular areoles densely to sparsely covered with minute stellate hairs; midrib very thick and prominent beneath; lateral nerves 12–15 pairs, slightly incurved, some of them bifurcate near the apex, nerves and bifurcations reaching the margin of the leaf; veins and reticulation very prominent and forming deep cavities; petiole c. 2 cm. long, very thick. Inflorescences, flowers and fruits as in *M. katangensis*.

N. Rhodesia. W: Mwinilunga Distr., Matonchi Farm, 7.ii.1938, *Milne-Redhead* 4463 (K); Mwinilunga, vii.1955, *Edmonds* E 10/55 (NDO).
Also in the Belgian Congo (Western Katanga).
In *Brachystegia* and *Marquesia* woodlands, on sandy or loamy soils.
Timber brittle.

Vernacular name: Chimpampa (ChiLunda).
All the characters tend to show that this species is probably a hybrid population geographically isolated from its parents *M. katangensis* and *M. magnificus*.

7. **Monotes glaber** Sprague in Kew Bull. **1909**: 305 (1909).—Eyles in Trans. Roy. Soc. S. Afr. **5**: 421 (1916).—De Wild., Pl. Bequaert. **4**, 2: 170 (1927).—Hutch. in Kew Bull. **1931**: 253 (1931).—Steedman, Trees etc. S. Rhod.: 51, t. 49 (1933).—Bancroft in Bol. Soc. Brot., Sér. 2, **13**: 372 (1939).—Duvign. in Lejeunea, **13**: 43 (1949).
Monotes africanus var. *denudans* sensu Eyles, loc. cit.
Vatica africana var. *glabra* Oliv., F.T.A. **1**: 173 (1868). Type: N. Rhodesia, Batoka Highlands, *Kirk* (K, holotype).

Tree 3–10 (20) m. high; branchlets glabrous. Leaf-lamina 5–9·5 × 2–5 cm., elliptic to oblong or obovate-oblong, obtuse to truncate at the apex, slightly cordate at the base; upper surface finely reticulate, shining, glabrous; lower surface concolorous, smooth, glabrous; midrib and lateral nerves slightly depressed above and prominent beneath; lateral nerves in 8–11 pairs, progressively fading out well before reaching the margin of the leaf; reticulation scarcely prominent, not conspicuous to the naked eye. Inflorescences axillary, 2–4 cm. long, slender, lax, 4–10-flowered, brownish, glabrescent, on a slender peduncle 1–1·5 cm. long; pedicels 5 mm. long. Sepals 3 mm. long, densely cottony-tomentose. Petals 7·5 mm. long, shortly greyish- or yellowish-sericeous-tomentose. Stamens with anthers produced into a very short mucro. Fruit 0·7–1·2 cm. in diam., subglobose, sericeous, rounded or slightly conical at the apex; wings 2–3 × 1–1·5 cm., yellow or brownish, elliptic to obovate.

Bechuanaland Prot. N: Sitengu Pan, fr. viii.1937, *Miller* 143 (BM; FHO). **N. Rhodesia.** B: Sesheke, fl. 19.xii.1952, *Angus* 966 (BRLU; FHO; K). C: Lusaka, 3.v.1932, *Stevenson* 393/32 (NDO). S: Batoka Highlands, fr. vii.1860, *Kirk* (K). **S. Rhodesia.** N: Mazoe, fr. i.1906, *Eyles* 242 (BM; SRGH). W: 128 km. from Bulawayo on road to Victoria Falls, fr. vii.1930, *Hutchinson & Gillett* 3409 (K). C: Beatrice, fl. 27.xii.1924, *Eyles* 4441 (K; SRGH); Salisbury, fl. 23.xii.1931, *Trapnell* 691 (K).
Widespread in sandy *Brachystegia* woodland in S. Rhodesia, extending to the *Baikiaea* woodland on sand in Barotseland.

8. **Monotes africanus** A.DC. in DC., Prodr. **16**, 2: 624 (1868) emend. excl. specim. Kirk.—Gilg in Engl., Bot. Jahrb. **28**: 137 (1899) pro parte excl. specim. *Welwitsch* 1077.—De Wild., Pl. Bequaert. **4**, 2: 173 (1927).—Bancroft in Exell & Mendonça, C.F.A. **1**, 1: 370 (1937); in Bol. Soc. Brot., Sér. 2, **13**: 372 (1939).—Duvign. in Lejeunea, **13**: 42 (1949).—Exell & Mendonça, C.F.A. **1**: 370 (1951).—Milne-Redh. in Mem. N.Y.Bot. Gard. **3**, 3: 222 (1953). Type from Angola (Huila).
Vatica africana Welw. & Kirk ex A.DC., loc. cit., *in synon.*—Welw. ex Oliv., F.T.A. **1**: 173 (1868) pro parte excl. vars. *hypoleuca, glomerata* et *glabra*. Type as above.
Vatica africana var. *denudans* Welw. in Trans. Linn. Soc. **27**: 16, t. 5 fig. 1 (1869) pro parte excl. specim. Kirk. Type from Angola.
Monotes rufotomentosus Gilg in Engl., Bot. Jahrb. **28**: 138 (1899).—Milne-Redh., loc. cit. Type from Tanganyika (Uhehe).

Shrub or small tree up to 8 m. high; branchlets greyish or brownish, glabrous. Leaf-lamina 5–10 × 2·5–5·5 cm., elliptic to oblong or obovate, obtuse to emarginate at the apex, rounded to slightly cordate at the base ; upper surface finely reticulate, typically bullate in the angles made by the lateral nerves and midrib (with a corresponding depression on the lower face), glabrous except sometimes for the midrib and lateral nerves, viscid from numerous punctiform yellow glands ; lower surface concolorous, puberulous with very minute curled or coiled meso-pachydermatous hairs mixed with some straight ones on the nerves and reticulation, the interreticular areoles generally glabrous ; midrib and lateral nerves slightly depressed above and prominent beneath ; lateral nerves in 8–13 pairs (with a strong tendency to the formation of short subsidiaries), nearly straight but incurved at the apex without reaching the margin of the leaf ; veins and reticulation not very prominent but quite conspicuous. Inflorescences 1·5–3 cm. long, axillary, ± condensed or lax, few-flowered, fulvous-tomentellous ; peduncle 1·5–3 cm. long, slender ; pedicels 3–8 mm. long. Sepals 2 mm. long, densely woolly-rufous-tomentose. Petals 8 mm. long, densely greyish-subsericeous-tomentose. Stamens with anthers produced into a large emarginate or triangular lobe. Fruit 1–1·5 cm. in diam., subglobose, rounded to conical at the apex, sericeous ; wings 2·5–3·5 × 1·2–1·6 cm., yellowish or brownish, elliptic to obovate.

N. Rhodesia. N : Mpika, fl. 3.ii.1955, *Fanshawe* 1955 (K ; NDO ; SRGH). W : Kitwe, fl. 19.i.1955, *Fanshawe* 1810 (K ; NDO ; SRGH) ; Mwinilunga Distr., fr. 21.viii.1920, *Milne-Redhead* 938 (K). E : Mbozi, 29.viii.1933, *Greenway* 3640 (NDO) ; Lundazi, fr. 27.iv.1952, *White* 2484 (BRLU ; FHO ; K). Nyasaland. N : Nyika Plateau, fr. vii.1896, *Whyte* (K). C : Nchisi Mt., fr. 5.viii.1946, *Brass* 17133 (BM ; K ; SRGH). S : Zomba, fl. xii.1915, *Purves* 240 (K) ; Zomba Distr., fl. i.1932, *Clements* 188 (FHO). Mozambique. N : Massangulo, fr. ii.1953, *Gomes e Sousa* (K) ; R. Msalu, fl., *Allen* 143 (FHO ; K). T : Macanga, fr. 25.viii.1941, *Torre* 3334 (BM).

Also in Angola, Tanganyika and the Belgian Congo. A small tree widespread in *Brachystegia* woodlands, on all kinds of soils, sometimes dominant and forming pure stands on compact heavy soils.

A most variable species in form and dimensions of leaves. The eastern and southeastern specimens tend to form larger leaves, with a more irregular, obovate and convex lamina with more distant lateral nerves, but all transitions occur. A more constant character is a more developed pilosity on the midrib and lateral nerves, and even on the lamina on the upper surface. As already stated by Bancroft (tom. cit.: 373, 376), the eastern specimens of the *M. rufotomentosus* type are probably hybrids with a possible introgression from *M. adenophyllus* ; good examples of this are *Allen* 143 and *Clements* 188 (described as " large, strong timber "). At the other end of the scale of variation, some western specimens (Mwinilunga, *Milne-Redhead* 938, Solwezi, *Milne-Redhead* 576) are completely glabrous on the two surfaces of the leaves, and may be confused with *M. glaber* (but nerves and reticulation are quite different).

9. **Monotes discolor** R.E.Fr., Wiss. Ergebn. Schwed. Rhod.-Kongo-Exped. **1** : 153, t. 12 fig. 13–14 (1914).—De Wild., Pl. Bequaert. **4**, 2 : 172 (1927).—Bancroft in Exell & Mendonça, C.F.A. **1**, 1 : 141 (1937) ; in Bol. Soc. Brot., Sér. 2, **13** : 378 (1939).—Duvign. in Lejeunea, **13** : 50 (1949).—Exell & Mendonça, C.F.A. **1**, 2 : 371 (1951). Type : N. Rhodesia, Mporokoso, *Fries* 1175 (S, holotype ; UPS).

A species with large discolorous leaves, falling into 3 distinct varieties according to the length of the hairs on the lower surface of the leaves.

Lower surface of leaves tomentose, the hairs of nerves and veins hiding the reticulation and the interreticular areoles :
Lower surface of leaves with short cottony hairs   -   -   -   -   var. *discolor*
Lower surface of leaves with long woolly hairs   -   -   -   -   var. *lanatus*
Lower surface of leaves tomentellous, the hairs of nerves and veins not hiding the reticulation and the interreticular areoles, which are nevertheless covered with minute stellate hairs   -   -   -   -   -   -   -   -   -   -   var. *cordatus*

The other characters, here given under the type variety, are the same for all three varieties.

Var. **discolor**

Small tree up to 10 m. high ; branchlets pubescent, soon becoming glabrous. Leaf-lamina 13–19 × 7–10 mm., broadly elliptic to obovate or suborbicular, never twice as long as wide, rounded, truncate or emarginate at the sometimes apiculate apex, rounded or more often cordate at the base ; upper surface yellowish-green, finely reticulate, somewhat shining, viscid from numerous punctiform yellow glands, glabrous except for the midrib ; lower surface whitish, shortly cottony-tomentose with relatively short dense curled or coiled mesodermatous hairs on the nerves and reticulation and hiding the reticulation and interreticular areoles, which are densely covered with minute stellate hairs; midrib and lateral nerves slightly depressed above and very prominent beneath; lateral nerves in 16–19 pairs (without short subsidiaries), curving towards the apex without reaching the margin of the leaf. Inflorescences axillary on young short leafy shoots; peduncles 3–8 cm. long, relatively few-flowered and lax, densely rufous-tomentose; pedicels 3 mm. long. Sepals 4–5 mm. long, densely cottony-greyish-tomentose. Petals 10 mm. long, densely fulvous-sericeous-tomentose. Stamens with anthers produced at the apex into a long triangular lobe. Fruit 2 cm. in diam., subglobose, ± truncate and apiculate at the apex; wings 4·5–5·5 ×1–2 cm., brownish, spathulate, with sparse minute penicillate hairs outside and inside.

N. Rhodesia. N: Abercorn, fl. 13.xii.1949, *Bullock* 2092 (K; SRGH); Mporokoso, fl. & fr. 31.x.1911, Fries 1175 (UPS). W: Ndola, fl. 1.x.1937, *Duff* 240 (NDO). C: Serenje, fl. & fr. 27.x.1938, *Miller* 259 (NDO).

Var. **lanatus** Duvign. in Bol. Soc. Brot., Sér. 2, **33**: 102 (1959). Type: N. Rhodesia, Mwinilunga, *Holmes* 1330 (K, holotype).

Differing from var. *discolor* by the much longer simple hairs on nerves and veins of the lower surface of the leaves which is whitish-woolly-tomentose instead of whitish-cottony-tomentose. Veins and reticulation completely hidden by this woolly tomentum and only the principal and lateral nerves visible.

N. Rhodesia. W: Mwinilunga, fl. & fr. 14.xi.1955, *Holmes* 1330 (K; NDO).

Var. **cordatus** (Hutch.) Duvign., loc. cit. Type: N. Rhodesia, Serenje, *Hutchinson & Gillett* 3703 (K, holotype).
Monotes cordatus Hutch. in Kew Bull. **1931**: 246 (1931). Type as above.
Monotes lukuluensis Hutch. tom. cit.: 247 (1931). Type: N. Rhodesia, Lukulu R., *Hutchinson & Gillett* 3741 (K, holotype).

Differing from var. *discolor* by the shorter indumentum of simple hairs on nerves and veins of the lower surface of the leaves which is greyish cottony-tomentellous instead of whitish-cottony-tomentose and as a result, reticulation, although densely covered by stellate hairs, just visible. Reticulation on the upper face not so prominent as in the type variety.

N. Rhodesia. N: Abercorn, fr. 3.vi.1936, *Burtt* 5923 (BM; K); Lukulu R., fr. 16.vii.1930, *Hutchinson & Gillett* 3741 (K). W: Ndola, fl. 26.x.1953, *Fanshawe* 453 (K; NDO); Ndola, fr. 25.i.1933, *Duff* 88/33 (FHO; K; NDO); Solwezi, Mutanda Bridge, fr. 26.vi.1930, *Milne-Redhead* 598 (K). C: Chitambo Mission, fr. 7.vi.1931, *Stevenson* 317/31 (BRLU; FHO; NDO); Serenje, 15.vii.1930, *Hutchinson & Gillett* 3703 (K).

Distribution as for the type variety.

10. **Monotes angolensis** De Wild., Pl. Bequaert. **4**, 2: 168 (1927).—Bancroft in Exell & Mendonça, C.F.A. **1**, 1: 140 (1937); in Bol. Soc. Brot., Sér. 2, **13**: 379 (1939). —Duvign. in Lejeunea, **13**: 46 (1949). Type from Angola (Bié).
Monotes oblongifolius Hutch. in Kew Bull. **1931**: 248 (1931). Type: N. Rhodesia, Kaloswe, *Hutchinson & Gillett* 3765 (K, holotype).

Tree up to 17 m. high; branchlets tomentellous, soon becoming glabrous. Leaf-lamina 4·5–8 ×2–4 cm., discolorous, elliptic to oblong, obtuse to truncate and mucronulate at the apex, rounded to slightly cordate at the base; upper surface smooth with minute flexuous penicillate scattered hairs, becoming glabrous, sometimes viscid from small punctiform glands; lower surface smoothly tomentellous, becoming yellowish, with a few straight hairs on the glabrescent brown nerves and very short woolly coiled yellow hairs covering the interreticular areoles; midrib prominent beneath; lateral nerves in 10–13 pairs (with a very strong tendency to the formation of subsidiaries), nearly straight but incurved just before and without reaching the margin of the leaf; petiole 1–2 cm. long, slender. Inflorescences up to 2·5 cm. long, axillary, relatively lax, few-flowered, greyish-tomentose, on a long slender peduncle; pedicels 3 mm. long. Sepals woolly-tomentellous. Petals densely sericeous-tomentellous. Stamens with connective produced into a triangular appendage. Fruit 9–12 mm. in diam., subglobose, woolly-tomentellous, generally conical at the apex; wings shining red or purple, 3–5 ×0·5–1 cm., generally narrowly sublanceolate.

Tab. 81. A.—MONOTES KATANGENSIS. A1, fruiting branch (× ⅔) *Angus* 120 ; A2, vertical section of flower (× 7) *Hutchinson & Gillett* 3595; A3, anther and part of filament (× 20) *Hutchinson & Gillett* 3595; A4, under surface of leaf (× 25). B.—MONOTES ENGLERI. B1, flowering branch (× ⅔); B2, anther and part of filament (× 12); B3, fruit (× ⅔) all from *Eyles* 4646.

N. Rhodesia. B: Mankoya, fl. 23.ii.1952, *White* 2128 (FHO; K). N: Abercorn, fr. 19.vi.1950, *Bullock* 2946 (K; SRGH); Kaloswe, fr. 16.vii.1930, *Hutchinson & Gillett* 3765 (K). W: Mwinilunga, Matonchi Farm, fl. 7.ii.1938, *Milne-Redhead* 4458 (K); Solwezi, Butanda Bridge, fr. 5.vii.1930, *Milne-Redhead* 673 (K).

Also in Angola and the Belgian Congo. *Brachystegia* woodlands, especially on light soils.

This tree is readily distinguished by its leaves, which become yellow with age and by its very brittle stems and branches.

11. **Monotes elegans** Gilg in Engl., Bot. Jahrb. **41**: 291 (1908).—Bancroft in Bol. Soc. Brot., Sér. 2, **13**: 376 (1939).—Duvign. in Lejeunea, **13**: 47 (1949). Type from Tanganyika.

    Monotes caloneurus sensu R.E.Fr., Wiss. Ergebn. Schwed. Rhod.-Kongo-Exped. **1**: 154 (1914).

Tree up to 12 m. high; branchlets tomentellous, becoming glabrous. Leaf-lamina 6–11 × 3·5–6 cm., elliptic to oblong or obovate-oblong, rounded to slightly emarginate and occasionally mucronate at the apex, rounded to slightly cordate at the base; upper surface smooth, persistently tomentellous, with minute, flexuous, geminate or penicillate hairs; lower surface discolorous, greyish, with relatively short cottony hairs on the nerves and reticulation and with the interreticular areoles densely covered with minute stellate hairs; midrib prominent beneath; lateral nerves in 12–16 pairs (with a strong tendency to the formation of short subsidiaries) curving towards the apex just before and without reaching the margin of the leaf; veins conspicuous; reticulation moderately conspicuous or not conspicuous, ± hidden by the indumentum; petiole 1–1·5 cm. long, rather slender. Inflorescences up to 6 cm. long, axillary, relatively lax, many-flowered, greyish- or fulvous-tomentose; pedicels 2–3 mm. long. Sepals 3 mm. long, with a dense cottony tomentum. Petals 10 mm. long, densely rufous-sericeous-tomentose. Stamens with anthers produced into a rounded appendage. Fruit 1–1·5 cm. in diam., subglobose, rounded or slightly subconical at the apex; wings reddish or brownish, very variable in shape, suborbicular (2·5–2 cm. in diam.) to linear (5·5 ×0·8 cm.).

N. Rhodesia. N: half-way to Kalambo Falls from Abercorn, fl. 13.v.1936, *Burtt* 5926 (BM; K; NDO), Abercorn Distr., fr. 30.v.1936, *Burtt* 5911 (BM; K). W: Ndola to Nkana, fl. 3.ii.1933, *Duff* 104/33 (FHO); Mufulira, fr. 2.vi.1934, *Eyles* 8249 (K; SRGH); Mwinilunga Distr., fr. 22.viii.1930, *Milne-Redhead* 942 (K); Bwana Mkubwa, fr. 17.viii.1911, *Fries* 368 (UPS).

Also in Tanganyika and the Belgian Congo.

12. **Monotes engleri** Gilg in Engl., Bot. Jahrb. **41**: 291, fig. K (1908).—De Wild., Pl. Bequaert. **4, 2**: 172 (1927).—Hutch. in Kew Bull. **1931**: 253 (1931).—Bancroft in Bol. Soc. Brot., Sér. 2, **13**: 376 (1939). Duvign. in Lejeunea, **13**: 45 (1949). TAB. **81** fig. B. Type: S. Rhodesia, near Umtali, *Engler* 3159 (B, holotype †).

    Monotes sp. cf. *M. engleri* sensu Eyles in Trans. Roy. Soc. S. Afr. **5**: 421 (1916).

    Monotes hypoleucus sensu Eyles, loc. cit.

    Monotes tomentellus Hutch. in Kew Bull. **1931**: 248 (1931). Type: S. Rhodesia, Odzani River Valley, *Teague* 431 (K, holotype).

Shrub or small tree up to 7 m. high; branchlets greyish- to rufous-tomentose, very soon glabrous. Leaf-lamina 4–8 × 2·5–4 cm., elliptic to oblong or obovate-oblong, obtuse to emarginate at the apex, rounded to slightly cordate at the base; upper surface finely reticulate, somewhat shining, tomentellous when young, soon becoming subglabrous, with minute simple or penicillate scattered hairs and numerous punctiform glands; lower surface discolorous, greyish- or fulvous-silvery-tomentellous with relatively short dense curled hairs on the nerves and reticulation hiding the interreticular areoles, which are densely covered with minute stellate hairs; midrib prominent beneath; lateral nerves in 9–15 pairs, curving towards the apex just before and without reaching the margin of the leaf; veins and reticulation not very prominent but quite conspicuous; petiole 1–2 cm. long, slender. Inflorescences 2·5–3 cm. long, axillary, more or less condensed, few-flowered, greyish- or fulvous-tomentose, on a peduncle 1–3 cm. long; pedicels 1–3 mm. long. Sepals 3 mm. long, densely cottony-rufous-tomentose. Petals 10 mm. long, densely greyish-subsericeous-tomentose. Stamens with connective produced into a triangular appendage. Fruit 1·5 cm. in diam., subglobose, rounded at the apex, cottony-tomentellous; wings 3–4·5 × 1·5–2·5 cm., yellowish, broadly oblong to obovate, with penicillate hairs outside and inside.

N. Rhodesia. N: Mpika, fl. 6.ii.1955, *Fanshawe* 2007 (NDO). **S. Rhodesia.** N: Mazoe, 28.vi.1933, *Pardy* 112/33 (FHO; NDO). C: Salisbury, fl. 8.xii.1926, *Eyles* 4555 (K); Marandellas, fl. 22.xii.1948, *Wild* 2708 (SRGH). E: Umtali Distr., Dora R., fl. 14.xii.1949, *Chase* 1914 (BM; SRGH). S: Fort Victoria, fr. 17.iv.1947, *Acheson* 15/47 (FHO). **Nyasaland.** C: Domira Bay, fr. 2.vii.1936, *Burtt* 5925 (FHO; K). S: main road west of Lake Nyasa from Zomba to the north of the lake, fr. vii.1935, *Smuts* 2086 (K; NH; PRE). **Mozambique.** N: Mutuali, fl. 4.ii.1954, *Gomes e Sousa* 4173 (K; SRGH); Mpanda Mt., fl. viii.1940, *Lewis* 105 (FHO). Z: Guruè, fr. 1.vii.1943, *Torre* 5651 (BM; LISC). MS: Manica, Mavita, fr. 20.ii.1948, *Mendonça* 3801 (BM; L); Cheringoma, fr. 4.v.1942, *Torre* 4041 (BM; LISC); Beira, fr. 15.iv.1957, *Gomes e Sousa* 4359 (K).

Specimens from the most southern part of the area are not so hairy and sometimes have a shorter and more dispersed tomentellum, with only scattered stellate hairs on the inter-reticular areoles. They show, therefore, some convergence with *M. africanus*.

28. MALVACEAE

By A. W. Exell

(Genera 10–16 by A. D. J. Meeuse)

Herbs, shrublets, shrubs or small trees, usually with stellate hairs or bristles, sometimes aculeate, more rarely lepidote. Leaves alternate, usually petiolate, often palmately divided. Stipules present. Flowers bisexual (rarely unisexual), actinomorphic. Epicalyx present or absent. Calyx (3–4) 5-lobed, truncate or occasionally 5- or 10-toothed; lobes valvate. Petals 5, free but often slightly adnate to the staminal tube, contorted or imbricate. Stamens numerous, united in, a tube surrounding the style; anthers 1-thecous. Ovary superior, (1) 2-multi-locular (carpels rarely in vertical rows); style simple at the base, often branched towards the apex, branches the same number as or twice as many as the carpels; ovules 1–numerous in each loculus; placentation axile. Fruit a loculicidally dehiscent capsule or composed of follicles, achenes or pseudo-achenes arranged around a central columella, or indehiscent and woody or fleshy. Seeds usually with some endosperm; cotyledons often folded.

In addition to the genera dealt with here, *Lagunaria patersonii* G. Don, an Australian species, is grown as an ornamental shrub.

Fruit indehiscent, woody or ± fleshy:
  Ovules 3–4 in each loculus; fruit not lobed, woody or somewhat fleshy
                                                         1. Thespesia
  Ovules 1 in each loculus; fruit (4) 5-lobed, fleshy - - - **2. Thespesiopsis**
Fruit a loculicidally dehiscent capsule or composed of follicles, achenes or pseudo-achenes arranged around a central columella:
  Fruit a loculicidally dehiscent capsule:
    Calyx splitting laterally and deciduous with the corolla - **3. Abelmoschus**
    Calyx 5-lobed or 5- or 10-toothed, persistent:
      Ovules 2 or more in each loculus:
        Style not distinctly branched (rarely slightly divided at the tip), clavate at the apex with coherent stigmas:
          Oil-glands in a double row along each nerve of the calyx; bracts of epicalyx 3–15, linear, sometimes minute - - - - **4. Cienfuegosia**
        Oil-glands irregularly scattered on the calyx (and elsewhere) or absent:
          Epicalyx of 3 persistent cordate-ovate and incised-dentate foliaceous bracts (in our species) - - - - - - - **5. Gossypium**
          Epicalyx fused with the calyx, cupuliform with 5–15 caducous teeth or lobes
                                           7. Azanza
      Style 3–5-branched:
        Style 3(4)-branched; bracts of epicalyx 3, large, persistent, cordate, incised-dentate, foliaceous - - - - - - **6. Gossypioides**
        Style 5-branched; bracts of epicalyx not as above (rarely absent)
                                          8. Hibiscus
      Ovules 1 in each loculus - - - - - - **9. Kosteletzkya**

Fruit composed of septicidally or loculicidally dehiscent follicles, achenes or pseudo-achenes arranged around a central columella and separating from it when loculicidally dehiscent:
Style-branches equal in number to the carpels :
  Epicalyx absent :
    Carpels not contracted in the middle :
      Loculi 1-ovulate ; mericarps up to 10 (rarely more), always 1-seeded, not dehiscent by apical slits - - - - - - - 10. **Sida**
      Loculi pluri-ovulate, mericarps 10 or more (rarely fewer) ; usually (2) 3 (8)-seeded, rarely 1-seeded, dehiscent by apical slits - - 11. **Abutilon**
    Carpels contracted in the middle, upper halves stellately spreading ; loculi 2-ovulate - - - - - - - - - - 12. **Wissadula**
  Epicalyx of 3 bracts (in our species) :
    Style-branches with longitudinally extended stigmatic surfaces on their adaxial surfaces - - - - - - - - - 13. **Malva**
    Style-branches with capitate or shortly subclavate stigmas - 14. **Malvastrum**
Style-branches twice the number of the carpels :
  Carpels with hooked spines ; bracts of the epicalyx connate at the base and adnate to the calyx - - - - - - - - - 15. **Urena**
  Carpels without hooked spines though sometimes setose ; bracts of the epicalyx free or nearly so - - - - - - - - 16. **Pavonia**

1. THESPESIA Soland. ex Correa

Thespesia Soland. ex Correa in Ann. Mus. Par. **9** : 290, t. 8 fig. 2 (1807)
nom. conserv.

Small trees or shrubs. Leaves lepidote. Flowers solitary, axillary or forming terminal racemes or panicles by reduction of the upper leaves. Epicalyx of 3–5 bracts. Calyx cupuliform, truncate, persistent. Ovary 5-locular, loculi 3–4-ovulate ; style not branched, clavate. Fruit woody or somewhat fleshy, not lobed, indehiscent. Seeds sericeous or glabrous. Chromosome number : n=13.

Leaves entire, up to 15 ×12 cm.; flowers up to 8 cm. in diam., buds lepidote; seeds sericeous-tomentellous - - - - - - - - 1. *populnea*
Leaves 3-lobed (more rarely entire), up to 7 ×6 cm.; flowers up to 4·5 cm. in diam., buds lepidote and tomentellous ; seeds glabrous or nearly so - - 2. *acutiloba*

1. **Thespesia populnea** (L.) Soland. ex Correa, loc. cit.—Mast. in Oliv., F.T.A. **1** : 209 (1868).—Sim, For. Fl. Port. E. Afr.: 16, t. 10 (1909).—Exell & Hillcoat in Contr. Conhec. Fl. Moçamb. **2** : 60 (1954).—Keay, F.W.T.A. ed. 2, **1**, 2 : 342 (1958). Type from India.
  Hibiscus populneus L., Sp. Pl. **2** : 694 (1753). Type as above.

Small tree or shrub ; branchlets densely lepidote. Leaf-lamina up to 15 ×12 cm., entire, ovate to broadly ovate, densely lepidote on both surfaces, apex somewhat acuminate, base cordate 7-nerved ; petiole up to 9 cm. long, lepidote ; stipules 5–7 mm. long, lanceolate to almost subulate. Flowers up to 8 cm. in diam., yellow ; peduncles up to 9 cm. long, not obviously articulated. Epicalyx of 3 bracts ; bracts 2–7 (11) mm. long, lanceolate-triangular to triangular, very early caducous. Calyx c. 15 ×7 mm., densely lepidote, 5-dentate. Petals 4 ×3·5 cm., obliquely obovate, lepidote outside, glabrous inside. Staminal tube c. 17 mm. long ; free parts of filaments 3–5 mm. long. Fruit up to 25 ×30 mm., depressed-globose, lepidote. Seeds 10 ×6 mm., sericeous-tomentellous, somewhat glabrescent.

Mozambique. N: Palma, bank of R. Rovuma, fl. & fr. 21.x.1942, *Mendonça* 1011 (LISC). Z : Quelimane, fl. 1908, *Sim* 21096 (PRE). MS : Beira, fl. ii.1924, *Honey* 815 (K ; PRE). LM: Marracuene, fl. 30.iii.1955, *Lemos* 67A (LMJ).
Widespread in the tropics and in S. Africa. Edges of mangrove swamps and along tidal waters.

2. **Thespesia acutiloba** (Bak. f.) Exell & Mendonça in Contr. Conhec. Fl. Moçamb. **2** : 63 (1954). TAB. **82**. Type : Mozambique, Delagoa Bay, *Monteiro* (K, holotype).
  Thespesia populnea var. *acutiloba* Bak. f. in Journ. of Bot. **35** : 51 (1897). Type as above.

Small tree or shrub 2–5 m. tall ; branchlets lepidote. Leaf-lamina up to 7 ×6 cm., broadly ovate to suborbicular in outline, bluntly or rather acutely shallowly 3-lobed (more rarely entire), lepidote on both surfaces, apex somewhat

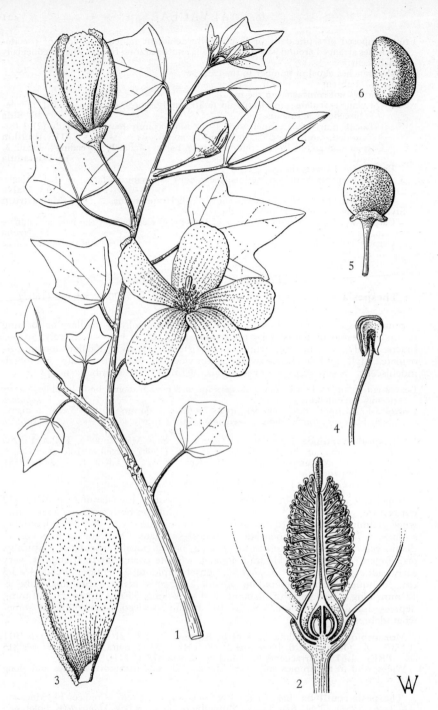

Tab. 82. THESPESIA ACUTILOBA. 1, flowering branch (× ⅔) *Torre* 2285; 2, vertical section of androecium and gynoecium (× 2) *Torre* 2120; 3, petal (× 1) *Torre* 2120; 4, stamen (× 6) *Torre* 2120; 5, fruit (× ⅔) *Myre & Carvalho* 1134; 6, seed (× 2) *Myre & Carvalho* 1134.

acuminate, margins entire, base truncate to very shallowly cordate and 5-nerved; petiole up to 4·5 cm. long, slender, lepidote; stipules not seen. Flowers 3–4·5 cm. in diam., solitary, axillary or forming terminal racemes or panicles by reduction of the upper leaves; peduncle 2–4 mm. long, 1–3-flowered, lepidote. Epicalyx of 3–5 bracts c. 3 mm. long, linear to subulate, very early caducous. Calyx 12 ×6–7 mm., lepidote outside, glabrous inside. Petals up to 4·5 ×3 cm., obovate. Staminal tube 12–13 mm. long; free parts of filaments 3–3·5 mm. long. Fruit subglobose, 15–16 mm. in diam., red, somewhat fleshy, glabrous. Seeds 9 ×5–6 mm., irregularly lunate, glabrous or almost so.

Mozambique. SS: Vila João Belo, fr. 23.ii.1941, *Torre* 2610 (BM; LISC). LM: Maputo, between Lagoa dos Patos and Catuane, fl. 16.xi.1944, *Mendonça* 2886 (BM; LISC); Inhaca I., fl. & fr. 14.iv.1944, *Torre* 6408 (BM; LISC). Also in Natal. In woodlands and thickets on recent sands near the coast.

2. THESPESIOPSIS Exell & Hillcoat

Thespesiopsis Exell & Hillcoat in Contr. Conhec. Fl. Moçamb. **2**: 55, t. 8 (1954).

Small trees or shrubs. Leaves lepidote. Flowers in terminal racemes or panicles. Epicalyx of 3 small caducous bracts. Calyx truncate, persistent. Ovary (4) 5-locular, loculi 1-ovulate, style not branched, clavate. Fruit (4) 5-lobed, indehiscent, fleshy. Chromosome number unknown.

Thespesiopsis mossambicensis Exell & Hillcoat, loc. cit. TAB. **83.** Type: Mozambique, Mocimboa da Praia, *Mendonça* 1050 (BM; K; LISC, holotype).

Small tree or shrub up to 6 m. tall; branchlets greyish-brown, appressed-peltate-lepidote when young. Leaf-lamina 2–8·5 ×2·5–8·5 cm., suborbicular, broadly ovate or cordate-acuminate, sparsely lepidote or glabrous above, closely appressed-peltate-lepidote below, eventually glabrescent, apex acuminate acute or rounded, margins entire, base usually cordate sometimes truncate or slightly rounded 5–7-nerved; petiole 1–6 cm. long, lepidote; stipules not seen. Flowers 4–4·5 cm. in diam., yellow; peduncle very short, 1–3-flowered; pedicels 5–8 mm. long, lepidote. Epicalyx of 3 bracts; bracts c. 2·5 mm. long, linear, lepidote, caducous. Calyx 8–10 ×4–5 mm., cupuliform, with 5 minute teeth, sericeous within. Petals 5, 4·5 ×3·5–4 cm., obliquely obovate, stellate-pubescent outside, glabrous inside. Staminal tube 20–25 mm. long; free parts of filaments 4–5 mm. long, sometimes partly connate in pairs. Style not branched, clavate. Fruit 12 ×15 mm., depressed-globose, red, glabrous. Seeds 12–13 ×7–8 mm., trigonous, silky-tomentose.

Mozambique. N: Mecufi, mouth of R. Lúrio, fr. 21.viii.1948, *Barbosa in Mendonça* 1842 (BM; LISC; SRGH). Endemic. On wooded hills near the coast.

J. B. Hutchinson (Appl. Genet. Cotton Improv.: 5 (1959)) disapproves of the separation of *Thespesiopsis* from *Thespesia*. He describes *Thespesia*, however, as having dry, woody or leathery fruits and 2–6 ovules per loculus; while *Thespesiopsis* has fleshy fruits and 1 ovule per loculus. If he wishes to include *Thespesiopsis* in *Thespesia* he should at least widen his conception of the latter genus and modify his description accordingly.

3. ABELMOSCHUS Medic.

Abelmoschus Medic., Malv.-Fam.: 45 (1787).

Annual or perennial herbs. Leaves palmately lobed or divided (rarely almost entire). Flowers solitary, axillary or in terminal racemes by reduction of the upper leaves. Epicalyx of 6–numerous filiform or linear bracteoles or much reduced and very caducous. Calyx thin, splitting laterally, circumscissile, slightly joined to the base of the corolla and deciduous with it. Staminal tube as in *Hibiscus*. Ovary 5-locular, loculi pluriovulate; style not manifestly branched, with 5 sessile or subsessile capitate stigmas. Capsule elongated, oblong or ellipsoid. Seeds with minute stellate hairs and sometimes also pilose.

Abelmoschus esculentus (L.) Moench, Meth. Pl.: 617 (1794).—Hochr., Fl. Madag., Malvac.: 7 (1955). TAB. **84.** Type from India.

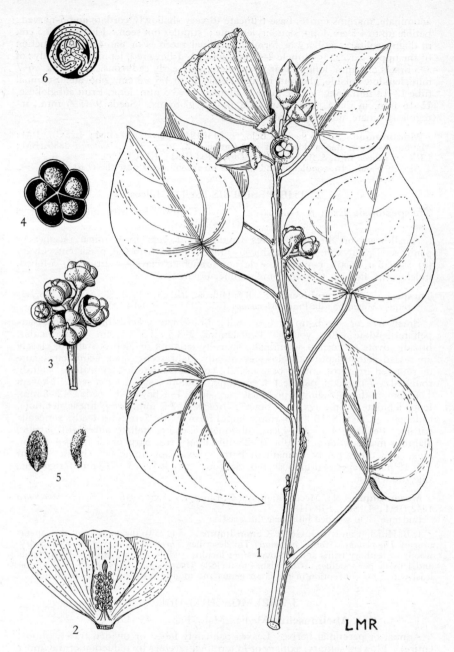

LMR

Tab. 83. THESPESIOPSIS MOSSAMBICENSIS. 1, flowering and fruiting branch (× ⅔) *Barbosa* 2023 ; 2, flower with calyx and 2 petals removed (× ⅔) *Barbosa* 2023 ; 3, infructescence (× ⅔) *Barbosa* 1842 ; 4, cross-section of fruit (× 1⅓) *Barbosa* 1842 ; 5, seed (× ¾) *Barbosa* 1842 ; 6, cross-section of seed (× 2) after a drawing by Miss D. Hillcoat.

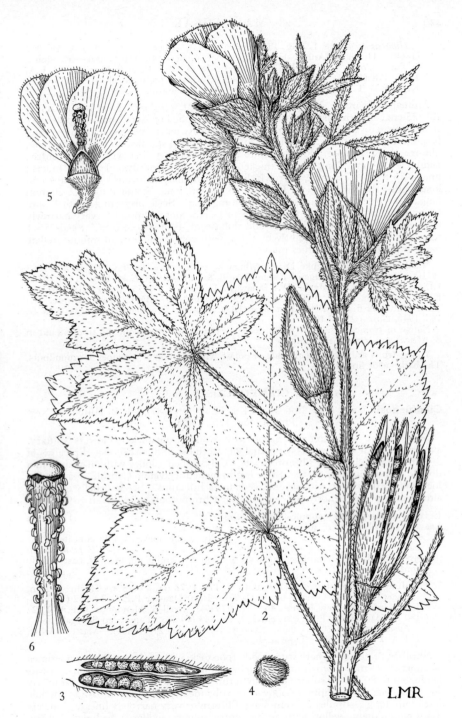

Tab. 84. ABELMOSCHUS ESCULENTUS.   1, flowering and fruiting shoot (× ⅔) *Welwitsch* 5278 ;
2, mature leaf (× ⅔) *Gilliland* 1567 ; 3, longitudinal section of fruit (× ⅔) *Gossweiler*
9241 ; 4, seed (×2) *Gossweiler* 9241 ; 5, flower with 2 petals, calyx and epicalyx
removed (× ⅔) *Pieton* 112 ; 6, staminal tube and stigma (×2) *Pieton* 112.

Hibiscus esculentus L., Sp. Pl. **2**: 696 (1753).—Mast. in Oliv., F.T.A. **1**: 207 (1868).—Hochr. in Ann. Conserv. Jard. Bot. Genève, **4**: 150 (1900).—Mendonça & Torre, Contr. Conhec. Fl. Moçamb. **1**: 18 (1950).—Exell & Mendonça C.F.A. **1**, 2: 178 (1951).—Williamson, Useful Pl. Nyasal.: 66, cum tab. p. 77 (1955).— Keay, F.W.T.A. ed. 2, **1**, 2: 348 (1958). Type as above.

Annual herb up to 2 m. tall; stems succulent, setulose. Leaf-lamina up to 25 × 25 cm., suborbicular in outline, palmatifid, -lobed or -sect, sparsely to densely setulose or setose-pilose on both surfaces especially on the nerves, margins serrate, base cuneate to cordate; petiole up to 30 cm. long; stipules up to 15 mm. long, filiform, densely pilose. Flowers up to 8 cm. in diam., yellow with purple centre; peduncle 1–4 cm. long, stout, thickened in fruit. Epicalyx of 10–12 bracts; bracts up to 25 × 2·5 mm., narrowly linear-triangular, caducous. Calyx 3–4 cm. long, with 5 short linear teeth. Petals up to 7–8 cm. long. Staminal tube 12–20 mm. long; free parts of filaments up to 0·5 mm. long. Style projecting up to 1 mm. beyond the staminal tube. Capsule up to 14 cm. long, ellipsoid to very narrowly ellipsoid, at first appressed-setose and pubescent, later glabrescent. Seeds 5 × 4 mm., depressed-globose, slightly humped, with concentric lines of minute stellate hairs or scales and sometimes pilose.

N. Rhodesia. E: Petauke Distr., Minga Forest Reserve, fl. 26.v.1952, *White* 2883 (FHO). S: Mazabuka, edge of Kafue Flats, fl. & fr. 7.iv.1955, *E.M. & W.* 1434 (BM; LISC; SRGH). **S. Rhodesia.** W: Bulawayo, 1370 m., fr. v.1955, *Miller* 2819 (SRGH). E: Nyumquarara Valley, 1070 m., fr. ii.1935, *Gilliland* 1567 (BM). **Mozambique.** T: Tete, Matunde, fl. 15.v.1948, *Mendonça* (LISC). MS: Chimoio, Amatongas, fl. & fr. 14.ii.1948, *Garcia* 232 (LISC).

Native of tropical Asia now widespread in cultivation throughout the tropics (and in S. Africa) and sometimes naturalized.

Young fruits are cooked with soda or potash with addition of tomatoes or groundnuts. The bark yields fibre.

4. CIENFUEGOSIA Cav.

Cienfuegosia Cav., Diss. Bot. **3**: 174, t. 72 fig. 2 (1787).—J. B. Hutch. in New Phytol. **46**: 125 (1947).

Small shrubs or perennial herbs with woody rootstock. Flowers solitary, axillary or in few-flowered sympodial or racemose inflorescences. Epicalyx of 3–12 (or more?) linear or filiform bracts. Calyx lobed; lobes with oil-glands in a double row along each nerve. Style not branched, clavate. Capsule 3–5-locular, loculicidally dehiscent, with several seeds in each loculus. Seeds usually ellipsoid, almost glabrous to densely silky or woolly. Chromosome number: n=10 or 11.

Leaves 3–5-palmatisect; stipules not foliaceous - - - - - 1. *digitata*
Leaves shallowly lobed to subentire; stipules foliaceous:
  Leaves flabelliform, minutely toothed especially at the apex, broadly cuneate at the base - - - - - - - - - - - 2. *hildebrandtii*
  Leaves suborbicular in outline, shallowly 3–5-lobed, margins not toothed, base usually cordate - - - - - - - - - - 3. *gerrardii*

1. **Cienfuegosia digitata** Cav., Diss. Bot. **3**: 174, t. 72 fig. 2 (1787).—Hochr. in Ann. Conserv. Jard. Bot. Genève, **6**: 55 (1902).—Burtt Davy, F.P.F.T. **2**: 282 (1932).— J. B. Hutch. in New Phytol. **46**: 128 (1947).—Exell & Mendonça, C.F.A. **1**, 2: 180 (1951).—Keay, F.W.T.A. ed. 2, **1**, 2: 343 (1958). Type from Senegal.
    Fugosia digitata (Cav.) Pers., Synops. Sp. Pl. **2**: 240 (1806).—Mast. in Oliv., F.T.A. **1**: 209 (1868). Type as above.

Shrublet or perennial herb with woody rootstock, up to c. 1 m. tall but sometimes as short as 9 cm.; stems minutely stellate-pubescent or nearly glabrous. Leaf-lamina up to 5 × 5 cm., suborbicular in outline, 3–5-palmatisect, glabrous or nearly so; lobes narrowly elliptic to obovate or subcuneate, sometimes 3–5-fid; petiole 8–25 mm. long; stipules 2·5 mm. long, filiform or very narrowly linear. Flowers 3·5–4 cm. in diam., yellow with purple centre, solitary, axillary; peduncle 2–6 cm. long. Epicalyx of up to 12 bracts; bracts 2–3 mm. long, filiform to very narrowly linear, caducous. Calyx 10–15 mm. long; lobes elongate-triangular, nigro-punctate, somewhat foliaceous at the tips, joined for 3–5 mm. at the base. Staminal tube 8 mm. long; free parts of filaments 1–1·5 mm. long. Capsule 9 mm. in diam., subglobose, glabrous. Seeds with silky floss.

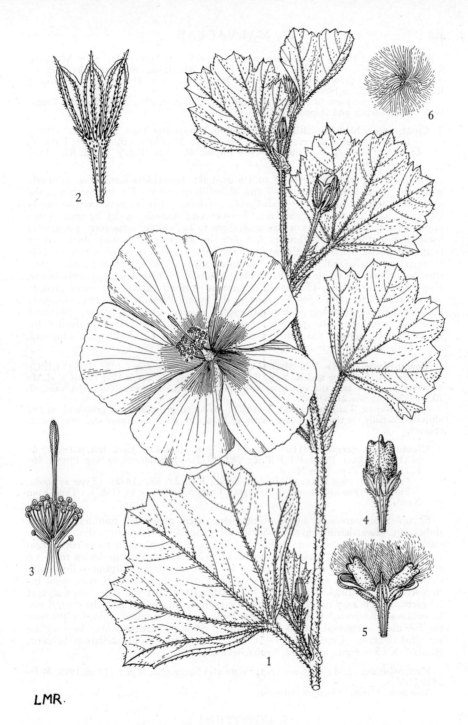

LMR.

Tab. 85. CIENFUEGOSIA HILDEBRANDTII. 1, flowering branch (×⅔) *Mendonça* 616 ; 2, calyx showing glands (×2) *Mendonça* 4023 ; 3, androecium, style and stigma (×2) *Mendonça* 4023; 4, capsule (×1) *Kassner* 439; 5, dehisced capsule (×1) *Kassner* 439; 6, seed (×1) *Kassner* 439.

428 28. MALVACEAE

Bechuanaland Prot. N: Francistown, fl. ii.1926, *Rand* 60 (BM). SE: Mochudi, fl. & fr. iii.1914, *Rogers* 6481 (BM; K; SRGH). **N. Rhodesia.** S: Mazabuka, 1035 m., fl. & fr. xii.1936, *Bebbington* 1693 (K). **S. Rhodesia.** W: Matobo Distr., Prospect Ranch, fl. & fr. 22.iii.1950, *Orpen* 08/50 (SRGH).
Also in the drier parts of western tropical Africa, Angola, SW. Africa and the Transvaal. Thorn-scrub and steppes.

2. **Cienfuegosia hildebrandtii** Garcke in Jahrb. Königl. Bot. Gart. Berl. **2**: 337 (1883). —Hochr. in Ann. Conserv. Jard. Bot. Genève, **6**: 57 (1902).—Burtt Davy, F.P.F.T. **2**: 282 (1932).—J. B. Hutch. in New Phytol. **46**: 126 (1947). TAB. **85**. Type from Kenya.

Shrublet or perennial herb up to 1·5 m. tall; branchlets tomentose to nearly glabrous. Leaf-lamina up to 6 ×7 cm., flabelliform, toothed at the margin in the upper half and occasionally very shallowly 3–5-lobed, densely pubescent to nearly glabrous, base broadly cuneate and 3–5-nerved; petiole up to 20 mm. long; stipules up to 10 ×6 mm., cordate-acuminate to lanceolate, foliaceous, sometimes somewhat amplexicaul. Flowers 6–8 cm. in diam., yellow with dark red or purplish centre, solitary in the axils of the upper leaves or more rarely in 2–3-flowered sympodial inflorescences; peduncle up to 25 mm. long, articulated near the base. Epicalyx of up to c. 10 bracts; bracts 2–3·5 mm. long, narrowly linear. Calyx 10–13 mm. long, 15-nerved, pubescent or nearly glabrous; lobes broadly ovate, mucronate, joined for 8–9 mm. at the base. Petals c. 4 ×4 cm., obliquely broadly obovate, minutely stellate-pubescent outside, glabrous inside. Staminal tube 7–8 mm. long; free parts of filaments 4–6 mm. long. Style projecting for up to 10 mm. beyond the staminal tube. Capsule 10–12 ×8–10 mm., ellipsoid, appressed-pubescent. Seeds with brownish silky floss.

Mozambique. MS: Vila Machado, fl. 18.iv.1948, *Mendonça* 4023 (BM; LISC). SS: between Mambone and R. Buzi, fl. & fr. 4.ix.1944, *Mendonça* 1995 (LISC). LM: between Moamba and the junction with the Lourenço Marques-Namaacha road, fl. 20.v.1957, *Carvalho* 219 (BM).
Also in Kenya, Tanganyika and S. Africa. Wooded grassland and grassland, at low altitudes, usually on the crests and slopes of hills with good soil-cover and near watercourses.

3. **Cienfuegosia gerrardii** (Harv.) Hochr. in Ann. Conserv. Jard. Bot. Genève, **6**: 56 (1902).—Burtt Davy, F.P.F.T. **2**: 282 (1932).—J. B. Hutch. in New Phytol. **46**: 127 (1947). Type from Natal.
Fugosia gerrardii Harv. in Harv. & Sond., F.C. **2**: 588 (1862). Type as above.
Thespesia rehmannii Szyszyl., Polypet. Thalam. Rehm.: 44 (1887). Type from Natal.

Shrublet or perennial herb; branchlets often prostrate, minutely stellate-pubescent. Leaf-lamina up to 5 ×5 cm., suborbicular to oblate in outline, sparsely stellate- or simple-pubescent, shallowly 3–5-lobed, apex blunt or rounded, margin entire, base cordate and 3-nerved; petiole up to 20 mm. long; stipules up to 15 ×6 mm., elliptic, foliaceous. Flowers up to 5 cm. in diam., sulphur-yellow with dark red or purplish centre, solitary, axillary or in 2–3-flowered sympodial inflorescences; peduncle 10–20 mm. long, articulated near the base. Epicalyx of 3 bracts; bracts very caducous and very variable in shape and size, the central one sometimes foliaceous. Calyx 8–12 mm. long; lobes triangular, joined at the base for 5–6 mm. Staminal tube 6–8 mm. long. Style projecting 6 mm. beyond the staminal column. Capsule c. 10 ×10 mm., subglobose-ovoid, stellate-pubescent. Seeds 7–8 ×5–6 mm., ellipsoid, brown-sericeous.

Mozambique. LM: between Impamputo and Namaacha, fl. & fr. 18.xii.1949, *Pedro* 3828 (LMJ).
Also in S. Africa. On rock outcrops.

5. GOSSYPIUM L.

Gossypium L., Sp. Pl. 2: 693 (1753); Gen. Pl. ed. 5: 309 (1754).—J. B. Hutch. in Hutch., Silow & Stephens, Evol. Gossyp.: 14 (1947).

Small trees, shrubs, shrublets or perennial herbs dotted everywhere with black oil-glands. Flowers in 1–several-flowered sympodial inflorescences. Epicalyx of 3 usually foliaceous and persistent bracts. Calyx cupuliform, truncate. Ovary

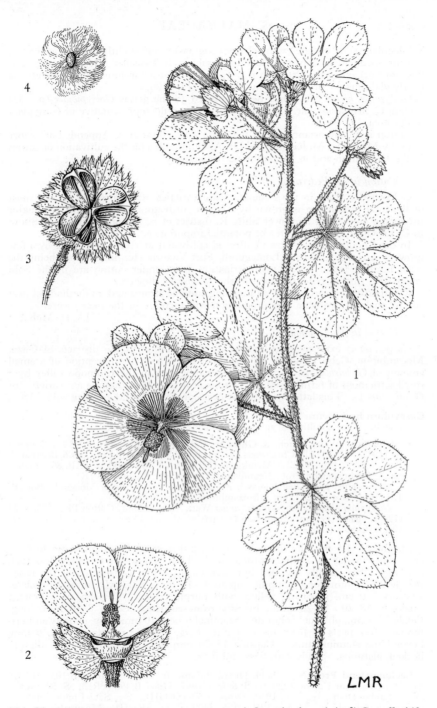

Tab. 86. GOSSYPIUM HERBACEUM VAR. AFRICANUM. 1, flowering branch ($\times \frac{2}{3}$) *Carvalho* 143 ; 2, flower with 1 bract and 3 petals removed ($\times \frac{2}{3}$) *Carvalho* 143 ; 3, dehisced capsule ($\times \frac{2}{3}$) *Pole Evans* 3311 ; 4, seed ($\times \frac{2}{3}$) *Pole Evans* 3311.

3–5-locular; style clavate (with coherent stigmas) rarely slightly divided at the tip. Capsule loculicidally dehiscent; loculi several (rarely 2)-seeded. Seeds usually with floss (or lint) and often also with fuzz (a shorter denser indumentum), sometimes nearly glabrous. Chromosome number: n=13 or 26.

Gossypium kirkii Mast. has been removed to the genus *Gossypioides* (p. 432) leaving *G. herbaceum* var. *africanum* as the only wild representative of *Gossypium* in our area.

Cotton is an important crop in the *Flora Zambesiaca* area. Appended are short notes by Mr. A. H. McKinstry and Dr. A. Quintanilha on the cultivation of cotton in the Federation and in Mozambique respectively.

1. *Cotton in the Federation.*

Cultivated cottons in the Federation are varieties of *G. hirsutum* L. and most of the total production is grown by African cultivators in Nyasaland, the major producing area being the lower Shire R. district of S. Province with Lake Shore in C. Province as another area of potential importance.

In the Rhodesias, rain-grown cotton is cultivated at c. 1050–1350 m. by a few growers at Hartley, Mazoe, Lomagundi, Fort Victoria etc., insect pests being the main limiting factor. The main irrigation areas under cotton are in the Sabi valley with further potential areas in the Zambezi valley etc.

Results from research work on breeding suitable strains and on methods of pest control make it probable that production will increase in the near future.

[A. H. McK.]

2. *Cotton in Mozambique.*

Gossypium is extensively cultivated in Mozambique in the districts of Gaza, Moçambique, Cabo Delgado, Niassa and Tete. With the exception of a small amount of *G. barbadense* L., grown under irrigation in the Limpopo valley by a small settlement of farmers, all the cottons grown in Mozambique are varieties of *G. hirsutum* L. This latter species may occasionally be found growing wild. [A.Q.]

Gossypium herbaceum L., Sp. Pl. 2: 693 (1753).—Mast. in Oliv., F.T.A. **1**: 211 (1868). Type a cultivated specimen probably from Asia Minor.

Var. **africanum** (Watt) Hutch. & Ghose in Ind. Journ. Agric. Sci. **7**: 34, t. 5 upper fig.(1947).—O.B.Mill., Check-lists For.Trees and Shrubs Brit. Emp.No.6, Bechuanal. Prot.: 39 (1948).—Exell & Mendonça, C.F.A. **1**, 2: 183 (1951). TAB. **86**. Lectotype: Bechuanaland Prot., Ngamiland, Kwebe Hills, *Lugard* 198 (K).
  ? *Gossypium puberulum* Klotzsch ex Garcke in Peters, Reise Mossamb. Bot. **1**: 128 (1861) *nom. nud.* Type: Mozambique, Sena, *Peters* (B†).
  Gossypium obtusifolium var. *africanum* Watt, Wild & Cult. Cotton Pl.: 153, t. 23 (1907). Many syntypes from various parts of Africa; lectotype as for *G. herbaceum* var. *africanum* (above).

Shrublet 1–1·5 m. tall; branches terete, pubescent. Leaf-lamina up to 6 ×6 cm., suborbicular in outline, cordate at the base, 5–7-lobed to rather less than half-way; lobes usually rounded but occasionally acute, constricted at the base; petiole 20–25 mm. long; stipules up to 9 ×15 mm., narrowly linear-lanceolate. Flowers 5–6 cm. in diam., yellow with purple centre. Epicalyx of 3 bracts; bracts c. 18–20 ×18–20 mm., broadly ovate-cordate. Calyx 8–10 mm. long. Petals 3 ×3 cm., obliquely obovate. Staminal tube c. 10 mm. long, densely antheriferous; free parts of filaments c. 1 mm. long. Style projecting for c. 6 mm. beyond the staminal tube. Capsule 15–20 mm. in diam., subglobose, slightly beaked, glabrous. Seeds with floss and fuzz.

Bechuanaland Prot. N: Chobe Distr., 915 m., fr. vi.1910, *Miller* B/1057 (PRE). SE: Pharing, near Kanye, 1220 m., fl. & fr. ii.1947, *Miller* B/498 (PRE). **S. Rhodesia.** W: Matobo Distr., fr. 23.iii.1950, *Orpen* 7/50 (SRGH). S: Sabi-Lundi Junction, Chitsa's Kraal, 245 m., fr. 8.vi.1950, *Wild* 3400 (BM; SRGH). **Mozambique.** T: Zumbo, between Maguè and Carinde, fr. 10.ix.1949, *Pedro & Pedrógão* 8251 (SRGH). MS: Nova Sofala, fr. 3.viii.1949, *Pedro & Pedrógão* 7945 (SRGH). SS: between Guijà and the Manuel Rodrigues estate, fl. 3.v.1957, *Carvalho* 143 (BM; LM). LM: Maputo, Goba, fl. & fr. 23.xii.1944, *Mendonça* 3471 (LISC).
  Also in S. Africa. Woodland, bush and grassland, mainly at low altitudes.

Wild Cotton.

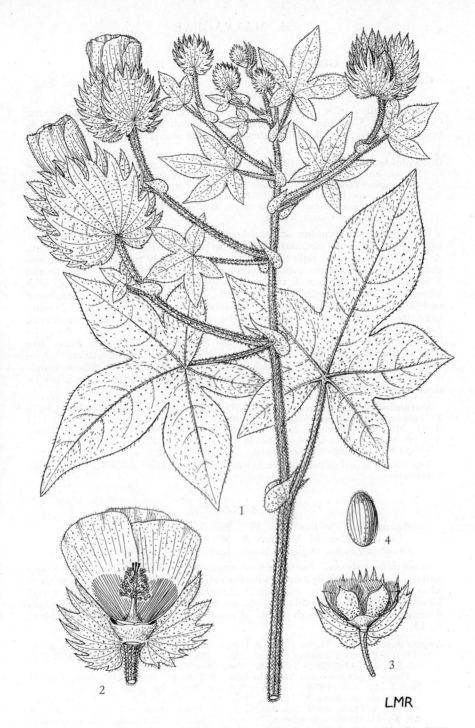

Tab. 87. GOSSYPIOIDES KIRKII. 1, flowering branch (×1) composite from *Vaughan* 2777 and *Schlieben* 2596; 2, flower with 1 bract and 2 petals removed (×1) *Drummond & Hemsley* 3580; 3, dehisced capsule with 1 bract removed (×⅔) *Schlieben* 5369; 4, seed (×2) *Schlieben* 5369.

6. GOSSYPIOIDES Skovsted ex J. B. Hutch.

Gossypioides Skovsted [in Journ. Genet. **31**: 287 (1935) *sine diagn. lat.*] ex J. B. Hutch. in New Phytol. **46**: 131 (1947).

Subscandent shrubs. Leaves lobed. Epicalyx of 3 incised-dentate bracts. Calyx shortly 5-dentate. Ovary 3–5-locular, loculi 2–∞ - ovulate; style 3–4-branched. Capsule loculicidally dehiscent. Seeds ovoid. Chromosome number : $n = 12$.

Gossypioides kirkii (Mast.) [Skovsted, loc. cit. *nom. illegit.*] J. B. Hutch. in New Phytol. **46**: 131 (1947). TAB. **87**. Type from Tanganyika.
    Gossypium kirkii Mast. in Journ. Linn. Soc., Bot. **19**: 214 (1882).—Watt, Wild & Cult. Cotton Pl.: 316, t. 51 (1907). Type as above.
    Gossypium microcarpum sensu Garcia in Bol. Soc. Brot., Sér. 2, **20**: 42 (1946).
    Gossypium sp.—Garcia, loc. cit.

Much-branched climbing or sprawling shrub; branchlets ribbed, angled or winged, tomentose to glabrous, gland-dotted. Leaf-lamina up to 10 × 10 cm., ± orbicular in outline, rather deeply 3–5-lobed, densely pubescent to glabrous, gland-dotted, apex acute usually acuminate, margin entire, base cordate and 3–5-nerved; petiole up to 8 cm. long; stipules up to 15 × 5 mm., falcate with a broadened base. Flowers c. 5 cm. in diam., pale yellow with dark centre, borne on 1–3-flowered sympodial branches. Epicalyx of 3 bracts; bracts c. 25 × 20 mm., large, free, ovate, persistent, incised-dentate. Calyx c. 6 mm. long, cupuliform, 5-dentate, with numerous scattered oil-glands. Petals 2·5–3 × 2·5–3 cm., obliquely obovate, stellate-pubescent outside. Staminal tube 6–7 mm. long; free parts of filaments 1–1·5 mm. long. Ovary 3–4-locular; loculi 2-ovulate; style-branches 3 (4), 3·5–5 mm. long. Capsule 12–15 mm. in diam., subglobose, gland-dotted, nearly glabrous. Seeds 6 × 3·5 mm., narrowly ellipsoid, with a brown woolly floss but no fuzz.

Mozambique. N: Moma, fl. & fr. 6.viii.1948, *Barbosa in Mendonça* 1772 (LISC). Z: between the beach and Maganja da Costa, fl. 27.ix.1949, *Barbosa & Carvalho* 4215 (LM). MS: Cheringoma, fr. 22.x.1949, *Pedro & Pedrógão* 8859 (LMJ; PRE; SRGH). SS: Inharrime, fr. 10.xii.1944, *Mendonça* 3370 (LISC; PRE).
Also in Kenya, Tanganyika and Natal. On coastal sands.

This species occurs in two apparently well-marked varieties : tomentose and glabrous respectively.

7. AZANZA Alef.

Azanza Alef. in Bot. Zeit. **19**: 298 (1861).—Exell & Hillcoat in Contr. Conhec. Fl. Moçamb. **2**: 58 (1954).
    Shantzia Lewton in Journ. Wash. Acad. Sci. **18**: 10, fig. 1–2 (1928).

Small trees or shrubs. Leaves stellate-tomentose or stellate-pubescent. Flowers solitary, axillary. Epicalyx fused with the calyx and with 5–15 caducous teeth or lobes. Calyx persistent, ± cupuliform, almost entire or toothed or lobed. Ovary 5-locular; loculi 2–pluriovulate; style not branched, clavate. Fruit loculicidally dehiscent by 5 valves. Seeds woolly or nearly glabrous. Chromosome number : $n = 13$.

Azanza garckeana (F. Hoffm.) Exell & Hillcoat in Contr. Conhec. Fl. Moçamb. **2**: 59 (1954).—Palgrave, Trees of Central Africa : 208 cum tab. (1956).—Keay, F.W.T.A. ed. 2, **1**, 2 : 342 (1958). TAB. **88**. Type from Tanganyika.
    Thespesia lampas sensu Mast. in Oliv., F.T.A. **1**: 209 (1868).—Sim, For. Fl. Port. E. Afr.: 16 (1909).—Garcia in Bol. Soc. Brot., Sér. 2, **20**: 41 (1946).
    Thespesia garckeana F. Hoffm., Beitr. Fl. Centr.-Ost-Afr.: 12 (1889).—Monro in Proc. & Trans. Rhod. Sci. Ass. **8**, 2 : 75 (1908).—Eyles in Trans. Roy. Soc. S. Afr. **5**: 417 (1916).—Steedman, Trees etc. S. Rhod.: 48, frontisp. et t. 47 (1933).—O. B. Mill., Check-lists For. Trees & Shrubs Brit. Emp. No. 6, Bechuanal. Prot.: 39 (1948).—Suesseng, in Proc. & Trans. Rhod. Sci. Ass. **43**: 104 (1951).—Pardy in Rhod. Agr. Journ. **49**: 74 cum phot. (1952).—Wild, Guide Fl. Vict. Falls : 147 (1953).—Williamson, Useful Pl. Nyasal.: 117 (1955). Type as for *Azanza garckeana*.
    Thespesia trilobata Bak. f. in Journ. of Bot. **35**: 52 (1897). Type from Tanganyika.
    Thespesia rogersii S. Moore in Journ. of Bot. **56**: 5 1918).—Burtt Davy & Hoyle,

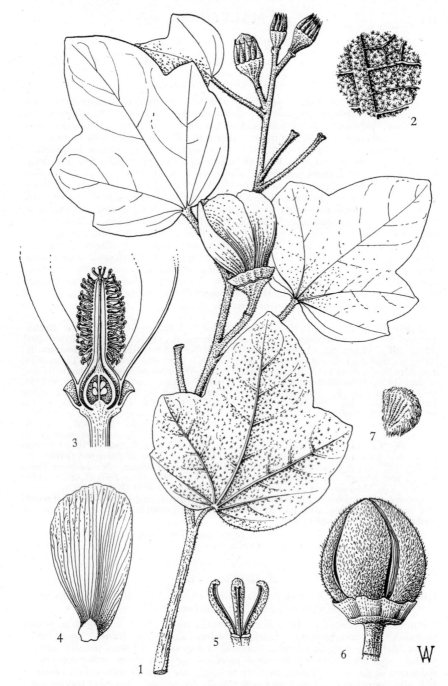

Tab. 88. AZANZA GARCKEANA.   1, flowering branch (× ⅔) *Norlindh & Weimarck* 4436 ;   2, part of under surface of leaf (× 4) *Norlindh & Weimarck* 4436 ;   3, vertical section of flower (× 1) *Goodwin* 51 ;   4, petal (× ⅔) *Goodwin* 51 ;   5, stigmas pulled apart (× 2) *Goodwin* 51 ;   6, fruit (× 1) *Hutchinson & Gillett* 3347 ;   7, seed (× 1) *Hutchinson & Gillett* 3347.

N.C.L.: 49 (1936).—Hutch., Bot. in S. Afr.: 468 cum fig. (1946).—O. B. Mill., loc. cit. Type: S. Rhodesia, Bulawayo, *Rogers* 5839 (BM, holotype).

Shantzia garckeana (F. Hoffm.) Lewton in Journ. Wash. Acad. Sci. **18**: 15, fig. 1–2 (1928). Type as for *Azanza garckeana*.

Thespesia sp. (*Pole Evans* 4178).—O. B. Mill., loc. cit.

Tree up to 10 m. tall or shrub of 3–4 m.; branchlets stellate-tomentose, ± floccose when young, glabrescent. Leaf-lamina up to 20 × 20 cm., suborbicular in outline, palmately 3–5-lobed, stellate-pubescent to nearly glabrous above, stellate-tomentose to stellate-pubescent beneath, apex usually blunt or rounded, margin entire, base cordate and 5–7-nerved, midrib usually with a longitudinal fissure*; petiole up to 13 cm. long, stellate-tomentellous; stipules 6–7 mm. long, linear to lorate, sometimes ± falcate. Flowers c. 6 cm. in diam., yellow or purplish with dark purple or dark red centre; peduncle 2–7 cm. long, stellate-tomentellous, articulated above the middle. Epicalyx cupuliform, fused with the calyx, with 9–10 teeth; teeth up to 12 mm. long, caducous, linear, leaving abscission scars on the rim of the persistent base. Calyx c. 10 mm. long, cupuliform, stellate-tomentose, with 5 short teeth. Petals c. 6 × 3·5–4 cm., narrowly obliquely obovate, stellate-pubescent outside, glabrous within, the outside overlapping margins thickened. Staminal tube 10–12 mm. long; free parts of filaments 2–5 mm. long, longer towards the apex of the tube. Style pilose, clavate apex bearing the stigmatic surfaces 3·5 mm. long. Capsule up to 4 × 3 cm., red, subglobose to broadly ellipsoid or ovoid, appressed-setulose or tomentellous, glutinous. Seeds c. 10 × 7 mm., hemispherical, with a brownish woolly floss.

Caprivi Strip. Mpilila I., fl. 15.i.1959, *Killick & Leistner* (PRE; SRGH). **Bechuanaland Prot.** N: between Tutumi and Sebena, fl. 14.xii.1929, *Pole Evans* 2601 (PRE). SE: 13 km. SE. of Serowe, fr. i.1941, *Miller* B/270 (PRE). **N. Rhodesia.** B. Kabompo mouth, fr. 27.v.1954, *Gilges* 384 (K). N: Abercorn Distr., Lake Tanganyika, near Mpulungu, fl. xi.1952, *Angus* 772 (BM; K). W: Solwezi Distr., Mutando Bridge, fr. 25.vi.1930, *Milne-Redhead* 599 (K). C: Mt. Makulu, near Lusaka, fr. 25.vii.1956, *Simwanda* 23 (K). E: Petauke, Luangwa R. above Beit Bridge, fl. 17.iv.1952, *White* 2697 A (FHO; K). S: Mazabuka, 1000 m., st. 6.xi.1930, *Veterinary Officer* CRS 105 (PRE). **S. Rhodesia.** N: Sebungwe Distr., Sengwa Camp, 460 m., st. ix.1955, *Davies* 1472 (BM; K; SRGH). W: Bulawayo, fl. ii.1903, *Eyles* 1196 (BM; SRGH). C: Salisbury, fl. i.1918, *Eyles* 906 (BM; K; SRGH). E: Umtali Commonage, fl. 20.i.1949, *Chase* 1328 (BM; SRGH). S: Zimbabwe, fr. vii.1929, *Smuts* (PRE). **Nyasaland.** N: 4·8 km. N. of Ekwendeni, fr. 9.vi.1938, *Pole Evans & Erens* 665 (PRE). S: Limbe, 1220 m., fl. 20.i.1948, *Goodwin* 51 (BM). **Mozambique.** N: between Porto Amélia and Mecufi, T: Montes de Zobuè, fr. 17.vi.1941, *Torre* 2870 (LISC). MS: between Vila Machado and Muda, fl. 19.iv.1948, *Mendonça* 4024 (BM; LISC).

Eastern and southern tropical Africa. Very widely distributed in many kinds of woodland from sea-level to c. 2000 m. Apparently never dominant but scattered among other species. On a wide range of soils.

A glutinous substance from the inside of the rind of the fruit is eaten by Africans and the timber is used for axe handles and knife sheaths.

8. HIBISCUS L.

Hibiscus L., Sp. Pl. **2**: 693 (1753); Gen. Pl. ed. 5: 310 (1754).—Hochr. in Ann. Conserv. Jard. Bot. Genève, **4**: 23 (1900).

Annual or perennial herbs (sometimes with annual shoots arising from woody rootstocks or underground stems), shrublets, shrubs or (rarely) small trees. Leaves petiolate, simple, lobed or digitately compound. Flowers usually solitary, axillary and often forming terminal racemose or corymbose inflorescences by reduction of the upper leaves, medium to large, often yellow with a dark centre or red, pink, purplish or white; peduncle usually articulated. Epicalyx (occasionally absent) of 5–20 bracts very variable in shape and length, free or adnate to the base of the calyx. Calyx 5-merous, usually with 5 lobes (or rarely 5 or 10 teeth) joined at the base or more rarely almost to the apex, persistent. Staminal tube truncate at the apex; free parts of filaments very variable in length, sometimes whorled.

---

* *Azanza lampas* (Cav.) Alef. (*Hibiscus lampas* Cav.) also has, at least sometimes, a similar longitudinal fissure on the midrib, which is confirmatory evidence of its relationship with *A garckeana*.

Ovary 4–5-locular; loculi 3–∞ -ovulate; style usually 5-branched.   Fruit a loculi-cidally dehiscent capsule.

The classification elaborated by Hochreutiner (loc. cit.) was considerably modified by Ulbrich as regards the African sections of the genus (see Engl., Pflanzenw. Afr. **3**, 2: 391 (1921)). It seems unwise to tamper with the sub-divisions of a large cosmopolitan genus, such as *Hibiscus*, without a study of the worldwide range and in the absence of such a study (since Hochreutiner) the correct nomenclature of the sections remains in doubt.   I have placed the species of our area in 13 series for convenience of keying without giving these series any nomenclatural status; but as they correspond very nearly to sections recognized by Hochreutiner and/or Ulbrich and as the names are very familiar to all workers on the genus, I have indicated, in parentheses, the names of the sections to which my series approximately correspond.

Key to the series

Capsule neither winged nor with prominent angles :
  Epicalyx cupuliform with 10 short teeth (TAB. **89** fig. 8) ; stipules foliaceous
                                          Series 1 (*Azanza*)
  Epicalyx with bracts free or joined only towards the base or obsolete or absent :
    Calyx with 10 principal nerves, a median one to each calyx-lobe and a commissural one to each sinus, the latter nerve bifurcating with a branch to the margin of each of the adjacent calyx-lobes (TAB. **89** fig. 1) :
      Calyx becoming coriaceous or fleshy at maturity, nerves very prominent ; capsule setose (TAB. **89** fig. 4) - - - - - Series 2 (*Furcaria*)
      Calyx thin, nerves not prominent; capsule glabrous or pubescent; epicalyx absent (in our species) (TAB. **89** fig. 1) - - - Series 3 (*Solandra*)
    Calyx variously nerved but not as above :
      Calyx inflated or lobes much enlarged, especially in fruit :
        Calyx scarious, lobes joined to above half-way, often nearly to the apex (TAB. **89** fig. 3) - - - - - - - - - Series 4 (*Trionum*)
        Calyx not scarious, lobes joined to about ⅓ - - - Series 5 (*Venusti*)
      Calyx neither inflated nor with much enlarged lobes in fruit :
        Seeds with silky or cottony floss at maturity (TAB. **90** fig. C2)
                                          Series 6 (*Bombycella*)
        Seeds glabrous, pubescent or tomentose :
          Staminal tube shorter than the corolla :
            Bracts of epicalyx 5 (TAB. **89** fig. 7) - - - Series 7 (*Calyphylli*)
            Bracts of epicalyx more than 5 :
              Bracts of epicalyx rigid and pungent (TAB. **89** fig. 9)
                                          Series 8 (*Ketmia* pro parte)
              Bracts of epicalyx not as above :
                Valves of capsule sometimes acute but not abruptly aristate :
                  Bracts of epicalyx filiform (TAB. **89** fig. 6) Series 9 (*Ketmia* pro parte)
                  Bracts of epicalyx linear to lanceolate or spathulate :
                    Bracts of epicalyx 10 mm. long or longer ; flowers yellowish, cream, white or purple :
                      Flowers with dark centre; bracts of epicalyx ± spathulate (TAB. **89** fig. 10) - - - Series 9 (*Ketmia* pro parte)
                      Flowers concolorous ; bracts of epicalyx linear to linear-oblong
                                          Series 10 (*Trichospermum* pro parte)
                    Bracts of epicalyx up to 6 mm. long :
                      Flowers red - - - - - Series 6 (*Bombycella*)
                      Flowers white, pale yellow or yellow with a dark centre
                                          Series 9 (*Ketmia* pro parte)
                Valves of capsule aristate, awns 2–3 mm. long (TAB. **89** fig. 5) :
                  Seeds appressed-pubescent - - Series 11 (*Aristivalves*)
                  Seeds lepidote - - - - Series 9 (*Ketmia* pro parte)
          Staminal tube exceeding the corolla, 3–7 cm. long Series 12 (*Lilibiscus*)
Capsule winged or with very prominent angles (TAB. **89** fig. 2) Series 13 (*Pterocarpus*)

Series 1 (*Azanza*)

Trees.   Epicalyx cupuliform, toothed.

1. **Hibiscus tiliaceus** L., Sp. Pl. **2**: 694 (1753).—Bertol. f., Ill. Piant. Mozamb. **2**: 17 (1852).—Mast. in Oliv., F.T.A. **1**: 207 (1868).—Hochr. in Ann. Conserv. Jard. Bot. Genève, **4**: 62 (1900).—Sim, For. Fl. Port. E. Afr.: 15, t. 9 (1909).—Engl., Pflanzenw. Afr. **3**, 2: 391, fig. 186E (1921).—Garcia in Bol. Soc. Brot., Sér. 2, **20**: 41 (1946).—Mendonça & Torre, Contr. Conhec. Fl. Moçamb. **1**: 18 (1950).—

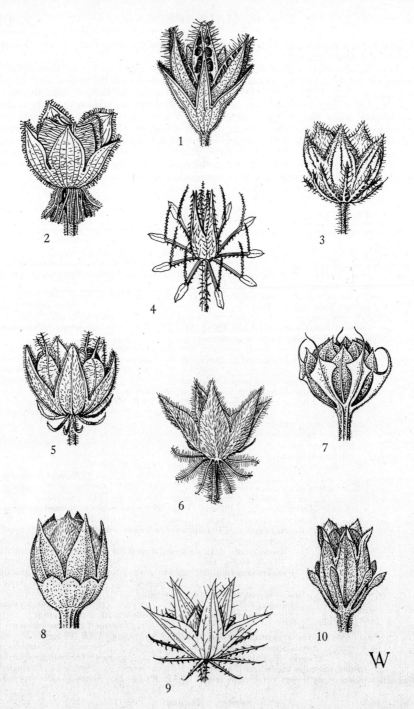

Tab. 89. Structure of epicalyx and calyx in HIBISCUS. 1. H. LOBATUS; 2. H. VITIFOLIUS SUBSP. VULGARIS; 3. H. TRIONUM; 4. H. SURATTENSIS; 5. H. PALMATUS; 6. H. PHYSALOIDES; 7. H. CALYPHYLLUS; 8. H. TILIACEUS; 9. H. CAESIUS; 10. H. PANDURIFORMIS. Fig. 1 (×2), fig. 6 (×⅔), the rest (×1).

Exell & Mendonça, C.F.A. **1**, 2 : 177 (1951).—Keay, F.W.T.A. ed. 2, **1**, 2 : 348 (1958). TAB. **89** fig. 8. Type from India.
Paritium tiliaceum (L.) St.-Hil., Fl. Bras. Merid. **1** : 256 (1825).—Harv. in Harv. & Sond., F.C. **1** : 177 (1860).—Garcke in Peters, Reise Mossamb. Bot. **1** : 128 (1861). Type as above.

Tree up to 5 m. tall or large bush; young branches stellate-tomentellous. Leaf-lamina 3–15 cm. in diam., suborbicular, densely and finely stellate-tomentellous but glabrescent on the upper surface, apex abruptly acuminate, margin entire or minutely dentate, base deeply cordate 5–9-nerved, nerves impressed above retaining some indumentum, prominent beneath and with longitudinal pore-like slits near the base (sometimes only on the median nerve but usually on 3 or 5); petiole up to 12 cm. long, stellate-tomentose or stellate-tomentellous, sometimes glabrescent; stipules up to c. 30 × 15 mm., ovate-cordate, semi-amplexicaul, caducous, each pair leaving a circular scar round the stem. Flowers up to c. 8 cm. in diam., yellow with a dark centre, in the axils of ovate bracts and forming an apparently terminal panicle when the bracts have fallen; pedicels c. 10 mm. long, stout, tomentellous. Epicalyx cupuliform with 10 short teeth. Calyx up to 3 cm. long, tomentellous; lobes joined nearly to the middle, lanceolate. Petals 6–7 × 4–6 cm., obovate, contorted. Staminal tube up to 25 mm. long; free parts of filaments 0·5–1 mm. long. Style-branches 5–6 mm. long, densely glandular. Capsule 2–2·5 cm. in diam., subglobose, acute at the apex, tomentose. Seeds 4·5 × 3 mm., subreniform, glabrous.

Mozambique. N: Fernão Veloso, fr. 23.x.1952, *Barbosa & Balsinhas* 5175 (LM). Z: between Régulo Morla and Maganja da Costa, fl. & fr. 22.ix.1949, *Barbosa & Carvalho* 4207 (LM). MS: Beira, between Manga and Macuti, fl. & fr. 25.x.1949, *Pedro & Pedrógão* 8942 (LMJ; SRGH). SS: Inhambane, fr. 30.x.1938, *Torre* 1588 (COI; LISC). LM: Inhaca I., fl. ix.1919, *Breyer* (PRE; SRGH); Marracuene, fl. & fr. 11.iv.1955, *Lemos* 68A (LMJ).
Widespread in the tropics and subtropics. Along the sea-shore, usually along the margins of water-courses or in tidal zone thickets.

The inflorescence seems to be cymose but the apparently terminal flowers are really axillary, the caducous bracts being the stipules of much reduced or suppressed leaves. After the bracts have fallen the inflorescences give the impression of being terminal cymose panicles. The stipules and bracts are paired and opposite and each pair occupies the complete circumference of the stem leaving a circular scar reminiscent of *Ficus*.

This is a very aberrant species of *Hibiscus*, providing a link with the genera *Azanza* and *Thespesia*, and a good case can be made for separating it generically, as St.-Hilaire (loc. cit.) and Harvey (loc. cit.) did, as *Paritium tiliaceum*.

Series 2 (*Furcaria*)

Bracts of epicalyx forked, or with a small appendage, or simple. Calyx 10-nerved, a median one to each lobe and a commissural one to each sinus, the latter nerve bifurcating.

Bracts of epicalyx clearly forked (TAB. **89** fig. 4) :
 Stipules broad, auriculate and ± amplexicaul   -   -   -   -   2. *surattensis*
 Stipules narrow, linear to subulate, not amplexicaul :
  Plant not aculeate   -   -   -   -   -   -   -   3. *acetosella*
  Plant ± aculeate :
   Peduncle up to 0·8 cm. long :
    Prickles stout, usually unbranched with swollen bases ; calyx-lobes acute
                                                            4. *torrei*
    Prickles less developed, usually forked or stellate and only slightly swollen at the base ; calyx-lobes acuminate :
     Leaves membranous, usually obtuse at the base (rarely slightly cordate), up to 18 × 18 cm.; plant erect, annual   -   -   -   5. *mastersianus*
     Leaves chartaceous, usually somewhat cordate at the base, up to c. 4 × 4 cm.; plant ± prostrate, usually perennial   -   -   -   -   6. *hiernianus*
   Peduncle 1–6 cm. long   -   -   -   -   -   -   7. *altissimus*
Bracts of epicalyx not clearly forked (occasionally with a small appendage on the adaxial surface near the apex) :
 Calyx fleshy, edible ; plant not aculeate   -   -   -   -   8. *sabdariffa*
 Calyx not fleshy ; plant aculeate :
  Stem (below the flowers) pilose, pubescent or puberulous (occasionally almost glabrous) as well as setose or aculeate ; flowers white or yellow with dark centres ; plant annual :
   Calyx up to 2·5 cm. long, with a ± conspicuous gland on the median vein of

each lobe ; stems pubescent to puberulous (occasionally almost glabrous) as well as aculeate ; leaf-lamina suborbicular to ovate-lanceolate in outline :
Flowers white, greyish or pale yellow, with dark purple centre ; calyx-lobes acuminate or long-acuminate ; gland on calyx-lobe conspicuous, usually at least 1·5 mm. in diam. (rarely absent) ; leaf-segments usually narrowly elliptic to linear (occasionally broader) ; stems puberulous (rarely pubescent) to nearly glabrous (as well as aculeate) - - - - - 9. *cannabinus*
Flowers yellow with dark red centre ; calyx-lobes usually ovate, only shortly acuminate ; gland on calyx-lobe less conspicuous, usually up to 1 mm. in diam. ; leaf-segments rhombic to oblanceolate ; stems shortly stellate-tomentose (as well as aculeate) - - - - - 10. *meeusei*
Calyx 2·5–4 cm. long, devoid of glands on the median veins of the lobes ; stems with long flexuous hairs (as well as aculeate) ; leaf-lamina ± flabelliform in outline - - - - - - - - - - 11. *mechowii*
Stem (below the flowers) tomentose, tomentellous or densely pubescent and with conspicuous short stout conical prickles the bases of which remain as small knobs ; flowers purple, reddish or pink (rarely white), with a darker centre ; calyx densely setose ; plant perennial, usually a shrub or small tree
12. *diversifolius* subsp. *rivularis*

2. **Hibiscus surattensis** L., Sp. Pl. **2** : 696 (1753).—Harv. in Harv. & Sond., F.C. **1** : 177 (1860).—Garcke in Peters, Reise Mossamb. Bot. **1** : 127 (1861).—Mast. in Oliv., F.T.A. **1** : 201 (1868).—Hochr. in Ann. Conserv. Jard. Bot. Genève, **4** : 110 (1900) pro parte.—Bak. f. in Journ. Linn. Soc., Bot. **40** : 27 (1911).—Eyles in Trans. Roy. Soc. S. Afr. **5** : 416 (1916).—Engl. Pflanzenw. Afr. **3**, 2 : 400, fig. 187 A (1921).—Burtt Davy, F.P.F.T. **2** : 280 (1932).—Exell & Mendonça, C.F.A. **1**, 1 : 166 (1937).—Garcia in Bol. Soc. Brot., Sér. 2, **20** : 39 (1946).—Brenan in Mem. N.Y. Bot. Gard. **8**, 3 : 225 (1953).—Wild, Guide Fl. Vict. Falls : 147 (1953).—Keay, F.W.T.A. ed. 2, **1**, 2 : 346 (1958). TAB. **89** fig. 4. Type from India.
Hibiscus hypoglossus E. Mey. ex Harv. in Harv. & Sond., loc. cit. *nom. nud.*
Hibiscus surattensis var. *genuinus* Hochr., tom. cit. : 111 (1900). Type as for the species.
Hibiscus surattensis var. *villosus* Hochr., tom. cit. : 112 (1900). Type : Nyasaland, Whyte (K).

Annual herb, prostrate or climbing or scrambling up to 3 m. ; stems hispid and/or pubescent and sparsely to rather densely aculeate. Leaf-lamina up to 10 × 10 cm., broadly ovate to suborbicular in outline, sparsely hispid on both surfaces and aculeate on the nerves beneath, palmately 3–5-lobed (sometimes only shallowly) to palmatipartite ; segments elliptic, somewhat constricted towards the base, rather coarsely toothed ; petiole up to 11 cm. long, hairy like the stems ; stipules up to 15 × 10 mm., broad, auriculate, ± amplexicaul. Flowers up to 10 cm. in diam., yellow with purple centre, solitary, axillary ; peduncle up to 6–7 cm. long, hispid and aculeate, articulated 3–7 mm. below the apex and often more densely hispid above the articulation. Epicalyx of 8–9 bracts ; bracts 10–15 mm. long, forked with the inner branch linear and inflexed and the outer branch spathulate. Calyx 10-veined, becoming scarious ; lobes up to 25 × 10 mm., ovate-lanceolate, usually somewhat rigid with long sharp apices, setose and aculeate on the median and marginal nerves. Petals up to 5 × 3·5 cm., obovate, glabrous or nearly so. Staminal tube up to 15 mm. long ; free parts of filaments 1·5–2·5 mm. long. Style branching 5 mm. above the apex of the staminal tube ; branches 2–2·5 mm. long. Capsule 15 × 12 mm., globose-ovoid, densely appressed-pilose, apex acute. Seeds 3–3·5 × 2·5 mm., subreniform, with a few minute white appressed-stellate hairs.

N. Rhodesia. N : Fort Rosebery, fr. 23.viii.1952, *Angus* 301 (BM ; FHO). S : Long I., fl. iv.1918, *Eyles* 1264 (BM ; SRGH). **S. Rhodesia.** N : confluence of Sanyati and Zambezi, 425 m., fr. ix.1955, *Davies* 1510 (BM ; SRGH). E : Umtali, fl. & fr. xi.1948, *Chase* 1665 (BM ; SRGH) ; Vumba, 1680 m., fl. & fr. 3.v.1957, *Chase* 6470 (BM ; K ; SRGH). **Nyasaland.** S : Mt. Mlanje, Great Ruo Gorge, 760 m., fl. & fr. 29.viii.1956, *Newman & Whitmore* 640 (BM). **Mozambique.** N : Palma, Ingoane, fl. 12.ix.1948, *Barbosa* 2071 (LISC ; LMJ). Z : Maganja da Costa, fl. & fr. 31.vii.1943, *Torre* 5722 (LISC). MS : Cheringoma, Inhaminga, fl. & fr. 24.v.1948, *Mendonça* 4362 (LISC). SS : between Chongoene and Vila João Belo, fl. 2.v.1957, *Carvalho* 136 (BM ; LM). LM : Umbeluzi, fr. 27.v.1949, *Myre & Balsinhas* 723A (BM ; LISC).
Widespread in the tropics of the Old World. Cultivated ground and waste places.

3. **Hibiscus acetosella** Welw. [ex Ficalho in Bol. Soc. Geogr. Lisb., Sér. 2 : 608 (1881) *nom. nud* .] ex Hiern, Cat. Afr. Pl. Welw. **1** : 73 (1896).—Exell & Mendonça, C.F.A.

1, 1: 167 (1937).—Williamson, Useful Pl. Nyasal.: 65 (1955). Type from Angola (Cuanza Norte).

    Hibiscus cannabinus sensu Hochr. in Ann. Conserv. Jard. Bot. Genève, **4**: 114 (1900) pro parte quoad syn. *H. acetosella.*

Annual or perennial herb, sometimes suffruticose; stems crisped-pubescent or nearly glabrous, devoid of prickles. Leaf-lamina up to 8 × 8 cm., ovate to suborbicular in outline, varying from almost entire to irregularly 3–5-lobed or palmatipartite, slightly fleshy, somewhat glaucous, often tinged with red; petiole up to 10–11 cm. long, minutely pubescent; stipules up to 15 mm. long, linear. Flowers up to 10 cm. in diam., purple-red or lemon-yellow, with a purple centre, solitary, axillary; peduncle up to 8–10 mm. long, articulated near the middle. Epicalyx of 9–10 bracts; bracts up to 15 mm. long, forked, the inner branch linear, the outer branch spathulate. Calyx 20 mm. long, 10-nerved, nearly glabrous; lobes ovate-lanceolate to triangular, rigid, sharp-pointed. Petals up to 5·5 × 3·5 cm., obovate, glabrous. Staminal tube 10 mm. long; free parts of filaments 1 mm. long. Style-branches (probably immature) 1·5 mm. long. Capsule 17 × 9 mm., ovoid-acuminate, hispid. Seeds 3 × 2·5 mm., subreniform, glabrous or nearly so.

N. Rhodesia. W: Mwinilunga, *Marks* 107 (K). **S. Rhodesia.** C: Salisbury, fl. 23.vi.1953, *Bertram* in GHS 43476 (BM; SRGH).

Also in S. Tomé, Principe, Belgian Congo, Angola and Tanganyika. Perhaps wild in Angola but probably cultivated elsewhere. Native plantations and roadsides.

The leaves have an acid flavour recalling that of *Rumex acetosella*, and the plant is cultivated as a vegetable or salad. Although very similar to *H. cannabinus* L., under which it was placed by Hochreutiner (loc. cit.), it is even nearer to *H. noldeae* Bak. f. (a wild species in Angola) from which it perhaps originated. The Africans may have isolated it from its prickly and hence inedible relatives by a process of selection or it could conceivably have arisen as a result of hybridization between *H. cannabinus* and *H. sabdariffa* (also devoid of prickles).

4. Hibiscus torrei Bak. f. in Journ. of Bot. **75**: 101 (1937).—Garcia in Bol. Soc. Brot.,   Sér. 2, **20**: 40 (1946). Type: Mozambique, Niassa, Vila Cabral, *Torre* 435 (BM;   COI, holotype; K).

Plant said to be shrubby but probably a much-branched perennial herb; stems rather sparsely pilose or setose-pilose, with one or two longitudinal lines of pubescence changing their radial position at each node, rather densely aculeate with stout usually unbranched deflexed prickles. Leaf-lamina up to 10 × 10 cm., suborbicular in outline, rather shallowly palmately 3–5-lobed, scabrous, setulose on both surfaces with some prickles on the nerves of the lower surface, base obtuse to subcordate; petiole up to 4·5 cm. long, hairy like the stems; stipules c. 7 × 1 mm., linear. Flowers 4–5 cm. in diam., yellow with a violet centre, solitary, axillary; peduncle 4–8 mm. long, hispid. Epicalyx of usually 9 bracts; bracts up to 10 mm. long, forked, the inner branch linear, the outer branch ovate to ovate-lanceolate. Calyx up to 22 mm. long, 10-veined, hispid or setose-pilose; lobes ovate-acute, aculeate on the margins, joined for up to 4 mm. at the base. Petals 4–5 cm. long, pubescent. Capsule up to 2 × 1·5 cm., subreniform, surface minutely faveolate.

Mozambique. N: Vila Cabral, fl. & fr. vii.1934, *Torre* 435 (BM; COI; K). Known only from the type-collection. Damp places.

Owing to paucity of material it was considered better not to demolish a flower to obtain the androecium and style dimensions.

5. Hibiscus mastersianus Hiern, Cat. Afr. Pl. Welw. **1**: 71 (1896) emend. excl.   specim. angol.—Milne-Redh. in Kew Bull. **1935**: 272 (1935). Lectotype: Mozam-   bique, Lupata, *Kirk* (K).

    Hibiscus furcatus sensu Mast. in Oliv., F.T.A. **1**: 201 (1868) pro parte excl. specim. ex Gambia.

    Hibiscus surattensis sensu Hochr. in Ann. Conserv. Jard. Bot. Genève, **4**: 110 (1900) pro parte quoad syn. *H. mastersianus* Hiern.

Erect annual herb up to 2 m. tall with irritant hairs; stems with short, sometimes sparse, usually stellate prickles with a swollen base, and one or more longitudinal lines of crisped pubescence or pile. Leaf-lamina 10(18) × 8(18) cm., usually suborbicular in outline, sometimes varying to narrowly oblong, stellate-pubescent on both surfaces, not lobed or shallowly 3–5-palmatilobed, lobes acute to rounded, sinuses shallow rounded, margin irregularly serrate, base rounded to truncate or slightly cordate 5–7-nerved, midrib with a longitudinal fissure 2–3 mm. long near the base; petiole 4–7 (18) cm. long, hairy like the young stems; stipules up to 8 mm. long, linear to filiform. Flowers up to 7 cm. in diam., yellow to orange with purple or maroon centre, solitary, axillary; peduncle 4–8 mm. long, hispid. Epicalyx of 9–10 bracts; bracts 8 × 0·8 mm., linear, forked. Calyx up to 2 cm. long, 10-nerved, pilose, becoming scarious; lobes lanceolate, acuminate, setose or aculeolate on the margins. Petals 3–5 × 2–3 cm., almost glabrous. Staminal tube 15 mm. long; free parts of filaments 0·5–1 mm. long. Style-branches 1 mm. long. Capsule up to 15 × 10 mm., ovoid-acuminate, densely appressed-setose.

Caprivi Strip. E. of Cuando R., 945 m., fr. x.1945, *Curson* 963 (PRE). **Bechuana-land Prot.** N: Chobe-Zambezi confluence, fl. & fr. 11.iv.1955, *E.M. & W.* 1458 (BM; LISC; SRGH). **N. Rhodesia.** N: Mbulu Is., Lake Tanganyika, 730 m., fl. & fr. 11.iv.1955, *Richards* 5389 (K; SRGH). C: Lusaka, N. of Barclays Bank, fr. 1.iv.1951, *Hodge* A 107 (K). S: Bombwe, fl. & fr. 25.iv.1932, *Martin* 204/32 (FHO; K). **S. Rhodesia.** W: Nyamandhlovu, Ngamo, fl. & fr. 4.iv.1950, *Orpen* 21/50 (SRGH). **Mozambique.** T: Lupata, fr. iv.1860, *Kirk* (K).
Also in the Belgian Congo, Tanganyika and SW. Africa. Roadside weed.

H. mastersianus was a new name proposed by Hiern (loc. cit.) for the species named *H. furcatus* Roxb. by Masters (loc. cit.). Masters cited two specimens (*Ingram* from Gambia and *Kirk* from Tete) and these are clearly the syntypes of *H. mastersianus*, although Hiern cited Angolan specimens later considered by Exell and Mendonça (in Journ. of Bot. **74**: 136 (1936)) to belong to different species. Milne-Redhead (loc. cit.) has dealt at length with the typification.

6. **Hibiscus hiernianus** Exell & Mendonça in Journ. of Bot. **74**: 136 (1936); C.F.A. **1**, 1: 167 (1937). Type from Angola (Huila).

Prostrate perennial (? sometimes annual) herb with prostrate-ascendent branches; stems stellate-setulose to stellate-tomentellous, scabrous, with a few scattered prickles. Leaf-lamina up to 4 × 4 cm., chartaceous, elliptic to suborbicular in outline, not lobed to 3–5-lobed, scabrous, densely stellate-setulose on both sur-faces, apex obtuse, margin crenate-serrulate, base slightly cordate; petiole up to 4 cm. long, hairy like the stems; stipules 3 mm. long, linear to subulate. Flowers 4–5 cm. in diam., yellow, solitary, axillary; peduncle up to 3·5 cm. long, tomentose or stellate-setose. Epicalyx of 10–11 bracts; the bracts 8–10 mm. long, linear, forked. Calyx 12–13 mm. long, 10-nerved; lobes triangular-acuminate, joined for 3–4 mm. at the base, usually with reddish-purple pigment on the nerves. Petals up to 3·5 cm. long. Staminal tube 11–12 mm. long; free parts of filaments 0·5 mm. long. Style-branches 1–2 mm. long. Capsule (immature) densely ap-pressed-setose.

N. Rhodesia. W: Ndola, fl. 11.iv.1954, *Fanshawe* 1075 (K; SRGH).
Also in Angola.

7. **Hibiscus altissimus** Hornby in Proc. & Trans. Rhod. Sci. Ass. **41**: 55 (1946). Type: Mozambique, Lusa River Valley, W. of Guruè, *A. J. W. Hornby* 4563 (K; LM; PRE).
Hibiscus furcatus sensu Harv. in Harv. & Sond., F.C. **1**: 176 (1860).

A woody scrambler over bushes; stems nearly 10 m. long, aculeate with prickles recurved and swollen at the base, otherwise nearly glabrous or with longitudinal lines or areas of crisped pubescence. Leaf-lamina 3–12 × 2–15 cm., ovate to broadly suborbicular in outline, usually nearly glabrous except for prickles on the nerves of the under surface but sometimes fairly densely stellate-pubescent (*Pedro & Pedrógão* 4057), usually 3–5-lobed (rarely not lobed), lobes acute, margin ir-regularly serrate, sinuses acute, base 5–7-nerved, midrib (and often the 2 adjacent nerves) with a longitudinal fissure up to 3 mm. long near the base; petiole up to 7·5 cm. long, hairy like the young stems; stipules 6–7 mm. long, acicular or fili-form. Flowers up to 10 cm. in diam., yellow or terracotta, sometimes with a dark centre, solitary, axillary; peduncle 1–6 cm. long. Epicalyx of about 7 bracts; bracts c. 15 mm. long, forked, the outer branch somewhat reflexed. Calyx 2–3 cm. long; lobes 13–22 × 6–10 mm., lanceolate, acute, joined at the base for 4–10 mm., setose on the midrib and margins, transversely ribbed in fruit. Petals up to

6 × 5 cm., obovate. Staminal tube c. 14 mm. long; free parts of filaments 0·5–1 mm. long. Style-branches 4–5 mm. long. Capsule 20 × 18 mm., densely appressed-setose. Seeds 4 × 3 mm., subreniform, with whitish irregularly discoid scales 0·1–0·2 mm. in diam.

Mozambique. N: Maniamba, S. Geci, fl. & fr. 29.v.1948, *Pedro & Pedrógão* 4057 (LMJ). Z: Bajone, fl. & fr. 2.x.1949, *Barbosa & Carvalho* 4273 (LM). MS: Cheringoma, Inhaminga, fl. & fr. 24.v.1948, *Mendonça* 4363 (BM; LISC). SS: between Bilene and Vila de João Belo, fl. & fr. 1.v.1957, *Carvalho* 131 (BM; LM). LM: Rikatla, fr. iv.1918, *Junod* 238 (BM; G; PRE).
Also in Natal and the Transvaal. In semi-evergreen riverine woodland.

Very near to *H. rostellatus* Guill. & Perr. but less hairy and the leaves usually more deeply lobed with narrower sinuses. *Pedro & Pedrógão* 4057 shows tendencies towards *H. rostellatus* and differences may break down when more material is available.

8. **Hibiscus sabdariffa** L., Sp. Pl. 2: 695 (1753).—Mast. in Oliv., F.T.A. 1: 204 (1868).—Hochr. in Ann. Conserv. Jard. Bot. Genève, 4: 116 (1900).—Ulbr. & Fr. in R.E.Fr., Wiss. Ergebn. Schwed. Rhod.-Kongo-Exped. 1: 145 (1914).—Engl., Pflanzenw. Afr. 3, 2: 402, fig. 187 B-D (1921).—Exell & Mendonça, C.F.A. 1, 1: 168 (1937).—Williamson, Useful Pl. Nyasal.: 66 (1955).—Keay, F.W.T.A., ed. 2, 1, 2: 347 (1958). Type from Ceylon.

Annual herb up to 1·5 m. tall; stems glabrous or almost so. Leaf-lamina up to 15 × 15 cm., suborbicular to elliptic in outline, usually rather deeply digitately 3–5-lobed, usually glabrous or nearly so, lobes narrowly to very narrowly elliptic; petiole up to 10 cm. long, glabrous or sparsely pubescent; stipules up to 10 mm. long, very narrowly triangular. Epicalyx of 9–10 bracts, glabrous or sparsely pubescent; bracts up to 18 mm. long, elliptic, joined at the base. Calyx up to 3(5) cm. long, red, fleshy, edible; lobes ovate to ovate-lanceolate, joined to nearly half-way, usually with a conspicuous gland on the median vein on the outside of each lobe. Flowers 3–5 cm. in diam., pale yellow with purplish-brown centres, solitary, axillary or in racemose inflorescences by reduction of the upper leaves; peduncle up to 2·5 cm. long, glabrous, pubescent or pilose, articulated below the middle. Capsule up to 20 × 18 mm., ovate-acuminate, glabrous or somewhat pilose. Seeds up to 5 × 4 mm., subreniform, minutely stellate-pubescent, glabrescent.

N. Rhodesia. B: Balovale Distr., 1040 m., fr. vii.1933, *Trapnell* 1262 (K). S: Mazabuka, 1070 m., fr. iv.1934, *Trapnell* 1414 (K). **S. Rhodesia.** N: Lomagundi, fr. vi.1931, *Pardy* in GHS 4839 (SRGH). S: Fort Victoria, Kukumere Mission, fl. & fr. 5.v.1949, *Gerstner* 7033 (PRE; SRGH). **Nyasaland.** C: Dowa Distr., Lake Nyasa Hotel, fr. 27.vii.1951, *Chase* 3892 (BM; SRGH). S: Chibisa, Lower Shire Valley, fr. v.1861, *Kirk* (K). **Mozambique.** Z: Chamo, fr. iii.1859, *Kirk* (K). T: Furancungo, fr. 30.ix.1947, *Pimenta* 67 (SRGH).
Cultivated throughout the tropics. Probably introduced in our area and found as an escape from cultivation.

Roselle; Guinea or Jamaica Sorrel.
The fleshy calyx and occasionally the leaves are cooked and eaten; good jelly can be made from the calyces.
H. sabdariffa is usually glabrous and devoid of prickles but, judging from herbarium specimens, it seems to tend to revert to the prickly condition of a wild species from which it evolved. Its affinity with *H. mechowii* is very close.

9. **Hibiscus cannabinus** L., Syst. Nat. ed. 10, 2: 1149 (1759).—Harv. in Harv. & Sond., F.C. 1: 176 (1860).—Mast. in Oliv., F.T.A. 1: 204 (1868).—Hochr. in Ann. Conserv. Jard. Bot. Genève, 4: 114 (1900).—Bak. f. in Journ. Linn. Soc., Bot. 40: 27 (1911).—Ulbr. & Fr. in R.E.Fr., Wiss. Ergebn. Schwed. Rhod.-Kongo-Exped. 1: 145 (1914).—Engl., Pflanzenw. Afr. 3, 2: 400, fig. 188 (1921).—Burtt Davy, F.P.F.T. 2: 280 (1932) pro parte.—Exell & Mendonça, C.F.A. 1, 1: 168 (1937).—Garcia in Bol. Soc. Brot., Sér. 2, 20: 39 (1946).—Mendonça & Torre, Contr. Conhec. Fl. Moçamb. 1: 12 (1950) pro parte.—Suesseng. in Proc. & Trans. Rhod. Sci. Ass. 43: 102 (1951) pro parte.—Wild, Guide Fl. Vict. Falls: 147 (1953); Common Rhod. Weeds: fig. 41 (1955).—Martineau, Rhod. Wild Fl.: 49 (1954) pro parte.—Williamson, Useful Pl. Nyasal.: 65 (1955).—Keay, F.W.T.A. ed. 2, 1, 2: 347 (1958). Type a specimen cultivated at Uppsala, probably of Indian origin.
Hibiscus henriquesii P. Lima in Brotéria, Sér. Bot., 19: 138 (1921). Type: Mozambique, Niassa, near Palma, fl. & fr. 14.viii.1916, *Lima* 24 (PO, holotype).

Hibiscus sp.—Garcia, tom. cit.: 41 (1946) quoad specim. coll. Dr. Braga.
Hibiscus sabdariffa sensu Mendonça & Torre, tom. cit.: 11 (1950).

Annual herb up to c. 2 m. tall; stems aculeate with small rather sparse prickles usually pointing upwards, otherwise nearly glabrous or with a longitudinal line of crisped pubescence changing its radial position at each node. Leaf-lamina up to c. 15 × 15 cm., usually suborbicular in outline, scaberulous or almost glabrous with a few minute prickles on the nerves, 3–7-palmatisect to palmate-lobed, often somewhat pedate (usually nearly entire in seedlings and sometimes near the apex of the stems), apex acute, margin serrate or dentate, rarely subentire, base broadly cuneate to shallowly cordate, usually with a prominent gland on the under surface near the base of the midrib, lobes elliptic to entire; petiole up to 22 cm. long, hairy like the stems; stipules 4·5 mm. long, narrowly linear to filiform, very caducous (rarely seen on dried specimens). Flowers up to 10 cm. in diam., usually pale yellow, whitish or greyish, with purple centre, solitary, axillary or in racemes by the reduction of the upper leaves; peduncle 2–6 mm. long, aculeate or setose. Epicalyx of 7–8 bracts, bracts up to 18 mm. long, linear, joined for about 2 mm. at the base. Calyx up to 2·5 cm. long, 10-nerved, setose; lobes long, acuminate (sometimes subcaudate), joined for up to 5 mm. at the base, aculeate or setose outside especially near the margin, margin sometimes with a woolly tomentum, usually with a prominent gland 1·5–2 mm. in diam. on the midrib. Petals up to 6 × 4·5 cm., obovate, pubescent outside, glabrous within. Staminal tube up to 23 mm. long; free parts of filaments 1–2 mm. long. Style-branches 2–3·5 mm. long. Capsule up to 20 × 15 mm., ovoid-acuminate, appressed-setose. Seeds 3–3·5 × 1·5–2·5 mm., irregularly subreniform, minutely faveolate.

Bechuanaland Prot. N: Kwebe Hills, fl. 7.iii.1898, *Lugard* 213 (K). **N. Rhodesia.** N: Abercorn, 1675 m., fl. & fr. 12.v.1952, *Siame* 24A (BM). W: Mufulira, 1220 m., fl. & fr. 18.iv.1948, *Cruse* 257 (K). C: Chilanga Distr., Quien Sabe Ranch, 1100 m., fr. ix.1929, *Sandwith* 190 (K). S: Mapanza Mission, fl. & fr. 26.iv.1953, *Robinson* 184 (K). **S. Rhodesia.** C: Hartley Distr., Poole Farm, fl. & fr. 15.iv.1950, *Hornby* 370 (BM; SRGH). E: Odzani R., fl. iii.1935, *Gilliland* 1721 (BM). S: Belingwe Distr., Mataga sale pens, fl. & fr. iv.1958, *Davies* 2472 (SRGH). **Nyasaland.** N: Nyika Plateau, 1830–2130 m., *Whyte* (K). C: Mua-Livulezi Forest, fr. 26.vii.1956, *Adlard* 215 (K). S: Zomba, fl. ix.1891, *Whyte* (BM). **Mozambique.** N: between Mogincual and Quixaxe, fl. & fr. 27.vii.1948, *Pedro & Pedrógão* 4688 (LMJ). Z: between Ile and Mugeba, fl. & fr. 26.vi.1943, *Torre* 5576 (LISC). T: Km. 148 on the railway, fr. 18.v.1948, *Mendonça* 4308 (LISC). MS: Vila Machado, fl. 18.iv.1948, *Mendonça* 4008 (BM; LISC). SS: between Chissano and Chibuto, Maniquenique, fl. & fr. 8.vii.1948, *Myre* 46 (BM; LM). LM: between Umbeluzi and Makola, fl. & fr. 27.xii.1955, *Lemos* 126 (LMJ).

Widespread in tropical and subtropical regions. Usually found as a weed of arable and waste land.

Grown as a fibre-plant, especially in India, and commercially in our area as a jute substitute under the name *kenaf.* The leaves and flowers are sometimes cooked and eaten as vegetables.
One specimen (N. Rhodesia, Solwezi, *Milne-Redhead* 472A (K; PRE)), named *H. furcatus* Roxb. at Kew, may be an introduction from India, as suggested; it appears to be *H. cannabinus* in nearly every respect except for having forked epicalyx-bracts.

10. **Hibiscus meeusei** Exell in Bol. Soc. Brot., Sér. 2, **33**: 165 (1959). Type from the Transvaal.
Hibiscus cannabinus sensu Ulbr. & Fr. in R.E.Fr., Wiss. Ergebn. Schwed. Rhod.-Kongo-Exped. **1**: 145 (1914) pro parte quoad specim. *Fries* 954.—Burtt Davy, F.P.F.T. **2**: 280 (1932) pro parte.—Suesseng. in Proc. & Trans. Rhod. Sci. Ass. **43**: 102 (1951).—Martineau, Rhod. Wild Fl.: 49 pro parte, t. 18 fig. 1 (1954).
Hibiscus cf. *diversifolius.*—Verdoorn & Collett in Farming in S. Afr. **34**: 2, fig. 2 (1947).
Hibiscus sabdariffa sensu Mendonça & Torre, Contr. Conhec. Fl. Moçamb. **1**: 11 (1950) pro parte quoad specim. *Mendonça* 332 (LISC).

Annual or biennial herb up to 1 (1·5) m. tall; stems ± woody, pale green or brownish-purple, shortly stellate-tomentose and sparsely setose when young, branched from the base. Leaves dark green, usually with a purplish margin, minutely pubescent above with stellate or simple hairs, usually minutely stellate-pubescent beneath; lamina of lower leaves 12 × 12 cm., ovate-orbicular in outline, base truncate or slightly cordate, 5-lobed or -angled; lamina of middle leaves ovate in outline, 3–5-palmatisect to -palmatipartite, lobes rhombic, obovate or oblanceolate; lamina of upper leaves 3–6 cm. long, scarcely lobed, lanceolate, margin serrate, base cuneate; petiole up to 12 cm. long, hairy like the stems; stipules c. 10 mm. long, subulate-setose. Flowers 3–6 cm. in diam., yellow with dark red centre, solitary, axillary or forming racemes at the apices of the stems; peduncle up to c. 5 mm. long, hairy like the stems, articulated near the base. Epi-calyx of 8–12 bracts; bracts c. 10 mm. long, linear, apices sometimes shortly appendiculate but not manifestly forked, joined at the base. Calyx c. 15 mm. long; lobes ovate-lanceolate or triangular, acute, margin tomentellous, midrib with a median gland less conspicuous than that of *H. cannabinus*, accrescent in fruit and becoming c. 20 mm. long and somewhat woody. Petals 2–4 cm. long, obovate. Staminal tube c. 6 mm. long; free parts of filaments 0·5 mm. long. Style-branches c. 1 mm. long, stigmas hidden among the anthers. Capsule 14–15 × 11–12 mm., ovoid-conical, rostrate, appressed-setose; valves woody. Seeds 3–4 × 3 mm., angular-subreniform, with minute pectinate scales.

Caprivi Strip. Bagani Pontoon, fl. 19.i.1956, *de Winter* 4341 (PRE). **N. Rhodesia.** N: Bangweulu, Kamindas, fl. & fr. 9.x.1911, *Fries* 954 (UPS). W: Ndola, fl. & fr. 20.iii.1954, *Fanshawe* 979 (K). C: Lusaka, fl. & fr. iv.1957, *Noak* 197 (BM; SRGH). S: Dundwa, near Mapanza, 1100 m., fl. & fr. 6.iv.1953, *Robinson* 161a (K). **S. Rhodesia.** N: Sinoia, Umboe Valley, 1130–1190 m., fl. & fr. 28.iii.1950, *Colvile* 108 (BM; SRGH). W: Victoria Falls, fl. v.1915, *Rogers* 13157 (BM). C: Marandellas, fl. 26.iii.1942, *Dehn* 201 (M; SRGH). S: Zimbabwe, fl. 17.iii.1958, *Leach* 8221 (SRGH). **Nyasaland.** C: Dowa Distr., near Lake Nyasa Hotel. 425 m., fl. & fr. 31.vii.1951, *Chase* 3911 (BM; SRGH). **Mozambique.** MS: Baruè, fr. 18.ix.1942, *Mendonça* 332 (LISC). Also in S. Africa. A ruderal species of roadsides, fallow and waste land.

Martineau (loc. cit.) clearly recognizes what he calls two " varieties " or " forms " of *H. cannabinus*. The one he figures, his " small form ", is almost certainly *H. meeusei*.

11. **Hibiscus mechowii** Garcke in Linnaea, **43**: 121 (1881).—Exell & Mendonça, C.F.A. **1**, 1: 169 (1937). Type from Angola (Cuanza Norte).
    Hibiscus cannabinus sensu Hochr. in Ann. Conserv. Jard. Bot. Genève, **4**: 114 (1900) pro parte quoad syn. *H. mechowii*.

Annual (or perennial?) herb up to 1·3 m. tall; young stems with small rather sparse prickles and sometimes with an additional longitudinal line of crisped pubescence changing its radial position at each node. Leaf-lamina up to 15 × 15 cm., obovate to broadly obovate in outline, rather sparsely pilose or setose on both surfaces, cuneate to obtuse at the base, midrib with a longitudinal fissure 2–3 mm. long near the base, 3–5-palmatipartite; lobes up to 12 × 2 cm., very narrowly elliptic, margins serrate; petiole up to 12 cm. long, hairy like the young stems; stipules 10–18 mm. long, linear. Flowers 5–6 cm. in diam., yellow with reddish centre, solitary, axillary; peduncle 8–25 mm. long, articulated just below the middle, usually densely setose or setose-aculeate above the articulation, less densely so and more pubescent beneath it. Epicalyx of 7–8 bracts; bracts 12–20 × 1–1·5 mm., linear, free. Calyx 2–3·5 cm. long; lobes triangular, acute, pilose-setose and aculeolate especially on the margins and midrib, joined at the base. Petals c. 4 × 3 cm., almost glabrous. Staminal tube 20 mm. long; free parts of filaments c. 0·5 mm. long. Style-branches c. 1 mm. long. Capsule up to 35 × 25 mm., ovoid, appressed-setose. Seeds c. 3·5 × 2 mm., subreniform, with concentric rings of minute scales, often glabrescent.

N. Rhodesia. B: Lealui Distr., Mongu Township, fr. 11.ii.1952, *White* 2049 (K). N: Abercorn Distr., Chilongowelo, fl. 10.v.1952, *Richards* 1697 (K). S: Livingstone, 915 m., fl. 6.iv.1956, *Robinson* 1450 (K; SRGH). Also in Angola, Belgian Congo, SW. Sudan, Tanganyika and SW. Africa. Ruderal.

Very close to *H. sabdariffa* and it could well be the wild species from which the latter originated.

12. **Hibiscus diversifolius** Jacq., Coll. Bot. **2**: 307 (1789); Ic. Pl. Rar. **3**: t. 551 (1792).—Edwards, Bot. Reg. **5**: t 381 (1819).—Harv. in Harv. & Sond., F.C. **1**: 171 (1860).—Mast. in Oliv., F.T.A. **1**: 198 (1868).—Hochr. in Ann. Conserv. Jard. Bot. Genève, **4**: 119 (1900).—Bak. f. in Journ. Linn. Soc., Bot. **40**: 27 (1914).—Eyles in Trans. Roy. Soc. S. Afr. **5**: 415 (1916).—Engl., Pflanzenw. Afr. **3**, 2: 402, fig. 187 E–G (1921).—Burtt Davy, F.P.F.T. **2**: 280 (1932).—Exell & Mendonça, C.F.A. **1**, 1: 173 (1937).—Garcia in Bol. Soc. Brot., Sér. 2, **20**: 40

(1946).—Mendonça & Torre, Contr. Conhec. Fl. Moçamb. **1**: 14 (1950).—
Wild, Guide Fl. Vict. Falls: 147 (1953).—Brenan in Mem. N.Y. Bot. Gard., **8**, 3:
225 (1953).—Williamson, Useful Pl. Nyasal.: 65 (1955). Type a plant cultivated
in Vienna.

    Hibiscus ficulneus sensu Cav., Diss. **3**: 148, t. 151 fig. 2 (1787).

    Hibiscus macularis E. Mey. ex Harv. in Harv. & Sond., F.C. **1**: 171 (1860) *nom.
nud.*

Small tree, shrub or scrambling perennial herb up to 10 m. tall; young stems
stellate-tomentose or densely stellate-pubescent and aculeolate. Leaf-lamina up
to 16 × 16 cm., suborbicular in outline, ± distinctly 3–7-palmatilobed or -palmati-
partite, stellate-pubescent above, stellate-tomentellous beneath, apex acute or
rounded, margin irregularly serrate or crenate-serrate, base truncate to cordate,
usually with a linear or elliptic fissure or a suborbicular to elliptic gland with a
central fissure on the lower surface near the base of the midrib; petiole up to 12
cm. long, hairy like the young stems; stipules 3–4 mm. long, linear to filiform.
Flowers c. 7–8 cm. in diam., yellow or purplish with dark red or purple centre,
solitary, axillary or forming racemes at the ends of the branches by reduction of
the subtending leaves; peduncle 5–7 mm. long, densely setose, articulated near
the base. Petals 4–5 × 3–4 cm., obovate, stellate-pubescent on the outside. Epicalyx of 7–8 bracts; bracts 8–12 mm. long, free, linear. Calyx up to 2·5–3 cm.
long, densely setose; lobes lanceolate, joined for c. 5 mm. at the base, with an
elliptic gland on the midrib. Staminal tube up to 20 mm. long; free parts of
filaments up to 0·5 mm. long. Style protruding c. 6 mm. beyond the staminal
tube before branching, branches 1·5 mm. long. Capsule 20 × 15 mm., ovoid-acute,
densely setose. Seeds 4 × 2·8 mm., subreniform, with minute scales in concentric
rings.

Subsp. **diversifolius**

Stems usually with 1 or more longitudinal lines of pubescence, sometimes nearly
glabrous and aculeolate with short stout conical prickles. Flowers yellow with
maroon centre.

Tropical and S. Africa, Madagascar, India, Australia, New Caledonia and Pacific Is.
Not yet recorded from our area, but occurring both to the north and south of it.

Subsp. **rivularis** (Bremek. & Oberm.) Exell, stat. nov. Type: Bechuanaland Prot.,
    Chobe R., Kabulabula, fl. vii.1930, *van Son* in Herb. Transv. Mus. 28936 (BM;
    PRE, holotype).

    Hibiscus rivularis Bremek. & Oberm. in Ann. Transv. Mus. **16**, 3: 424 (1935).
Type as above.

Stems more densely hairy, ± uniformly stellate-pubescent to stellate-tomentose.
Leaves densely pubescent to tomentose. Flowers reddish to purple with darker
centre.

    Caprivi Strip. Lisikili, 24 km. E. of Katima Mulilo, 975 m., fl. 17.vii.1952, *Codd*
7101 (BM; PRE). **Bechuanaland Prot.** N: Chobe R., 13 km: N. of Kachikau, fl.
9.vii.1937, *Erens* 366 (PRE; SRGH). **N. Rhodesia.** N: Fort Rosebery, Lake Bangweulu, near Samfya Mission, fl. & fr. 20.viii.1952, *White* 3097 (FHO). C: Kafue R.
gorge, fl. x.1947, *Benson* (BM). S: Livingstone Distr., Katombora, fl. & fr. 26.viii.1947,
Brenan & Greenway 7759 (FHO; K). **Nyasaland.** C: Kota Kota Distr., Chia,
470 m., fl. & fr. 2.ix.1946, *Brass* 17480 (BM; K; PRE; SRGH). S: Cholo Mt.,
1200 m., fl. 21.ix.1946, *Brass* 17707 (K; SRGH). **Mozambique.** N: edge of Lake
Nyasa, fr. 11.x.1942, *Mendonça* 758 (LISC). Z: Maganja da Costa, fl. 14.ix.1944,
Mendonça 2057 (LISC). T: Lupata, fl. iii.1859, *Kirk* (K). MS: Gaza, fr. viii.1919,
Junod 366 (PRE). SS: Inhambane, Mulamba, fl. & fr. vii.1956, *Gomes e Sousa* 1756
(COI). LM: Catembe, fl. 10.vi.1957, *Carvalho* 277 (BM; LMJ).

    Also in Uganda, Tanganyika and Angola. In damp places along the margins of rivers
and lakes and in thickets.

Series 3 (*Solandra*)

Annual herbs. Flowers white or yellowish. Epicalyx absent in our species
(present in *H. upingtoniae*). Calyx 5-lobed, 10-nerved. Carpels often awned.

Calyx-lobes longer than the capsule; stipules 1–2 mm. long   -   -   13. *sidiformis*
Calyx-lobes shorter than the capsule; stipules 4–8 mm. long   -   -   14. *lobatus*

13. **Hibiscus sidiformis** Baill. in Bull. Soc. Linn. Par. **1** : 518 (1885) (" sidaeformis ").—
Hochr., Fl. Madag., Malvac.: 44, t. 12 fig. 4–5 (1955). Type from Madagascar.
 Solandra ternata Cav., Diss. Bot. **5** : 279, t. 136 fig. 2 (1788). Type from Senegal.
 Hibiscus ternatus (Cav.) Mast. in Oliv., F.T.A. **1** : 206 (1868) non *H. ternatus*
Cav. (1787).—Hochr., in Ann. Conserv. Jard. Bot. Genève, **4** : 126 (1900).—
Eyles in Trans. Roy. Soc. S. Afr. **5** : 416 (1916).—Engl. in Pflanzenw. Afr. **3**,
2 : 403, fig. 187 L–N (1921).—Burtt Davy, F.P.F.T. **2** : 280 (1932).—Mendonça
& Torre, Contr. Conhec. Fl. Moçamb. **1** : 15 (1950).—Martineau, Rhod. Wild
Fl. : 51 (1954). Type as above.
 Lagunaea schinzii Gürke in Verh. Bot. Verein. Brand. **30** : 180 (1888). Type
from SW. Africa.
 Hibiscus schinzii (Gürke) Hochr., tom. cit.: 127 (1900) non *H. schinzii* Gürke
(1889). Type as above.
 Hibiscus ternifoliolus F. W. Andr., Fl. Pl. Anglo-Egypt. Sudan, **2** : 31 (1952).
Type as for *Solandra ternata* Cav.

Annual herb from 6 cm. to 1 m. tall, erect or with arcuate-ascending branches
from near the base ; young stems densely pubescent or tomentellous with occasional longer weak or stiffish hairs. Leaf-lamina 1–5 × 1–5 cm., broadly ovate to
suborbicular, often not or scarcely lobed towards the base of the stem, becoming
shallowly to deeply 3-lobed towards the apex, hairy like the young stems on both
surfaces, lobes broadly elliptic to almost linear, usually narrower in the upper
leaves, apex acute to rounded, margin entire or coarsely serrate, base usually
cordate ; petiole up to 4·5 cm. long, hairy like the young stems ; stipules 1–2 mm.
long, filiform, caducous. Flowers 3–3·5 cm. in diam., white or pale yellow,
solitary, axillary ; peduncle up to 4 cm. long, slender, pubescent, articulated c.
2 mm. below the apex. Epicalyx absent. Calyx 8–13 mm. long, becoming somewhat scarious in fruit, hairy like the young stems ; lobes narrowly lanceolate,
joined for c. 3 mm. at the base. Petals up to 18 × 8 mm., narrowly obovate, glabrous. Staminal tube up to 12 mm. long, antheriferous only in the upper half ;
free parts of filaments 1·5–2 mm. long. Style-branches c. 0·5 mm. long. Capsule
5–8 × 5–6 mm., ellipsoid, carpels pointed or with short awns 0·5–1 mm. long.
Seeds 2 × 1·5 mm., irregularly subreniform, densely verruculose otherwise glabrous.

Bechuanaland Prot. N : Ngamiland, *Curson* 405 (PRE). **N. Rhodesia.** S : Livingstone, fl. & fr. iv.1909, *Rogers* 7064 (SRGH). **S. Rhodesia.** N : Urungwe Distr.,
Mauora R., fl. 26.ii.1958, *Phipps* 892 (SRGH). W : Wankie Distr., Matetsi, 850 m.,
fl. & fr. iii.1918, *Rand* 1289 (BM). E : Umtali, Maranka Reserve, 760 m., fl. & fr.
10.ii.1953, *Chase* 4772 (BM ; SRGH). S : Gwanda Distr., Shashi R., fl. & fr. 18.xii.1956,
Davies 2366 (SRGH). **Nyasaland.** N : 13 km. W. of Karonga, 490 m., fl. & fr. 24.ii.1953,
Williamson 182 (BM). C : near Lake Nyasa, near Senga Bay, 470 m., fl. & fr. 17.ii.1959,
Robson 1621A (BM ; K ; LISC ; SRGH). **Mozambique.** T : Lupata, fl. & fr.
20.iv.1862, *Kirk* (K). MS : Meringua, fl. & fr. 30.vi.1950, *Chase* 2472 (SRGH).
 Widespread in tropical and S. Africa and in Madagascar. In river valleys up to c.
1000 m., often on alluvial silt and in mopane woodland and *Terminalia-Commiphora*
thickets.

14. **Hibiscus lobatus** (Murr.) Kuntze, Rev. Gen. Pl. **3**, 2 : 19 (1898).—Exell & Mendonça, C.F.A. **1**, 1 : 176 (1937).—Mendonça & Torre, Contr. Conhec. Fl.
1 : 16 (1950).—Hochr., Fl. Madag., Malvac.: 42, t. 12 fig. 1–3 (1955).—Keay,
F.W.T.A., ed. 2, **1**, 2 : 346 (1958). TAB. **89** fig. 1. Type a cultivated plant of
unknown origin.
 Solandra lobata Murr. in Comment. Soc. Reg. Sci. Gotting. **6** : 20, t. 1 (1785).
Type as above.
 Hibiscus solandra L'Hérit., Stirp. Nov. **1** : 103, t. 49 (1788) *nom. illegit.*—Mast.
in Oliv., F.T.A. **1** : 206 (1868).—Hochr. in Ann. Conserv. Jard. Bot. Genève, **4** :
128 (1900).—Eyles in Trans. Roy. Soc. S. Afr. **5** : 416 (1916).—Wild, Guide Fl.
Vict. Falls : 147 (1953). Type as above.

Annual herb from 0·5–1·3 m. tall ; young stems pubescent, pilose or tomentellous. Leaf-lamina 2–12 × 1·5–10 cm., suborbicular to ovate in outline, varying on
the same individual from not lobed to deeply 3–5-palmatilobed or incised, usually
pubescent on both surfaces with additional longer simple hairs on the veins and
2–4-pronged hairs mainly on the lower surface, lobes sometimes secondarily
pinnately lobed or incised, apex acute to subcaudate, margin bluntly toothed or
crenate or irregularly lobed, base slightly to distinctly cordate 3–5-nerved ; petiole
up to 8 cm. long, pubescent or pilose or tomentellous ; stipules 4–8 × 0·5–1 mm.,
filiform to narrowly linear. Flowers 1–2 cm. in diam., white to yellowish, in

few-flowered terminal racemes ; peduncle 10–20 mm. long, pubescent or pilose or tomentellous, articulated near the apex. Epicalyx absent. Calyx pubescent; lobes 5–7 × 1·5–2·5 mm., narrowly lanceolate, joined near the base. Petals 8–12 × 4–6 mm., obovate, thin-textured and transparent. Staminal tube 5 mm. long; free parts of filaments 0·5 mm. long. Style-branches 1 mm. long. Capsule 10 × 7 mm., oblong-ellipsoid, pubescent, with awns 1·5–2 mm. long. Seeds 1·5 × 1·3 mm., irregularly prismatic, minutely verruculose otherwise glabrous.

N. Rhodesia. N: Lake Mweru, 1080 m., fl. & fr. 10.iv.1957, *Richards* 9124 (K). W: Kitwe, fr. 6.iii.1955, *Fanshawe* 2119 (K). E: Nyamadzi R., fr. 25.iii.1955, *E.M. & W.* 1178 (BM; LISC; SRGH). S: 5 km. E. of Mapanza, 1070 m., fl. & fr. 1.iv.1954, *Robinson* 657 (BM; K). **S. Rhodesia.** N: Miami, fl. & fr. v.1926, *Rand* 136 (BM). W: Victoria Falls, 940 m., fl. & fr. iv.1918, *Eyles* 1297 (BM; K). C: Salisbury Distr., Makabusi R., fr. 31.iii.1947, *Wild* 1882 (BM; SRGH). E: Umtali Distr., Maranka Reserve, 820 m., fl. & fr. 27.ii.1953, *Chase* 4792 (BM; SRGH). **Nyasaland.** N: Njakwa Gorge, fl. & fr. 26.iii.1954, *Jackson* 1278 (K). S: Zomba, near Misanji R., fl. & fr. 27.iv.1955, *Jackson* 1649 (BM; K; SRGH). **Mozambique.** Z: Massingire, Montes de Metaloa, fr. 24.v.1943, *Torre* 5380 (LISC). T: Lupata, fl. & fr. iv.1860, *Kirk* (K). MS: Vila Machado, Montes de Siluvo, fl. & fr. 11.ii.1948, *Garcia* 985 (BM; LISC).

Tropical Africa, Madagascar and tropical Asia. Woodland, river valleys and roadsides from c. 300–1400 m.

Series 4 (*Trionum*)

Annual herbs. Calyx inflated, scarious, enclosing the capsule ; lobes connate nearly to the apex.

15. **Hibiscus trionum** L., Sp. Pl. **2**: 697 (1753).—Curt., Bot. Mag.: t. 209 (1793).— Harv. in Harv. & Sond., F.C. **1**: 176 (1860).—Mast. in Oliv., F.T.A. **1**: 196 (1868).—Hochr. in Ann. Conserv. Jard. Bot. Genève, **4**: 144 (1900).—Eyles in Trans. Roy. Soc. S. Afr. **5**: 416 (1916).—Engl., Pflanzenw. Afr. **3**, 2: 403, fig. 187, O (1921).—Burtt Davy, F.P.F.T. **2**: 280 (1932).—Arwidss. apud Norlindh & Weim. in Bot. Notis. **1934**: 99 (1934).—Exell & Mendonça, C.F.A. **1**, 1: 176 (1937).—Mendonça & Torre, Contr. Conhec. Fl. Moçamb. **1**: 16 (1950).— Suesseng. in Proc. & Trans. Rhod. Sci. Ass. **43**: 29 (1951).—Martineau, Rhod. Wild Fl.; 51, t. 19 (1954).—Wild, Common Rhod. Weeds: fig. 42 (1955).—Keay, F.W.T.A. ed. 2, **1**, 2: 346 (1958). TAB. **89** fig. 3. Syntypes cultivated plants originating from Italy and S. Africa.

Annual herb up to 1·5 m. tall; young stems stellate-pubescent to stellate-tomentose or hispid. Leaf-lamina 2–6 × 2–6 cm., suborbicular to ovate in outline (or hastate but not apparently in our area), varying from not lobed to deeply 3–5-palmatilobed or incised, hispid on the nerves, apex rounded, base truncate to slightly cordate, 3–5-nerved, lobes pinnately incised ; petiole 1–4 cm. long, hairy like the young stems ; stipules 4–8 mm. long, linear. Flowers 2·5–4 cm. in diam., white, cream or yellow, with purple centres, solitary, axillary ; peduncle up to 5·5 cm. long, hairy like the young stems. Epicalyx of 12 bracts ; bracts 7–14 mm. long, filiform, often hispid. Calyx up to 2·5 cm. long, scarious with purplish longitudinal veins, stellate-pilose or hispid, becoming inflated ; lobes united nearly to the apex. Petals 2–3 × 1·5–3 cm., obovate, nearly glabrous. Staminal tube 3–4 mm. long, somewhat glandular ; free parts of filaments 3–4 mm. long. Style-branches 2 mm. long. Capsule 10–14 mm. in diam., subglobose, hispid, enclosed in the inflated calyx. Seeds 2 × 1·5 cm., subreniform, tuberculate.

Bechuanaland Prot. N: Ngamiland, fl., *Curson* 540 (PRE). SE: Kanye, fl. & fr. 1895, *Marloth* 2165 (PRE). **N. Rhodesia.** S: Mazabuka, 1040 m., fl. & fr. ii.1934, *Trapnell* 1364 (K). **S. Rhodesia.** W: Nyamandhlovu, fl. & fr. 17.iii.1953, *Plowes* 1562 (BM; SRGH). C: Salisbury, 1460 m., fl. i.1932, *Brain* 9051 (SRGH). E: Umtali Distr., Engwa, 1600 m., fl. & fr. 1.ii.1955, *E.M. & W.* 18 (BM; LISC; SRGH). **Mozambique.** SS: near Xai-xai, fl. & fr. 24.x.1947, *Barbosa* 488 (LM). LM: Sabiè, Moamba, R. Incomati, fl. & fr. 29.iv.1944, *Torre* 6527 (LISC).

Widespread in the warmer regions of the Old World from S. Europe to S. Africa, Asia and Australia. Upland grasslands up to 1700 m. and a common weed of cultivated land and waste places.

Many varieties of this species have been described (see Harvey, loc. cit., and Hochreutiner, loc. cit.) based on differences in the indumentum, the lobing of the leaves and the size of the flowers but great variation can be seen on an individual specimen. The species is far more variable in S. Africa than in our area where, in S. Rhodesia at least, it never seems to differ markedly from Wild's figure (loc. cit.); but in S. Mozambique it is more variable in leaf-shape.

Series 5 (*Venusti*)

Large shrubs. Flowers large. Calyx much enlarged in fruit, not scarious; lobes united for about ⅓ their length. One introduced species in our area.

16. **Hibiscus mutabilis** L., Sp. Pl. **2**: 694 (1753).—Harv. in Harv. & Sond., F.C. **1**: 172 (1860).—Hochr. in Ann. Conserv. Jard. Bot. Genève, **4**: 147 (1900).—Engl., Pflanzenw. Afr. **3**, 2: 397, fig. 186, P–R (1921).—Brenan, T.T.C.L.: 303 (1949). Syntypes cultivated plants from China and India.

A large shrub with suborbicular 5–7-angled leaves and large flowers 8–10 cm. in diam., opening white and turning red by evening.

S. Rhodesia. N: Shamva, fl. 22.x.1933, *Eyles* 7656 (K; SRGH). C: Salisbury Distr., Govt. Forest Nursery, fl. 23.iii.1950, *Shepherd* in GHS 27283 (SRGH). S: Victoria Distr., fl. & fr. 1909–12, *Monro* 1957 (BM).
Probably a native of China, now cultivated throughout the tropics and subtropics. Sometimes established as an escape from cultivation.

Series 6 (*Bombycella*)

Shrubs, shrublets or perennial (rarely annual?) herbs. Flowers usually red, pink or white (more rarely yellow), concolorous (except in *H. pusillus*). Epicalyx of more than 5 bracts, sometimes very small (absent in *H. elliottiae* Harv., not yet recorded from our area). Seeds usually with a cottony or silky floss (glabrous or minutely pubescent in *H. nyikensis* and *H. allenii*) when mature.

A natural group (except perhaps for *H. pusillus* and possibly *H. subreniformis*) easily recognizable even when the seeds are not ripe.

Note

Hibiscus ebracteatus Mast. (in Oliv., F.T.A. **1**: 206 (1868)) is the same as *H. elliottiae* Harv. (1863). The type (*Chapman & Baines* (K)) was apparently collected in SW. Africa just outside our area and there is as yet no record from within it. It is easily distinguishable from all the other species in this series by the absence of an epicalyx.
Seeds with silky or cottony floss when mature (TAB. **90** fig. C2):
  Flowers yellow or yellow with a dark centre:
    Flowers yellow with a dark centre; petals up to 4·5 cm. long; leaves varying from
      not lobed to deeply lobed, margin serrate or toothed - - - 17. *pusillus*
    Flowers yellow, concolorous; petals up to 1·5 cm. long; leaves subreniform, densely
      stellate-tomentose, not lobed, margin crenate or bluntly serrate - 18. *subreniformis*
  Flowers red, pink, lilac or white:
    Petals 3 cm. long or longer:
      Flowers usually with a dark centre; staminal tube 4–8 mm. long   17. *pusillus*
      Flowers concolorous; staminal tube 20–30 mm. long   -   19. *pedunculatus*
    Petals up to 2·5 cm. long:
      Indumentum partly of fuscous or dark tawny hairs, especially on the calyx,
        epicalyx and peduncle (see also *H. nyikensis*):
        Flowers white; calyx 1·5–2·5 cm. long in fruit; hairs 1·5–2·5 mm. long
                                                                              20. *fuscus*
        Flowers red; calyx less than 1·5 cm. long; hairs 0·5–1 mm. long:
          Epicalyx shorter than the calyx (TAB. **90** fig. B):
            Stipules 5–8 mm. long; bracts of epicalyx up to 1 mm. wide   21. *shirensis*
            Stipules 2–3 mm. long; bracts of epicalyx 1·5–2 mm. wide   22. *rupicola*
          Epicalyx equalling or exceeding the calyx (TAB. **90** fig. C1)   23. *debeerstii*
      Indumentum not as above (the stellate hairs may be golden-brown when dried):
        Petals 5–9 mm. long; epicalyx 1–3 mm. long;
          Leaves not or rarely lobed; flowers white or pink; staminal tube 2–2·5 mm.
            long:
            Young stems with an indumentum of stellate-setulose hairs projecting up
              to 0·5 mm. (TAB. **90** fig. D) - - - - - 25. *micranthus*
            Young stems with an indumentum of stellate-hispid hairs projecting 1–2
              mm. (TAB. **90** fig. E) - - - - - - 26. *sabiensis*

Leaves tending to be deeply 3-lobed; flowers red; staminal tube 4–4·5 mm.
long - - - - - - - - - -   27. *migeodii*
Petals 10–25 mm. long:
   Free parts of filaments 0·5–1·5 mm. long; plant a bush, shrublet or virgate
     herb:
      Style-branches 2–3 mm. long; calyx-lobes up to 5 mm. long; indu-
      mentum of closely appressed hairs (TAB. **90** fig. A4) -   28. *meyeri*
      Style-branches 4–15 mm. long; calyx-lobes at least 6 mm. long:
       Hairs on young stems closely appressed; leaves sparsely stellate-pubes-
       cent or stellate-hispid; flowers red -   -   -   29. *praeteritus*
       Hairs on young stems somewhat patent, some at least projecting 0·5 mm.:
        Leaves not lobed or with only a slight tendency to become 3-lobed;
        flowers red, pink or white:
         Flowers white, fading to red; peduncles 1 cm. long   30. *gwandensis*
         Flowers red or pink; peduncles up to 3–4 cm. long:
          Indumentum tawny; calyx-lobes elongate-triangular
                                  31. *aponeurus*
          Indumentum silvery-grey; calyx-lobes lanceolate   32. *richardsiae*
        Leaves mostly 3-lobed (some at least on each specimen); flowers
        opening white, turning pink then red -   -   -   33. *mutatus*
   Free parts of filaments 1·5–4 mm. long; plant sending up annual shoots from
   a woody rootstock (except perhaps *H. rupicola*):
     Leaves 3-nerved at the base - - - - - -   34. *rhodanthus*
     Leaves 5–7-nerved at the base:
      Plant stellate-hispid, somewhat rough to the touch; leaves ovate with a
      tendency to become 3-lobed; bracts of epicalyx linear-oblong, up to
      1 mm. broad - - - - - - -   35. *barbosae*
      Plant densely stellate-tomentose, soft to the touch; leaves very broadly
      ovate to suborbicular; bracts of epicalyx linear-oblong, up to 2 mm.
      broad - - - - - - - - -   22. *rupicola*
Seeds glabrous or minutely pubescent:
   Leaves not lobed; stems with fuscous appressed hairs -   -   24. *nyikensis*
   Leaves (at least some of them) deeply 3-lobed; stems hispid with patent simple or
   branched hairs - - - - - - - - -   36. *allenii*

17. **Hibiscus pusillus** Thunb., Prodr. Pl. Cap. **2**: 118 (1800).—Harv. in Harv. & Sond.,
F.C. **1**: 175 (1860).—Hochr. in Ann. Conserv. Jard. Bot. Genève, **4**: 79 (1900).—
Eyles in Trans. Roy. Soc. S. Afr. **5**: 416 (1916).—Burtt Davy, F.P.F.T. **2**: 281
(1932).—Exell & Mendonça, C.F.A. **1**, 1: 166 (1937).—Martineau, Rhod. Wild
Fl.: 51, t. 19 fig. 3 (1954). Type from S. Africa.
   Hibiscus gossypinus Thunb., loc. cit. Type from S. Africa.
   Hibiscus atromarginatus Eckl. & Zeyh., Enum. Pl. Afr. Austr. Extratrop. **1**: 38
(1834).—Harv. in Harv. & Sond., loc. cit.—Burtt Davy, tom. cit.: 280 (1932).—
Mendonça & Torre, Contr. Conhec. Fl. Moçamb. **1**: 16 (1950). Type from S.
Africa.

Perennial herb 5–30 cm. high with ± prostrate annual shoots from a woody
rootstock; stems hispid or with appressed stellate hairs or a mixture of patent and
appressed hairs. Leaf-lamina 5–6 × 2–4·5 cm., suborbicular to elliptic in outline,
not lobed to deeply 3–5-palmatisect, nearly glabrous to hispid, especially on the
nerves of the lower surface; lobes narrowly elliptic to sharply toothed or pinnatifid; petiole 2–15 mm. long; stipules 5–7 mm. long, subulate. Flowers up to
6 cm. in diam., yellow (rarely pink) with a purple or crimson centre, usually
solitary, axillary; peduncle 1·5–12 cm. long, inconspicuously articulated near the
apex. Epicalyx of 8–11 bracts; bracts 3–12 mm. long, acicular. Calyx up to
25 mm. long; lobes 7–25 × 3–8 mm., narrowly lanceolate-triangular, acute. Petals
1–4·5 cm. long, obovate, nearly glabrous or with a few scattered stellate hairs.
Staminal tube 4–8 mm. long; free parts of filaments 4–8 mm. long. Style-
branches 2·5–4 mm. long. Capsule 10–12 × 10–12 mm., hispid on the margins
of the valves. Seeds with a well-developed white, brownish or reddish silky
floss.

Bechuanaland Prot. N: between Mumpswe and Sigara pans. fl. 25.iv.1957, *Drum-
mond & Seagrief* 5215 (CAH; K; PRE; SRGH). SE: near Lobatsi, fl. & fr. 28.xii.1911,
Rogers 6077 (SRGH). **S. Rhodesia.** W: Matopo Hills, 1370 m., fl. iii.1903, *Eyles* 110
(BM; SRGH). C: Salisbury, 1460 m., fl. 27.xii.1933, *Brain* 9793 (SRGH). S: Ndanga
Distr., Triangle Sugar Estate, fl. 25.i.1949, *Wild* 2734 (SRGH). **Mozambique.** LM:
Goba, fr. 19.xii.1952, *Myre & Carvalho* 1395 (K; LMJ).
   Also in S. Africa. Mopane woods and among rocks.

18. **Hibiscus subreniformis** Burtt Davy, F.P.F.T. **1**: 43 (1926). Type from the Transvaal (Messina).

Perennial herb or shrublet, up to 1 m. high; stems densely stellate-tomentose. Leaf-lamina 1–5 × 1–4 cm., broadly ovate to subreniform, densley stellate-tomentose, apex rounded, margin irregularly crenate, base truncate to cordate 7-nerved; petiole 3–20 mm. long; stipules c. 2 mm. long, subulate. Flowers c. 2·5 cm. in diam., pale yellow or orange-yellow, solitary, axillary; peduncle 8–25 mm. long, densely stellate-tomentose, articulated near the middle. Epicalyx of c. 10 bracts; bracts 3–7 mm. long, acicular to subulate. Calyx up to 11 mm. long, stellate-tomentose; lobes 7–11 × 2–4 mm., narrowly elliptic to lanceolate. Petals 10–15 mm. long, obovate, glabrous. Staminal tube 10 mm. long; anthers spirally arranged; free parts of filaments 1 mm. long. Style-branches 3–3·5 mm. long. Capsule 8 × 7 mm., subglobose, stellate-pubescent. Seeds with long white silky hairs.

Bechuanaland Prot. SE: NW. of Gaberones, 1040 m., fl. & fr. 1.xii.1954, *Codd* 8910 (PRE; SRGH). **S. Rhodesia.** W: Matobo Distr., Besna Kobila Farm, 1430 m., fl. ii.1954, *Miller* 2206 (BM; SRGH).
Also in S. Africa. In open woodland and on rock outcrops.

Burtt Davy (loc. cit.) describes the petals as " pallide rubra " and the flowers of the type-specimen certainly look somewhat pink but whenever collectors have given information the flower-colour is said to be yellow.

19. **Hibiscus pedunculatus** L.f., Suppl. Pl.: 309(1781).—Thunb., Prodr. Pl. Cap. **2**: 118 (1800).—Edwards, Bot. Reg. **3**: t. 231 (1817).—Harv. in Harv. & Sond., F.C. **1**: 173 (1860).—Hochr. in Ann. Conserv. Jard. Bot. Genève, **4**: 78 (1900).—Burtt Davy, F.P.F.T. **2**: 281 (1932). Type from S. Africa.

Perennial herb 1–2 m. tall; stems densely appressed-stellate-pilose. Leaf-lamina 2–8·5 × 2–8 cm., suborbicular to ovate in outline, 3-lobed, with stiffish simple or branched hairs on both surfaces, apex rounded, margin bluntly toothed, base obtuse to slightly cordate 5–7-nerved; lobes rounded at the apex; petiole 0·6–4·5 cm. long; stipules 4 mm. long, subulate. Flowers up to 7 cm. in diam., pale lilac or pink, solitary, axillary; peduncles 4–12 cm. long, appressed-stellate-setulose, articulated 3–10 mm. below the apex. Epicalyx of 7–8 bracts, usually slightly exceeding the calyx; bracts 14–18 × 1–3·5 mm., linear to linear-elliptic. Calyx up to 16 mm. long, setulose-hispid; lobes 10–16 × 3–5 mm., lanceolate. Petals 3–4·5 cm. long, obovate, stellate-pubescent outside, glabrous within. Staminal tube 20–30 mm. long; anthers borne on the upper half of the tube; free parts of filaments 0·5–1 mm. long. Style-branches 2–5 mm. long. Capsule 10–12 × 8–10 mm., subglobose, hispid, acuminate at the apex. Seeds with a white or brownish fuzz.

Mozambique. LM: Namaacha, 500 m., fl. & fr. 27.iii.1956, *Barbosa & Lemos* 7510 (COI; LMJ).
Also in S. Africa. Clearings in riverine forest.

20. **Hibiscus fuscus** Garcke in Bot. Zeit. **7**: 854 (1849).—Ulbr. in Notizbl. Bot. Gart. Berl. **8**: 679 (1924).—Cufod. in Ann. Naturh. Mus. Wien, **56**: 40 (1948). Type from S. Africa.
Hibiscus gossypinus sensu Harv. in Harv. & Sond., F.C. **1**: 175 (1860) non Thunb. —Mast. in Oliv., F.T.A. **1**: 205 (1868).—Bak. f. in Journ. Linn. Soc., Bot. **40**: 28 (1911).—Brenan in Mem. N.Y. Bot. Gard. **8**, 3: 224 (1953).
Hibiscus ferrugineus sensu Hochr. in Ann. Conserv. Jard. Bot. Genève, **4**: 84 (1900) pro parte, non *H. ferrugineus* Cav.

Shrub up to 3 m. tall or shrublet; stems densely fuscous-stellate-hispid and often with an underlying tomentum of shorter stellate hairs. Leaf-lamina 2–7 × 1–5 cm., narrowly ovate to suborbicular, sometimes slightly lobed, brownish- or tawny-tomentose, apex rounded to acute, margin rather bluntly to sharply serrate, base obtuse rounded or nearly truncate usually 5-nerved; petiole 2–33 mm. long; stipules up to 10 mm. long, filiform. Flowers 1·5–3 cm. in diam., white or yellowish with orange anthers, congested towards the tips of the branches, forming leafy somewhat corymbose panicles; peduncles up to 6 cm. long, hairy like the stems, articulated near the apex. Epicalyx of 10–12 bracts; bracts 5–11 mm. long,

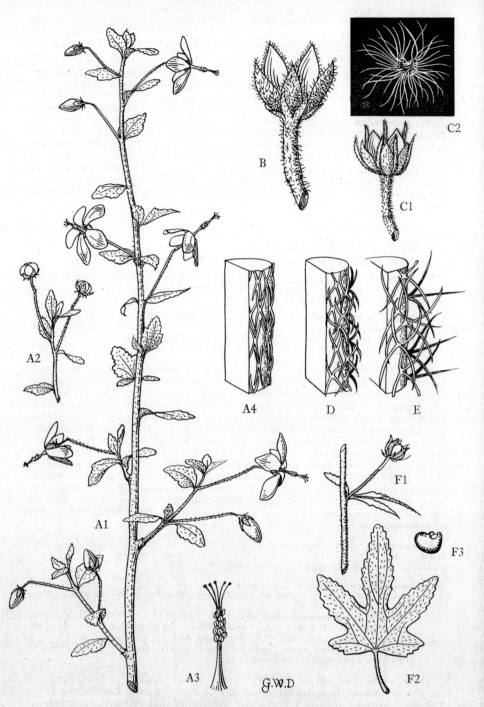

Tab. 90. A.—HIBISCUS MEYERI SUBSP. MEYERI. A1, flowering stem (×⅔); A2, fruiting stem (×⅔); A3, staminal tube and style-branches (×2); A4, indumentum of young stem (×12) all from *Mendonça* 3523. B.—H. SHIRENSIS. Flower (×2) *Torre* 432. C.—H. DEBEERSTII. C1, flower (×2); C2, seed (×2) both from *Gamwell* 59. D.—H. MICRANTHUS. Indumentum of young stem (×12) *Gilges* 5013. E.—H. SABIENSIS. Indumentum of young stem (×12) *Wild* 2312. F.—H. ALLENII. F1, stem and young fruit (×⅔); F2, leaf (×⅔); F3, seed (×2) all from *Simons* s.n.

filiform to subulate. Calyx up to 22 mm.; lobes 6–20 × 3–6 mm., narrowly lanceolate, acute, fuscous-hispid especially on the midrib and near the margins, shortly connate at the base. Petals up to 18 mm. long, narrowly obovate, stellate-setulose outside, glabrous inside. Staminal tube 8–12 mm. long; free parts of filaments c. 0·5 mm. long. Style-branches 2–3·5 mm. long. Capsule c. 10 mm. in diam., subglobose, minutely pubescent or nearly glabrous. Seeds with a white or brownish silky floss.

S. Rhodesia. E: Vumba Distr., Witchwood Farm, fl. 7.vii.1948, *Chase* 742 (BM; K; LISC; SRGH). **Nyasaland.** N: Misuku Hills, Mugesse Forest Reserve, 1830 m., fl. ix.1953, *Chapman* 213 (BM). S: Cholo Distr., Cholo Mt., 1200 m., fl. & fr. 19.ix.1946, *Brass* 17646 (BM; K; PRE; SRGH). **Mozambique.** Z: Serra de Milange, fl. & fr. 11.ix.1949, *Barbosa & Carvalho* 4041 (LM). T: near Zobuè, fl. & fr. 27.viii.1943, *Torre* 5798 (LISC). MS: Mavita, fl. & fr. 14.iv.1949, *Pedro & Pedrógão* 6500 (LMJ; SRGH).
Also from Ethiopia to S. Africa and in the Belgian Congo. Rain-forest, forest margins and upland grasslands.

Brenan (loc. cit.) rightly pointed out that Hochreutiner (loc. cit.) was wrong in considering *H. gossypinus* sensu Harv. to be a synonym of *H. ferrugineus* Cav. but it should be added that the latter author had already changed his mind (see Candollea, **12**: 169 (1949)). This species has long been known as *H. gossypinus* Thunb. but the true *H. gossypinus* is the same as *H. pusillus* (q.v.).

21. **Hibiscus shirensis** Sprague & Hutch. in Kew Bull. **1907**: 47 (1907).—Bak. f. in Journ. Linn. Soc., Bot. **40**: 28 (1911).—Eyles in Trans. Roy. Soc. S. Afr. **5**: 416 (1916).—Garcia in Bol. Soc. Brot., Sér. 2, **20**: 39 (1946).—Cufod. in Ann. Naturh. Mus. Wien, **56**: 57 (1948).—Brenan in Mem. N.Y. Bot. Gard. **8**, 3: 224 (1953). TAB. **90** fig. B. Type: Nyasaland, banks of Likangola R., *Buchanan* 385 (K).
  Hibiscus heterochlamys Ulbr. in Notizbl. Bot. Gart. Berl. **7**: [367] (1920). Syntypes from Uganda, Tanganyika and Nyasaland, Shire Highlands, *Buchanan* (B†).

Shrub or perennial herb, up to 2–4(5) m. tall; stems with fuscous stellate hairs and an underlying pubescence of finer, pale green, stellate hairs. Leaf-lamina up to 10 × 7 cm., ovate to ovate-lanceolate with a tendency to become 3-lobed, stellate-pubescent to stellate-tomentose on both surfaces with occasional larger fuscous stellate hairs, apex acute, margin irregularly serrate, base obtuse to cordate 5–7-nerved. Flowers 2–3 cm. in diam., red, solitary, axillary and forming terminal racemes (sometimes corymbose) at the apices of the stems by reduction of the upper leaves; peduncles up to 6 cm. long but usually much shorter. Epicalyx of 6–8 bracts; bracts 2–7 mm. long, filiform to linear, silvery-green when dried and contrasting strongly with the fuscous calyx. Calyx 8–14 mm. long, fuscous-pilose; lobes 5–13 × 2–5 mm., narrowly lanceolate, acute. Petals 7–20 mm. long, obovate, with a few scattered setulose stellate hairs. Staminal tube 4–5 mm. long; free parts of filaments 2–3(5) mm. long. Style-branches 3–5 mm. long. Capsule 10 mm. in diam., subglobose, puberulous. Seeds with long white silky hairs and a short white pubescence.

N. Rhodesia. N: Fort Rosebery, Lake Bangweulu, fl. & fr. 23.viii.1952, *White* 3143 (FHO). W: Kitwe, fl. & fr. 10.v.1958, *Fanshawe* 2272 (K). **S. Rhodesia.** E: Chirinda, 1160 m., fl. & fr. iv.1906, *Swynnerton* 298 (BM; K). **Nyasaland.** N: Rumpi Distr., near Nchena-Chena, fl. & fr. 12.v.1952, *White* 2841 (FHO). C: Dedza, Golomoti Escarpment, fl. & fr. 13.vi.1950, *Wiehe* N. 586 (SRGH). S: Zomba, Mulunguzi R., fl. & fr. 17.vi.1954, *Banda* 14 (BM). **Mozambique.** N: Vila Cabral, Lichinga, fl. & fr. 8.i.1955, *Torre* 432 (BM; COI; K; LISC). Z: 16 km. SW. of Guruè, fl. & fr. 6.vii.1942, *Hornby* 4565 (PRE). MS: Manica, Serra de Vumba, 1000 m., fl. & fr. 25.iii.1948, *Garcia* in Mendonça 716 (BM; LISC).
Also in Tanganyika and the Belgian Congo (Katanga). In *Brachystegia* woodland, wooded grassland, secondary thickets and cultivated ground up to about 1250 m.

Barbosa 1110 (LISC) from Mozambique, Serra do Garuso (MS), has exceptionally large flowers but appears to be this species.

22. **Hibiscus rupicola** Exell in Bol. Soc. Brot., Sér. 2, **33**: 177 (1959). Type: Mozambique, Tete, Serra de Zobuè, *Mendonça* 579 (LISC, holotype).

Shrub or perennial herb 1·5 m. tall; stems densely tomentose. Leaf-lamina up to 5 × 4·5 cm., broadly ovate to suborbicular, stellate-tomentose on both surfaces, apex rounded, margin crenate-dentate, base rounded or truncate 5–7 nerved; petiole up to 1 cm. long; stipules 2–3 mm. long, subulate. Flowers 1·5 cm. in diam., red, solitary in the axils of the upper leaves and leaf-like bracts; peduncle 1·5 cm. long, articulated 5 mm. below the apex. Epicalyx of c. 7 bracts; bracts 5–6 × 1·5–2 mm., linear or linear-oblong. Calyx 5–12 mm. long, stellate-tomentose; lobes up to 12 × 4·5 mm., lanceolate-triangular, shortly connate at the base. Petals 10–20 × 5–10 mm., obovate. Staminal tube 2 mm. long; free parts of filaments 3–3·5 mm. long. Style-branches 5–6 mm. long. Capsule 8–10 mm. in diam., subglobose, minutely pubescent. Seeds with silky floss.

Mozambique. T: Serra de Zobuè, fl. & fr. 3.x.1942, *Mendonça* 579 (LISC). Known only from the locality stated. Among rocks.

It was difficult to decide whether to describe this as a new species. The leaves and indumentum are very similar to those of *H. aponeurus* from which it differs, however, in the length of the free parts of the filaments of the stamens (only 0·5–1 mm. long in *H. aponeurus*). The closest affinity is undoubtedly with *H. shirensis* from which it can be separated by the leaf-shape, indumentum (more densely tomentose) and the shorter staminal tube. There are some fuscous hairs in the indumentum but these are far less conspicuous than in *H. shirensis*. Future collections will show whether these differences can be maintained.

23. **Hibiscus debeerstii** De Wild. & Dur. in Bull. Soc. Bot. Belg. **38**: 21 (1899).— Hochr. in Ann. Conserv. Jard. Bot. Genève, **4**: 91 (1900).—Cufod. in Ann. Naturh. Mus. Wien, **56**: 36 (1948) pro parte excl. *H. micranthus* var. *macranthus* et specim. Rhod.—Garcia in Bol. Soc. Brot., Sér. 2, **20**: 39 (1946). TAB. **90** fig. C. Type from the Belgian Congo.

Perennial herb up to 1·5 m. tall; stems with fuscous-stellate hairs and a finer pale green stellate pubescence. Leaf-lamina 2–6 × 0·3–2 cm., lanceolate to linear-oblong, stellate-pubescent to stellate-pilose on both surfaces, somewhat scabrid, apex acute, margin serrate, base obtuse to rounded 3–5-nerved; petiole up to 15 mm. long; stipules 4–6 mm. long, filiform. Flowers 1·5–3 cm. in diam., red, solitary, axillary and forming racemes or panicles at the apices of the stems by reduction of the upper leaves; peduncle up to 3·5 cm. long, hairy like the stems, articulated 5–8 mm. below the apex. Epicalyx of 6–7 (10) bracts, equalling or exceeding the calyx; bracts 4–7 mm. long, filiform. Calyx up to 5 mm. long, fuscous-pilose; lobes 4–5 × 2–2·5 mm., ovate-lanceolate. Petals 12–18 mm. long, obovate. Staminal tube 3 mm. long; free parts of filaments 1–2 mm. long. Style-branches 2–5 mm. long. Capsule 8–10 × 10–12 mm., depressed-subglobose, puberulous. Seeds with white silky floss.

N. Rhodesia. N: Abercorn Distr., 1220–1520 m., fl. & fr. 22.vii.1931, *Gamwell* 59 (BM). **Nyasaland.** N: Katowa, fr. 5.x.1952, *Chapman* 26 (BM). **Mozambique.** N: Mecaloja, Posto de Sanga, fr. 28.i.1934, *Torre* 433 (BM; COI; K; LISC). T: Zobué, fl. & fr. 24.viii.1943, *Torre* 5785 (LISC; SRGH). Also in Tanganyika and the Belgian Congo. River banks.

24. **Hibiscus nyikensis** Sprague in Kew Bull. **1908**: 56 (1908).—Cufod. in Ann. Naturh. Mus. Wien, **56**: 52 (1948). Syntypes: Nyasaland, Nyika Plateau, *Whyte* 226 (K); between Mpata and the Tanganyika Plateau, *Whyte* (K). *Whyte* 226 (K) is chosen as lectotype.

Perennial herb 1–1·5 m. tall; stems with somewhat appressed-fuscous-stellate hairs. Leaf-lamina up to 10 × 2·5 cm., narrowly oblong-lanceolate, stellate-pubescent on both surfaces, apex acute, margin serrate, base truncate to slightly cordate 5-nerved; petiole up to 2·5 cm. long; stipules 3 mm. long, filiform. Flowers 2·5–3·5 cm. in diam., scarlet, solitary, axillary, and forming racemes at the apices of the stems by reduction of the upper leaves; peduncle up to 3·5 cm. long. Epicalyx of 7–8 bracts; bracts 3–6 mm. long, filiform. Calyx up to 8 mm. long, fuscous-pilose; lobes up to 8 × 2·5 mm., lanceolate, acute. Petals 15–18 mm. long, obovate, with scattered stellate hairs. Staminal tube 5–6 mm. long; free parts of filaments 1–2 mm. long. Style-branches 4·5–6 mm. long. Capsule 8–9 mm. in diam., puberulous. Seeds nearly glabrous.

N. Rhodesia. N? " South of Lake Tanganyika ", fl. & fr. 1914, *Clark* (BM). **Nyasaland.** N: Nyika Plateau, 1800–1200 m., *Whyte* 226 (K). Also in Tanganyika (Mt. Mbeya, *MacInnes* 244 (BM)). Upland grasslands, 1300–2100 m.

25. **Hibiscus micranthus** L.f., Suppl. Pl.: 308 (1781).—Mast. in Oliv., F.T.A. **1**: 205 (1868).—Hochr. in Ann. Conserv. Jard. Bot. Genève, **4**: 82 (1900) pro parte excl. specim. *Rand* 438 et 591.—Burtt Davy, F.P.F.T. **2**: 282 (1932).—Exell & Mendonça, C.F.A. **1**, 1: 163 (1937).—Cufod. in Ann. Naturh. Mus. Wien, **56**: 46 (1948).—Wild, Guide Fl. Vict. Falls: 147 (1953).—Keay, F.W.T.A. ed. 2, **1**, 2: 346 (1958). TAB. **90** fig. D. Type from India.

Virgate shrub up to c. 2·5 m. tall; stems stellate-setulose, hairs somewhat projecting giving a slightly rough appearance. Leaf-lamina 1–5 × 0·5–2·5 cm., narrowly elliptic, oblong-elliptic to lanceolate-elliptic or suborbicular, stellate-hispid on both surfaces, apex rounded to acute, margin serrate, base obtuse to rounded; petiole 1–5 mm. long; stipules 3–5 mm. long, setaceous. Flowers up to c. 1 cm. in diam. (small for the genus), white, pinkish or purplish, solitary, axillary, and forming racemes or panicles at the apices of the stems by reduction of the upper leaves; peduncle up to 2 cm. long, articulated near the middle. Epicalyx of 5–7 bracts; bracts 1·5–3 mm. long, filiform. Calyx up to 4·5 mm. long, setulose; lobes up to 4 × 2 mm., lanceolate, acute. Petals 5–8 mm. long, obovate, pubescent outside, glabrous inside. Staminal tube 2–2·5 mm. long; free parts of filaments up to 0·5 mm. long. Style-branches 1–1·5 mm. long. Capsule 7–8 × 8–12 mm., depressed-globose, minutely pubescent. Seeds with white silky floss.

Bechuanaland Prot. N: Francistown, fl. ii.1926, *Rand* 59 (BM). SE: Kanye Distr., Pharing, fl. x.1949, *Miller* B/974 (PRE). **N. Rhodesia.** S: Victoria Falls, fl. 24.xi.1949, *Wild* 3199 (K; SRGH). **S. Rhodesia.** N: near Binga, fl. 6.xi.1958, *Phipps* 1373 (SRGH). W: Wankie, 915 m., fl. x–xi.1935, *Levy* 160 (PRE). E: Melsetter Distr., Hot Springs, fr. 23.vii.1943, *Hopkins* in GHS 10275 (SRGH). S: Sabi-Lundi Junction, 240 m., *Wild* 3458 (BM; SRGH). **Nyasaland.** "Lukoma, Lake Nyasa", *Bellingham* (BM). **Mozambique.** T: Tete, fl. & fr. ii.1859, *Kirk* (K). SS: between Guijá and Mabalane, fl. & fr. 3.vi.1959, *Barbosa & Lemos* in *Barbosa* 8575 (LMJ; K). LM: Lourenço Marques, fl. & fr. 12.iii.1947, *Pedro & Pedrógão* 126 (LMJ).
Tropical and S. Africa, Arabia and India. Woodlands, often ruderal.

For a subdivision of this polymorphic species into varieties see Cufodontis (loc. cit.). Garcke (in Peters, Reise Mossamb. Bot. **1**: 127 (1861)) cited several specimens (now destroyed) under *H. micranthus* but it is uncertain whether any or all of them belong to this species. *H. micranthus* has been treated by several authors in a very broad sense including in it several taxa (even the very distinct *H. rhodanthus*) here considered to be separate species.

26. **Hibiscus sabiensis** Exell in Bol. Soc. Brot., Sér. 2, **33**: 169 (1959). TAB. **90** fig. E. Type: S. Rhodesia, lower Sabi R., *Wild* 2312 (SRGH, holotype).
Hibiscus micranthus var. *hispidus* Cufod. in Ann. Naturh. Mus. Wien, **56**: 48 (1948) pro parte excl. specim. *Dinter* 83. Syntypes from the Transvaal and SW. Africa; lectotype *Schlechter* 4591 from the Transvaal.

Perennial herb up to 100 cm. tall; young stems patent-stellate-hispid, hairs 1–1·5 mm. long. Leaf-lamina up to 5 × 3 cm., ovate, somewhat sparsely to rather densely stellate-hispidulous on both surfaces, apex rather blunt to somewhat acute, margin serrate, base rounded 5–7-nerved; petiole up to 18 mm. long; stipules 5–7 mm. long, subulate, stellate-hispid. Flowers up to 1 cm. in diam. (small for the genus), white or pinkish, solitary, axillary; peduncle up to 7 mm. long, not conspicuously articulated. Epicalyx of 5–6 bracts; bracts 2 mm. long, filiform. Calyx 3–3·5 mm. long; lobes 3 × 1·5 mm., ovate-triangular, shortly connate at the base. Petals 5 × 2·5 mm., obovate-elliptic, stellate-setulose outside, glabrous inside. Staminal tube 2·5 mm. long; free parts of filaments less than 0·5 mm. long. Style-branches 1·5 mm. long. Capsule 6 mm. in diam., subglobose, minutely pubescent. Seeds with white silky floss.

S. Rhodesia. E: Lower Sabi, east bank, fl. & fr. 28.i.1948, *Wild* 2312 (SRGH).
Also in the Transvaal. In *Colophospermum mopane* woods on alluvium.

27. **Hibiscus migeodii** Exell in Journ. of Bot. **68**: 83 (1930).—Cufod. in Ann. Naturh. Mus. Wien, **56**: 52 (1948). Type from Tanganyika.
Hibiscus micranthus sensu Bak. f. in Journ. Linn. Soc., Bot. **40**: 28 (1911) pro parte quoad specim. *Swynnerton* 2052.—Garcia in Bol. Soc. Brot., Sér. 2, **20**: 39 (1946).
Hibiscus sp.—Garcia, tom. cit.: 41 (1946) quoad specim. Carvalho.

Hibiscus zanzibaricus Cufod., tom. cit.: 58 (1948) pro parte quoad specim. Mozamb.

Annual or perennial herb up to c. 1 m. tall; stems ± patent-stellate-setose with additional smaller simple or branched hairs. Leaf-lamina up to 7 × 5 cm., broadly ovate to narrowly oblong, sometimes hastate, with a tendency to become shallowly or deeply 3-lobed, sparsely to densely stellate-setose on both surfaces, apex usually acute, margin irregularly bluntly toothed, base rounded to cordate 5–7-nerved; petiole up to 4·5 cm. long; stipules 2–4 mm. long, setaceous. Flowers up to 1·5 cm. in diam., red, solitary, axillary and forming terminal racemes or panicles by reduction of the upper leaves; peduncle up to 3·5 cm. long, articulated 3–5 (10) mm. below the apex. Epicalyx of 6–7 bracts; bracts 2–3 mm. long, linear. Calyx 6 mm. long; lobes 4–5 mm. long, lanceolate. Petals 8–9 mm. long, obovate, stellate-setose outside, glabrous inside. Staminal tube 4–5 mm. long; free parts of filaments 0·5–1 mm. long. Style-branches 2–2·5 mm. long. Capsule 6–8 × 8–9 mm., subglobose, minutely pubescent. Seeds with a white silky floss.

Nyasaland. N: 20 km. W. of Karonga, Nyanja Hill, 640 m., fl. 14.iv.1953, *Williamson* 247 (BM). **Mozambique.** N: Metangula, fl. & fr. 25.v.1948, *Pedro & Pedrógão* 3904 (LMJ). Z: between Mocuba and Régulo Mataia, fr. 22.v.1949, *Barbosa & Carvalho* 2891 (LM). MS: Vila Machado, R. Mucuzi, fl. & fr. 23.iv.1948, *Mendonça* 4032 (BM; LISC). SS: Inhambane, Mongwe, fl. 20.vii.1954, *Eccles* 7 (BM).
Also in Tanganyika. Open woodland, wooded grassland and roadsides, up to 650 m.

28. **Hibiscus meyeri** Harv. in Harv. & Sond., F.C. **1**: 173 (1861). Syntypes from S. Africa.
    Hibiscus microphyllus E. Mey. in Drège, Zwei Pflanz.-Docum.: 158, 192 (1843) *nom. nud.*, non *H. microphyllus* Vahl (1790). Type from S. Africa.
    Hibiscus mendoncae Exell in Bol. Soc. Brot., Sér. 2, **33**: 171 (1959). Type: Mozambique, Lourenço Marques, Maputo, Goba, *Mendonça* 3019 (LISC, holotype).

Subsp. **meyeri.** TAB. **90** fig. A.

Shrublet or perennial herb up to 2–2·5 m. tall: branchlets appressed-stellate-hispid. Leaf-lamina 1–3 × 0·8–2·5 cm., suborbicular to broadly ovate in outline, often 3-lobed, sparsely stellate-pubescent or stellate-hispid, apex acute to rounded, margin serrate, base obtusely cuneate to nearly truncate 3–5-nerved; petiole up to 8 mm. long; stipules 2–4 mm. long. Flowers 1·5–2·5 cm. in diam., white or pinkish or opening white and turning pink or purplish, solitary, axillary; peduncle up to 3·5 cm. long, stellate-pubescent, articulated 7–10 mm. below the apex. Epicalyx of usually 7 bracts; bracts 2–3 mm. long, filiform. Calyx up to 6·5 mm. long; lobes 5 × 2 mm., elliptic-triangular, acute, pubescent, joined at the base for c. 1·5 mm. Petals 15 × 10 mm., reflexed, obovate, stellate-pilose outside, glabrous inside. Staminal tube 6–8 mm. long; free parts of filaments 0·5 mm. long, ± in 4 whorls. Style-branches 2–3 mm. long. Capsule 10 mm. in diam., subglobose, minutely pubescent. Seeds with a white or slightly brownish silky floss.

Mozambique. LM: Goba, near the bridge over the Umbeluzi, Maputo, fl. 18.iii.1945, *Sousa* 106 (LISC).
Also in Zululand and Natal. In *Androstachys* woodland and on cultivated ground.
Subsp. *transvaalensis* (Exell) Exell* (tom. cit.: 172 (1959)) is fairly widespread in the Transvaal; it is less densely hairy and the bracts of the epicalyx are usually longer.

29. **Hibiscus praeteritus** R. A. Dyer, Fl. Pl. S. Afr. **11**: t. 436 (1931).—Burtt Davy, F.P.F.T. **2**: 282 (1932).—Cufod. in Ann. Naturh. Mus. Wien, **56**: 54 (1948).—Brenan in Mem. N.Y. Bot. Gard. **8**, 3: 224 (1953).—Wild, Guide Fl. Vict. Falls: 147 (1953).—Martineau, Rhod. Wild Fl.: 50, t. 18 fig. 2 (1954). Type from the Transvaal (Soutpansberg).
    Hibiscus micranthus sensu Bak. f. in Journ. Linn. Soc., Bot. **40**: 28 (1911) pro parte quoad specim. *Swynnerton* 2051.
    Hibiscus hirtus sensu Ulbr. & Fr. in R.E.Fr., Wiss. Ergebn. Schwed. Rhod.-Kongo-Exped. **1**: 146 (1914).
    Hibiscus hirtus subsp. *africanus* Cufod., tom. cit.: 44 (1948).

* **Hibiscus meyeri** subsp. **transvaalensis** (Exell) Exell, comb. nov. (*H. mendoncae* subsp. *transvaalensis* Exell in Bol. Soc. Brot., Sér. 2, **33**: 172 (1959)).

Virgate perennial herb up to 2 m. tall; stems appressed-stellate-setulose. Leaf-lamina up to 6 × 3 cm., ovate to narrowly lanceolate, sparsely to fairly densely stellate-hispid on both surfaces, apex acute to rather blunt, margin serrate, base obtuse to rounded or truncate 3–5-nerved; petiole up to 3 cm. long; stipules 5 mm. long, subulate. Flowers 2–3 cm. in diam., red or pink? (perhaps hybrids), solitary, axillary; peduncle up to 8 cm. long, articulated 4–5 mm. below the apex. Epicalyx of 6–7 bracts; bracts 5–10 mm. long, linear. Calyx up to 9–10 mm. long, stellate-setulose; lobes up to 8–9 × 1·5 mm., linear-lanceolate. Petals 10–20 mm. long, narrowly obovate, sparsely stellate-setose outside, glabrous inside. Staminal tube 7–12 mm. long; free parts of filaments 0·5–1 mm. long. Style-branches 7–15 mm. long. Capsule 9 × 9 mm., subglobose, minutely pubescent. Seeds with white silky floss.

Bechuanaland Prot. N: Ngamiland, *Curson* 766 (PRE). **S. Rhodesia.** W: Victoria Falls, 850 m., fl. iii.1918, *Eyles* 1273 (BM; SRGH). E: Melsetter Distr., Cashel, fl. 13.i.1951, *Crook* 329 (BM; SRGH). **Nyasaland.** S: Chikwawa, *Banda* 129 (BM). **Mozambique.** T: Kabankangywa Kraal, fl. & fr. 22.ix.1948, *Wild* 2589 (K; SRGH). MS: Chirinda, 60 m., fl. & fr. 19.xii.1906, *Swynnerton* 2051 (BM).
Also in S. Angola and S. Africa. Margins of rivers.
The specimens cited under *H. hirtus* L. by Garcke (in Peters, Reise Mossamb. Bot. **1**: 127 (1861)), now destroyed, were perhaps *H. praeteritus*.

30. **Hibiscus gwandensis** Exell in Bol. Soc. Brot., Sér. 2, **33**: 174 (1959). Type: S. Rhodesia, Marungudzi, *Drummond* 5754 (BM, holotype; SRGH).

Erect shrublet or perennial herb; branchlets subpatent-stellate-setulose. Leaf-lamina up to 3 × 2·5 cm. (and probably larger), ovate, stellate-setulose above, stellate-hispidulous below, apex acute, margin rather coarsely serrate (for the size of the lamina), base obtuse to truncate 5-nerved; petiole up to 11 mm. long; stipules 4 mm. long. Flowers 1·5–2 cm. in diam., white, fading to red, in 2–5-flowered cymose inflorescences, axillary; peduncle up to 1 cm. long, rather inconspicuously articulated near the apex. Epicalyx of 7–9 bracts; bracts 2·5–4·5 mm. long, subulate to filiform. Calyx 6–8 mm. long; lobes 4–6 × 1·5–2 mm., ovate-lanceolate to elongate-triangular, slightly joined at the base. Petals 12–15 mm. long, obovate, sparsely stellate outside, glabrous inside. Staminal tube 6 mm. long; free parts of filaments 0·5 mm. long, ± in 4 whorls. Style-branches 5 mm. long. Capsule 8 × 9 mm., subglobose, pubescent. Seeds immature, probably producing a floss.

S. Rhodesia. S: Marungudzi, on rim of ancient volcano, 700 m., fl. & fr. 10.v.1958, *Drummond* 5754 (BM; SRGH).
Known only from the locality cited. Among syenite rocks.

Near *H. meyeri* and *H. okavangensis* but geographically far removed from both and differing in the length of the style-branches.

31. **Hibiscus aponeurus** Sprague & Hutch. in Kew Bull. **1908**: 54 (1908).—Sprague in Curt. Bot. Mag.: t. 8231 (1908).—Cufod. in Ann. Naturh. Mus. Wien, **56**: 32 (1948). Syntypes from Kenya and Tanganyika of which *Grant* 215 (K) from Tanganyika is chosen as lectotype.

Shrublet or perennial herb up to 1 m. tall; stems somewhat scabrous, stellate-setulose, hairs mainly appressed but often somewhat patent. Leaf-lamina 2–4 × 2–3 cm., ovate to elliptic with a tendency to become 3-lobed, densely stellate-pubescent or stellate-pilose, apex rounded or obtuse, margin serrate, base rounded to obtuse 5-nerved; petiole 1–2 cm. long; stipules 5–9 mm. long, acicular. Flowers 1·5–2 cm. in diam., red or pink, solitary, axillary; peduncle up to 3 cm. long (often much shorter). Epicalyx of 8–13 bracts; bracts 3·5–8 mm. long, linear-subulate. Calyx up to 9 mm. long, hispid; lobes 5–8 × 1·5–3 mm., triangular to narrowly triangular-lanceolate. Petals 13–15 mm. long, narrowly obovate, stellate-setose outside, glabrous inside. Staminal tube 7–10 mm. long; free parts of filaments 0·5–1 mm. long. Style-branches 4–5 mm. long. Capsule 8–12 × 8–12 mm., ellipsoid to subglobose, pubescent. Seeds with white silky floss.

Mozambique. N: Serra de Ribauè, fl. & fr. 15.x.1948, *Pedro & Pedrógão* 5514 (K; LMJ).

Sudan, Uganda, Kenya, Belgian Congo, Tanganyika and N. Mozambique. Bush and grassland.

Specimens from S. Mozambique and the Transvaal which have been named *H. aponeurus* are mostly *H. barbosae* (q.v.).

32. Hibiscus richardsiae Exell in Bol. Soc. Brot., Sér. 2, **33**: 176 (1959). Type: N. Rhodesia, Lake Tanganyika, Sumba, fl. & fr. 5.iv.1957, *Richards* 9033 (BM; K, holotype).

Perennial herb c. 0·6 m. tall; stems erect, ± patent-stellate-pubescent. Leaf-lamina 1–2·5 × 0·8–1·5 cm., ovate, ± densely pubescent above, rather sparsely setose below, hairs 2–3-pronged, apex acute, margin serrate, base obtuse or rounded; petiole up to 1·5 cm. long; stipules 3–5 mm. long, brown, subulate. Flowers c. 2·5 cm. in diam., red, solitary, axillary and forming corymbose clusters at the apices of the stems; peduncles up to 3–4 cm. long, articulated near the apex. Epicalyx of 8–9 bracts; bracts 3–5 mm. long, ciliate. Calyx 10–11 mm. long, pilose outside, glabrous inside; lobes up to 10 × 3 mm., lanceolate, acute. Petals 1·5–2 × 1–1·3 cm., obovate, stellate-pilose outside, glabrous inside. Staminal tube 9–10 mm. long; free parts of filaments 1–1·5 mm. long. Style-branches 6·5 mm. long. Capsule 9–10 mm. in diam., subglobose, minutely pubescent. Seeds (immature) with a floss.

N. Rhodesia. N: Mporokoso Distr., Lake Tanganyika, Sumba, escarpment road, fl. & fr. 5.iv.1957, *Richards* 9033 (BM; K).
Also in Tanganyika (Ufipa Distr.). Among rocks (N. Rhodesia) and in dense bush (Tanganyika).

Very close to *H. aponeurus* but strikingly different when dried, having a silvery-grey appearance in contrast with the tawny indumentum of *H. aponeurus* and more closely resembling *H. praeteritus* from which it differs by its more patent indumentum and more densely pubescent leaves.

33. Hibiscus mutatus N.E.Br. in Kew Bull. **1906**: 99 (1906).—Eyles in Trans. Roy. Soc. S. Afr. **5**: 415 (1916).—Arwidss. apud Norlindh & Weim. in Bot. Notis. **1934**: 96 (1934).—Cufod. in Ann. Naturh. Mus. Wien, **56**: 52 (1948).—Martineau, Rhod. Wild Fl.: 51 (1954). Type: S. Rhodesia, Matopos, *Cecil* 108 (K, holotype).
Hibiscus micranthus forma.—Gibbs in Journ. Linn. Soc., Bot. **37**: 431 (1906).
Hibiscus micranthus sensu Eyles in Trans. Roy. Soc. S. Afr. **5**: 415 (1916) pro parte.

Shrub up to 2–2·5 (5) m. tall; branchlets ± patent-stellate-hispid, some hairs projecting to c. 0·5 mm. Leaf-lamina up to 3·5 × 3 cm., elliptic or ovate to sub-orbicular in outline with tendency to become 3-lobed, rather densely stellate-hispid on both surfaces, apex obtuse or rounded, margin irregularly crenate-serrate, base obtuse to slightly cordate 5–7-nerved; petiole 4–15 mm. long; stipules 4 mm. long, acicular. Flowers 2–2·5 cm. in diam., opening pure white and turning rose-pink, solitary, axillary; peduncle up to 3 cm. long, articulated 5 mm. below the apex. Epicalyx of 5–7 bracts; bracts 4–7 mm. long, linear. Calyx up to 10 mm. long, hispid; lobes 5–8 × 1–2 mm., linear-triangular. Petals 1–2 cm. long, obovate to elliptic, stellate-setose outside, glabrous inside. Staminal tube 6–8 mm. long; free parts of filaments 0·5 mm. long. Style-branches 5–7 mm. long. Capsule 8–9 × 8–10 mm., subglobose, pubescent. Seeds with a white silky floss.

Bechuanaland Prot. N: between Tutumi and Sebena, fl. 14.xii.1925, *Pole Evans* 2606 (PRE). SE: Pharing, 900 m., fl. x.1946, *Miller* B/466 (PRE; SRGH). **S. Rhodesia.** W: Matopos, World's View, fl. & fr. 13.iv.1955, *E.M. & W.* 1486 (BM; LISC; SRGH); Wankie, 1370 m., fl. iv.1953, *Davies* 527 (SRGH). S: Belingwe Distr., Mnene, fl. & fr. 26.ii.1931, *Norlindh & Weimarck* 5157 (BM; PRE). **Nyasaland.** S: Fort Johnston, fl. 21.iv.1955, *Jackson* 1629 (K).
Also in SW. Africa (Okovango Native Territory). Open woodland and bush.

This beautiful shrub with pure white flowers turning pink with age is common near Rhodes's Grave. Martineau's statement (loc. cit.) that the flowers open pink and fade to white is incorrect.
The specimens from Pharing and Wankie are both poor and their identification somewhat uncertain. The single specimen recorded from Nyasaland (*Jackson* 1629) is from a bush up to 5 m. tall, according to the collector, and has less distinctly lobed leaves but seems to be identical in other respects.

34. **Hibiscus rhodanthus** Gürke apud Schinz in Bull. Herb. Boiss. **3**: 405 (1895).—Hochr. in Ann. Conserv. Jard. Bot. Genève, **4**: 91 (1900).—Ulbr. & Fr. in R.E.Fr., Wiss. Ergebn. Schwed. Rhod.-Kongo-Exped. **1**: 146 (1914).—Engl., Pflanzenw. Afr. **3, 2**: 395 (1921).—Exell & Mendonça, C.F.A. **1, 1**: 164 (1937).—Garcia in Bol. Soc. Brot., Sér. 2, **20**: 39 (1946).—Cufod. in Ann. Naturh. Mus. Wien, **56**: 26 (1948).—Brenan in Mem. N.Y. Bot. Gard. **8**, 3: 224 (1953).—Martineau, Rhod. Wild. Fl.: 51 (1954). FRONTISPIECE *to Part* 2. Type from Angola (Malange).

Hibiscus carsonii Bak. in Kew Bull. **1897**: 244 (1897). Syntypes: Nyasaland, Fort Hill, *Whyte* (K); N. Rhodesia, Fwambo, *Carson* (K). The latter is chosen as lectotype.

Hibiscus micranthus var. *sanguineus* (Franch.) Hochr., tom. cit.: 83 (1900) pro parte quoad specim. *Rand* 438 et 591.

Hibiscus micranthus forma *macranthus* Bak. f. in Journ. of Bot. **37**: 424 (1899).—Eyles in Trans. Roy. Soc. S. Afr. **5**: 415 (1916).

Hibiscus debeerstii sensu Cufod. in Ann. Naturh. Mus. Wien, **56**: 36 (1948) pro parte quoad synon. et specim. Rhod.

Perennial herb 5 cm.–1 m. tall, producing annual shoots from a woody rootstock; stems stellate-setose or stellate-setulose. Leaf-lamina 2–13 × 0·8–3 cm., oblong or oblong-elliptic to very narrowly oblong, sparsely to fairly densely stellate-setose, apex acute or rounded, margin rather distantly serrate, base cuneate to rounded usually 3-nerved; petiole 1–14 mm. long; stipules 2–4 mm. long, acicular. Flowers 2·5–4 cm. in diam., red, solitary, axillary and often forming terminal racemes or panicles by reduction of the upper leaves; peduncle up to 5 cm. long, stellate-hispid, rather inconspicuously articulated usually near the middle. Epicalyx of 6–7 bracts; bracts 1–4 mm. long, linear to subulate. Calyx up to 8 mm. long, stellate-hispid outside, glabrous inside; lobes 3–6 × 1·5–3 mm., lanceolate to triangular, slightly connate or joined nearly to half-way. Staminal tube 6–8 mm. long; free parts of filaments 1·5–4 mm. long. Style projecting up to 10 mm. beyond the staminal tube, branches 3·5–4·5 mm. long. Capsule 8–10 × 10–12 mm., subglobose, usually densely pubescent. Seeds with a white silky floss.

N. Rhodesia. B: Sesheke Distr., *Macaulay* 88 (K). N: Abercorn Distr., road to Kalambo Falls, fl. 29.vii.1951, *Nash* 7A (BM). W: Mwinilunga Distr., above Kalenda dambo, fl. 8.x.1937, *Milne-Redhead* 2641 (BM; K). C: 16 km. S. of Lusaka, fl. 6.iv.1955, *E.M. & W.* 1411 (BM; LISC; SRGH). E: Nyika Plateau, 2130 m., fl. 24.ix.1956, *Benson* NR 162 (BM). S: between Livingstone and Kalomo, fl. & fr. 10.vii.1930, *Pole Evans* 2802 (PRE; SRGH). **S. Rhodesia.** N: Mazoe, fl. viii.1905, *Eyles* 183 (BM; SRGH). C: Salisbury, fl. ix.1898, *Rand* 591 (BM). **Nyasaland.** N: Nyika Plateau, 2290 m., fl. & fr. 10.x.1947, *Benson* 1438 (BM). C: near Lilongwe, fl. 5.x.1952, *Kantikana* 8 (BM). **Mozambique.** N: Vila Cabral, fl. viii–x.1934, *Torre* 563 (BM; COI; LISC). T: between Furancungo and Vila Coutinho, fl. 29.ix.1942, *Mendonça* 165 (LISC).

Also in Angola and Tanganyika. Open woodland and grassland from 1150–2300 m.

35. **Hibiscus barbosae** Exell in Bol. Soc. Brot., Sér. 2, **33**: 174 (1959). Type: Mozambique, Namaacha, *Barbosa & Lemos* 7545 (LMJ, holotype).

Hibiscus aponeurus sensu Burtt Davy, F.P.F.T. **2**: 281 (1932) pro parte quoad specim. *Breyer* 17961.—Cufod. in Ann. Naturh. Mus. Wien, **56**: 32 (1948) pro parte quoad distrib. Transv.

Perennial herb up to 1·5 m. tall, producing annual shoots from a woody rootstock; stems somewhat scabrous, densely stellate-setose. Leaf-lamina 1·5–6 × 0·6–4 cm., ovate to narrowly oblong, ± deeply 3-lobed, densely stellate-hispid on both surfaces, apex obtuse or rounded, margin irregularly crenate-serrate, base usually truncate 5–7-nerved; petiole 3–4·5 cm. long; stipules 2 mm. long, acicular. Flowers 3–3·5 cm. in diam., red, solitary, axillary and forming terminal racemes by reduction of the upper leaves. Epicalyx of 5–7 bracts; bracts 4–5 mm. long, linear. Calyx 12 mm. long, densely pilose; lobes 8–10 × 2·5–3·5 mm., lanceolate, acute. Petals 2 × 1·5–1·8 cm., stellate-pilose outside, glabrous inside. Staminal tube 5–6 mm. long; free parts of filaments 2–4 mm. long. Style-branches 4–4·5 mm. long. Capsule 8–10 × 8–10 mm., subglobose, pubescent. Seeds with a white silky floss.

Mozambique. LM: between Maputo and Goba, fl. 8.i.1947, *Pedro & Pedrógão* 404 (LMJ; PRE).

Also in the Transvaal. Grassland and fallow land, up to c. 600 m.

Geographically separated from *H. aponeurus* (extending from Kenya to the Niassa Province of Mozambique), with which it has been misidentified and easily distinguishable from it by the much longer free parts of the filaments.

36. **Hibiscus allenii** Sprague & Hutch. in Kew Bull. **1907**: 45 (1907).—Eyles in Trans. Roy. Soc. S. Afr. **5**: 414 (1916).—Cufod. in Ann. Naturh. Mus. Wien, **56**: 32 (1948).—Wild, Guide Fl. Vict. Falls: 147 (1953). TAB. **90** fig. F. Syntypes: Victoria Falls, *Allen* 103 (K) and 113 (K) of which *Allen* 103, already labelled " type " at Kew, is confirmed as the lectotype.

Shrublet or perennial herb; stems stellate-hispid or strigose with yellowish patent simple or branched hairs. Leaf-lamina up to 8 × 8 cm., ovate to sub-orbicular in outline, usually deeply 3–5-lobed, strigose to hispid above, stellate-hispidulous beneath, apex acute, base obtuse to truncate; lobes narrowly oblong lanceolate to elliptic, margins rather coarsely serrate; petiole up to 3·5 cm. long; stipules 4–8 mm. long, filiform. Flowers 1·5–2 cm. in diam., red, solitary, axil-lary; peduncle up to 4·5 cm. long, articulated above the middle. Epicalyx of 6–7 bracts; bracts 2–4 mm. long, narrowly linear. Calyx up to 9 mm. long, stellate-hispidulous; lobes 7–8 × 1·5–2 mm., narrowly triangular, acute. Staminal tube 7–9 mm. long; free parts of filaments 0·5 mm. long. Style-branches 3–5 mm. long. Capsule 8–9 × 9–10 mm., subglobose, pubescent. Seeds glabrous or minutely pubescent.

Caprivi Strip. Mpilila I., fl. 13.i.1959, *Killick & Leistner* 3356 (PRE; SRGH).

Bechuanaland Prot. N : Ngamiland, fl. & fr. 1930, *Curson* 416 (PRE). **N. Rhodesia.** B : Sesheke Distr., fl. & fr. rec. 1911, *Macaulay* 427 (K). S : Mazabuka, 72 km. SE. of Choma, 610 m., fr. 17.xii.1956, *Robinson* 1778 (K; SRGH). **S. Rhodesia.** N : Urungwe Distr., near Chavaru R., fl. & fr. 13.iv.1956, *Phelps* 151 (BM; PRE; SRGH). W : Bulawayo Distr., fl. & fr. i.1951, *Orpen* 20/51 (PRE; SRGH). C : Marandellas Distr., between Sabi R. and Macheke R., fl. & fr. 8.i.1950, *Munch* 219 (K; SRGH). E : Melsetter Distr., Hot Springs, 610 m., fl. & fr. 9.xii.1951, *Chase* 4242 (BM; SRGH). **Nyasaland.** " Lake Nyasa ", fl. & fr. 1876, *Simons* (BM). **Mozambique.** N : 32 km. E. of Mandimba, fl. & fr. 18.iv.1942, *Hornby* 3739 (K; PRE). MS : Gorongoza, fl. & fr. 11.x.1946, *Simão* 1053 (LMJ).
Also in S. Tanganyika. Woodland and grassland up to c. 1400 m.

Series 7. (*Calyphylli; Ketmia* pro parte)

Epicalyx of 5 bracts.

Leaves glabrous or nearly so (sometimes with a few bristles on the nerves beneath) coarsely
    serrate or crenate-serrate; bracts of epicalyx lorate or linear   -   37. *dongolensis*
Leaves pubescent or setulose and pubescent at least on the undersurface:
    Calyx-lobes ± lorate or caudate at the apex, joined to about half-way; bracts of epi-
      calyx elongate-triangular to linear-triangular, usually broadest at or near the base
                                                       38. *lunarifolius*
    Calyx-lobes ovate to lanceolate:
      Bracts of the epicalyx variously shaped but usually broadened somewhere near the
        middle (see TAB. **89** fig. 7) and narrowed towards the base:
        Leaves ± concolorous (at least when dried), sparsely to densely pubescent or
          stellate-setulose below, indumentum rarely entirely covering the surface, sub-
          orbicular to broadly ovate in outline, 1–1½ times as long as broad, usually
          manifestly 3-lobed - - - - - - - - - 39. *calyphyllus*
        Leaves discolorous (at least when dried), greyish-tomentose below with the in-
          dumentum entirely covering the surface, ovate to narrowly ovate (rarely sub-
          orbicular), usually twice as long as broad, not or obscurely 3-lobed, margin
          sometimes entire - - - - - - - - 40. *platycalyx*
      Bracts of the epicalyx ovate-lanceolate to narrowly linear-lanceolate, broadest at or
        near the base - - - - - - - - 41. *ludwigii*

37. **Hibiscus dongolensis** Del., Cent. Pl. Afr.: 59 (1826).—Garcke in Peters, Reise Mossamb. Bot. **1**: 126 (1861).—Exell & Mendonça, C.F.A. **1**, 1 : 175 (1937).—Mendonça & Torre, Contr. Conhec. Fl. Moçamb. **1**: 15 (1950). Type from the Sudan.
    Hibiscus lunarifolius var. *dongolensis* (Del.) Hochr. in Ann. Conserv. Jard. Bot. Genève, **4**: 161 (1900). Type as above.

Shrub up to 2 m. tall or somewhat woody perennial herb; stems nearly glab-rous. Leaf-lamina up to 10 × 7 cm., ovate to narrowly ovate or ovate-lanceolate, glabrous or nearly so (with occasional hairs on the veins of the lower surface), rarely obscurely 3–5-lobed, apex acute or somewhat rounded, margin coarsely serrate, serrate, or crenate, base obtuse, truncate or slightly cordate; petiole up to 6 cm. long, usually nearly glabrous except for a longitudinal line of pubescence on the adaxial surface; stipules 5–6 mm. long, filiform to narrowly linear, usually glabrous. Flowers up to c. 8 cm. in diam., yellow with dark centre, solitary, axillary; peduncle up to 10 mm. long, glabrous or puberulous, not conspicuously articulated. Epicalyx of (usually) 5 bracts; bracts up to 20 × 2 mm., lorate, glabrous or nearly so, approximately equalling the calyx-lobes, joined for 1–2 mm. at the base. Calyx up to 20 mm. long, nearly glabrous; lobes ± lorate or caudate with 1 prominent central nerve, nearly glabrous, joined for up to c. 7 mm. at the base. Petals up to 7·7 × 4·5 cm., obovate, pubescent outside where not overlapping, densely tomentellous in bud. Staminal tube 20–25 mm. long; free parts of filaments 2·5–3 mm. long. Style-branches 3–4 mm. long. Capsule 15–20 mm. in diam., subglobose, stellate-setulose, valves gradually attenuated into awns up to 6 mm. long. Seeds 2·5 × 3 mm., subreniform, minutely stellate-puberulous.

Bechuanaland Prot. N: Ngami, Motlhatlogo, fl. 10.v.1930, *van Son* in Herb. Transv. Mus. 28938 (BM; PRE; SRGH). **N. Rhodesia.** S: Mazabuka, edge of Kafue Flats, fl. & fr. 7.iv.1955, *E.M. & W.* 1437 (BM; LISC; SRGH). **S. Rhodesia** W: Nyamandhlovu, fl. & fr. 25.i.1956, *Plowes* 1917 (BM; SRGH). E: Inyanga, near Cheshire, alt. 1300 m., fl. & fr. 15.i.1931, *Norlindh & Weimarck* 4351 (BM; PRE; SRGH). S: Beitbridge Distr., Chibeza Hot Springs Camp, fl. & fr. v.1955, *Davies* 1244A (BM; SRGH). **Mozambique.** T: Tete, *Peters* (fide Garcke, loc. cit.). LM: Magude, fl. & fr. 12.vii.1948, *Torre* 7984 (LISC).
Widespread in tropical and S. Africa. In woodland (especially *Colophospermum mopane*) and on alluvial flats.

The relationship between this species and *H. lunarifolius* requires study: there are some intermediates.

38. **Hibiscus lunarifolius** Willd. in L., Sp. Pl. ed. 4, **3**: 811 (1800).—Mast. in Oliv., F.T.A. **1**: 202 (1868) pro parte.—Hochr. in Ann. Conserv. Jard. Bot. Genève, **4**: 161 (1900) pro parte.—Burtt Davy, F.P.F.T. **2**: 281 (1932).—Martineau, Rhod. Wild Fl.: 50 (1954).—Keay, F.W.T.A. ed. 2, **1**, 2: 347 (1958). Type from India.

Erect herb about 1 m. tall; stems densely patent- or subappressed-stellate-setose and pubescent. Leaf-lamina up to 10 × 9 cm., ovate to suborbicular, shallowly or obscurely 3-lobed or not lobed, pubescent (at least on the nerves) above and often sparsely stellate-setose, sparsely to densely pubescent to stellate-setose below, sometimes tomentose when young, indumentum often ± floccose (perhaps due to some pathological condition), apex acute or somewhat rounded, margin crenate-serrate, base shallowly cordate (more rarely obtuse); petiole up to 8 cm. long, pubescent; stipules c. 5 mm. long, lorate to filiform. Flowers 3–5 cm. in diam., yellow (usually without a dark centre, at least in our area), solitary, axillary and often forming rather congested terminal racemes by reduction of the upper leaves; peduncle usually up to 5 mm. long, stellate-setose. Epicalyx of 5 bracts; bracts 16 × 3 mm., very narrowly lanceolate to linear-lanceolate, acute, joined for 1–1·5 mm. at the base. Calyx 5–6 mm. long, cupuliform, usually slightly shorter than the epicalyx; lobes very narrowly triangular with 1 prominent central nerve and broad sinuses. Petals 5–6 × 4–5 cm., obliquely obovate, tomentose outside when in bud. Staminal tube 15–20 mm. long; free parts of filaments 1–1·2 mm. long. Style-branches 2–3 mm. long. Capsule 15 × 12–14 mm., subglobose to broadly ellipsoid; awns 2·5–3 mm. long. Seeds 2·5 × 2·5 mm., angular subreniform, sparsely and minutely lepidote.

S. Rhodesia. W: Matopos, fl. iii.1918, *Eyles* 970 (BM; K; SRGH). C: Salisbury, fl. 30.iii.1924, *Eyles* 4342 (K). E: Umtali, Marakwa's Hill, 1430 m., fl. & fr. 6.iii.1952, *Chase* 4396 (BM; SRGH).
India and tropical Africa. Hillsides and rocky places.

39. **Hibiscus calyphyllus** Cav., Diss. Bot. **5**: 283, t. 140 (1788).—Hochr. in Ann. Conserv. Jard. Bot. Genève, **4**: 99 (1900).—Engl., Pflanzenw. Afr. **3**, 2: 397, fig. 186 U–W (1921).—Burtt Davy, F.P.F.T. **2**: 281 (1932).—Exell & Mendonça, C.F.A. **1**, 1: 174 (1937).—Mendonça & Torre, Contr. Conhec. Fl. Moçamb. **1**: 15 (1950).—Martineau, Rhod. Wild Fl.: 50 (1954). TAB. **89** fig. 7. Type from Réunion.

Hibiscus calycinus Willd. in L., Sp. Pl. ed. 4, **3**: 817 (1800) *nom. illegit.*—Harv. in
Harv. & Sond., F.C. **1**: 170 (1860).—Mast. in Oliv., F.T.A. **1**: 203 (1868).—
Eyles in Trans. Roy. Soc. S. Afr. **5**: 415 (1916). Type as above.
Hibiscus wildii Suesseng. in Proc. & Trans. Rhod. Sci. Ass. **43**: 103 (1951).
Type: S. Rhodesia, Marandellas, fl. 12.iii.1941, *Dehn* 202 (M, holotype).

Shrub or perennial herb up to 3 m. tall; stems tomentose or pubescent when
young, later glabrescent. Leaf-lamina up to 12 × 12 cm., suborbicular in outline,
obscurely or distinctly 3–5-lobed, stellate-pubescent or stellate-pilosulose above,
stellate-pubescent or stellate-pilose or stellate-tomentose beneath, apex acute,
margin serrate, base cordate or subcordate; petiole usually up to 9 (18) cm. long,
stellate-pilose; stipules up to 15 mm. long, filiform or subsetaceous, somewhat
expanded at the base. Flowers up to 12 cm. in diam., yellow usually with a brownish or dark red centre, solitary, axillary; peduncle c. 7 (10) mm. long, stellate-
pubescent, usually rather inconspicuously articulated near the base. Epicalyx
of 5 bracts, stellate-pubescent; bracts up to 18 mm. long, varying greatly in shape
but nearly always broadest near the middle then narrowed suddenly to a caudate
tip which is a prolongation of the midrib, joined for 3–4 mm. at the base. Calyx
up to 16 mm. long, stellate-tomentellous; lobes ovate to ovate-lanceolate, usually
3-nerved, joined to nearly half-way. Petals up to 6 × 4·5 cm., obliquely obovate,
pubescent outside, glabrous within. Staminal tube up to 15 mm. long; free parts
of filaments 1·5–3 mm. long. Style-branches 4–5 mm. long. Capsule 25 × 15
mm., ellipsoid; valves aristate. Seeds 3 × 2·5 mm., subreniform, tomentellous.

Bechuanaland Prot. N: Ngamiland, fl., *Curson* 257 (PRE). SE: Pharing, 1200 m.,
fl. v.1948, *Miller* B/610 (PRE). **N. Rhodesia.** N: Chilongowelo, 1490 m., fl. & fr.
4.iv.1952, *Richards* 1539 (K). W: Ndola, fr. 3.vi.1954, *Fanshawe* 1264 (K). C:
Lusaka, fl. 7.iii.1957, *Noak* 146 (BM; SRGH). **S. Rhodesia.** N: Trelawney Distr.,
fr. 29.iii.1944, *Jack* 241 (SRGH). W: Bulalima-Mangwe, fl. & fr. 6.v.1942, *Feiertag* in
GHS 45502 (SRGH). C: Maidstone, near Rusape, c. 1450 m., fl. 20.ii.1931, *Norlindh
& Weimarck* 5118 (BM). E: Umtali Distr., near Colonel Peacock's Farm, fl. 21.ii.1954,
Chase 5195 (BM; SRGH). S: 13 km. SE. of Tuli, fl. & fr. 22.iii.1959, *Drummond* 5918
(SRGH). **Nyasaland.** N: Mbawa, fl. & fr. 7.iv.1955, *Jackson* 1616 (SRGH). S:
Ncheu Distr., Bawi, 750 m., fl. & fr. 16.iii.1955, *E.M. & W.* 924 (BM; LISC; SRGH).
Mozambique. MS: Chimoio, Vila Pery, Monte Chizombero, fl. & fr. 26.iii.1948,
Garcia 740 (BM; LISC). SS: Guijà, between Massingire and Mpulanguene, fr.
8.v.1957, *Carvalho* 172 (BM). LM: between Umbeluzi and Matola, fl. & fr. 27.xii.1958,
Lemos 123 (LISC; LMJ).
Widespread in tropical Africa, S. Africa, Madagascar and Mascarene Is. In rain-
forest, riverine forest, thickets and grassland.

There seems to be a transition between this species and *H. dongolensis*. Specimens
occur with the bracts of the epicalyx only slightly broader in the middle and with a
sparser indumentum on the lower surface of the leaf. *Garcia* 740, cited above, is one of
these. *Pedro & Pedrógão* 6330 (LMJ; SRGH) from near Dombe (Manica e Sofala)
seems to be intermediate between *H. calyphyllus* and *H. platycalyx*.

40. **Hibiscus platycalyx** Mast. in Oliv., F.T.A. **1**: 202 (1868).—Hochr. in Ann.
  Conserv. Jard. Bot. Genève, **4**: 100 (1900) pro parte.—O. B. Mill., Check-lists
  For. Trees & Shrubs Brit. Emp. No. 6, Bechuanal. Prot.: 39 (1948). Syntypes:
  Mozambique, near Tete, fl. 1.ii.1860, *Kirk* (K).; Rovuma, *Kirk* (K). The former
  is chosen as lectotype.
    Hibiscus seineri Ulbr. ex. Engl., Pflanzenw. Afr. **3**, 2: 397 (1921); in Notizbl.
  Bot. Gart. Berl. **8**: 161 (1922). Type: Bechuanaland Prot., Mabeleapudi, *Seiner*
  II.310 (B†).
    Hibiscus rogersii Burtt Davy, F.P.F.T. **1**: 43 (1926). Type from the Transvaal.

Shrub up to 2 m. tall; branches glabrous to greyish-pubescent. Leaf-lamina
5–14 × 2–10 cm., papyraceous, narrowly ovate to broadly ovate (rarely suborbicular), rarely obscurely 3-lobed, stellate-pubescent above, greyish-tomentose
beneath, apex acute to rounded sometimes slightly acuminate sometimes mucronate, margin subentire to crenate or serrate, base rounded to cordate 3–7-nerved;
petiole up to 8 cm. long, greyish-tomentellous or nearly glabrous; stipules 4–10
mm. long, subulate with expanded base. Flowers up to c. 7 cm. in diam., yellow
with dark reddish centre, solitary in the axils of the upper leaves and forming
short terminal racemes at the ends of the branches; peduncle up to 2 cm. long,
greyish-pubescent, 5-angled at the apex, articulated near the base. Epicalyx of

5 bracts, tomentellous; bracts 15–35 mm. long, very variable in shape as in *H. calyphyllus* and, while less obviously broadened towards the middle, usually clearly if slightly narrowed towards the base, sometimes with long caudate tips. Calyx up to 20 mm. long, accrescent, pubescent; lobes ovate-acute, joined for up to 5–6 mm. at the base. Petals up to 6 × 5 cm., obovate, pubescent outside except where overlapping, glabrous within. Staminal tube 12–18 mm. long; free parts of filaments up to 3 mm. long. Style-branches 2–4 mm. long. Capsule 2·5 × 2 cm., broadly ovoid, stellate-setulose; valves acute. Seeds 5 × 5 mm., reniform, silky-tomentose.

Bechuanaland Prot. N: Ngami, fl. 5.v.1930, *van Son* in Herb. Transv. Mus. 28941 (BM; PRE; SRGH). SW: Mabeleapudi, *Seiner* II.310 (B†). **S. Rhodesia.** N: Sebungwe Distr., Kariangwe Hill, fl. 8.xii.1951, *Plowes* 214 (BM; SRGH). W: Wankie Game Reserve, 915 m., fl. 20.ii.1956, *Wild* 4782 (BM; PRE; SRGH). E: Hot Springs, fl. 16.iv.1948, *Chase* 1682 (SRGH). S: Ndanga Distr., Chidumo Clinic, fl. i.1959, *Farrell* 18 (SRGH). **Mozambique.** N: Rovuma, *Kirk* (K). T: near Tete, fl. 1.ii.1860, *Kirk* (K). MS: between Inhamitanga and Lacerdónia, fl. & fr. 7.iv.1942, *Torre* 4089 (LISC).
Also in S. Africa and probably in Tanganyika. In forest and in *Combretum* and *Colophospermum mopane* woodlands up to c. 1000 m.

Torre 4089 (LISC) cited above, and *Simão* 745 (LM) from Inhamitanga were both collected in forest and differ from the rest of the material in having larger more remotely dentate leaves with longer mucros and much longer epicalyx-bracts with caudate tips. There is insufficient evidence to decide whether this is only a shade form or a taxon requiring nomenclatural recognition. On the other hand, *H. seineri* appears from the description and from *van Son* in Herb. Transv. Mus. 28941, likely to be identical with it, to be a dry-climate form. There is evidence of a cline from Bechuanaland Prot. in the west to the forest form in the east represented by *Torre* 4098. The latter form approaches *H. owariensis* Beauv. from W. Africa.

41. **Hibiscus ludwigii** Eckl. & Zeyh., Enum. Pl. Afr. Austr. Extratrop. **1**: 39 (1834).—Harv. in Harv. & Sond., F.C. **1**: 171 (1860).—Mast. in Oliv., F.T.A. **1**: 203 (1868).—Hochr. in Ann. Conserv. Jard. Bot. Genève, **4**: 161 (1900).—Burtt Davy, F.P.F.T. **2**: 281 (1932).—Brenan in Mem. N.Y. Bot. Gard. **8**, 3: 225 (1953). Type from S. Africa.
    Hibiscus macranthus Hochst. ex A. Rich., Tent. Fl. Abyss. **1**: 55 (1847).—Arwidsson apud Norlindh & Weimarck in Bot. Notis. **1934**: 97 (1934). Type from Ethiopia.

Shrub or perennial herb up to 2–3 m. tall; stems sparsely to very densely setose. Leaf-lamina up to 8 × 8 cm., suborbicular in outline, obscurely or distinctly 3–5-lobed, setose-pilosulose on both surfaces, especially on the nerves, apex acute, margin crenate-serrate, base cordate or subcordate; petiole up to 7 cm. long; stipules 4–5 mm. long, filiform. Flowers up to 7 cm. in diam., yellow with purple centre, solitary in the axils of the upper leaves and forming racemes at the apices of the stems; peduncle up to 15 mm. long, stellate-setose, articulated near the base. Epicalyx of 5 bracts, pubescent; bracts up to 25 × 9 mm., ovate-lanceolate to narrowly triangular, acute, joined for c. 4–5 mm. at the base. Calyx shorter than or equalling the epicalyx, pubescent; lobes ovate to ovate-lanceolate, acute, joined to nearly half-way. Petals up to 7 × 6 cm., obovate, pubescent. Staminal tube 30 mm. long; free parts of filaments 2–3 mm. long. Style-branches 4 mm. long, pubescent. Capsule 20 × 12 mm., ellipsoid, densely stellate-setose. Seeds 3 × 2·5 mm., subreniform, glabrous.

S. Rhodesia. E: Inyanga, c. 1700 m., 26.i.1931, *Norlindh & Weimarck* 4603 (BM; PRE). **Nyasaland.** C: Mt. Nchisi, 1650 m., *Brass* 17063 (BM; K; PRE; SRGH). From Ethiopia to S. Africa. Forest margins.

There is a considerable range of variation linking *H. macranthus* and *H. ludwigii*. *Norlindh & Weimarck* 4603, cited above, has narrowly triangular epicalyx-bracts differing quite considerably from the more usual ovate-lanceolate form. It could conceivably be *H. ludwigii × lunarifolius*.

Series 8 (*Ketmia* pro parte)
Bracts of epicalyx 7–9, rigid and pungent. Seeds minutely lepidote.

42. **Hibiscus caesius** Garcke in Bot. Zeit. **7**: 850 (1849); in Peters, Reise Mossamb. Bot. **1**: 125 (1861).—Ulbr. & Fr. in R.E.Fr., Wiss. Ergebn. Schwed. Rhod.-

Kongo-Exped. **1** : 145 (1914).—Garcia in Bol. Soc. Brot., Sér. 2, **20** : 40 (1946)
TAB. **89** fig. 9.   Type : Mozambique, Tete, *Peters* (B†).
   Hibiscus pentaphyllus F. Muell., Fragm. Phyt. Austr. **2** : 13 (1860) non Roxb.
(1832).—Mast. in Oliv., F.T.A. **1** : 198 (1868).—Eyles in Trans. Roy. Soc. S. Afr.
5 : 415 (1916).—Wild, Guide Fl. Vict. Falls : 147 (1953).—Martineau, Rhod.
Wild Fl. : 51, 19 fig. 2 (1954).   Type from Australia.
   Hibiscus gibsonii Stocks ex Harv. in Harv. & Sond., F.C. **2** : 587 (1862).—Mast.
in Hook., Fl. Brit. Ind. **1** : 339 (1874).   Type from S.W. Africa.
   Hibiscus caesius var. *genuinus* Hochr. in Ann. Conserv. Jard. Bot. Genève, **4**,
160 (1900).—Exell & Mendonça, C.F.A. **1**, 1 : 169 (1937).   Type as for *H. caesius*.

Shrub up to 2 m. tall, sometimes subscandent ; stems glabrous or sparsely setose.
Leaf digitately 3–5-foliolate or palmatisect ; leaflets 2–10 × 1–3 cm., with sparse
2–3-pronged hairs mainly on the lower surface, apex acute, margin serrate, base
cuneate ; petiole up to 8 cm. long, pilose or setose-pilose ; stipules 10–12 mm.
long, linear to filiform.   Flowers up to 10 cm. in diam., yellow or cream with
purple or reddish-brown centre, solitary, axillary ; peduncle up to 12 cm. long,
sparsely setose, articulated near the apex.   Epicalyx of 7–9 bracts ; bracts 1·5–3·5
cm. long, stiff, pungent, slightly joined at the base.   Calyx up to 3 cm. long ;
lobes up to 3 × 1 cm., ovate-lanceolate, acute to acuminate, 3–5-nerved, free almost
to the base, sparsely pubescent on the margins.   Petals up to 5 × 4 cm., obovate,
very sparsely setose or glabrous.   Staminal tube 15 mm. long ; free parts of filaments 1·5–2 mm. long.   Style projecting 1·5 mm. above the apex of the staminal
tube before branching ; branches 2 mm. long.   Capsule c. 15 × 13 mm., ovoid,
sparsely setose.   Seeds 5 × 3·5 mm., lune-shaped, very minutely scaly, with a few
scattered hairs, ± glabrescent.

Bechuanaland Prot. N : Chobe Airstrip, fl. 29.vii.1950, *Robertson & Elffers* 79
(PRE ; SRGH).   SE : Bamangwato Territory, fl. iii.1895, *Holub* 1190 (K).   **N. Rhodesia.**
B : Nangweshi, 1040 m., fl. & fr. 19.vii.1952, *Codd* 7122 (BM ; PRE ; SRGH).   C :
Lusaka, Mt. Makulu, fl. & fr. 4.iv.1955, *E.M. & W.* 1404 (BM ; LISC ; SRGH).
S : Victoria Falls, fl. & fr. 30.vii.1911, *Fries* 150 (UPS).   **S. Rhodesia.** N : Urungwe
Distr., Rifa R., 1000 m., 24.ii.1953, *Wild* 4102 (BM ; SRGH).   W : Victoria Falls,
915 m., fl. & fr. 1.ii.1934, *Saunders Davies* (BM).   C : Gwelo, 1750 m., fl. i.1920, *Walters*
2700 (BM ; SRGH).   E : Chipinga, Sabi Valley, fl. 28.iii.1956, *Whellan* 1022 (BM ;
SRGH).   S : Gwanda, Mbezi R., 610 m., fl. & fr. v.1955, *Davies* 1313 (BM ; SRGH).
Mozambique. T : Muatize, Inhantoto, fl. & fr. 7.v.1948, *Mendonça* 4106 (BM ; LISC).
MS : Maringuè, fl. 30.vi.1950, *Chase* 2480 (BM ; SRGH).   **Nyasaland.** S : Chikwawa
Distr., S. of Lilanje R., fl. 25.iii.1960, *Phipps* 2710 (BM ; PRE ; SRGH).
Tropical and subtropical regions of the Old World.   Woodland and riverine fringes.

Series 9 (*Ketmia* pro parte; *Panduriformes*)

   Epicalyx of 7–20 filiform to linear or linear-lanceolate bracts, usually ± flexuous.
Seeds usually minutely lepidote.

Bracts of the epicalyx linear, linear-lanceolate or spathulate, up to 2 mm. broad :
   Leaves almost glabrous or with sparse hairs on the nerves beneath ; bracts of the
      epicalyx linear to linear-lanceolate -   -   -   -   -   43. *articulatus*
   Leaves tomentose beneath ; bracts of the calyx usually spathulate (broader at the tips)
                                                            44. *panduriformis*
Bracts of the epicalyx threadlike, ± flexuous, usually not more than 0·5 mm. broad :
   Chalky incrustations present towards the base of the under surface of the leaf-lamina ;
      leaves ± deeply palmatilobed -   -   -   -   -   -   45. *physaloides*
   Chalky incrustations absent from the leaves :
      Flowers pink or white ; plant densely fulvous-setose especially on the young parts
                                                            46. *burtt-davyi*
      Flowers yellow (rarely white), usually with a dark centre ; plant usually with a
         greyish-green indumentum :
         Plant prostrate or semiprostrate -   -   -   -   -   47. *schinzii*
         Plant erect :
            Calyx-lobes elongate-triangular, acuminate ; bracts of the epicalyx extending
               beyond the sinuses formed by the calyx-lobes ; leaves not lobed, stellate-
               pubescent beneath -   -   -   -   -   -   48. *rhabdotospermus*
            Calyx-lobes deltoid to triangular ; bracts of the epicalyx usually not or only just
               reaching the sinuses formed by the calyx-lobes ; leaves usually (but not
               always) lobed :
               Leaves tomentose beneath and sparsely setose, indumentum completely con-
                  cealing the lower surface of the lamina, smooth or slightly rough :

Plant robust, up to 1·5 m. tall; flowers 6–8 cm. in diam.   49. *kirkii*
Plant slender, up to 0·5 m. tall; flower up to c. 2·5 cm. in diam.
50. *jacksonianus*
Leaves pubescent beneath, indumentum usually not completely concealing the
lower surface of the lamina, and also stellate-setose with yellow irritant
hairs, rough   -   -   -   -   -   -   -   -   51. *engleri*

43. **Hibiscus articulatus** Hochst. ex A. Rich., Tent. Fl. Abyss. **1**: 60 (1847).—Mast.
in Oliv., F.T.A. **1**: 200 (1868).—Oliv. in Trans. Linn. Soc. **29**: 36, t. 13 (1872).—
Hochr. in Ann. Conserv. Jard. Bot. Genève, **4**: 159 (1900).—Arwidss. apud
Norlindh & Weim. in Bot. Notis. **1934**: 98 (1934).—Exell & Mendonça, C.F.A. **1**,
1: 172 (1937).—Wild, Guide Fl. Vict. Falls: 147 (1953).—Martineau, Rhod.
Wild Fl.: 51 (1954).—Keay, F.W.T.A. ed. 2, **1**, 2: 347 (1958). Type from
Ethiopia.
Hibiscus rhodesicus Bak. f. in Journ. of Bot. **37**: 424 (1899).—Eyles in Trans.
Roy. Soc. S. Afr. **5**: 416 (1916). Type: S. Rhodesia, Bulawayo, *Rand* 28 (BM,
holotype).
Hibiscus eburneopetalus Bak. f., op. cit. **75**: 100 (1937). Type: Mozambique,
Mucobezi, *Le Testu* 911 (P, holotype).

Shrublet or perennial herb, often prostrate, producing annual shoots from a
woody rootstock; stems sparsely pilose, pilose or hirsute. Leaf-lamina up to
10 × 6 cm., broadly elliptic in outline, varying from not lobed to 3–5-partite,
almost glabrous or with simple or stellate hairs on the nerves, obtuse to cuneate
at the base; lobes linear to broadly elliptic, central one often much longer than
the lateral ones, margins varying from almost entire to coarsely toothed; petiole
3–15 mm. long; stipules up to 8 mm. long, linear. Flowers up to 5 cm. in diam.,
white or pale yellow, solitary, axillary; peduncle up to 6 cm. long, densely minutely
pubescent, articulated near the apex. Epicalyx of 7–9 bracts; bracts 3·5–6 mm.
long, linear, usually free almost to the base but occasionally some of them united
in an irregular fashion sometimes almost to the apex. Calyx up to c. 18 mm. long,
minutely pubescent and sparsely pilose; lobes up to 15 × 3·5 mm., narrowly
lanceolate to elliptic, acute, often somewhat acuminate, joined for 3–4 mm. at the
base. Petals up to 3 × 2 cm., obovate to narrowly obovate, glabrous or nearly so.
Staminal tube 10–12 mm. long; free parts of filaments 2–3 mm. long. Style
sometimes projecting 2–3 mm. beyond the staminal tube before branching;
branches 2–2·5 mm. long, pubescent. Capsule 18 × 12 mm., ellipsoid, minutely
pubescent; margins of valves pubescent. Seeds 2·5 × 2 mm., angular-subreniform,
glabrous or minutely pubescent.

Caprivi Strip. Caprivi side of the river near Andara Mission, fl. 23.ii.1956, *de Winter
& Marais* 4813 (PRE). **Bechuanaland Prot.** N: Leshumo Valley, fl. i.1876, *Holub* (K).
N. Rhodesia. C: N. of Lusaka, 3·2 km. S. of Chisamba Station, Concord Ranch, fl.
& fr. 6.i.1958, *Benson* N.R. 204 (BM). E: Ft. Jameson, 1030 m., fl. 26.xii.1935, *Winter-
bottom* 31 (K). S: Livingstone, 910 m., fl. i.1906, *Rogers* 7240a (K; SRGH). **S.
Rhodesia.** N: Mazoe, 1310 m., fl. & fr. xii.1905, *Eyles* 225 (BM; SRGH). W:
Victoria Falls, 880 m., fl. i.1944, *Martineau* 61 (SRGH). C: Beatrice, fl. 23.i.1959, *Drewe*
in GHS 92540 (SRGH). E: Inyanga, 1300 m., fl. & fr. 15.i.1931, *Norlindh & Weimarck*
4402 (BM; SRGH). S: Victoria Distr., 1200 m., fl. & fr. xii.1920, *Mainwaring in Eyles*
2786 (PRE; SRGH). **Nyasaland.** N: Karonga, 460 m., fl. & fr. 18.ii.1953, *Williamson*
165 (BM). C: Kasungu, 1000 m., fl. & fr. 14.i.1959, *Robson* 1171 (BM; K; LISC;
SRGH). S: Shire R., near Liwonde Ferry, 475 m., fl. 13.iii.1955, *E.M. & W.* 851 (BM;
SRGH). **Mozambique.** N: Chamba, fl. & fr. 20.viii.1948, *Pedro & Pedrógão* 5586
(LMJ; SRGH). Z: Mocuba, 120 m., fl. & fr. i.1943, *Faulkner* 155 (K; PRE; SRGH).
Widespread in tropical Africa from Ethiopia to the Transvaal. Woodland and road-
sides from sea-level to about 1500 m.

44. **Hibiscus panduriformis** Burm.f., Fl. Ind.: 151, t. 47 fig. 2 (1768).—Garcke in Peters,
Reise Mossamb. Bot. **1**: 127 (1861).—Mast. in Oliv., F.T.A. **1**, 203 (1868).—
Hochr. in Ann. Conserv. Jard. Bot. Genève, **4**: 95 (1900).—Eyles in Trans. Roy.
Soc. S. Afr. **5**: 415 (1916).—Engl., Pflanzenw. Afr. **3**, 2: 397, fig. 186 S–T
(1921).—Exell & Mendonça, C.F.A. **1**, 1: 173 (1937).—Mendonça & Torre,
Contr. Conhec. Fl. Moçamb. **1**: 14 (1950).—Martineau, Rhod. Wild Fl.: 50
(1954).—Keay, F.W.T.A. ed. 2, **1**, 2: 346 (1958). TAB. **89** fig. 10. Type from
India (probably destroyed).
Hibiscus multistipulatus Garcke in Bot. Zeit. **7**: 849 (1849). Type from " East
Africa " (no specimen cited).
Hibiscus friesii Ulbr. [apud Ulbr. & Fr. in R.E.Fr., Wiss. Ergebn. Schwed.

Rhod.-Kongo-Exped. **1**: 146 (1914) *nom. nud.*] in Fedde Repert. **13**: 521 (1915). Type: N. Rhodesia, Chirukutu, near Broken Hill, *Fries* 239 (UPS, holotype).

Shrub or perennial herb with woody base, up to 2·5 m. tall; stems stellate-tomentose and often pilose or setose. Leaf-lamina up to 12 × 10 cm., suborbicular in outline, shallowly 3–5-lobed, tomentose to pilose above, tomentose beneath, apex somewhat acuminate, margin shallowly and irregularly toothed, base cordate to truncate; petiole up to 12 cm. long, hairy like the stems; stipules 5–10 mm. long, filiform, often 2–3 side by side at a node. Flowers 3·5–10 cm. in diam., yellow with a dark red or purple centre, solitary, axillary; peduncle 5–12 mm. long, tomentose, articulated near the middle. Epicalyx of 8–10 bracts; bracts up to 10 × 2 mm., linear-spathulate (slightly broader at the apex), joined for c. 2 mm. at the base. Calyx 18 mm. long, tomentose; lobes triangular, often conspicuously 3-nerved, joined to about half-way. Petals up to 3–5 × 2·5–4 cm., tomentose outside except where overlapping, nearly glabrous within. Staminal tube 10–18 mm. long; free parts of filaments 1·5–3 mm. long. Style-branches 3–5 mm. long. Capsule c. 15 mm. in diam., subglobose, densely setose. Seeds 3 × 2 mm., lune-shaped to subreniform, densely fulvous-pubescent.

N. Rhodesia. C: Chirukutu, near Broken Hill, fl. 7.viii.1911, *Fries* 239 (UPS). S: between Choma and the Kafue R., fl. & fr. 11.viii.1930, *Hutchinson & Gillett* 3567 (BM; K; SRGH). **S. Rhodesia.** N: Mazoe, 1340–1460 m., fl. i.1906, *Eyles* 531 (BM; SRGH). C: Salisbury, fl. 1921, *Godman* 25 (BM). E: Umtali, fl. ii.1959, *Head* (BM). **Nyasaland.** C: Lilongwe, fl. 5.x.1952, *Kantikana* 5 (BM). S: Zomba Distr., Chilwa I., fl. 2.iii.1955, *Jackson* (BM; SRGH). **Mozambique.** N: between Ancuabeana and Metage, fl. & fr. 7.ix.1948, *Barbosa* 2013 (LISC; LMJ). T: Sisitso, Ulere Stream, R. Zambezi, fl. & fr. 11.vii.1950, *Chase* 2638 (BM; SRGH). MS: Chemba, fl. & fr. 15.iv.1960, *Lemos & Macuácua* 113 (BM; LMJ).
Widespread in tropical Africa, Madagascar, tropical Asia and Australia. Dry places, old cultivation, sandy soil.

Specimens of this species differ in three pairs of characters: (1) stems tomentose or hirsute; (2) flowers larger (petals c. 5 mm. long) or smaller (petals c. 3–5 cm. long): (3) seeds fulvous-pubescent or glabrous. These features occur, however, in various combinations and it seems ill-advised to attempt an infraspecific classification. *H. friesii* Ulbr. (*Fries* 239) has a distinct facies and, if eventually found to be separable, *Kantikana* 5 from Lilongwe would probably belong to the same taxon.
For a fuller synonymy see Hochreutiner (loc. cit.).

45. **Hibiscus physaloides** Guill. & Perr. in Guill., Perr. & Rich., Fl. Senegamb. Tent. **1**: 52 (1831).—Harv. in Harv. & Sond., F.C. **1**: 172 (1860).—Mast. in Oliv., F.T.A. **1**: 199 (1868) pro parte.—Hochr. in Ann. Conserv. Jard. Bot. Genève, **4**: 161 (1900) pro parte.—Eyles in Trans. Roy. Soc. S. Afr. **5**: 415 (1916).—Burtt Davy, F.P.F.T. **2**: 281 (1932).—Exell & Mendonça, C.F.A. **1**, 1: 171 (1937).—Garcia in Bol. Soc. Brot., Sér. 2, **20**: 40 (1946).—Wild, Guide Fl. Vict. Falls: 147 (1953).—Williamson, Useful Pl. Nyasal.: 66 (1955).—Keay, F.W.T.A. ed. 2, **1**, 2: 346 (1958). TAB. **89** fig. 6. Type from Senegambia.
  Hibiscus variabilis Garcke in Peters, Reise Mossamb. Bot. **1**: 126 (1861). Type: Mozambique, Querimba I., *Peters* (B†).
  Hibiscus physaloides var. *genuinus* Hochr., tom. cit.: 162 (1900). Type as for *H. physaloides.*
  Hibiscus cf. *kirkii.*—Garcia in Bol. Soc. Brot., Sér. 2, **20**: 40 (1946).

Annual (or perennial?) herb up to 2 m. tall; stems setose-pilose with irritant hairs and densely pubescent to tomentellous. Leaf-lamina up to 20 × 15 cm., suborbicular in outline, shallowly to rather deeply digitately 3–7-lobed, stellate-pubescent to stellate-tomentose on both surfaces and with several irregular chalky concretions on the under surface near the base, apex acute, margin irregularly crenate to toothed, base cordate; lobes usually triangular, acute; petiole up to 17 cm. long, hairy like the stems; stipules 4 mm. long, filiform. Flowers c. 9 cm., in diam., yellow or yellowish-orange with purplish or reddish centres, solitary, axillary; peduncle up to c. 7 cm. long, hairy like the stems, articulated towards the apex. Epicalyx of 7–10 bracts; bracts 6–10 mm. long, filiform. Calyx 15–30 mm. long, pilose and pubescent; lobes up to 20 × 12 mm., ovate-elliptic, acute, joined at the base for 3–6 mm. Petals up to 5 × 3·5 cm., obovate, almost glabrous or with a few sparse dark hairs. Staminal tube 15–18 mm. long; free parts of filaments 1–2 mm. long. Style-branches c. 1·5 mm. long. Capsule 15 × 10 mm., ovoid, setose-pilose. Seeds 2–2·5 × 1·8–2 mm., angular-subreniform, sparsely scaly or verruculose, often glabrescent.

N. Rhodesia. N : Abercorn Distr., 850 m., fl. 6.iv.1952, *Richards* 1366 (K). E : Luangwa R., 520 m., fl. & fr. 25.iii.1955, *E.M. & W.* 1188 (BM ; LISC ; SRGH). **S. Rhodesia.** E : Nyamkwarara Valley, 1065 m., fl. ii.1935, *Gilliland* 1381 (BM ; K). S : Belingwe Distr., fl. iv.1958, *Davies* 2474 (SRGH). **Nyasaland.** N : Nyika Plateau, 2350 m., fl. & fr. ii.1903, *McClounie* 182 (K). S : Fort Johnston, fl. & fr. 23.iv.1955, *Banda* 78 (BM). **Mozambique.** N : Mandimba, fl. 15.ii.1942, *Hornby* 3702 (PRE). Z : Mocuba Distr., fr. 9.iv.1948, *Faulkner* 249 (COI ; K ; SRGH). MS : Msusa, 215 m., fl. & fr. 25.vii.1950, *Chase* 2797 (BM ; COI ; LISC ; SRGH). SS : Inhambane, Mambone, fl. & fr. 27.v.1941, *Torre* 2761 (LISC). LM : Rikatla, fl. iv.1919, *Junod* 554 (PRE).

Widespread in tropical and S. Africa, Madagascar and Seychelles. Dry stony places.

Yields a short white fibre of good strength (Williamson).

46. **Hibiscus burtt-davyi** Dunkley in Kew Bull. **1935** : 256 (1935).—Burtt Davy & Hoyle, N.C.L. : 49 (1936). Type : Nyasaland, Mt. Mlanje, *Burtt Davy* 22097 (FHO ; K, holotype).

Small tree up to 7 m. tall, or shrub ; stems densely fulvous-stellate-tomentose, sometimes viscid. Leaf-lamina up to 9–10 × 7–8 cm., suborbicular or very broadly elliptic to elliptic in outline, shallowly to deeply 3–5-lobed, stellate-setose on both surfaces, apex acute, margin irregularly dentate or serrate, base obtuse to truncate ; petiole up to 5 cm. long, fulvous-stellate-setose ; stipules 6–8 mm. long, filiform. Flowers up to 5 cm. in diam., pink or white, solitary, axillary or in short axillary 2–5-flowered racemes often clustered at the apices of the stems ; peduncle up to 3·5 cm. long, densely fulvous-setose, articulated near the apex. Epicalyx of up to c. 20 bracts, often exceeding the calyx ; bracts up to 16 mm. long, filiform or very narrowly linear. Calyx c. 15 mm. long, densely setose ; lobes lanceolate, usually joined to about half-way. Petals 3–4·5 × 2·5 cm., obovate, stellate-setulose and stellate-pubescent outside except where overlapping, nearly glabrous inside. Staminal tube 16 mm. long ; free parts of filaments 1 mm. long. Style-branches 2 mm. long. Capsule 13 × 12 mm., subglobose, setose. Seeds 2·5 × 2·5 mm., angular-subreniform, minutely lepidote.

S. Rhodesia. E : Melsetter, 1830 m., fl. 7.vii.1950, *Thompson* 10 (PRE ; SRGH). **Nyasaland.** S : Tuchila Plateau, 1830 m., fl. 22.vii.1956, *Newman & Whitmore* 145 (B ; BM ; NY ; SRGH). **Mozambique.** MS : Mavita, fl. 6.vii.1949, *Pedro & Pedrógão* 7209 (LMJ).

Recorded only from S. Rhodesia, Nyasaland and Mozambique. In *Widdringtonia-Podocarpus* forest, forest margins and rocky hillsides up to 2130 m.

47. **Hibiscus schinzii** Gürke in Verh. Bot. Verein. Brand. **30** : 176 (1889).—Eyles in Trans. Roy. Soc. S. Afr. **5** : 416 (1916).—Burtt Davy, F.P.F.T. **2** : 281 (1932).— Mendonça & Torre, Contr. Conhec. Fl. Moçamb. **1** : 17 (1950).—Exell & Mendonça, C.F.A. **1,** 2 : 178 (1951).—Wild, Guide Fl. Vict. Falls : 147 (1953). Type from SW. Africa.

    Hibiscus physaloides var. *genuinus* forma *schinzii* (Gürke) Hochr. in Ann. Conserv. Jard. Bot. Genève, **4** : 162 (1900). Type as above.

    Hibiscus swynnertonii Bak. f. in Journ. Linn. Soc., Bot. **40** : 28 (1911).—Eyles, loc. cit. Type : S. Rhodesia, Sabi R., *Swynnerton* (BM, holotype).

    Hibiscus cordatus sensu Garcia in Bol. Soc. Brot., Sér. 2, **20** : 40 (1946).

Prostrate or semiprostrate perennial herb ; stems usually trailing along the ground, glandular-pubescent and with long weak or subsetose hairs. Leaf-lamina up to 4 (6) cm. in diam., suborbicular in outline, from very shallowly 3–5-lobed or 5-angular to 3–5-palmatipartite, stellate-pubescent above, stellate-tomentose beneath, apex subacute or somewhat rounded, margin crenate or subserrate, base cordate ; petiole up to 5 cm. long, pubescent ; stipules 3–4 mm. long, filiform, very caducous. Flowers up to 8 cm. in diam., yellow (rarely white) with reddish-purple centre, solitary, axillary ; peduncle up to 10 cm. long, with a somewhat glandular pubescence, articulated near the apex. Epicalyx of 7–9 bracts ; bracts 2·5 mm. long, filiform to very narrowly linear. Calyx 10–20 mm. long, densely pubescent, and stellate-setulose ; lobes elongate-triangular, acute, joined at the base for up to 4–5 mm. Petals 3–5 × 1·5–2·5 cm., narrowly obliquely obovate, pubescent outside where not overlapping, glabrous inside. Staminal tube 7–18 mm. long; free parts of filaments up to 0·5 mm. long. Style-branches 2–5 mm. long. Capsule 12 mm. in diam., subglobose, densely stellate-pubescent; margins of valves setose. Seeds 2 × 2 mm., subreniform, sparsely minutely lepidote.

Bechuanaland Prot. N: Nata R., fl. & fr. 21.iv.1957, *Drummond & Seagrief* 5147 (CAH; SRGH). SE: Mochudi, 940 m., fl. & fr. iii.1914, *Rogers* 6482 (BM; PRE). **N. Rhodesia.** B: Sesheke Distr., fl. & fr. 1911, *Macaulay* 472 (K). **S. Rhodesia.** W: Nymandhlovu, fl. & fr. 13.v.1953, *Plowes* 1598 (BM; SRGH). C: Salisbury, fl. & fr. 1916, *Craster* (K). E: Sabi-Lundi Junction, 245 m., fl. & fr. 5.vi.1950, *Chase* 2329 (BM; K; SRGH). S: Birchenough Bridge, fl. & fr. i.1938, *Obermeyer* 2469 (PRE). **Mozambique.** MS: between Dondo and R. Buzi, fl. & fr. 29.viii.1948, *Pedro & Pedrógão* 8949 (LMJ; SRGH). LM: Angoane, fl. & fr. 8.x.1954, *Lemos* 41 (LISC; LMJ; SRGH); Inhaca I., fl. & fr. ix.1919, *Breyer* 20481 (PRE; SRGH).
Also in Angola and SW. Africa. Coastal bush, coastal sand-dunes, woodlands and grasslands from sea-level to 1400 m.

H. swynnertonii differs in having white flowers (according to Swynnerton) and the lobes of the leaf slightly constricted towards the base. It is unlikely to be specifically distinct.

48. **Hibiscus rhabdotospermus** Garcke in Bot. Zeit. **7**: 839 (1849).—Mast. in Oliv., F.T.A. **1**: 200 (1868). Type from " Africa orientalis ".
　　Hibiscus cordatus Hochst. ex Webb, Fragm. Fl. Aethiop.-Aegypt.: 45 (1854) non *H. cordatus* D. Dietr. (1847).—Hochr. in Ann. Conserv. Jard. Bot. Genève, **4**: 164 (1900).—Burtt Davy, F.P.F.T. **2**: 281 (1932). Type from Ethiopia.
　　Hibiscus rhabdotospermus var. *mossamedensis* Hiern, Cat. Afr. Pl. Welw. **1**: 70 (1896). Type from Angola (Mossâmedes).
　　Hibiscus mossamedensis (Hiern) Exell & Mendonça in Journ. of Bot. **74**: 138 (1936); C.F.A. **1**, 1: 170 (1937). Type as above.

Herb 1-2 m. tall; stems crisped-pubescent or tomentose and with sparse to dense stellate or simple bristles. Leaf-lamina 3·5–12 × 2·5–10 cm., ovate, broadly ovate or suborbicular, pubescent to densely pubescent with few to numerous branched or simple bristles mainly on the lower surface, apex acute, margin serrate, base cordate, truncate or broadly cuneate; petiole up to 8·5 cm. long, hairy like the stems; stipules up to 5 mm. long, filiform. Flowers up to 5 cm. in diam., yellow with reddish or purple centre, solitary, axillary; peduncle up to 4 cm. long, articulated above the middle. Epicalyx of c. 10 bracts; bracts up to 12 mm. long, filiform, usually extending beyond the sinuses formed by the calyx-lobes. Calyx up to 25 mm. long; lobes elongate-acuminate, joined at the base for up to 8 mm. Petals up to 3 × 2 cm., obovate-elliptic, stellate-pubescent outside where not overlapping. Staminal tube c. 18 mm. long; free parts of filaments 0·5–1·5 mm. long. Style-branches 2 mm. long. Capsule 12 × 10 mm., subglobose, densely pubescent; awns of the carpels up to 5 mm. long. Seeds 2·3 × 1·8 mm., angular-subreniform, sparsely white-lepidote.

Bechuanaland Prot. N: Kwebe Hills, 1000 m., fl. & fr. 8.iii.1898, *Lugard* 215 (K). **S. Rhodesia.** N: Urungwe Distr., Zambezi valley, 520 m., fl. 24.ii.1953, *Wild* 4103 (BM; PRE; SRGH). E: Melsetter Distr., Hot Springs, 610 m., fl. & fr. 24.ii.1952, *Chase* 4380 (BM; SRGH). **Mozambique.** T: Tete, fl. & fr. 1859, *Kirk* (K). MS: Lupata, fr. 20.iv.1860, *Kirk* (K).
Also in Angola, SW. Africa, Sudan and Kenya.

Recent material has linked *H. rhabdotospermus* and *H. mossamedensis* by a series of intermediates. Specimens from the south and west of our area tend to be more setose until, in SW. Africa, separation from *H. engleri* becomes difficult. *Wild* 4103, cited above, is a very luxuriant specimen with much larger leaves and longer stipules than usual.

49. **Hibiscus kirkii** Mast. in Oliv., F.T.A. **1**: 199 (1868).—Hochr. in Ann. Conserv. Jard. Bot. Genève, **4**: 163 (1900).—Eyles in Trans. Roy. Soc. S. Afr. **5**: 415 (1916).—Hutch., Botanist in S. Afr.: 464 (1946).—Suesseng. in Proc. & Trans. Rhod. Sci. Ass. **43**: 102 (1951). Type: Mozambique, Sena, *Kirk* (K, holotype).

Erect perennial herb up to 1·5 m. tall; stems tomentose, sparsely setose and often somewhat glandular. Leaf-lamina up to 10 (15) × 7 (12) cm., lanceolate to suborbicular in outline, rarely not lobed, usually 3–5-lobed, palmatifid or palmatisect, central lobe or segment usually much the longest, tomentose and setose on both surfaces, apex acute, margin crenate to serrate, base cordate; petiole up to 10 cm. long; stipules up to 10 mm. long, filiform to narrowly linear. Flowers up to 8 cm. in diam., yellow with dark purple or red centre, solitary, axillary, forming terminal racemes by reduction of the upper leaves; peduncle up to 3 cm., long, tomentose, articulated from near the middle to almost at the base. Epicalyx of 7–9 bracts; bracts up to 8 mm. long, filiform. Calyx up to 15 mm. long, tomentose and setulose outside, appressed-pilose within; lobes ovate to triangular, joined at the base for 5–7 mm. Petals up to 4 × 3·5 cm., obliquely obovate, tomentose outside where not overlapped, glabrous inside. Staminal tube c. 20 mm. long; free parts of filaments 1–1·5 mm. long. Style-branches 1·5–2 mm. long (perhaps immature). Capsule up to 10 mm. in diam., broadly ellipsoid to subglobose, tomentose. Seeds 2 × 2 mm., angular-subreniform, sparsely lepidote.

Bechuanaland Prot. N: Kwebe Hills, 1000 m., fl. 8.iii.1898, *Lugard* 214 (K). **S. Rhodesia.** N: Concession Distr., fl. 23.ii.1938, *Hopkins* in GHS 6817 (SRGH). W: Matopo Hills, 1370 m., fr. 3.vii.1905, *Rattray* 6958 (PRE). C: Marandellas, fl. & fr. 6.iv.1950, *Wild* 3269 (K; M; SRGH). E: Umtali, 1220 m., fl. & fr. 11.ii.1951, *Chase* 3610 (BM; COI; LISC; SRGH). S: Lundi R., fl. & fr. 30.vi.1930, *Hutchinson & Gillett* 3309 (BM; K). **Mozambique.** Z: Montes do Ile, 1000 m., fl. & fr. 2.iv.1943, *Torre* 5040 (LISC). MS: Chimoio, fl. & fr. 2.iii.1948, *Garcia* in *Mendonça* 447 (BM; LISC). SS: Guijà, fl. 3.ix.1950, *Myre* 795 (BM; LM). LM: Umbeluzi, fl. 5.iv.1949, *Myre & Balsinhas* 409 (BM; LM).

Also from S. Africa. Grassland and roadsides.

50. **Hibiscus jacksonianus** Exell in Bol. Soc. Brot., Sér. 2, **33**: 180 (1959). Type: Nyasaland, Namwera Escarpment, *E.M. & W.* 912 (BM; SRGH, holotype).

Erect annual herb up to 50 cm. tall, branched from near the base; stems slender, arcuate-ascending, crisped-pubescent and rather sparsely stellate-setose with comparatively large 1–5-pronged bristles. Leaf-lamina up to 3·5 × 3 cm., ovate to suborbicular in outline, 3-sect or 3-lobed, greyish-tomentose and sparsely setose, margin crenate or coarsely serrate, base cordate; petiole up to 2 cm. long; stipules about 2 mm. long, filiform, very caducous. Flowers 1·5–2·5 cm. in diam., yellow with purple centre, in small 3–5-flowered terminal cymes; peduncle 4 mm. long. Epicalyx of 7–8 bracts; bracts 3–4 mm. long, narrowly linear to filiform. Calyx 10–12 mm. long, densely pubescent; lobes 2–3 mm. broad at the base, triangular, joined for 3–5 mm. Petals 2·5 cm. long, narrowly obovate. Staminal tube c. 6 mm. long; free parts of filaments 0·5 mm. long. Style-branches 1·5 mm. long (probably immature).

Nyasaland. C: Lilongwe, Bunda Hill, 1250 m., fl. 7.ii.1959, *Robson* 1491 (BM; LISC; K; SRGH). S: Namwera Escarpment, Jalasi, 1120 m., fl. 15.iii.1955, *E.M. & W.* 912 (BM; SRGH).

Not yet found elsewhere. *Brachystegia-Julbernardia-Cussonia* woodland and old cultivation.

51. **Hibiscus engleri** K. Schum. in Engl., Bot. Jahrb. **10**: 47 (1888).—Burtt Davy, F.P.F.T. **2**: 281 (1932).—Mendonça & Torre, Contr. Conhec. Fl. Moçamb. **1**: 14 (1950). Type from SW. Africa.
  Hibiscus cordatus Harv. in Harv. & Sond., F.C. **1**: 172 (1860) non D. Dietr. (1847) nec Hochst. ex Webb (1854).—Burtt Davy, loc. cit. Type from S. Africa.
  Hibiscus subphysaloides Hochr. in Ann. Conserv. Jard. Bot. Genève, **20**: 163 (1917).—Burtt Davy, loc. cit. Type from S. Africa (Komati Poort).
  Hibiscus microcalycinus Engl., Pflanzenw. Afr. **3**, 2: 399 (1921).—Ulbr. in Notizbl. Bot. Gart. Berl. **8**: 165 (1922). Type from S. Africa (Komati Poort).
  Hibiscus irritans R. A. Dyer, Fl. Pl. Afr. **27**: t. 1050 (1948). Type as for *H. cordatus* Harv.

Erect herb up to c. 15 m. tall; stems tomentose or pubescent and rough with yellow stellate-setose or simple irritant hairs. Leaf-lamina up to 9 × 8 cm., ovate to suborbicular in outline, shallowly to rather deeply 3–5-lobed or 3–5-angled (occasionally not lobed), tomentose to densely pubescent and with stellate, 2–3-

branched or simple bristles, apex acute to rounded, margin serrate, base cordate; petiole up to 8 cm. long, hairy like the stems; stipules 3–4 mm. long, filiform. Flowers 5–6 cm. in diam., yellow with maroon or dark red centres, solitary in the axils of the upper leaves and in corymbose terminal or lateral racemes or panicles; peduncle up to 6 cm. long, hairy like the stems, articulated near the apex. Epicalyx of 7–10 bracts; bracts 1·5–3·5 mm. long, filiform, usually extending less than halfway to the sinuses formed by the calyx-lobes. Calyx up to c. 16 mm. long; lobes triangular, joined at the base for 7–8 mm. Petals up to 5 × 4 cm., obovate, stellate-pubescent outside where not overlapped, glabrous within. Staminal tube 16 mm. long; free parts of filaments 1–1·5 mm. long. Style-branches 3–4 mm. long. Capsule up to 15 × 9 mm., ellipsoid, setose, with awns up to 3 mm. long. Seeds 2·5 × 2 mm., angular-subreniform, sparsely lepidote.

S. Rhodesia. W: Matobo Distr., 1435 m., fl. & fr. ii.1954, *Miller* 2170 (BM; PRE; SRGH). S: Beitbridge, fl. & fr. 16.ii.1955, *E.M. & W.* 451 (BM; SRGH). **Mozambique.** LM: Namaacha, 600 m., fl. & fr. 27.iii.1957, *Barbosa & Lemos in Barbosa* 7540 (COI; LMJ).
Also in S. Africa. Dry bush.

In our area there is no great difficulty with this species, but in S. Africa, and especially in SW. Africa, there seems to have been some interchange of genes with *H. kirkii* and *H. rhabdotospermus*.

Series 10 (*Trichospermum* pro parte)
Bracts of epicalyx 6 mm. long or longer, linear to lanceolate. Flowers yellowish, cream, purple or white. Seeds tomentose or pubescent (rarely almost glabrous) but with no floss.

Leaves suborbicular to narrowly elliptic, up to 4 times as long as broad, entire or rather coarsely toothed, especially towards the apex -   -   -   -   - 52. *aethiopicus*
Leaves linear to very narrowly oblong or linear-elliptic, 5–10 times as long as broad, usually with a few sharp teeth or deeply incised with narrow lobes 53. *malacospermus*

52. **Hibiscus aethiopicus** L., Mant. Pl. Alt.: 258 (1771).—Cav., Diss. Bot. **3**: 155, t. 61 fig. 1 (1787).—Harv. in Harv. & Sond., F.C. **1**: 174 (1860).—Hochr. in Ann. Conserv. Jard. Bot. Genève, **4**: 98 (1900).—Eyles in Trans. Roy. Soc. S. Afr. **5**: 414 (1916).—Engl., Pflanzenw. Afr. **3**, 2: 396, fig. 186 J–L (1921).—Burtt Davy, F.P.F.T. **2**: 281 (1932). Syntypes from S. Africa.

Var. **ovatus** Harv. in Harv. & Sond., F.C. **1**: 174 (1860).—Hochr., loc. cit.—Arwidss. apud Norlindh & Weim. in Bot. Notis. **1934**: 97 (1934).—Suesseng. in Proc. & Trans. Rhod. Sci. Ass. **43**: 102 (1951). Syntypes from S. Africa.
   Hibiscus ovatus Cav., Diss. Bot. **3**: 143, t. 50 fig. 3 (1787). Type from S. Africa.
   Hibiscus leiospermus Harv. in Harv. & Sond., tom. cit.: 73 (1860).—Burtt Davy, F.P.F.T., tom. cit.: 280 (1932). Type from S. Africa.

Perennial herb 14–35 cm. tall, with woody rootstock; stems upright or prostrate-ascending, stellate-hispid. Leaf-lamina 1–8 × 0·6–4·5 cm., usually entire (in our area) but occasionally toothed especially towards the apex, narrowly to broadly elliptic, lanceolate, ovate-lanceolate or suborbicular, nearly glabrous to densely stellate-hispid, apex blunt to rounded, base usually rounded 3–5-nerved; petiole 5–15 mm. long, stellate-hispid; stipules 5 mm. long, subulate. Flowers yellowish, cream or white (usually drying pinkish), solitary, axillary; peduncle up to 4·5 cm. long, stellate-hispid, articulated c. 5 mm. below the apex. Epicalyx of 7–9 bracts; bracts 12 × 1·5 mm., linear-oblong (more variable in S. Africa), hispid especially on the margins. Calyx 10–20 mm. long, lobes up to 18 × 3·5 mm., lanceolate or elongate-triangular, hispid on the margins, joined for about 3 mm. at the base. Petals up to 3·5 × 2 cm., obovate, glabrous. Staminal tube 10–11 mm. long; free parts of filaments 0·5–1 mm. long. Style-branches 4 mm. long. Capsule 9 mm. in diam., subglobose, sparsely hispid. Seeds 2·5 × 2 mm., subreniform, tomentellous, pubescent or nearly glabrous.

S. Rhodesia. W: Bulawayo, fl. & fr. 1903, *Chubb* 103 (BM; SRGH). C: Marandellas, fl. & fr. 31.xii.1941, *Dehn* 568 (M; PRE; SRGH). E: Umtali Distr., Odzani R. bridge, fl. & fr. 27.xi.1956, *Chase* 3227 (BM; LISC; SRGH). **Mozambique.** LM: Delagoa Bay, fl. 1906, *Junod* 1370 (G).
Also from Eritrea to Natal and in the Transvaal. Upland grasslands up to 1700 m.

A specimen from S. Rhodesia, Matobo Distr., Besna Kobila, *Miller* 2160 (SRGH) is very different from the other Rhodesian specimens seen, having relatively broader, toothed leaves and a more densely hispid indumentum. In the Transvaal and Natal, *H. aethiopicus* var. *ovatus* is extremely variable, much more so than in Rhodesia, and *Miller* 2160 comes well within the range of variation found there.

The varietal epithet *ovatus* though probably derived from *H. ovatus* Cav. must be considered as nomenclaturally independent of the latter and differently typified for Harvey included *H. ovatus* in his synonymy with a mark of interrogation.

The type-variety was described from Cape Province and is usually a smaller plant with rounder, usually toothed leaves, a more appressed indumentum and flowers of a deeper yellow (usually drying yellow). There is considerable intergrading between the varieties.

53. **Hibiscus malacospermus** (Turcz.) E. Mey. [in Drège, Zwei Pflanz.-Docum.: 47, 192 (1843) *nom. nud.*] ex Harv. in Harv. & Sond., F.C. **1**: 174 (1860).—Hochr. in Ann. Conserv. Jard. Bot. Genève, **4**: 98 (1900) pro parte excl. var. *palmatipartitus*. Type from S. Africa.

    Kosteletzkya malacosperma Turcz. in Bull. Soc. Nat. Mosc. **31**: 192 (1858). Type as above.

Perennial herb 6–24 cm. tall, with woody rootstock; stems stellate-setose. Leaf-lamina 2–8 ×0·1–1·5 cm., narrowly linear, linear-elliptic or narrowly oblong, entire or rarely with a few coarse teeth, appressed-stellate-setulose especially on the nerves and margins, apex acute (rarely rounded), base narrowly cuneate 3–5-nerved; petiole up to 1 mm. long; stipules up to 12 mm. long, linear to subulate. Flowers up to 8 cm. in diam., purple, solitary, axillary; peduncle up to 6 cm. long, stellate-setose or stellate-hispid, articulated 5–10 mm. below the apex. Epicalyx of 7–8 bracts; bracts up to 18 ×2 mm., linear, acute, sparsely stellate-setulose. Calyx 15–22 mm. long, usually slightly exceeding the epicalyx; lobes up to 20 × 6 mm., ovate-lanceolate, acute, slightly acuminate, stellate-setose, stellate-setulose or stellate-hispid especially on the margins, slightly joined at the base. Petals up to 5 × 3 cm., narrowly obovate to obovate, glabrous or nearly so. Staminal tube 10–12 mm. long; free parts of filaments 1–2 mm. long. Style-branches 4 mm. long. Capsule 10–12 × 10 mm., subglobose, sparsely setose. Seeds 3 × 2 mm., subreniform, tomentellous.

S. Rhodesia. W: Bulawayo, fl. 1908, *Chubb* 22 (BM). C: Umvuma, Mtao, fl. xii.1922, *Eyles* 6973 (K; SRGH).
Also in S. Africa. Grassland.

Series 11 (*Aristivalves*)

Bracts of epicalyx more than 5. Flowers yellow, often concolorous. Valves of capsule aristate with awns 2–3 mm. long. Seeds appressed-sericeous-pubescent.

Included by Hochreutiner in his Sect. *Trichospermum* (in Ann. Conserv. Jard. Bot. Genève, **4**: 94 (1900)).

54. **Hibiscus palmatus** Forsk., Fl. Aegypt.-Arab.: CXVII, 126 (1775). TAB. **89** fig. 5. Type from Arabia.

    Hibiscus intermedius A. Rich., Tent. Fl. Abyss. **1**: 58 (1847) non Bélanger (1834). —Mast. in Oliv., F.T.A. **1**: 198 (1868).—Hochr. in Ann. Conserv. Jard. Bot. Genève, **4**: 94 (1900).—Engl., Pflanzenw. Afr. **3**, 2: 397, fig. 186 M–O (1921). Type from Ethiopia.

    Hibiscus aristivalvis Garcke in Bot. Zeit. **7**: 849 (1849) (" aristaevalvis "); in Peters, Reise Mossamb. Bot. **1**: 124 (1861).—Exell & Mendonça, C.F.A. **1**, 1: 173 (1937).—Mendonça & Torre, Contr. Conhec. Fl. Moçamb. **1**: 13 (1950). Type: Mozambique, Sena, Peters (B†).

Perennial ± prostrate herb; stems rather sparsely to densely setose or setulose and often with a longitudinal line of pubescence changing its radial position at each node. Leaf-lamina up to 12 × 16 cm. in outline (usually much smaller), 3–5-palmatilobed or -sect, sparsely setulose mainly on the nerves, apex rounded, base cordate; lobes elliptic to narrowly lanceolate or oblong-lanceolate, margins irregularly toothed or ± pinnatifid; petiole up to 6 cm. long, sparsely setulose; stipules 3 mm. long, filiform or subulate. Flowers 3·5–5 cm. in diam., pale yellow or yellow, with or without a dark centre. Epicalyx of 7–8 bracts; bracts 12–15 mm. long, ± filiform or narrowly linear. Calyx 10–15 mm. long; lobes sub-scarious, triangular or narrowly triangular, joined for 2–3 mm. at the base. Petals

3·5 × 2 cm., obliquely obovate, glabrous. Staminal tube c. 6 mm. long; free parts of filaments 3·5 mm. long; style-branches 3·5–4 mm. long. Capsule 8–9 mm. in diam., subglobose, sparsely pubescent; valves with awns 2–3 mm. long. Seeds 3 × 2 mm., angular-subreniform, appressed-sericeous-pubescent.

S. Rhodesia. N: Urungwe Distr., near the Kessesse R., fl. 3–7.iii.1956, *Goodier* 40 (BM; SRGH). W: Bulawayo, fl. & fr. i.1898, *Rand* 74 (BM). **Mozambique.** T: Tete, fl. & fr. ii.1859, *Kirk* (K). SS: between Massingire and Mapulanguene, fr. 8.v.1957, *Carvalho* 173 (BM; LM). LM: between Umbeluzi and Catuene, fl. & fr. 16.ix.1944, *Mendonça* 2880 (BM; LISC).
India, Arabia and widespread in tropical and S. Africa. *Colophospermum mopane* and other woodlands from 0–1400 m., mainly on alluvial soils.

After examining the type of *H. palmatus* Forsk., by kind permission of the Director of the Copenhagen Botanical Museum, I have no doubt at all that it is the species usually known as *H. aristivalvis* Garcke. Some modern Arabian collections from the same region as the type match it very closely.

Series 12 (*Lilibiscus*)
Shrubs with large, ornamental flowers. Staminal tube exceeding the corolla.

Petals entire; epicalyx c. 10 mm. long - - - - - - 55. *rosa-sinensis*
Petals pinnatifidly laciniate; epicalyx minute - - - - - 56. *schizopetalus*

55. **Hibiscus rosa-sinensis** L., Sp. Pl. **2**: 694 (1753).—Hochr. in Ann. Conserv. Jard. Bot. Genève, **4**: 133 (1900).—Brenan, T.T.C.L.: 303 (1949).—Williamson, Useful Pl. Nyasal.: 66 (1955).

Large ornamental shrub with dark green ovate toothed leaves and large showy, usually red flowers. Staminal tube 3–7 cm. long, exceeding the corolla.
Native of tropical Asia. Much planted throughout the tropics, often as a hedge-plant. Forms with variously coloured flowers are in cultivation and also variegated and double forms (see Hochreutiner, loc. cit.).

56. **Hibiscus schizopetalus** (Mast.) Hook. f. in Curt. Bot. Mag.: t.᷾6524 (1880).—Brenan, T.T.C.L.: 302 (1949). Type a plant grown from seed from Kenya, near Mombasa.
   Hibiscus sp.—Kirk & Oliv. in Journ. Linn. Soc., Bot. **15**: 478, fig. 1 (1876).
   Hibiscus rosa-sinensis var. *schizopetalus* Mast. in Gard. Chron. **1879**: 272, fig. 45 (Aug. 1879). Type as above.

A shrub probably closely related to *H. rosa-sinensis* but with large pendent pink flowers, laciniate petals and a minute epicalyx.

Native of Kenya and Tanganyika. Not yet found wild in our area.
Cultivated in the Municipal Garden in Lourenço Marques (*Gomes e Sousa* 3484 (COI)).

The varietal epithet was first mentioned in print by Thiselton Dyer (Gard. Chron. **1879**: 568 (May, 1879)) but without sufficient description to validate it. His reference to Kirk and Oliver (loc. cit.) is also insufficient validation as he leaves the identification in doubt by saying that var. *schizopetalus* is only *apparently* the same as their plant. It seems best to attribute the authorship to Masters, as J. D. Hooker did, although the epithet was presumably given to the plant either by Veitch's Nursery or by Thiselton Dyer.
This species is of interest as the laciniate petals certainly give the impression that it is a cultivated form of some species presumably near *H. rosa-sinensis*; but its wild origin seems to be well-established. It may well be found wild in the north of Mozambique.

Series 13 (*Pterocarpus*)
Shrubs or perennial herbs. Epicalyx of c. 10 filiform to linear bracts. Calyx-lobes joined to the middle. Capsule 5-winged, scarious. Clearly transitional towards the genus *Kosteletzkya*.

57. **Hibiscus vitifolius** L., Sp. Pl. **2**: 696 (1753).—Garcke in Peters, Reise Mossamb. Bot. **1**: 127 (1861).—Mast. in Oliv., F.T.A. **1**: 197 (1868).—Hochr. in Ann. Conserv. Jard. Bot. Genève, **4**: 168 (1900).—Bak. f. in Journ. Linn. Soc., Bot. **40**: 27 (1911).—Ulbr. & Fr. apud R.E.Fr. in Wiss. Ergebn. Schwed. Rhod.-Kongo-Exped. **1**: 145 (1914).—Eyles in Trans. Roy. Soc. S. Afr. **5**: 416 (1916).—Burtt Davy, F.P.F.T. **2**: 281 (1932).—Exell & Mendonça, C.F.A. **1**, 1: 176 (1937).

—Mendonça & Torre, Contr. Conhec. Fl. Moçamb. **1**: 17 (1950).—Brenan in Mem. N.Y. Bot. Gard. **8**, 3: 225 (1953).—Wild, Guide Fl. Vict. Falls: 147 (1953).—Martineau, Rhod. Wild Fl.: 50 (1954).—Keay, F.W.T.A. ed. 2, **1**, 2: 346 (1958).—Brenan & Exell in Bol. Soc. Brot., Sér. 2, **32**: 69 (1958). Type from Ceylon.

Shrub or perennial herb up to 1·5 m. tall, sometimes scandent or scrambling; stems terete, nearly glabrous or with a very variable indumentum, sometimes tomentose or tomentellous, often glandular, sometimes hispid or stellate-hispid or aculeate and with these different types of hairs occurring in various combinations. Leaf-lamina 3–15 × 3–15 cm., ovate to suborbicular; not lobed to deeply 3–5 (7)-palmatilobed, with a variable indumentum (as on the stems) but stellate-hispid or bifurcate hairs, rather rare on the stems, are more frequent especially on the under surface, margin serrate, base truncate to cordate 5–9-nerved; lobes acute or bluntish; petiole 2–7 (18) cm. long, hairy like the stems; stipules 3–5 mm. long, filiform. Flowers 5–9 cm. in diam., yellow with a dark reddish or purple centre, solitary or forming terminal racemes by reduction of the upper leaves; peduncle 1–1·5 (5) cm. long, articulated above the middle. Epicalyx of c. 10 bracts; bracts 8–10 mm. long, filiform. Calyx up to 15 mm. long in flower, accrescent, somewhat scarious; lobes up to 15 × 10 mm., broadly elliptic, acuminate, joined to about half-way. Petals up to c. 5 × 4·5 cm., glabrous or nearly so. Staminal tube up to 15 mm. long; free parts of filaments up to 5 mm. long. Style-branches 1·5–3 mm. long, minutely glandular. Capsule 10–12 × 12–15 mm. winged, scarious, usually hispid; valves ± aristate. Seeds 2·5 × 1·5 mm., irregularly reniform, glabrous.

Widespread in the tropics and subtropics of the Old World; probably introduced in tropical America.

Key to the subspecies

Leaves and stems nearly glabrous to rather sparsely hairy, sometimes aculeate, young stems sometimes with a fine pubescence; under surface of the leaf-lamina hairy only on the veins or with a few sparse hairs on the reticulation, surface always visible and with glabrous areas; leaves drying dark green, not lobed to deeply 3–5 (7)-lobed (rain-forest habitats) - - - - - - - - - - subsp. *vitifolius*

Leaves and stems tomentose, tomentellous, densely pilose or pubescent or hispid, often glandular; leaves usually drying brownish-green, not lobed to shallowly 3–7-lobed (woodland and grassland habitats) - - - - - - subsp. *vulgaris*

Subsp. **vitifolius**

Hibiscus jatrophifolius A. Rich., Tent. Fl. Abyss. **1**: 57 (1847). Type from Ethiopia.

Hibiscus ricinoides Garcke in Bot. Zeit. **7**: 834 (1849). Type from S. Africa.

Hibiscus ricinifolius E. Mey. [in Drège, Zwei Pflanz.-Docum.: 192 (1843) *nom. nud.*] ex Harv. in Harv. & Sond., F.C. **1**: 171 (1860) *nom. illegit.* Type from S. Africa.

Hibiscus natalitius Harv. in Harv. & Sond., op cit., **2**: 587 (1862).—Bak. f. in Journ. Linn. Soc., Bot. **40**: 28 (1911). Type from S. Africa.

Hibiscus vitifolius var. *ricinifolius* Hochr. in Ann. Conserv. Jard. Bot. Genève, **4**: 170 (1900).—Exell & Mendonça, C.F.A. **1**, 2: 177 (1951).—Brenan in Mem. N.Y. Bot. Gard. **8**, 3: 225 (1953).—Keay, F.W.T.A. ed. 2, **1**, 2: 346 (1958). Type as for *H. ricinifolius*.

Hibiscus vitifolius var. *genuinus* forma *zeylanicus* Hochr. in Ann. Conserv. Jard. Bot. Genève, **4**: 169 (1900).

Young stems sparsely hairy, often with a longitudinal line of fine pubescence, sometimes with scattered weak simple hairs and often with small scattered prickles. Leaves usually drying dark green. Leaf-lamina usually rather deeply 3–5 (7)-lobed; lower surface with glabrous areas and hairs only on the veins or more rarely with some scattered hairs on the reticulation. Carpellary awns varying from nearly absent to 7 mm. long.

S. Rhodesia. E: Melsetter Distr., Chirinda Forest, fr. 26.x.1947, *Chase* 422 (BM; SRGH). **Nyasaland.** S: Cholo Distr., Cholo Mt., fl. 21.ix.1946, *Brass* 17720 (BM; K; SRGH). **Mozambique.** SS: Buzi, Madanda Forest, 120 m., fl. 5.xii.1906, *Swynnerton* 2054 (BM).

Rain-forests in Ceylon, Cameroons, Ethiopia, Uganda, Angola, Belgian Congo, Tanganyika, S. Rhodesia, Nyasaland, Mozambique and S. Africa.

Although widespread the type subspecies has rarely been collected and seems to be uncommon and sporadic. In the very few flowers I have been able to examine the style seems to branch almost at the apex of the staminal tube and does not extend for some millimetres beyond it before dividing, as it does in subsp. *vulgaris*. If this difference is real it might become necessary to raise the two subspecies to specific rank.

For an account of the typification of this species see Brenan and Exell (in Bol. Soc. Brot., Sér. 2, **32**: 69 (1958)). The type in Hermann's herbarium (BM) is this forest taxon (hence subsp. *vitifolius*): the plant commonly known as *H. vitifolius* is subsp. *vulgaris* Brenan & Exell (see below).

Subsp. **vulgaris** Brenan & Exell in Bol. Soc. Brot., Sér. 2, **32**: 73 (1958). TAB. **89** fig. 2. Type from Angola (Luanda).

    Hibiscus vitifolius var. *genuinus* Hochr. in Ann. Conserv. Jard. Bot. Genève, **4**: 169 (1900) pro parte excl. forma *zeylanicus*.

    Hibiscus vitifolius var. *adhaerens* Ulbr. in Engl., Bot. Jahrb. **48**: 377 (1912).—Arwidss. apud Norlindh & Weim. in Bot. Notis. **1934**: 98 (1934). Type from SW. Africa.

    Hibiscus vitifolius var. *vitifolius* sensu Keay, F.W.T.A. ed. 2, **1**, 2: 346 (1958).

Young stems tomentose or tomentellous, often glandular, sometimes hispid, prickles very infrequent. Leaves usually drying brownish-green. Leaf-lamina not lobed to rather shallowly 3–5 (7)-lobed; lower surface fairly densely to densely hairy.

Bechuanaland Prot. N: Chobe-Zambezi confluence, fl. 11.iv.1955, *E.M. & W.* 1480 (BM; LISC; SRGH). **N. Rhodesia.** E: Petauke–Old Petauke Boma road, fl. & fr. 4.xii.1958, *Robson* 835 (K; SRGH). S: Livingstone, fl. iv.1909, *Rogers* 7095 (SRGH). **S. Rhodesia.** N: Urungwe Distr., Nyanyanga R., 16 km. E. of Kariba, fl. 22.i.1951, *Goodier* 18 (BM; SRGH). W: Bulawayo, fl. x.1897, *Rand* 26 (BM). C: Hartley, Gatooma, fl. xii.1924, *Hoffe* 3 (PRE). E: Inyanga, near Cheshire, 1300 m., fl. & fr. 15.i.1931, *Norlindh & Weimarck* 4357 (BM). S: Ndanga, Triangle Sugar Estate, fl. & fr. 25.i.1949, *Wild* 2732 (BM; SRGH). **Nyasaland.** N: Mzimba, fr. 2.vi.1947, *Benson* 1287 (BM). C: Lilongwe, fl. & fr. 1.iv.1955, *Jackson* 1559 (BM; K; SRGH). S: Chikwawa, 180 m., fl. & fr. 5.x.1946, *Brass* 17983 (K; SRGH). **Mozambique.** N: between Chiure and the Lúrio Falls, fr. 19.viii.1948, *Pedro & Pedrógão* 4851 (SRGH). Z: Mocuba, fl. 1945, *Faulkner* 370 (SRGH). T: Benga, fl. & fr. 16.v.1948, *Mendonça* 4271 (BM; LISC). MS: Vila Machado, fl. & fr. 15.iv.1948, *Mendonça* 3917 (BM; LISC). LM: Maputo, fl. 18.i.1957, *Barbosa & Lemos in Barbosa* 7507 (LMJ).

Widespread in tropical and subtropical regions of the Old World; probably introduced in tropical America. Woodlands, grasslands and roadsides, often on alluvial sands.

Subsp. *vulgaris* is the plant which has commonly been known as *H. vitifolius*. It could be divided into varieties or forms (see Hochreutiner, loc. cit.) on the presence or absence of glandular hairs, hispid hairs, prickles etc. but it is doubtful whether these differences have much significance (see Brenan & Exell, loc. cit.).

9. KOSTELETZKYA C. Presl

Kosteletzkya C. Presl, Reliq. Haenk. **2**: 130, t. 70 (1835).

Shrubs or herbs. Flowers solitary, axillary or in axillary or terminal racemes. Epicalyx of 7–10 filiform to linear bracts. Calyx 5-lobed or 5-toothed. Ovary 5-locular; loculi 1-ovulate; style 5-branched, stigmas capitate. Capsule depressed-globose with 5 prominent angles, dehiscing loculicidally. Seeds reniform.

Upper leaves oblong-lanceolate to linear; flowers white - - - 1. *buettneri*
Upper leaves ovate to suborbicular; flowers purple or pink:
  Stems with subflexuous ± deflexed hairs - - - - 2. *adoensis*
  Stems with subsetose patent hairs - - - - - 3. *begoniifolia*

1. **Kosteletzkya buettneri** Gürke in Verh. Bot. Verein. Brand. 31: 92 (1889).—Eyles in Trans. Roy. Soc. S. Afr. **5**: 416 (1916).—Exell & Mendonça, C.F.A. **1**, **1**: 157 (1937).—Keay, F.W.T.A. ed. 2, **1**, 2: 349 (1958). TAB. **91**. Type from Angola (Zaire).

Erect annual or perennial herb with rhizomatous rootstock; stems ± appressed-stellate-setulose-pubescent, branched from the base. Leaf-lamina 4–10 (15) ×

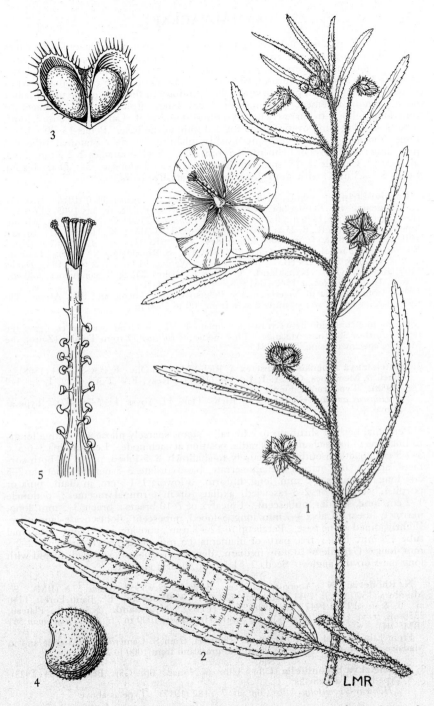

Tab. 91. KOSTELETZKYA BUETTNERI. 1, flowering and fruiting stem (×⅔) *Rand* s.n.; 2, leaf (×⅔) *Eyles* 224; 3, vertical section of capsule (×4) *Eyles* 224; 4, seed (×6) *Eyles* 224; 5, staminal tube and style-branches (×4) *Milne-Redhead* 3397.

0·4–1·5 (2·5) cm., narrowly ovate to lanceolate becoming linear-oblong to narrowly linear towards the apices of the stems, densely stellate-pubescent, apex rather blunt, margin serrate, base cuneate to obtuse; petiole 8–30 mm. long; stipules 3–10 mm. long, filiform, caducous. Flowers up to 4·5 cm. in diam., white sometimes tinged with pink, drying yellow, solitary in the axils of the upper leaves or in terminal racemes; peduncle 0·6–6 cm. long. Epicalyx of 7–10 bracts 3·5 mm. long, filiform to narrowly linear, setulose-pubescent. Calyx up to 8 mm. long; lobes 4–5 × 2–4 mm., ovate, joined for 2–3 mm. at the base. Petals 15–20 × 10–15 mm., obovate, stellate-pubescent outside, glabrous inside. Staminal tube 8–9 mm. long; free parts of filaments 0·5 mm. long. Style-branches 2–2·5 mm. long, glabrous. Capsule 8–15 mm. in diam., depressed-globose, 5-angled, hispid. Seeds 3·5 × 3 mm., with short dense brown scale-like hairs.

Caprivi Strip. Kakumba I., fl. 17.i.1959, *Killick & Leistner* 3480 (PRE; SRGH). **N. Rhodesia.** B: Barotseland (without precise locality), fl. & fr. 5.ix.1905, *Allen* 133 (K). N: Abercorn, 1520 m., fl. & fr. 15.i.1955, *Fanshawe* 2129 (K; NDO). W: Mwinilunga Distr., Lunga R., fl. 25.xi.1937, *Milne-Rehead* 3397 (BM; K; PRE). C: Broken Hill, 1220 m., fl. & fr. xi.1909, *Rogers* 8646 (K; SRGH). S: Livingstone Distr., Kasungula, 975 m., fl. 5.i.1957, *Gilges* 709 (SRGH). **S. Rhodesia.** N: Mazoe, 1310 m., fl. & fr. xii.1905, *Eyles* 224 (BM; SRGH). W: Wankie, 910 m., fl. & fr. iv.1955, *Davies* 1109 (BM; SRGH). **Nyasaland.** C: Bua R., below Mude R. confluence, 900 m., fl. & fr. 10.ii.1959, *Robson* 1542 (BM; K; LISC; SRGH).
Also from Gambia to Nigeria, in the Belgian Congo, Angola and SW. Africa. On river margins and in wet swampy places from 900 to 1600 m.

Owing to a confusion between two specimens, Exell and Mendonça (loc. cit.) gave the type as *Büttner* 381 from Malange. This should be *Büttner* 72 from Tondoa (Zaire), the Malange specimen being *Teucz* 381.

2. **Kosteletzkya adoensis** (Hochst. ex A. Rich.) Mast. in Oliv., F.T.A. **1**: 194 (1868).—Exell & Mendonça, C.F.A. **1**, 1: 158 (1937).—Keay, F.W.T.A., ed. 2, **1**, 2: 349 (1958). Type from Ethiopia.
    Hibiscus adoensis Hochst. ex A. Rich., Tent. Fl. Abyss. **1**: 54 (1847). Type as above.

Perennial herb or shrub up to 2 m. tall; stems sparsely pilose and with a longitudinal line of hairs changing its radial position at each node. Leaf-lamina 3–6·5 × 3–7·5 cm., ovate in outline, obscurely to shallowly 3–5-lobed, pilose on both surfaces, apex acute, margin crenate-serrate, base cordate 3–5-nerved; petiole 2–5 cm. long; stipules 4–5 mm. long, filiform. Flowers 1–1·5 cm. in diam., pink or purplish, solitary or 2–5-fascicled, axillary or in terminal racemes; peduncle 1–2 cm. long, slender, pubescent. Epicalyx of 7–10 bracts; bracts 3–5 mm. long, narrowly linear. Calyx 4–7 mm. long, 5-lobed, pubescent; lobes ovate-lanceolate, slightly joined at the base. Petals up to 12 × 8 mm., obovate, pubescent. Staminal tube 4·5 mm. long; free parts of filaments 0·5 mm. long. Style-branches 2–2·5 mm. long. Capsule 6–10 mm. in diam., depressed-globose, 5-angled, hispid with long hairs on the angles. Seeds 2 × 1·5 mm., nearly glabrous.

N. Rhodesia. N: Abercorn, 1675 m., fl. & fr. 22.v.1952, *Siame* 57A (BM). S: Mumbwa, 1500 m., fl. 1911, *Macaulay* 729 (K). **S. Rhodesia.** E: Banti Forest, 1740 m., fl. & fr. 19.viii.1947, *Farquhar* 21/47 (FHO). **Nyasaland.** N: Nyika Plateau, 2350 m., fr. vii.1896, *Whyte* (K). S: Shire Highlands, 1000 m., fr. 1891, *Buchanan* 561 (BM; K).
From Ethiopia and the Sudan to S. Rhodesia, from S. Cameroons to Angola and in Madagascar. In forest clearings, bush and grassland from 1000 to 2350 m.

3. **Kosteletzkya begoniifolia** (Ulbr.) Ulbr. in Notizbl. Bot. Gart. Berl. **8**: 684 (1923). Type from Tanganyika.
    Hibiscus begoniifolius Ulbr., op. cit. **7**: 182 (1917). Type as above.

Erect perennial herb or shrublet; stems patent-setose. Leaf-lamina 5–9 × 4–6 cm., suborbicular, densely hairy on the nerves of the under surface of the leaf, apex acute, margin serrate or crenate, base cuneate to slightly cordate 5–7-nerved; petiole 1·5–3·5 cm. long; stipules 3–7 mm. long, filiform to narrowly lorate. Flowers pink, solitary, axillary or in terminal or axillary racemes; peduncle up to 5·5 cm. long. Epicalyx of 8–10 bracts; bracts 4–10 mm. long, filiform to narrowly linear, setulose. Calyx up to 12 mm. long, 5-lobed, hispid; lobes 8–12 × 4–7 mm. Petals up to 25 mm. long, narrowly obovate, pubescent. Staminal tube 10–12 mm. long; free parts of filaments 0·5 mm. long. Style-branches 3 mm. long. Capsule 1·5 cm. in diam., with very prominent angles, hispid. Seeds 2·5 × 1·5 mm., subreniform, with short scattered hairs or nearly glabrous.

N. Rhodesia. W: Mwinilunga, fl. & fr. 1.xii.1937, *Milne-Redhead* 3466 (K). Also in Ethiopia, Kenya and Tanganyika. Marshy ground.

10. SIDA L.

Sida L., Sp. Pl. 2: 683 (1753); Gen. Pl. ed. 5: 306 (1754).

Annual or perennial herbs or suffrutices, erect or prostrate, glabrous or pubescent or pilose to densely tomentose or velutinous. Leaves usually undivided, serrate or crenate-serrate, occasionally 3-lobed, cuneate to cordate at the base, usually petiolate. Flowers small to medium-sized, usually cream or orange, long-pedicelled to subsessile or solitary, clustered or fasciculate in the leaf-axils or arranged in racemes (sometimes subcapitate or subumbellate) or spikes. Epicalyx absent. Calyx shallow-campanulate to saucer-shaped; lobes ovate or triangular, generally acute to acuminate. Petals not clawed. Staminal tube dilated at the base, divided at the apex into several to many free filaments; free parts of filaments terete. Ovary of 5 to many carpels, each with a single pendulous ovule; style terete or subclavate; stigmas capitate or truncate. Fruit of 5 to many mericarps ultimately separating from the torus, dehiscent at the apex (or rarely at the base) or indehiscent, smooth, transversely ribbed, glabrous or hairy, usually acute, beaked or awned. Seeds triangular; cotyledons folded; endosperm scanty or absent.

About 200 species, circumtropical, mainly in the New World.

The African species can be roughly divided into three groups, one including several circumtropical weeds and semi-weedy species with a wide distribution and often found in waste places (*S. acuta*, *S. rhombifolia*, *S. cordifolia*), the second including species more or less restricted to the tropical regions of Africa or of the Old World (*S. veronicifolia*, *S. urens*), and the third group which includes species of a restricted distribution, such as *S. chrysantha*, *S. dregei*, *S. hoepfneri*, *S. pseudo-cordifolia* (southern Africa) and *S. serratifolia* (tropical to S. Africa). The third group is usually not found in waste places but in a variety of other habitats. There is one species of *Sida*, viz. *S. ternata* L. f., found in the Ethiopian region and in S. Africa as far north as the Transvaal, of which I have not seen any records from our region although it is likely to occur in Nyasaland and the Eastern Border Mts. of Rhodesia. It is easily distinguishable by its 3-lobed leaves, small solitary flowers and long slender pedicels.

The flowers of *Sida* species open in the morning, while those of *Abutilon* are generally open only in the afternoon and evening.

Erect to prostrate plants with cordate leaves and with long spreading hairs on stems and petioles; mericarps 5, not dehiscing at the base:
  Often prostrate; stems weakly pilose; flowers solitary or few together; pedicels filiform, usually over 20 mm. long; calyx not pilose - - - 1. *veronicifolia*
Erect or decumbent, more or less densely pilose on stems, petioles, pedicels and calyx; flowers often clustered; pedicels mostly under 15 mm. long - - 2. *urens*
Erect or spreading plants, variously pubescent, but without long spreading hairs; mericarps 6–12, or, if 5, opening at the base or leaves not cordate:
Mericarps 5, dehiscing at the base and setaceously birostrate - - - 3. *alba*
Mericarps not dehiscing at the base:
  Flowers axillary, solitary, rarely fasciculate or terminal and corymbose; leaves generally oblong or elliptic to lanceolate, usually at least twice as long as broad, and usually under 3 cm. broad, with base rarely subcordate; mericarps, if awned, without reflexed hairs on the awns:
    Plant glabrous to sparsely or rarely more densely pubescent, but not velutinous or densely tomentose; leaves more or less lanceolate, sharply serrate; mericarps 5–8:
      Pedicels usually under 20 mm. long, often much shorter; mericarps rugose, sharply birostrate - - - - - - - 4. *acuta*
      Pedicels over 20 mm. long, often much longer; mericarps smooth (or wrinkled only through drying), usually muticous; pericarp thin and papery, usually more or less stellate-pubescent - - - - - 5. *dregei*

Plant more or less tomentose; leaves either not lanceolate or, if so, discolorous
  and tomentose-canescent on lower surface; mericarps 7–12:
  Pedicels usually shorter than the petioles; mericarps 7–8, birostrate but beaks
    more or less connivent, rarely erostrate; leaves ovate-elliptic, rounded
    at the apex - - - - - - - - - - -   6. *ovata*
  Pedicels longer than the petioles; mericarps 7–12:
    Mericarps distinctly birostrate with suberect often connivent awns, nearly
      glabrous - - - - - - - - -   7. *rhombifolia*
    Mericarps not birostrate, often more or less stellate-pubescent:
      Leaves usually distinctly discolorous, whitish- or greyish-tomentose
        beneath:
        Low much-branched suffrutex, up to 30 (60) cm. tall; leaves ± oblong,
          rounded to subacute at the apex, upper surface usually drying green
            9. *chrysantha*
        Erect subvirgate shrub, 0·7–3 m. tall, not much branched from the base;
          leaves ± lanceolate, attenuate and usually acute at the apex, upper
          surface often drying dark brown - - - -   8. *serratifolia*
      Leaves nearly concolorous; plant canescent or yellowish-grey- to grey-
        green-tomentose, usually over 50 cm. tall - - -   10. *hoepfneri*
Flowers usually clustered in the upper axils, sometimes solitary in the lower axils;
  pedicels short; leaves broadly ovate, subcordate to truncate at the base, usually
  not more than 1·5–2 times as long as broad, the larger ones over 3 cm. broad:
  Mericarps long-rostrate with erect awns covered with reflexed short stiff hairs
                                                11. *cordifolia*
  Mericarps muticous or acute but not awned - - -   12. *pseudocordifolia*

1. **Sida veronicifolia** Lam., Encycl. Méth. Bot. **1**: 5 (1783).—Exell & Mendonça,
    C.F.A. **1, 1**: 147 (1937).—Keay, F.W.T.A. ed. 2, **1, 2**: 338 (1958). Type from India.
    Sida humilis Cav., Diss. **5**: 277, t. 134 fig. 2 (1788).—Mast. in Oliv., F.T.A. **1**:
    179 (1868). Type from Asia.

Trailing annual or biennial herb, sometimes rooting at the nodes; stems up to
1 m. long, slender, with sparse weak long hairs. Leaf-lamina up to 5 × 4 cm.,
broadly ovate-cordate to suborbicular-cordate, papyraceous, apex usually acuminate, margin crenate-serrate or ± doubly serrate, base broadly cordate, sparsely
pubescent (hairs usually simple) on both surfaces; petiole about as long as the
lamina, hairy like the stems. Flowers pale yellow, solitary or in shortly peduncled
subumbellate few-flowered racemes in the axils of the upper leaves; pedicels up
to 35 mm. long, slender to filiform, usually with a few hairs, articulated near the
middle. Calyx 5–6 mm. long, lobed to about the middle, nearly glabrous; lobes
ciliate, triangular, acute. Petals 4–6 mm. long. Mericarps 5, c. 3 mm. long, not
reticulated, shortly bicuspidate, pubescent at the apex and on the subsetaceous
± connivent awns, slightly longitudinally keeled on the back and lateral edges.

Nyasaland. S: Upper Shire Valley, fl. & fr. viii.1861, *Kirk* (K). **Mozambique.**
N: Mocuba, Namagoa, fl. & fr. vii–viii.1945, *Faulkner* 121 (K; PRE).
Widely distributed in the tropics.

2. **Sida urens** L., Syst. Nat. ed. 10, **2**: 1145 (1759).—Mast. in Oliv., F.T.A. **1**: 179
    (1868).—Exell & Mendonça, C.F.A. **1, 1**: 148 (1937).—Keay, F.W.T.A. ed. 2, **1,**
    2: 339 (1958). Type from Jamaica.

Perennial herb up to c. 60 cm. tall, branched from the base, covered on stems,
petioles, pedicels and calyx with hispid to pilose, long-patent to somewhat curly
hairs; stems erect or trailing, slender, terete. Leaf-lamina up to 7 × 4 cm.,
cordate-ovate to triangular-cordate, usually acuminate or attenuate at the apex,
regularly and acutely serrate, more or less pubescent to strigose-pilose on both
surfaces; petiole usually about as long as the lamina, often somewhat abruptly
upturned at the apex. Flowers pale yellow or buff, sometimes with reddish centre,
in clusters, fasciculate or subcapitate on a short common peduncle, often subsessile;
pedicels slender to filiform, up to 12 mm. long. Calyx 5–7 mm. long, lobed to a
little beyond the middle, angular; lobes triangular, attenuate-aristate. Petals c.
9 mm. long, ± emarginate. Mericarps 5, c. 2·5 mm. long, dorsally not reticulated
and nearly smooth, laterally faintly flabellately striate, shortly beaked or muticous,
somewhat pubescent near the apex.
  N. Rhodesia. N: Abercorn Distr., Chilongowelo, fl. 13.v.1952, *Richards* 1803 (K).

W: Mukinge Hill Mission, Kasempa, fl. x.1934, *Trapnell* 1633 (K). **S. Rhodesia.** N: Miami, fl. iv.1926, *Rand* 8 (BM). C: Salisbury Experimental Station, fl. 24.iii.1958, *Phipps* 1060 (K; PRE; SRGH). **Nyasaland.** N: Kondawe-Karonga, fl. vii.1896, *Whyte* (K). S: Domasi Valley, fl. 25.iv.1949, *Wiehe* N/81 (K). **Mozambique.** N: between Balama and Nungo, *Pedro & Pedrógão* 4942 (LMJ).
Widespread in tropical and subtropical regions. Ruderal.

3. **Sida alba** L., [Fl. Jam.: 18 (1759) *nom. nud.*] Sp. Pl. ed. 2, **2**: 960 (1763).—Exell & Mendonça, C.F.A. **1**, 1: 148 (1937).—Brenan, T.T.C.L.: 306 (1949).—Keay, F.W.T.A. ed. 2, **1**, 2: 339 (1958). Type from the West Indies.
Sida spinosa L., Sp. Pl. **2**: 683 (1753) pro parte.—Harv. in Harv. & Sond., F.C. **1**: 167 (1860).—Mast. in Oliv., F.T.A. **1**: 180 (1868).—Ulbr. in Engl., Bot. Jahrb. **51**: 41 (1913).—R.E.Fr., Wiss. Ergebn. Schwed. Rhod.-Kongo-Exped. **1**: 144 (1914). —Burtt Davy, F.P.F.T. **2**: 276 (1932).—Arwidss. apud Norlindh & Weim. in Bot. Notis. **1934**: 94 (1934). Type from India.

Annual or perennial woody or suffruticose plant 30–75 cm. tall, often branched from the base, with somewhat virgate erect or occasionally somewhat trailing branches, covered with a dense short stellate pubescence; stems faintly angular to terete, often soon purplish-brown in colour, at length glabrescent. Leaf-lamina 2–5 × 0·5–3 cm., usually narrowly ovate but varying from oblong to lanceolate or ovate-lanceolate, apex usually narrowed and subacute but often somewhat rounded, margin sharply and regularly serrate, base usually truncate to rounded sometimes somewhat cordate, upper surface dark green drying brown, thinly and shortly pubescent, glabrescent, lower surface whitish to ash-grey, softly tomentellous; petiole up to c. 3 cm. long, slender, somewhat purplish in older leaves, shortly pubescent, with a callus at the base that is often shortly spinescent; stipules 6–10 mm. long, usually persistent, setaceous. Flowers white to yellow, solitary or fascicled in the leaf-axils; pedicels rarely more than 15 mm. long, slender to fili-form, articulated in the upper 3–5 mm.; Calyx c. 6 mm. long, cupular, 10-ribbed, tomentellous, lobed to about the middle; lobes deltoid, acute, minutely mucronate-apiculate. Petals 5–6 mm. long. Mericarps 5, 4–5 mm. long (including the awns), finely reticulate, with 2 setaceous awns, opening somewhat irregularly at the base. Seeds c. 1·5 mm. long, dark brown, almost smooth, glabrous.

N. Rhodesia. B: near Senanga, fl. & fr. 21.vii.1952, *Codd* 7285 (BM; PRE; SRGH). N: near Mpulungu, fl. & fr. 23.ii.1955, *Richards* 4634 (K). W: Mukinge Hill Mission, Kasempa, fr. x.1934, *Trapnell* 1632 (K). C: Lusaka, *Robinson* 1473 (K). S: Mapanza Mission, fl. & fr. 22.i.1953, *Robinson* 63 (K). **S. Rhodesia.** N: Trelawney, fl. 27.iii.1943, *Jack* 157 (SRGH). W: Matobo Distr., Besna Kobila, fl. & fr. iii.1958, *Miller* 5168 (K; PRE; SRGH). C: Salisbury, fl. & fr. 25.v.1931, *Brain* 4637 (BM; SRGH). E: Inyanga, 1700 m., fl. & fr. 15.i.1931, *Norlindh & Weimarck* 4419 (BM; PRE; SRGH). S: Nuanetsi, *Davies* 1698 (SRGH). **Nyasaland.** N: Nyika Plateau, ii–iii.1903, *McClounie* 164 (K). C: Lilongwe, fl. 1.iv.1955, *Jackson* 1555 (K; SRGH). **Mozam-bique.** T: between Tete and Boroma, *Mendonça* 345 (LISC). MS: between Manica and the frontier Customs Office, fl. & fr. 4.iv.1948, *Mendonça* 3868 (BM; LISC). LM: Maracuene, near Namaacha, fr. 24.iv.1947, *Pedro & Pedrógão* 675 (LMJ; PRE).
Throughout tropical Africa to SW. Africa, Transvaal and Natal; also in America. A widespread species, found in grassland and woodland types of vegetation on a variety of soils; also semi-ruderal, but apparently rarely a weed.

4. **Sida acuta** Burm. f., Fl. Ind.: 147 (1768).—Eyles in Trans. Roy. Soc. S. Afr. **5**: 413 (1916).—Arwidss. apud Norlindh & Weim. in Bot. Notis. **1934**: 96 (1934).—Keay, F.W.T.A. ed. 2, **1**, 2: 339 (1958). Type from India.
Sida carpinifolia sensu Mast. in Oliv., F.T.A. **1**: 180 (1868).

Erect suffrutex up to c. 1 (3) m. tall, glabrous or sometimes sparsely stellately hairy; stems usually much branched, green, woody only at the base. Leaf-lamina 2–6 (10) × 0·5–1·5 (3·8) cm., green, lanceolate or linear-lanceolate, occasionally ovate-lanceolate, apex acute, margin sharply and regularly serrate, base obtuse to rounded; petiole 2·5 mm. long, pubescent; stipules usually longer than the petiole, linear. Flowers yellow, solitary or paired, axillary; pedicels 4–12 mm. long. Calyx 6–8 mm. long, membranous, glabrous, saucer-shaped, somewhat angular, divided to about the middle; lobes deltoid, caudate-acuminate. Petals 6–8 mm. long, ciliate. Mericarps 5–6, c. 4 mm. long, birostrate with a groove between the awns, reticulately striate and sulcate. Seeds smooth, glabrous except for a pub-escent area around the hilum.

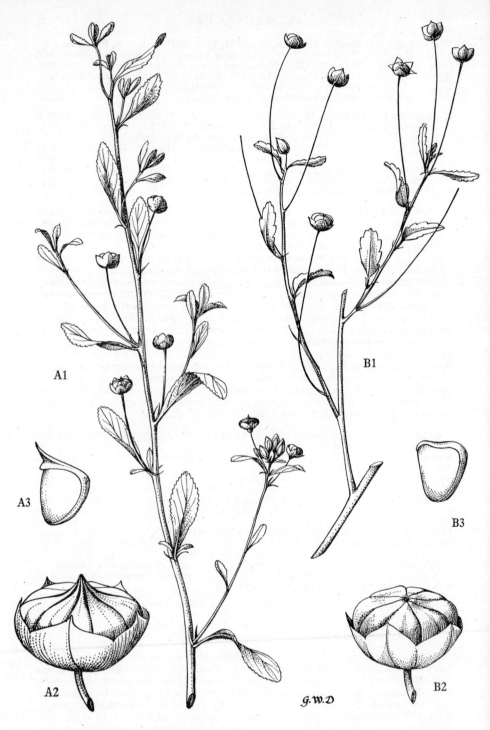

Tab. 92. A.—SIDA RHOMBIFOLIA. A1, stem with flowers and fruits (×1); A2, fruit (×5); A3, mericarp (×5). All from *Chase* 5857. B.—SIDA DREGEI. B1, stem with fruits (×1); B2, fruit (×5); B3, mericarp (×5). All from *Borle* 162.

S. Rhodesia. N: Trelawney, fl. & fr. 23.xii.1942, *Jack* 79 (K; PRE; SRGH). C: Salisbury, *Eyles* 5342 (SRGH). E: Umtali, fl. & fr. 13.xi.1955, *Chase* 5857 (BM; PRE; SRGH). **Nyasaland.** N: Rumpi Distr., Deep Bay, fr. 12.v.1952, *White* 2828 (FHO; K). C: Kota Kota, fl. & fr. 6.iv.1944, *Benson* 74 (PRE). S: Mlanje, Likabula, fl. & fr. 9.v.1958, *Chapman* 609 (SRGH). **Mozambique.** Z: Mocuba, Namagoa, fl. iv.1945, *Faulkner* 205 (K; PRE; SRGH). MS: Vila Pery, *Pedro & Pedrógão* 5985 (LMJ; SRGH). SS: R. das Pedras, fr. vii.1936, *Gomes e Sousa* 1777 (K). LM: Catembe, fr. 9.v.1920, *Borle* 499 (K; PRE; SRGH).

Widespread in the tropics and extending into SW. Africa, Transvaal and Natal. In fairly high-rainfall areas with practically no winter frost. In cultivation and along roadsides, often locally frequent, from sea-level to c. 1000 m.

5. **Sida dregei** Burtt Davy, F.P.F.T. **1**: 49 (1926); op. cit. **2**: 276 (1932). TAB. **92** fig. B. Type from Cape Prov.

Sida capensis sensu Eckl. & Zeyh., Enum. Pl. Cap.: 40 (1834).—Suesseng. in Proc. & Trans. Rhod. Sci. Ass. **43**: 104 (1951).

Sida longipes E. Mey. ex Harv. in Harv. & Sond., F.C. **1**: 167 (1860) non A. Gray (1852).—Ulbr. in Engl., Bot. Jahrb. **51**: 41 (1913).—Eyles in Trans. Roy. Soc. S. Afr. **5**: 413 (1916) pro parte. Type from Cape Prov.

Sida lancifolia Burtt Davy, op. cit. **1**: 50 (1926); op. cit. **2**: 276 (1932). Type from the Transvaal.

Sida pilosella Arwidss. apud Norlindh & Weim. in Bot. Notis. **1934**: 94 (1934). Type: S. Rhodesia, Inyanga, *F.N. & W.* 2858 (LD, holotype).

Erect, somewhat virgate or low and much-branched (after burning or grazing) annual or biennial suffrutex up to c. 1 m. tall, usually sparsely and finely stellate-pubescent on vegetative parts and calyx, occasionally either nearly glabrous or more densely pubescent; stems usually stiff or wiry when young, green, terete, ultimately glabrescent. Leaf-lamina 2–7 × 0·5–3·5 cm., ovate-lanceolate to lanceolate or ovate, often somewhat rhombic, apex somewhat narrowed and acute, margin sharply and rather coarsely serrate, base rounded to obtuse; petiole up to 10 (15) mm. long, stellate-pubescent and often with longer soft hairs at the apex; stipules setaceous to linear, caducous. Flowers yellow to orange, axillary; pedicels usually 4–6 (8) cm. long, occasionally shorter, articulated in the upper 9 mm. Calyx 5–8 mm. long, angular owing to the raised commissural lines of the lobes, thinly papyraceous, green, divided to about the middle; lobes triangular, acute to apiculate, finely stellate-pubescent or stellate-puberulous mainly along the margins and on the median veins. Petals 9–11 mm. long. Mericarps (6) 7 (8), c. 2·5 mm. long, truncate at the apex, muticous, usually with a short shallow longitudinal groove at the ventral apical angle, smooth (somewhat wrinkled when immature at time of drying), usually with sparse many-rayed hairs especially at the apex. Seeds c. 2 mm. long, black, ovoid-reniform, smooth, glabrous except for the area of the hilum.

S. Rhodesia. W: Matopos, fl. & fr. i.1957, *Miller* 4075 (PRE; SRGH). C: Selukwě, 915 m., fl. & fr. 8.xii.1953, *Wild* 4302 (BM; PRE; SRGH). E: Umtali, *Hopkins* in GHS 9742 (SRGH). S: Fort Victoria, fr. 17.iv.1946, *Greatrex* in GHS 14767 (K; SRGH). **Mozambique.** SS: Gaza, Chifucaze, R. Limpopo, fl. & fr. 6.x.1945, *Pedro* 194 (K; LMJ; PRE). LM: Goba, *Mendonça* 1649 (BM; LISC).

Also in the wetter areas of the Transvaal, Natal and E. Cape Prov. Usually in forest margins, in light shade, among scrub or among herbaceous plants along streams, rivers, etc., often on rocky and acid soils, from sea-level to about 1500 m. in areas with a fairly high to high rainfall and little or no frost.

The roots of this species survive bush-fires but the regrowth is often stunted and more or less atypical. The type gathering of *S. pilosella* is from a specimen that had been burnt and regenerated from the rootstock. Pedro and Pedrógão collected a specimen in Mozambique, Nampula, which differs in having rather dense clusters of flowers in the upper axils. It may belong here but more material is needed to decide its status.

6. **Sida ovata** Forsk., Fl. Aegypt.-Arab.: CXVI, 124 (1775). Type from Arabia.

Sida grewioides Guill. & Perr. in Guill., Perr. & Rich., Fl. Senegamb. Tent. **1**: 71 (1831).—Garcke in Peters, Reise Mossamb. Bot. **1**: 128 (1861).—Mast. in Oliv., F.T.A. **1**: 182 (1868).—Burtt Davy, F.P.F.T. **2**: 277 (1932).—Brenan, T.T.C.L.: 306 (1949).—Keay, F.W.T.A. ed. 2, **1**, 2: 339 (1958). Type from Senegambia.

Sida spp.—Eyles in Trans. Roy. Soc. S. Afr. **5**: 414 (1916) pro parte quoad specim. *Chubb* 304.

Suberect suffrutex up to c. 50 cm. tall, densely grey- or grey-green-tomentose; branches sometimes decumbent; stems terete or somewhat angular when young, wiry, at length glabrescent and somewhat woody. Leaf-lamina 10–35 × 6–25 mm., elliptic, ovate, oblong or somewhat obovate to suborbicular, apex obtuse to rounded, margin rather coarsely serrate-crenate, base usually cuneate to rounded, both surfaces densely grey-green-tomentose, the lower surface slightly paler and usually with longer hairs; petiole 2–10 mm. long, tomentose; stipules filiform, often somewhat patent, tomentose. Flowers yellow, axillary, solitary or occasionally 2–3-fasciculate, subsessile or pedicellate; pedicels up to 5 (9) mm. long. Calyx c. 6 mm. long, subglobose-campanulate, 10-ribbed, tomentose, lobed to about the middle; lobes broad, subacute to shortly apiculate. Petals not much exceeding the calyx. Mericarps 7–8, c. 4 mm. long, birostrate, separating at dehiscence, dorsally rugulose and very faintly cross-veined, laterally reticulate-striate; awns 2, c. 1 mm. long, often connivent, setaceous, pungent. Seeds c. 2 × 1·5 mm., rather tumid, nearly smooth, glabrous.

S. Rhodesia. W: Wankie, *Plowes* 1844 (SRGH). E: near Sabi Valley Experiment Station, fl. & fr. 12.i.1960, *Whellan* 1261 (SRGH). **Mozambique.** T: Metunde, fl. & fr. 15.iv.1948, *Mendonça* 4265 (LISC; SRGH). MS: Sena, fl. & fr. iv.1860, *Kirk* (K). SS: Gaza, Chibuto, *Torre* 6759 (LISC). LM: Magude, fr. 11.v.1957, *Carvalho* 193 (LM; PRE).
Arid and semi-arid regions from Senegal to Ethiopia and southwards; in SW. Africa and the Transvaal and also in Arabia and Socotra. Widespread but apparently nowhere common. Usually in the drier areas.

Confusion of this species with short-pedicelled forms of *S. hoepfneri* is possible. If there are no ripe mericarps distinction is difficult, but the awned mericarps of *S. ovata* and the muticous or minutely apiculate mericarps of *S. hoepfneri* always provide a clear-cut distinguishing character.

7. **Sida rhombifolia** L., Sp. Pl. **2**: 684 (1753).—Harv. in Harv. & Sond., F.C. **1**: 167 (1860).—Mast. in Oliv., F.T.A. **1**: 181 (1868).—Ulbr. in Engl., Bot. Jahrb. **51**: 51 (1913).—R.E.Fr., Wiss. Ergebn. Schwed. Rhod.-Kongo-Exped. **1**: 144 (1914).—Eyles in Trans. Roy. Soc. S. Afr. **5**: 413 (1916).—Burtt Davy, F.P.F.T. **2**: 277 (1932).—Arwidss. apud Norlindh & Weim. in Bot. Notis. **1934**: 96 (1934).—Exell & Mendonça, C.F.A. **1**, 1: 149 (1937).—Brenan, T.T.C.L.: 307 (1949).—Martineau, Rhod. Wild Fl.: 48 (1954).—Keay, F.W.T.A. ed. 2, **1**, 2: 339 (1958). TAB. **92** fig. A. Type from India.

Annual or biennial herb or suffrutex or shrublet, up to c. 1 m. tall; erect, often ± virgate, usually not much branched; stems terete, grey or brown, stellate- pubescent to stellate-tomentose, ultimately glabrescent. Leaf-lamina 2–6 × 0·5–2·5 cm., oblong to lanceolate-oblong or somewhat rhombid, shortly and sparsely stellate-pubescent or almost glabrous on the darker upper surface, shortly stellate-pubescent to velutinous on the pale lower surface, apex usually obtuse to rounded or rarely acute, margin finely crenate or subserrate, base broadly cuneate to rounded; petiole usually less than 5 mm. long; stipules c. 5. mm. long. Flowers pale yellow to cream, axillary, solitary; pedicels rarely exceeding 20 mm. long in flower and 35 mm. long in fruit. Calyx 4–5 mm. long, saucer- to cup-shaped, divided to about ⅓, stellate-velutinous or more sparsely pubescent, 10-nerved; lobes triangular, acute, sometimes mucronate. Petals c. 8 mm. long. Staminal tube glabrous or sparsely glandular-papillose. Mericarps 8–12, c. 4 mm. long, birostrate, with awns usually connivent until dehiscence of the fruit, glabrous except for the usually shortly pubescent awns. Seeds smooth and glabrous except for the pubescent area around the hilum.

N. Rhodesia. C: Mt. Makulu, near Lusaka, fl. 14.iv.1956, *Robinson* 1473 (K; SRGH). S: Mumbwa, fl. 1911, *Macaulay* 755 (K). **S. Rhodesia.** N: Mazoe, fl. iii.1906, *Eyles* 272 (BM; SRGH). W: Umgusa Forest Reserve, fl. & fr. x.1953, *Orpen* 11/53 (SRGH). C: Marandellas, *Newton* 70 (SRGH). E: Inyanga, fl. & fr. 20.i.1931, *Norlindh & Weimarck* 4468 (BM; PRE). **Nyasaland.** C: Namitete R., Lilongwe-Fort Jameson road, 1150 m., fl. & fr. 5.ii.1959, *Robson* 1458 (BM; K; LISC; SRGH). **Mozambique.** MS: Vumba Mts., *Pedro & Pedrógão* 6771 (LMJ; SRGH). SS: Vila de João Belo, fl. & fr. 8.x.1953, *Balsinhas* 1 (BM; LMJ). LM: Namaacha, fl. & fr. 21.v.1957, *Carvalho* 236 (LM; PRE).
Pantropical. Ruderal, usually near settlements.

8. **Sida serratifolia** Wilczek & Steyaert in Bull. Jard. Bot. Brux. **22**: 105 (1952). Type from the Belgian Congo.

Suffrutex or shrub, 0·75–2·5 m. tall; stems erect, ± virgate, not much branched, often only in upper portion, the younger parts densely floccose-stellate-pubescent. Leaf-lamina 4–10 × 1·3–3·5 cm., narrowly lanceolate to rhombic-lanceolate, gradually attenuate into an acute or obtuse apex or somewhat acuminate, margin regularly and finely serrate, base cuneate, upper surface dark green often drying dull dark grey-brown, sparsely covered with stellate hairs to almost glabrous, the paler canescent lower surface shortly and densely stellate-tomentose to subvelutinous particularly on the prominent veins; petiole up to c. 12 mm. long, densely tomentose, sometimes floccose; stipules up to c. 12 mm. long, filiform, stellate-hairy. Flowers yellow or golden-yellow, sometimes with a reddish centre, solitary or fascicled, axillary mainly in upper leaf-axils and forming pseudo-racemes and terminal corymbs (sometimes a solitary flower and a corymbose inflorescence in the same axil); peduncle up to 1·5 cm. long; pedicels up to 3·5 cm. long, slender to filiform, articulated at or above the middle. Calyx 5–7 mm. long, cupuliform-campanulate, finely stellate-pubescent outside, 10-ribbed near the base, lobed to about the middle; lobes triangular, apiculate to acuminate. Petals 7–10 mm. long, somewhat asymmetrical and ± bilobed. Staminal tube stellate-pilose. Mericarps 8–9, c. 2·5 × 2 mm., muticous or minutely apiculate, glabrous or very nearly so, dorsally somewhat rugose and faintly flabellately striate-reticulate, laterally thin to membranous. Seeds c. 1·5 mm. × 1·25 mm., dark brown, glabrous.

S. Rhodesia. C: Marandellas, *Davies* 980 (SRGH). E: Chirinda, fr. 27.v.1906, *Swynnerton* (BM; K). **Nyasaland.** S: Nampingudya Stream, Che Pira Village, fl. & fr. 22.iv.1955, *Banda* 76 (BM; SRGH). **Mozambique.** MS: Chimoio, Serra de Garuso, *Mendonça* 3891 (LISC; SRGH).
Also in the Belgian Congo, Uganda, Tanganyika, Transvaal and Swaziland. Forest edges, clearings and scrub, and on mountains up to 2000 m. in areas with a fairly high rainfall.

The habit is very similar to forms of the polymorphic *S. rhombifolia*, and without ripe mericarps it is sometimes very difficult to distinguish between the two species.

9. **Sida chrysantha** Ulbr. in Engl., Bot. Jahrb. **51**: 46, t. 2 fig. K–T (1913).—Burtt Davy, F.P.F.T. **2**: 277 (1932). Type from SW. Africa.
   Sida spp.—Eyles in Trans. Roy. Soc. S. Afr. **5**: 414 (1916) pro parte quoad specim. *Chubb* 355.

Low perennial suffrutex, much branched and woody at the base, up to 30 cm., rarely up to 60 cm. tall; stems usually branched throughout, terete or somewhat angled, usually greyish-green or greyish (rarely brownish), shortly and densely tomentose-velutinous, the upper portions usually slender and wiry. Leaf-lamina 2–4 × 1–2·5 cm., usually ovate-oblong to oblong but occasionally narrower or broader, apex subacute, margin irregularly rather coarsely serrate, base generally cuneate or more or less rounded, the upper surface dark green or brownish-green, usually sparsely and minutely stellate-pubescent, glabrescent, lower surface usually much paler, greyish- or yellowish-green, densely velutinous-tomentose and with prominent veins; petiole 3–10 mm. long, tomentose-velutinous; stipules 4 × 1 mm., ovate-lanceolate to lanceolate, velutinous, sometimes somewhat falcate, rather persistent. Flowers deep yellow to orange, occasionally paler, solitary, axillary; pedicels 3–6 cm. long, slender, terete, velutinous or tomentose, articulated near the apex. Calyx c. 8 mm. long, obconical-campanulate to saucer-shaped, tomentose outside, and along the margin inside, incised to about the middle; lobes triangular, somewhat narrowed or acuminate at the apex, 5 mm. broad at the base, with the median vein pinnately branching at about the middle of the lobe. Petals c. 10 × 8 mm., finely pubescent outside with simple hairs. Staminal tube with stellate hairs (usually sparse). Mericarps 7–9, c. 4 × 3 mm., muticous or subacute at the apex, usually stellate-hairy at the apex only, reticulate-veined on the flat lateral sides. Seeds brown, glabrous except for a small tomentose area around the hilum.

Bechuanaland Prot. SE: Kanye, fl. xi.1948, *Miller* B/794 (PRE; SRGH). **S. Rhodesia.** W: Bulawayo, fl. & fr. xii.1897, *Rand* 30 (BM).

Also in S. Africa and SW. Africa. On acid rocky slopes and in soils derived from acid rocks, among grasses and low growing plants or in woodland in light shade, usually at 1000–1500 m.

This species is closely related to *S. hoepfneri* and differs consistently only in the characters mentioned in the key, i.e. habit and discolorous leaves. Other characters used to separate the two species, such as the size of the flower, the pubescence on the staminal tube and the fruit, and the length of the pedicels, are not invariably constant. Some herbarium specimens, if not properly pressed or dried, or if in poor condition when collected, are difficult to place. There can be no doubt, however, that *S. hoepfneri* and *S. chrysantha* are quite distinct species, differing in ecology and hence with a different, though partly overlapping distribution.

10. **Sida hoepfneri** Gürke in Bull. Herb. Boiss. **3** : 404 (1895).—Ulbr. in Engl., Bot. Jahrb. **51** : 43 (1913).—Burtt Davy, F.P.F.T. **2** : 277 (1932).—Exell & Mendonça, C.F.A. **1**, 1 : 150 (1937). Syntypes from SW. Africa and Angola.

    Sida longipes var. *canescens* Szyszyl., Polypet. Thalam. Rehm. : 127 (1884).— Eyles in Trans. Roy. Soc. S. Afr. **5** : 413 (1916). Type from the Transvaal.

    Sida dinterana Hochr. in Bull. Herb. Boiss., Sér. 2, **2** : 1001 (1902).—Ulbr., tom. cit. : 45, t. 2 fig. A–J (1913). Type from SW. Africa.

    Sida chionantha Ulbr., tom. cit. : 44 (1913). Type from SW. Africa.

    Sida aurescens Ulbr., tom. cit. : 48 (1913). Syntypes from SW. Africa and Angola.

    Sida longipes sensu Eyles, loc. cit. pro parte quoad specim. *Rogers* 7168.

    Sida flexuosa Burtt Davy, op. cit. **1** : 49 (1926) ; op. cit. **2** : 277 (1932). Type from the Transvaal.

Erect annual or perhaps perennial herb, 0·5–1·75 m. tall, woody at the base, branched throughout but not much branched at the base, densely and shortly tomentose-subvelutinous with a whitish, yellowish, greenish-white, greyish-yellow or pale greyish-green indumentum on the younger parts of the stems, stipules, petioles, pedicels and outside of calyx ; stems terete, wiry, purplish-grey to brownish, longitudinally fissured, ultimately glabrous. Leaf-lamina 6 (7) × 0·5–3·5 cm., usually oblong or ovate-oblong to broadly elliptic, apex rounded or obtuse or occasionally acute to somewhat acuminate, margin distinctly and sometimes rather coarsely serrate or crenate-serrate or ± doubly serrate, shortly and densely tomentose on both surfaces with an indumentum coloured like that of the stems, upper surface occasionally ± glabrescent, veins usually ± impressed on the upper surface and prominent on the lower surface ; petiole 5 (8)–15 (25) mm. long. Flowers white to pale yellow (occasionally golden), solitary, axillary ; pedicels 1·5–5 (7) cm. long, usually slender, terete, articulated at or above the middle. Calyx (6) 8–11 mm. long, cupuliform, glabrous inside except for a narrow marginal tomentose zone, divided to about the middle ; lobes triangular to ovate-triangular, acute to ± acuminate, sometimes distinctly 3-nerved but more usually with only the raised median nerve conspicuous. Petals (14) 18–22 mm. long, with the narrow base ± fimbriate, glabrous or finely pubescent inside (mainly towards apex). Staminal tube glabrous or with sparse rather coarsely stellate hairs. Mericarps (7) 8–10, c. 6 mm. long, brown, muticous, rounded or minutely apiculate but not awned at the apex, usually stellate-pubescent at least in the upper part, dorsally finely rugose, laterally conspicuously reticulate with usually yellowish slightly raised ridges. Seeds ovoid, brown, glabrous except for the often tomentose or velutinous area around the hilum.

Caprivi Strip. Without precise locality, fr. x.1945, *Curson* 921 (PRE). **Bechuanaland Prot.** N : Ngamiland, Kwebe Hills, fl. & fr. 9.ii.1898, *Lugard* 167 (GRA ; K). SE : Mochudi, *Harbor* in *Rogers* 6488 (BM ; BOL ; K ; PRE ; SRGH). **N. Rhodesia.** B : Sesheke, fl. i.1876, *Holub* 390 (K). S : Livingstone, fl. iv.1909, *Rogers* 7168 (SRGH). **S. Rhodesia.** W : Victoria Falls, *Rogers* 5627 (BOL ; PRE ; SRGH). S : Beitbridge, fl. & fr. 16.ii.1955, *E.M. & W.* 443 (BM ; LISC ; SRGH). **Mozambique.** SS : Gaza, Guijà, fl. 16.xi.1957, *Barbosa & Lemos* in *Barbosa* 8174 (K ; LM).

Also in southern Angola, SW. Africa, Cape Prov. (Griqualand-W.), Transvaal and northern Natal.

Found mostly in rather arid woodland, usually on deep sandy soils (in association with *Terminalia sericea* Burch., etc.), but also sometimes on somewhat brackish or alkaline soils (in pans and periodically dry river beds, etc.) or on lower mountain slopes, usually between 600–1200 m.

This species is variable in some characters (hence the rather numerous synonyms) but constant in the short pale dense indumentum of the vegetative parts, the scarcely discolorous leaves and the muticous to acute, laterally flabellately streaked mericarps. The differences between *S. hoepfneri, S. dinterana, S. aurescens* and *S. chionantha* mentioned by Ulbrich in his key (tom. cit.: 40 (1913)) and in his descriptions, viz. colour of petals, pubescent or glabrous staminal tube and glabrous or hairy fruits, are not constant and I failed to find any other characters showing correlated differences. *S. flexuosa* is merely a depauperate small-leaved and small-flowered form.

11. **Sida cordifolia** L., Sp. Pl. **2**: 684 (1753).—Harv. in Harv. & Sond., F.C. **1**: 168 (1860).—Mast. in Oliv., F.T.A. **1**: 181 (1868).—Ulbr. in Engl., Bot. Jahrb. **51**: 49 (1913).—R.E.Fr., Wiss. Ergebn. Schwed. Rhod.-Kongo-Exped. **1**: 144 (1914).— Eyles in Trans. Roy. Soc. S. Afr. **5**: 413 (1916).—Burtt Davy, F.P.F.T. **2**: 277 (1932).—Arwidss. apud Norlindh & Weim. in Bot. Notis. **1934**: 96 (1934).—Exell & Mendonça, C.F.A. **1**, 1: 149 (1937).—Suesseng. in Proc. & Trans. Rhod. Sci. Ass. **43**: 104 (1951).—Martineau, Rhod. Wild Fl.: 47 (1954).—Wild, Common Rhod. Weeds: fig. 44 (1955). Type from India.

Erect annual or perhaps sometimes perennial suffrutex 0·5–1 m. tall; stems stellate-pubescent and usually with additional patent simple hairs up to 3 mm. long, ultimately glabrous.   Leaf-lamina 3–7 × 1–4 cm., broadly ovate- to sub-orbicular-subcordate, apex usually obtuse to rounded less often acute to acuminate, margin irregularly crenate, base somewhat cordate to rounded, ± densely stellate-pubescent on both surfaces, lower surface paler and with a denser indumentum of longer hairs, upper surface sometimes glabrescent; petiole 1–3 cm. long, stellate-tomentose and often with additional longer patent hairs; stipules c. 5 mm. long, filiform, weak.   Flowers bright yellow, fasciculate, mainly towards the tips of the branches or, on terminal and lateral branches bearing small leaves, in a subspicate inflorescence, or sometimes solitary in the axils of the lower leaves; pedicels 4–20 mm. long, the longer ones articulated near the apex. Calyx 4–6 × 8 mm., cam-panulate-cupuliform, densely and softly tomentose, lobed to about the middle; lobes triangular. Petals c. 7 mm. long.   Staminal tube hirsute. Mericarps 8–12 (normally 10), usually c. 7 mm. long, somewhat strigose and birostrate at the apex; awns 3–4 (6) mm. long, exceeding the calyx, erect, finely retrorse-hispid. Seeds c. 2 × 2 mm., glabrous except for some hairs around the hilum.

Caprivi Strip. Without precise locality, fr. x.1945, *Curson* 929 (PRE). **Bechuanaland Prot.** N: Kwebe, fl. & fr. ii.1899, *Lugard* 235 (GRA; K). SE: Gaberones, *van Son* in Herb. Transv. Mus. 28928 (BM; PRE). **N. Rhodesia.** B: Balovale, fl. & fr. vi.1952, *Gilges* 96 (K; PRE; SRGH). N: Abercorn, fl. & fr. 15.iv.1952, *Richards* 1489 (EA; K). W: Kitwe, fl. & fr. 27.ii.1955, *Fanshawe* 2107 (K). C: Chisamba, fl. & fr. 23.vii.1956, *Clarke* 151 (PRE). S: Livingstone, fl. iv.1909, *Rogers* 7088 (GRA; K; SRGH). **S. Rhodesia.** N: Trelawney, fl. & fr. 29.iii.1944, *Jack* 156 (K; PRE; SRGH). W: Victoria Falls, *Eyles* 145 (PRE; SRGH). C: Marandellas, fl. & fr. 29.x.1949, *Corby* 526 (SRGH). E: Inyanga, 1700 m., fl. & fr. 24.xi.1930, *Norlindh & Weimarck* 3166 (BM; PRE; SRGH). S: Belingwe, *Norlindh & Weimarck* 5161 (PRE). **Nyasaland.** C: Namitete R., Lilongwe-Fort Jameson road, 1150 m., fl. 5.ii.1959, *Robson* 1472 (BM; K; LISC; SRGH). S: Shire R., 16° S., fl. & fr. 1863, *Kirk* (K). **Mozambique.** N: Nampula, *Pedro & Pedrógão* 3125 (EA; LMJ; PRE). Z: Quelimane, *Sim* 20549 (PRE). MS: Vila Pery, bud 14.xi.1946, *Pedro & Pedrógão* 230 (LMJ; PRE). SS: between Vilanculos and Nhachengo, *Barbosa & Bettencourt* 4974 (LM). LM: Namaacha, fl. & fr. 21.v.1957, *Carvalho* 234 (LM; PRE; SRGH).

Pantropical. A weed of cultivation. Usually growing at low altitudes in hot conditions and absent both from most areas with cold dry winters and from areas at a high altitude with a high rainfall (such as the Eastern Border Mts. of S. Rhodesia).

Cultivated on experimental farms as a potential fibre crop in S. Rhodesia and Mozambique.

12. **Sida pseudocordifolia** Hochr. in Ann. Conserv. Jard. Bot. Genève, **20**: 135 (1917). Type from the Transvaal.
    Sida hislopii Burtt Davy & Greenway in Burtt Davy, F.P.F.T. **1**: 49 (1926); op. cit. **2**: 277 (1932). Type: S. Rhodesia, without precise locality, *Hislop* 1 (K, holotype).

Small suffrutex or soft-stemmed shrub up to c. 1·25 m. tall, covered on stems, petioles, pedicels and calyces with a ± soft stellate or subvelutinous indumentum; stems terete, yellowish-green to fawn-green when young, ultimately olive-brown or deep purplish-brown and glabrous. Leaf-lamina 2–7 × 1·5–6 cm., broadly ovate to suborbicular, often slightly rhomboid, apex usually obtuse or rounded, margin slightly to distinctly crenate or serrate-dentate or doubly serrate, base broadly cuneate, truncate to rounded, upper surface sparsely and shortly appressed-pubescent to almost glabrous, dark green, lower surface paler and densely tomentose, nervation impressed above, prominent beneath ; petiole up to 25 mm. long, terete ; stipules up to 15 mm. long, filiform, tomentose. Flowers yellow, in small fascicles or clusters mainly towards the tips of the branches, often in subspicate inflorescences, some solitary in the lower axils ; pedicels usually very short but up to 15 mm. long in some of the lower solitary flowers. Calyx 5·5–7·5 mm. long, cupuliform-campanulate, densely velutinous, lobed to about the middle ; lobes triangular, subacute. Petals c. 9 mm. long, glabrous. Staminal tube glabrous. Mericarps 8–10, 3–4 mm. long, narrowly trapezoid in side view, truncate at the apex, faintly reticulate-veined, ± pubescent in the upper part. Seeds c. 2 × 2 mm., suborbicular-obpyriform, smooth, glabrous except for the area around the hilum.

S. Rhodesia. E : Chirinda, 1160 m., fl. 18.v.1906, *Swynnerton* 471 (BM). **Mozambique.** N : Nampula, fl. & fr. 7.viii.1948, *Pedro & Pedrógão* 4727 (EA ; K ; LMJ). MS : Chimoio, Missão de Amatongas, fl. & fr. 25.iv.1948, *Mendonça* 4050 (BM ; LISC).
Also in the Transvaal, Swaziland and Natal. Lower mountain slopes among scrub and tall herbs in forest edges and clearings.

A species closely resembling *S. cordifolia* in habit, leaf-shape, inflorescence and indumentum but with quite different mericarps and other minor morphological differences and also quite distinct in its ecology. It is a forest species while *S. cordifolia* is ruderal.

11. ABUTILON Mill.

Abutilon Mill., Gard. Dict. Abridg. ed. 4 (1754).

Biennial to perennial (rarely annual) erect or occasionally spreading herbs or shrubs, variously pubescent, usually with stellate hairs. Leaves petiolate, usually more or less ovate in outline with cordate base. Flowers generally yellow to orange, rarely white, mauve or purple, small to medium-sized, axillary, solitary or fascicled, rarely 2–4-nate on a common peduncle, sometimes on short axillary leafy side-shoots, sometimes aggregated in terminal and lateral leafy pseudo-panicles ; pedicels usually articulated in the upper half often near the apex. Epicalyx absent. Calyx with a cupular to campanulate tube ; lobes 5, distinct, semi-orbicular to lanceolate, usually acute to acuminate. Petals 5, connate at the base and adnate to the base of the staminal tube, usually conspicuously longer than the calyx and in open flowers usually spreading to rotate, generally obovate with a narrow subunguiculate often ciliate basal portion. Staminal tube divided at the apex into many filaments, dilated below, glabrous or stellate-pubescent ; free parts of filaments terete ; anthers reniform. Carpels 5 to c. 40, 3–9-ovulate, in a circle around a distinct torus and joined to form a subglobose gynoecium ; style-branches as many as the carpels, terete, filiform or clavate ; stigmas simple to somewhat capitate. Fruit subglobose or turbinate to hemispherical or almost disk-shaped, often truncate, depressed or umbilicate at the apex ; mericarps 5 to many, laterally compressed, follicular, (1) 2–3 (9)-seeded, separating from the ultimately conical or subcylindric and usually more or less produced or dilated to capitate torus and usually dehiscing by the ventral suture, ultimately grey or brown to black, oblong to subrectangular, reniform or more or less semi-orbicular, rounded to truncate at the base and rounded, truncate or acute at the apex, muticous to mucronate, apiculate or awned at the upper dorsal (outer) angle or at the apex, the ventral side with a usually distinct retrorse tooth which originally fitted over and against the apex of the torus. Seeds reniform, often unequally so and more or less comma-shaped, glabrous, puberulous or stellate-tomentose, smooth, finely pitted or minutely papillose to verruculose ; embryo curved ; cotyledons folded ; endosperm scanty.

Pantropical and also in most subtropical regions. Over 150 species, some ruderal, some in open, dry and sunny situations such as forest margins, savannas, exposed rocky slopes etc. and a few in moister habitats such as dense thickets and (mountain) forests. A few species not cited here (mainly from S. America and China) are ornamental garden shrubs. Most species contain strong bast fibres and *A. theophrasti* Medic. (*A. avicennae* Gaertn.) and perhaps other species are cultivated as fibre crops, mainly in China.

The flowers of the African species generally only open in the afternoon or evening (but *A. fruticosum* is said to be an exception). This is in sharp contrast with species of *Sida*, whose flowers open in the morning.

The best characters for distinguishing *Abutilon* spp. are provided by the morphology of the mericarps. Collectors are urged always to collect the fruits when possible as it is often almost impossible to identify specimens without them.

Apart from the species treated here, there is one record (from Nyasaland) of *A. theophrasti* Medic., characterised by large hemispherical fruits, mericarps with subulate-falcate awns incurved at the apex and large (3–7 mm.) seeds. *A. grandifolium* (Willd.) Sweet is cultivated in Mozambique as a potential fibre crop on experimental farms; this species is conspicuously pilose on stems, petioles and pedicels, and has mericarps awned at the dorsal apical angle (not with apical awns as in the indigenous *A. engleranum*, which is also pilose).

Pedicels solitary, very rarely in pairs in the same axil but, if so, fascicled and not on a common peduncle :
  Young parts viscid ; indumentum of stems stellate-velvety usually intermingled with sparse longer patent hairs ; fruit 12–15 mm. high, about 20 mm. in diam. ; mericarps numerous, muticous or occasionally mucronate, densely tomentose ; seeds (2) 3, usually distinctly hairy - - - - - - - - 1. *hirtum*
  Young parts either not viscid or, if so, mericarps not as above ; seeds glabrous or puberulous :
    Mericarps rounded at the top and at the back, slightly angled near the outer apical corner, or completely rounded, usually muticous, rarely slightly mucronate, over 6 mm. long and over 5 mm. broad, 1-seeded ; flowers numerous, in terminal and lateral pseudo-panicles :
      Flowers yellow or orange; older stems markedly angled by coarse longitudinal ridges
                                                2. *angulatum*
      Flowers mauve or lavender with purple centre ; older stems subterete, not conspicuously angled - - - - - - - - 3. *longicuspe*
    Mericarps distinctly mucronate to aristate, or if muticous, less than 7 × 5 mm. and/or 2–8-seeded :
      Mericarps 1-seeded, up to 7 mm. long :
        Indumentum of (young) branches, petioles, pedicels and calyx-tube tomentose with additional patent stiff long hairs, the latter more rarely scanty to almost absent ; mericarps obliquely truncate with the slanting upper edge ending in a point at the outer (dorsal) upper angle ; young parts of stems terete ; leaf-margins usually shallowly serrate or crenate - - 4. *austro-africanum*
        Indumentum without additional long patent hairs ; mericarps rounded to subtruncate at the apex with a subapical mucro ; young parts of stems more or less distinctly longitudinally ribbed to angular ; leaf-margins usually conspicuously serrate or crenate - - - - - - 5. *matopense*
      Mericarps 2–8-seeded :
        Mericarps usually c. 10, up to 8 × 4 mm., obliquely truncate at the apex with the outer (dorsal) apical angle acute or subacute to submucronate, but not awned, when immature densely and often somewhat floccosely grey-green stellate-tomentose dorsally and on the upper lateral faces, with the apical dorsal portion broadly keeled, thus giving the fruit a characteristic ribbed appearance - - - - - - - - - 6. *fruticosum*
        Mericarps not as above :
          Mericarps subrectangular in outline, much compressed, nearly glabrous and papery when mature, the horizontal upper edge truncate with the upper dorsal angle acute to shortly apiculate ; the ventral tooth usually conspicuously upcurved at the tip - - - - 7. *sonneratianum*
          Mericarps not subrectangular or, if so, with an apical (not dorsal) tooth or awn, or with upper edge obliquely truncate ; the ventral tooth not upcurved at the tip :
            Mericarps much compressed, papery and nearly glabrous when mature, c. 11 mm. long and 8–9 mm. broad near the base, the apical edge slanting upwards into the sharply acute to awned outer angle
                                              8. *grantii*
          Mericarps not as above, under 8 mm. broad near the base :
            Mericarps numerous (usually over 25), ultimately black and stellately spreading, with the long apical awn at least one-third of the total length of the mericarp - - - - - 9. *mauritianum*
            Mericarps either not stellately spreading or less numerous (16 or fewer), with the apical or dorsal awn less than one-third of the total length of the mericarp :
              Leaves slightly serrate or crenate-dentate to subentire ; short indu-

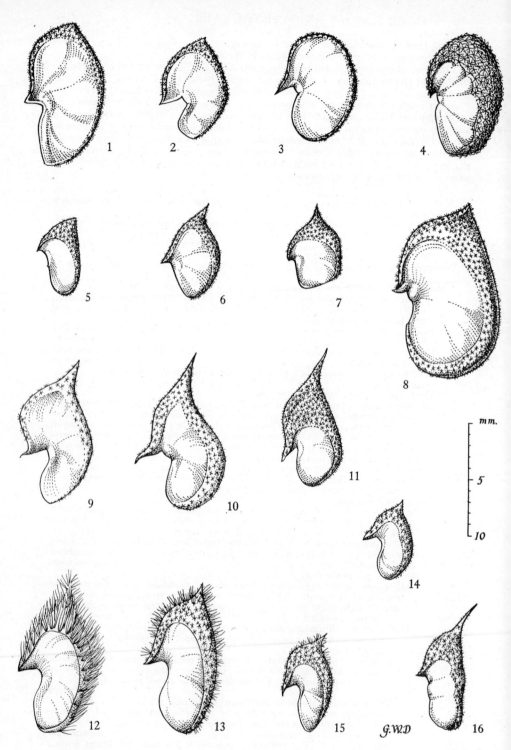

Tab. 93. Mericarps of ABUTILON species, lateral view (all ×3). 1 and 2. A. HIRTUM;
3. A. ANGULATUM; 4. A. LONGICUSPE; 5. A. FRUTICOSUM; 6. A. AUSTRO-AFRICANUM;
7. A. MATOPENSE; 8. A. SONNERATIANUM; 9. A. GRANTII; 10. A. MAURITIANUM;
11. A. LAURASTER; 12. A. GUINEENSE; 13. A. GRANDIFLORUM; 14. A. REHMANNII;
15. A. ENGLERANUM; 16. A. RAMOSUM.

mentum on stems, petioles, pedicels and calyx devoid of conspicuous long patent hairs :
Upper surface of leaf softly pubescent, not rough to the touch ;
mericarps 9–16, ultimately black and nearly glabrous, stellately
spreading, rather abruptly acuminate into a subulate apical awn
10. *lauraster*
Upper surface of leaf scabrid, rough to the touch ; ripe fruit densely
hirsute, nearly enclosed by the calyx ; mericarps produced into
a triangular-acuminate awn at the dorsal apical side, conspicuously
hirsute dorsally and on the lateral faces near the apex
11. *guineense*
Leaves (at least the young ones) distinctly and often irregularly
serrate-crenate or doubly serrate, or, if only slightly serrate-
crenate, plant pubescent on stems, petioles, pedicels and calyx
with additional conspicuous long patent hairs :
Indumentum velvety, without additional long hairs :
Fruit stellate-pilose and stellate-pubescent ; pedicels up to 8 cm.
long ; calyx 15–17 mm. long   -   -   - 12. *grandiflorum*
Fruit velutinous ; pedicels up to 3 (5) cm. long ; calyx 6–8 mm.
long -   -   -   -   -   -   -   - 13. *rehmannii*
Indumentum with additional (rather sparse) long white hairs
14. *engleranum*
Pedicels 2–4-nate on a common peduncle ; mericarps c. 10, with long subulate spreading
or recurved awns ; leaves usually ± 3-lobed   -   -   -   - 15. *ramosum*

1. **Abutilon hirtum** (Lam.) Sweet, Hort. Brit.: 53 (1826).—Garcke in Peters, Reise
Mossamb. Bot. **1** : 129 (1861).—Ulbr. in Engl., Bot. Jahrb. **51** : 33 (1913).—Exell
& Mendonça, C.F.A. **1**, 2 : 373 (1951).—Brenan in Kew Bull. **1953** : 91 (1953).—
Keay, F.W.T.A. ed. 2, **1**, 2 : 337 (1958). TAB. 93 fig. 1–2. Type from India.
Sida hirta Lam., Encycl. Méth. Bot. **1** : 7 (1783). Type as above.
Sida graveolens Roxb. [Hort. Beng.: 50 (1814) *nom. nud.*] ex Hornem., Suppl.
Hort. Bot. Hafn.: 77 (1819). Type from India.
Abutilon graveolens (Roxb. ex Hornem.) Wight & Arn. in Wight, Cat. Pl.: 13
(1833). Type as above.
Abutilon lugardii Hochr. & Schinz in Bull. Herb. Bois s., Sér.2, 3 : 825 (1903).—
Ulbr., tom. cit.: 32 (1913).—O. B. Mill. in Journ. S. Afr. Bot. **18** : 39 (1952).
Type : Bechuanaland Prot., Kwebe Hills, *Lugard* 171 (GRA ; K ; Z, holotype).

Large erect herb to soft-stemmed shrub up to about 1·5 m. tall, covered on
young parts, stems, petioles, pedicels and calyx with a dense usually somewhat
yellowish or brownish usually velutinous occasionally somewhat harsh stellate-
tomentose indumentum intermingled on young parts, especially the tips of the
branches, with short glandular hairs, and usually on younger parts of stems,
petioles and pedicels also with long patent white or yellow hairs, the latter more
rarely very scanty; stems terete, the older portions usually stout, at length glabrescent, slightly lignified with a large pith, covered with a thin greyish-brown
cortex with close lanceolate-rhomboid to linear markings (shallow fissures or
lenticels) often forming an almost continuous pattern. Leaf-lamina 4–20 cm. long
and up to 18 cm. broad, suborbicular-cordate or broadly ovate-cordate, generally
drying a yellowish or brownish-green colour, usually markedly acuminate with a
narrow mucronate acumen, irregularly and distinctly to coarsely serrate or serrate-
crenate to biserrate, occasionally very slightly 3-lobed ; finely and harshly stellate-
pubescent, glabrescent, somewhat rough to the touch, often also with sessile
glands (hence viscid) on the upper surface, more densely and more softly stellate-
pubescent and with prominent veins on the slightly paler lower surface ; petiole
generally as long as or longer than the corresponding lamina, terete. Flowers
yellow, often with a reddish centre, and/or the venation reddish towards the centre,
axillary, on main branches and sometimes also on short lateral shoots and often
forming a terminal leafy panicle ; pedicels up to about 5 cm. long, articulated in
upper 11 mm. Calyx 8–10 mm. long, campanulate, strigose-tomentose inside,
divided about half-way down ; lobes triangular-ovate, acute to shortly apiculate,
ciliate. Petals c. 16 mm. long. Staminal tube stellate-hairy. Fruit 12–15 × c. 20
mm., depressed-globose, broadly and shallowly umbilicate, densely and shortly
stellate-pubescent, nearly enclosed by the accrescent up to 14 mm. long appressed
fruiting calyx. Mericarps 20–30, dorsally rounded, usually bluntly angled and
muticous to shortly pointed near the dorsal (outer) upper side. Seeds 2–3, verruculose and usually distinctly but finely stellate-pubescent.

Caprivi Strip. Mpilila I., fl. & fr. 15.i.1959, *Killick & Leistner* 3405 (BM; K; PRE; SRGH). **Bechuanaland Prot.** N: Ngamiland, Kwebe Hills, fl. & fr. i.1897, *Lugard* 148 (GRA; K). SE: Francistown, *Rand* 31 (BM). **S. Rhodesia.** N: Chirundu, fl. & fr. ii.1958, *Drummond* 5466 (K; PRE; SRGH).

Widespread in the tropics and in SW. Africa and the Transvaal. In a variety of habitats from dry ones to riverine silty soils, but usually on the more alkaline types of soil; apparently infrequent in rocky places.

2. **Abutilon angulatum** (Guill. & Perr.) Mast. in Oliv., F.T.A. **1**: 183 (1868).—Eyles in Trans. Roy. Soc. S. Afr. **5**: 412 (1916).—Burtt Davy, F.P.F.T. **2**: 275 (1932).— Steedman, Trees etc. S. Rhod.: 47 (1933).—Arwidss. apud Norlindh & Weim. in Bot. Notis. **1934**: 93 (1934).—Burtt Davy & Hoyle, N.C.L.: 49 (1936).—Gomes e Sousa, Pl. Menyhart.: 80 (1936).—Exell & Mendonça, C.F.A. **1**, 1: 152 (1937); tom. cit., 2: 373 (1951).—Brenan, T.T.C.L.: 297 (1949); in Mem. N.Y. Bot. Gard. **8**, 3: 223 (1953).—Keay, F.W.T.A. ed. 2, **1**, 2: 337 (1958). TAB. **93** fig. 3. Type from Senegal.

Bastardia angulata Guill. & Perr. in Guill., Perr. & Rich., Fl. Senegamb. Tent. **1**: 65 (1831). Type as above.

Abutilon intermedium Hochst. ex Garcke in Schweinf., Beitr. Fl. Aethiop.: 49 (1867).—Bak. f. in Journ. Linn. Soc., Bot. **40**: 27 (1911).—Ulbr. in Engl., Bot. Jahrb. **51**: 20 (1913).—R.E.Fr., Wiss. Ergebn. Schwed. Rhod.-Kongo-Exped. **1**: 143 (1914). Type from Ethiopia.

An erect glaucous or yellowish or brownish shrub, usually 1–3 m. tall, occasionally taller, usually with only a few long side-branches in addition to the main stem but producing many short side-branches mainly in the flowering portions; stems up to 2–3 cm. in diam. near the base, greenish, terete or angled when young, shortly velutinous (as are all other vegetative parts, pedicels and calyces), occasionally with additional sparse small floccose tufts of stellate hairs, soon becoming angular owing to coarse longitudinal ridges originating below each node, glaucous-green to dull greyish-purple, semi-woody with a large pith, only at length glabrescent. Leaf-lamina up to 30 × 25 cm. (much smaller in upper leaves), rather dark glaucous-grey-green above, much paler below (usually conspicuously so and greyish), cordate-ovate to suborbicular-cordate or occasionally triangular-cordate, usually acuminate, with deep basal sinus, indistinctly serrate, crenate or minutely callous-dentate; petiole of larger leaves about as long as the corresponding lamina, rather stout, longitudinally sulcate, subpulvinate at the base, that of upper leaves thinner and often shorter to much shorter than the lamina. Flowers yellow, orange or apricot, numerous, on lateral and subterminal short shoots (often again branched), the flowering branchlets arranged in large lateral and terminal pseudopanicles, each flower solitary but the buds (and new side-branches) formed in such rapid succession that the buds are often apparently fasciculate; pedicels 3 (5) cm. long, articulated in the upper 10 mm. Calyx 10–15 mm. long, shallowly cupular, lobed to about the middle; lobes 3–4 (5) mm. long, suberect, ovate-triangular to triangular or ovate-lanceolate, minutely apiculate and densely but finely white-ciliate with inconspicuous or distinctly keeled midrib. Petals 9–12 mm. long. Staminal tube densely stellate-hairy. Fruit c. 9–12 × 8 mm., depressed-globose, umbilicate, densely stellate-tomentose, often ± floccose. Mericarps 20–30, rounded apically and dorsally but in the outer (dorsal) upper portion often with an obtuse (rarely subacute) angle, 1-seeded. Seeds verruculose or smooth, glabrous.

Calyx-lobes usually under 6 mm. long, triangular, not or inconspicuously veined; flower-buds not angular   -   -   -   -   -   -   -   -   -   -   var. *angulatum*
Calyx-lobes usually over 7 mm. long, ovate to ovate-lanceolate, usually distinctly acuminate with distinct median veins or ± keeled; flower-buds distinctly angular
var. *macrophyllum*

Var. **angulatum**

Indumentum glaucous. Flower-buds not angular. Calyx-lobes usually less than 6 mm. long, triangular, not or inconspicuously veined.

Bechuanaland Prot. N: Ngami, fl. & fr. 5.v.1930, *van Son* in Herb. Transv. Mus. 28931 (PRE). **N. Rhodesia.** B: Nangweshi, fl. & fr. 23.vii.1952, *Codd* 7172 (BM; K; PRE; SRGH). C: Chisamba, fl. & fr. 22.vii.1956, *Clarke* 144 (PRE). S: Livingstone, fl. & fr. 20.v.1956, *Gilges* 618 (PRE; SRGH). **S. Rhodesia.** N: Sebungwe, *Davies* 1446 (BM; SRGH). W: Wankie, fl. & fr. 11.v.1955, *Plowes* 1831 (PRE; SRGH). C:

Mondoro Reserve, fl. & fr. 31.viii.1943, *Hopkins* in GHS 10378 (SRGH). E: Umtali, *Eyles* 7300 (K; SRGH). S: Sabi-Lundi Junction, fl. & fr. 8.vi.1950, *Wild* 3399 (PRE; SRGH). **Nyasaland.** S: Mboma, Cholo, fr. 10.x.1943, *Hornby* 3835 (PRE). **Mozambique.** T: Boroma, *Chase* 2833 (BM; SRGH). MS: Maríngua, fl. & fr. 24.vi.1950, *Chase* 2529 (BM; PRE; SRGH). SS: Massingire, fl. & fr. 7.v.1957, *Carvalho* 167 (K; LM; PRE; SRGH).

Widespread in tropical Africa and in SW. Africa, Transvaal and Natal. In the drier regions below 1000 m. often on alkaline or somewhat brackish soils in open situations such as pans, river banks, roadsides, low scrub and bush.

Var. **macrophyllum** (Bak. f.) Hochr., Fl. Madag., Malvac.: 139 (1955). Type from Madagascar.

 Abutilon intermedium var. *macrophyllum* Bak. f. in Journ. of Bot. **31**: 72 (1893). Type as above.

 Abutilon pseudangulatum Hochr. in Ann. Conserv. Jard. Bot. Genève, **6**: 13 (1902). Type as above.

Indumentum more yellowish or brownish than glaucous. Inflorescences usually shorter and narrower and often more condensed than in var. *angulatum*. Flower-buds distinctly angular. Calyx-lobes usually more than 7 mm. long, ovate to ovate-lanceolate, with distinct median veins or ± keeled.

S. Rhodesia. E: Vumba Mts., fl. & fr. 7.vii.1947, *Chase* (PRE; SRGH). **Nyasaland.** N: Nyika Plateau, 4 km. SW. of Rest House, 2150 m., fr. 24.x.1958, *Robson* 308 (BM; K; LISC; SRGH). S: Cholo Mt., fl. 21.vii.1946, *Brass* 17715 (K; PRE; SRGH). **Mozambique.** N: Metuge, fl. 3.ix.1948, *Pedro & Pedrógão* 265 (PRE). LM: Maputo, fl. 16.vii.1948, *Gomes e Sousa* 3757 (K; LM; PRE; SRGH).

Also in SW. Africa, Transvaal, Natal and Madagascar. Usually in moister situations than var. *angulatum*; often in forest margins, forest regrowth, dense scrub and riverine bush, up to 2000 m.

3. **Abutilon longicuspe** Hochst. ex A. Rich., Tent. Fl. Abyss. **1**: 69 (1847).—Mast. in Oliv., F.T.A. **1**: 184 (1868).—Brenan, T.T.C.L.: 298 (1949); in Mem. N.Y. Bot. Gard. **8**, 3: 223 (1953).—Exell & Mendonça, C.F.A. **1**, 2: 373 (1951). TAB. **93** fig. 4. Type from Ethiopia.

 Abutilon cecilii N.E. Br. in Kew Bull. **1906**: 99 (1906) (" cecili ").—Eyles in Trans. Roy. Soc. S. Afr. **5**: 412 (1916). Type: S. Rhodesia, Inyanga, *Cecil* 196 (K, holotype).

 Abutilon longicuspe var. *epilosum* Exell in Journ. of Bot. **65**, Suppl. Polypet.: 33 (1927).—Exell & Mendonça, C.F.A. **1**, 1: 152 (1937). Type from Angola (Benguela).

Perennial shrub up to 5 m. tall; stems when young somewhat angular to sub-terete, densely covered with a short velutinous to somewhat harsh stellate indumentum, usually of a greyish-olive colour and as a rule with long whitish spreading hairs (very rarely absent) at least on the very young parts, when older terete, glabrescent, woody and ultimately with a grey cortex, faintly fissured by rather short longitudinal markings. Leaf-lamina up to 20 × 18 cm. (much smaller in upper leaves), suborbicular-cordate or broadly ovate-cordate, acuminate, with a usually long narrow acumen and a deep narrow basal sinus, somewhat irregularly but usually distinctly crenate or serrate; the upper surface dark green, very shortly stellately subvelutinous, the lower surface much paler, grey, densely stellate-velutinous, and sometimes in addition with few-rayed to simple stiff stellate hairs mainly on the distinct prominent reticulate nervation and with long patent hairs on the main veins near the base; petiole usually about as long as the lamina, velutinous, with or without long patent hairs. Flowers pale mauve, lavender or lilac with deep-red-purple centre and radiating purple veins, numerous, axillary, on main branches and also on short axillary shoots, aggregated in terminal and lateral, ultimately leafless, often large pseudo-panicles; pedicels usually under 3 cm. long, velutinous, with or without pilose hairs, articulated in upper 6 mm. Calyx 4–6 mm. long, cupuliform, tomentose and sometimes also pilose; lobes 4–7 mm. long, triangular or ovate-triangular, acute or mucronate, tomentose-velutinous, with inconspicuous median veins. Petals c. 12 mm. long, obovate-obcuneate. Staminal tube deep red-purple and glabrous in conical portion, stellate-hairy at the apex. Fruit c. 10 × 15 mm., subglobose to depressed-globose, umbilicate, densely stellate-hairy. Mericarps 12–25, c. 10 × 5–6 mm., semi-orbicular-reniform, rounded and muticous, papery and brittle when mature, 1-seeded. Seeds c. 3 × 2 mm.

N. Rhodesia. N : Abercorn, fl. 20.v.1955, *Richards* 5793 (K ; SRGH). **S. Rhodesia.** E : Inyanga, fl. & fr. 20.iv.1953, *Chase* 4915 (PRE ; SRGH). **Nyasaland.** N : Vipya, fl. & fr. 10.vii.1952, *Jackson* 966 (K). C : Kota Kota, Nchisi, fl. 26.vii.1946, *Brass* 16959 (BM ; K ; PRE ; SRGH). S : Manganja, Mt. Chiradzulu, fr. ix.1861, *Meller* (K). **Mozambique.** T : Serra de Zobuè, *Mendonça* 609 (LISC).

Eastern and southern tropical Africa. Mainly found on lower mountain slopes at 1000–2000 m. in forest regrowths, scrub and forest margins.

The pubescence appears to be extremely variable. Some specimens are conspicuously pilose on stems, petioles, pedicels, calyx-tubes and even on the main veins near the leaf-base on the lower surface, others scantily so or only on the youngest parts of the shoots. This variation in pubescence is found throughout the area so I do not think that var. *epilosum* can be upheld.

4. **Abutilon austro-africanum** Hochr. in Ann. Conserv. Jard. Bot. Genève, **7** : 25 (1902).—Ulbr. in Engl., Bot. Jahrb. **51** : 14 (1913).—Burtt Davy, F.P.F.T. **2** : 275 (1932). TAB. **93** fig. 6. Syntypes from Griqualand-W. and SW. Africa.

Shrublet 0·5–0·75 m. tall, often spreading, canescent to glaucous with a short greyish velvety indumentum and additional long soft patent white hairs (rarely very sparse or almost lacking) ; stems greyish- or yellowish-green when young, soon glabrescent and becoming pale-purplish-brown, ultimately woody with an ash-grey smooth or finely longitudinally fissured bark. Leaf-lamina 2–5 (8) × 1–3 (5) cm., cordate-triangular to ovate-cordate, apex acute or somewhat acuminate or rounded, margin crenate to crenate-serrate often with minutely callous-mucronate serrations, dark-greyish-green and velvety above, much paler, glaucous-grey and finely velvety beneath, venation of lower surface somewhat prominent and conspicuous owing to its whitish or pale yellow colour ; petiole usually shorter than the lamina, terete. Flowers yellow, solitary, axillary on main branches (not on condensed short axillary shoots) ; pedicels (10) 25–50 mm. long, slender, terete, articulated near the apex. Calyx 9–12 mm. long, widely campanulate, incised beyond the middle ; lobes 6–8 mm. long, triangular-ovate to ovate-lanceolate, acuminate-apiculate, with a prominent median vein and usually in addition with a faint longitudinal vein on either side. Petals 11–14 × 8 mm., conspicuously ciliate in basal narrowed portion, often marked with reddish spots at the base and reddish-veined in lower portion. Staminal tube rather shortly conical, sparsely stellate-hairy to glabrous except at the very base. Fruit c. 14 × 5 mm., discoid-subglobose, truncate at the apex and widely umbilicate in the centre, stellate-hairy. Mericarps 20–30, c. 7 × 5 mm., ultimately black, 1-seeded, the upper edge slanting upwards into the usually sharply pointed to shortly apiculate dorsal apical angle. Seeds c. 2·5 mm. long, punctate-verruculose.

Bechuanaland Prot. SE : Mochudi, fl. & fr. 1.iv.1914, *Harbor* in *Rogers* 6444 (BM ; BOL ; GRA ; K ; PRE). **S. Rhodesia.** W : Bulawayo, *Orpen* 80/50 (BM ; SRGH). **Mozambique.** LM : Sabiè, Moamba, *Pedrógão* 212 (LMJ ; PRE).

Also in SW. Africa, Cape Prov. (Griqualand-W.) and the Transvaal. Usually found in open vegetation on sandy to gravelly, often brackish soils, often in pans, generally in areas with a low rainfall.

This species is an example of a plant with an E–W distribution pattern found in a relatively small group of species in southern Africa. They occur from SW. Africa (and sometimes Angola) through Bechuanaland, S. Rhodesia and/or the Transvaal (sometimes also Griqualand-W.) to Mozambique. The explanation of this distribution pattern is most probably an ecological one ; these species are either more or less confined to dry sandy or gravelly, sometimes alkaline or brackish soils occurring throughout the area (*Barleria senensis* Klotzsch and *Abutilon austro-africanum* are good examples) or they are distributed by water in an E–W or W–E direction (the general directions of the rivers in this area). An example of the latter group is *Abutilon engleranum*, the only other species of the genus with this distribution pattern.

5. **Abutilon matopense** Gibbs in Journ. Linn. Soc., Bot. **37** : 431 (1902).—Eyles in Trans. Roy. Soc. S. Afr. **5** : 413 (1916). TAB. **93** fig. 7. Type : S. Rhodesia, Matopos, *Gibbs* 98 (BM, holotype ; BOL).

Abutilon betschuanicum Ulbr. in Engl., Bot. Jahrb. **51** : 15 (1913). Type from Cape Prov. (Kuruman).

Abutilon messinicum Burtt Davy, F.P.F.T. **1** : 36 (1926) ; op. cit. **2** : 275 (1932). Type from the Transvaal (Messina).

Perennial 0·75–2 m. tall, erect, much branched, shrubby, glaucous, usually densely leafy towards the tips of the branches, with an ash-grey (except on the leaves) appressed short velvety indumentum; stems herbaceous to wiry when young, usually with some longitudinally raised ridges decurrent from the leaf-bases or somewhat angular, later glabrescent and terete, ultimately woody and covered with a thin finely longitudinally fissured grey to brownish bark. Leaf-lamina 2–6 (10) × 4·5 (9) cm., thin but firm, glaucous or yellowish-glaucous above, slightly paler beneath, broadly ovate to ovate, apex shortly and usually bluntly acuminate, obtuse or acute (rarely rounded), margin usually rather coarsely crenate or crenate-serrate (occasionally crenulate to subentire), base cordate, venation on lower surface distinct and prominent; petiole 0·5–6 (12) cm. long, finely sulcate. Flowers yellow to apricot, solitary, one to few on short (6–8 cm. long) leafy axillary shoots; pedicels up to 5 cm. long, terete or in upper portion slightly angular, articulated at or above the middle. Calyx 7–9 mm. long, campanulate, incised to about the middle; lobes triangular-ovate, apiculate to acuminate, 1-nerved, finely ciliate. Petals 14–20 mm. long. Staminal tube hairy at the base. Fruit 14 mm. in diam., disk-shaped to depressed-hemispherical, nearly enclosed by the slightly accrescent calyx, umbilicate to truncate. Mericarps usually 18–24, 5–6 × 3–4·5 mm., blackish, shortly apiculate at the rounded apex, finely and somewhat floccosely stellate-tomentose dorsally and laterally in apical-dorsal region, 1-seeded. Seeds c. 2·5 × 1·5 mm., brown.

S. Rhodesia. W: Bulalima-Mangwe, *Feiertag* in GHS 45372 (BM; SRGH); Matobo, fl. & fr. ii.1953, *Miller* 1588 (PRE; SRGH). S: West Nicholson, fl. & fr. 23.iii.1953, *Plowes* 1575 (BM; PRE; SRGH); Beitbridge, *E.M. & W.* 448 (BM; LISC; SRGH).
Also in the Transvaal and Cape Prov. (Griqualand-W.). Occurs mostly scattered, apparently never gregariously, on granite, norite and sandstone slopes in fairly exposed to rather shady situations, more rarely in riverine vegetation and in this habitat with larger and softer leaves.

The seeds vary from quite glabrous and smooth to finely white-punctate or sometimes minutely stellate-lepidote.

6. **Abutilon fruticosum** Guill. & Perr. in Guill., Perr. & Rich., Fl. Senegamb. Tent. **1**: 70 (1831).—Mast. in Oliv., F.T.A. **1**: 187 (1868).—Eyles in Trans. Roy. Soc. S. Afr. **5**: 412 (1916).—Burtt Davy, F.P.F.T. **2**: 275 (1932).—Exell & Mendonça, C.F.A. **1**, 1: 153 (1937); tom. cit., 2: 373 (1951).—Brenan, T.T.C.L.: 298 (1949).—Keay, F.W.T.A. ed. 2, **1**, 2: 337 (1958). TAB. **93** fig. 5. Type from Senegal.

Shrub 0·5–1·25 (2) m. tall, much branched, canescent to glaucous-grey with a dense very short velvety indumentum; stems terete, slender, at length glabrescent, woody and ultimately covered with a light brown or greyish bark with short darker linear markings. Leaf-lamina 2–6 (10) × 1·5–4 (6) cm., apex obtuse to acute or somewhat acuminate, margin usually subentire to slightly crenate or serrate, less often more conspicuously serrate, upper surface grey-green with indistinct venation, lower surface paler and canescent with distinct somewhat prominent venation; petiole somewhat shorter or slightly longer than the corresponding lamina. Flowers solitary on the main branches and/or on short leafy axillary shoots; pedicels 0·5–4 (8) cm. long, slender, terete, articulated near the apex. Calyx 5–6 mm. long, broadly campanulate to cupular, divided to about the middle; lobes triangular or triangular-ovate, mucronate, minutely ciliate, with indistinct median vein. Petals 7–10 mm. long. Staminal tube stellate-hairy. Fruit 8 × 10–12 mm., broadly cylindric with rounded base, widely and shallowly umbilicate. Mericarps usually c. 10, broadly keeled in apical half (hence fruit in upper half characteristically furrowed between the mericarps), obliquely truncate-convex at the apex with the dorsal angle subacute to shortly mucronate but not awned, when not yet ripe densely and shortly greyish-green, ultimately grey-brown, stellate-tomentose dorsally and on the apical dorsal area of the flat sides, not turning black. Seeds usually 3, c. 1·5 × 1·5 mm., usually greyish-brown, minutely verruculose-punctate.

Bechuanaland Prot. N: Ngamiland, Kwebe Hills, fl. & fr. 4.i.1898, *Lugard* 85 (GRA; K). **S. Rhodesia.** S: Beitbridge, fr. 22.iii.1959, *Drummond* 6145 (K; SRGH).
Widely spread in all the drier areas of the western half of Africa from Senegambia to SW. Africa, in the Transvaal and also from Kenya to Egypt and from Arabia to India.

In xerophilous low scrub and in wooded grassland vegetation, often on alkaline soil, but also on rocky slopes and gravelly soil; rarely in riverine bush on silt; usually in regions with a low annual rainfall.

This is the only species of *Abutilon* in southern Africa the flowers of which open in the morning.

7. **Abutilon sonneratianum** (Cav.) Sweet, Hort. Brit.: 54 (1826).—Harv. in Harv. & Sond., F.C. **1**: 168 (1860).—Bak. f. in Journ. Linn. Soc., Bot. **40**: 27 (1911).— Ulbr. in Engl., Bot. Jahrb. **51**: 14 (1913).—Eyles in Trans. Roy. Soc. S. Afr. **5**: 413 (1916).—Burtt Davy, F.P.F.T. **2**: 275 (1932). TAB. **93** fig. 8. Type from Cape Prov.
　　Sida sonneratiana Cav., Diss. **1**: 29, t. 6 fig. 4 (1790). Type as above.
　　Abutilon umtaliense Bak. f. in Journ. of Bot. **74**: 194 (1936). Type: S. Rhodesia, Umtali, *Teague* 385 (BM, holotype; BOL; K; SRGH).

Shrubby perennial 0·5–1·5 (2) m. tall, usually with only one or a few main stems, with a very dense short soft velvety indumentum and additional sparse long white patent hairs (rarely completely absent), less often long-pilose; stems slender, terete or slightly angular when young, tough or wiry, olive-drab to purplish-brown, woody when older, and ultimately covered with a grey to dark brown thin bark densely marked with short longitudinal shallow grooves. Leaf-lamina 2–7 (10) × 1–5 (7) cm., usually triangular-cordate or ovate-cordate, sometimes somewhat 3-lobed with blunt lobes on either side of the lamina near the middle, apex acuminate, margin ± dentate, crenate or serrate, dark green, brownish-green when dry, dark grey-green or olive-green on upper surface, paler greyish- to light-glaucous-green on the lower surface, venation of lower surface distinct, fine but prominent; petiole lender, terete, with the long patent hairs often only at the very apex, that of ower leaves often longer, that of upper leaves often shorter than the lamina. Flowers yellow or orange-yellow, solitary in axils on developed branches; pedicels c. 6 cm. (in fruit 10 cm.) long, slender, terete, articulated in the upper 6–12 mm. Calyx 8–10 mm. long and 6 mm. in diam. at the throat, greyish to olive-drab, divided to or beyond the middle; lobes 4–10 mm. long, ovate-elliptic, ovate-lanceolate to lanceolate-triangular, attenuate-acuminate into an acute tip, median-veined. Petals c. 10 mm. long, glabrous except at the base. Staminal tube glabrous except at line of fusion with the petals. Fruit 12–15 × 20 mm., subcylindric-semiglobose, truncate, densely and finely stellate-pubescent. Mericarps 8–15, often 9–11, 10–14 × 7–9 mm., much compressed, papery, subrectangular in outline, with rounded base and nearly horizontal truncate apical edge produced at the dorsal apical angle into a point or subulate awn up to 2 mm. long, the ventral tooth often somewhat upturned at the apex, 3–8 (often 4 or 5)-seeded. Seeds c. 2 × 2 mm., finely verruculose-rugulose, glabrous.

S. Rhodesia. E: Chirinda, *Swynnerton* 504 (BM; K); Umtali, fl. 11.xii.1945, *Wild* 472 (K; SRGH). **Mozambique.** MS: Vila Pery, *Pedro & Pedrógão* 6030 (LMJ). LM: Namaacha, fl. & fr. 25.iv.1947, *Pedro & Pedrógão* 758 (LMJ; PRE).
　　Also in the Transvaal, Swaziland, Natal and Cape Prov.
　　In S. Rhodesia and in the Transvaal mountains a typical forest-edge plant on mountain slopes, usually from 1000–2000 m., on rocky or loamy soils, usually in light shade or growing among scrub or low herbaceous plants, often locally frequent, but in Natal and the Cape Prov. down to sea-level in coastal bush.
　　The flowers open after 4 p.m.

From observation of specimens growing wild in my garden in Pretoria it appears that this species loses much of its lower foliage in the cold (and dry) winter-season. The tips of the branches still produce some flowers late in the season and young shoots start flowering in spring again. When collected in these stages the specimens often have, apart from the small leaves, smaller flowers and fruit. The type gathering of *A. umtaliense* is apparently from such seasonally depauperate plants. The description given here applies to the Rhodesian specimens; plants from the Cape are often more loosely stellate-pubescent, but there are many intermediates and Rhodesian specimens can be matched with some Cape specimens.
　　A. sonneratianum is an example of the so-called " Cape Floral Element " (see H. Wild in Proc. & Trans. Rhod. Sci. Ass. **44**: 53 (1956)) mainly found in the Eastern Border Mts. of S. Rhodesia and consisting of genera and species with their main distribution in the south.

8. **Abutilon grantii** Meeuse, sp. nov.* TAB. **93** fig. 9. Type from Natal, " Port Natal " (Durban), *Grant* (K, holotype ; PRE).

   Abutilon indicum sensu Harv. in Harv. & Sond., F.C. **1** : 168 (1860).—Burtt Davy, F.P.F.T. **2** : 275 (1932).

Perennial or biennial shrublet up to 1 (1·5) m. tall ; stems slender, tough when young, somewhat angular or sulcate, soon terete, covered with a short stellate-tomentose, floccose or scabridulous but not velvety indumentum, sometimes somewhat glandular-viscid on the young parts, glabrescent, greyish usually turning brown to purplish-brown or almost black, ultimately woody and marked with longitudinal short shallow grooves. Leaf-lamina 1–4 (6) × 0·75–3 (5) cm., thin but firm, usually triangular-cordate with broadly rounded basal lobes and abruptly acuminate into a long narrowly-triangular acumen (often with lobules at the base of the acumen and the lamina thus appearing 3-lobed), sometimes ovate-cordate or cordate-triangular, margin varying from subentire or sinuous to ± crenate or bluntly and occasionally coarsely dentate, often discolorous but sometimes lower surface only slightly paler, upper surface dark greyish-green to brown, very shortly velutinous and smooth or with additional coarse stellate hairs, glabrescent, lower surface paler to almost ash-grey or white, velutinous ; petiole slender, longer or shorter than the lamina, more or less densely stellate-pubescent, sometimes somewhat viscid and often with pilose hairs at the apex (if so, often also long hairs on veins near leaf-base on lower surface). Pedicels on main shoots, slender, generally much longer than the petioles and often exceeding the leaves, finely stellate-pubescent, articulated in upper 10 mm. Calyx grey-green, velutinous-tomentose ; tube 3–4 mm. long and 6–7 mm. in diam. at the apex, cupuliform ; lobes 5–7 × 3–4 mm., ovate- or oblong-elliptic to elliptic, subacute, minutely apiculate. Petals 11–13 mm. long, yellow, glabrous except at the base. Staminal tube stellate-hirsute. Fruit c. 12 × 18 mm., hemispherical, umbilicate-truncate, finely stellate-pubescent. Mericarps 10–20, 10–11 × 8–9 mm., much compressed, papery, 2–3-seeded, obliquely truncate to somewhat convex at the slanting apical side and produced at the outer angle into a sharp point or a subulate awn up to 3 mm. long. Seeds c. 2·5 × 2 mm., dark brown, minutely verrucose-papillose.

Mozambique. LM : Matola, fl. & fr. 10.xii.1897, *Schlechter* 11688 (BM ; BOL ; GRA ; K ; L ; PRE).
Also in the coastal regions of Natal and the Eastern Cape Prov., rarely further inland. Lowland and coastal bush, usually below 300 m., in scrub, among herbaceous plants, in forest margins etc., sometimes close to the shore.

The plants I refer to this species are variable in leaf-shape, the degree and colour of the pubescence, the presence or absence of viscid glands on young stems and petioles, and the presence or absence of long hairs at the apex of the petiole ; but they agree in the characters of the calyx and the fruit (mericarps) and there are many intermediates between forms with discolorous leaves and non-discolorous leaves, densely pubescent and more glabrous plants, etc. An extreme form is the gathering *Carvalho* 257 (LM ; PRE) which has a somewhat different " look ". I tentatively referred this specimen to *A. grantii*, but it may prove to require varietal status. More material is needed.

9. **Abutilon mauritianum** (Jacq.) Medic., Künstl. Geschlecht. Malv.-Fam. : 28 (1787). —Burtt Davy, F.P.F.T. **2** : 275 (1932).—Exell & Mendonça, C.F.A. **1**, 1 : 153 (1937).—Brenan, T.T.C.L. : 298 (1949).—Keay, F.W.T.A. ed. 2, **1**, 2 : 337 (1958). TAB. **93** fig. 10. Type from Mauritius.

   Sida mauritiana Jacq., Misc. Austr. **2** : 352 (1781) ; Ic. Pl. : t. 137 (1781). Type as above.

   Abutilon zanzibaricum Boj. ex Mast. in Oliv., F.T.A. **1** : 186 (1868) pro parte excl. specim. ex Senna.—Bak. f. in Journ. Linn. Soc., Bot. **40** : 27 (1911).—Eyles in Trans. Roy. Soc. S. Afr. **5** : 413 (1916). Type from Zanzibar.

Soft-wooded shrub up to c. 1·5 m. tall, usually much branched, with a short greyish-drab soft velvety indumentum and sometimes with additional long soft patent hairs ; stems stoutish, tough when young, ultimately woody and glab-

---

* A. grantii Meeuse, sp. nov., *A. sonneratiano* arcte affinis sed columna staminali stellato-hirsuta mericarpiis brevioribus oblique truncatis differt. *Suffrutex* erectus annuus vel biennis. *Caules* minute tomentosi, junioribus interdum plus minusve viscidis. *Folia* velutina plerumque discoloria. *Mericarpia* 10–20, c. 11mm. longa et 8–9 mm. lata, 2–3-sperma.

rescent. Leaf-lamina up to 18 × 16 cm., suborbicular-cordate, apex acuminate into a usually obtuse to subacute, minutely mucronate acumen, margin usually slightly but distinctly serrate-crenate, ± discolorous, the upper surface dark green, usually smooth, the lower surface grey-green with distinct and prominent venation; petiole up to c. 18 cm. long, terete, sometimes longitudinally sulcate. Flowers axillary on main branches and on short axillary shoots; pedicels often exceeding the petioles. Calyx campanulate to cupular, 10–18 mm. long and 8–10 mm. in diam.; lobes 6–12 × 3–6 mm., ovate-lanceolate, lanceolate-linear or narrowly triangular-lanceolate, gradually acuminate or attenuate. Petals 14–20 mm. long, yellow, sometimes reddish at the base inside and/or with reddish veins in the basal portion. Staminal tube stellate-hairy. Fruit c. 15 × 20–25 mm., stellate-pubescent, but the mericarps ultimately stellately spreading. Mericarps 25–40, 2–3-seeded, ultimately glabrescent, black, produced at the apex into a long, pointed acumen about ⅓ of the total length of the mericarp. Seeds papillose-verruculose.

N. Rhodesia. B : Sesheke, fl., *Macaulay* 184 (K). N : Mpika, fl. 12.ii.1955, *Fanshawe* 2065 (K). C : Lusaka, fl. 2.ii.1957, *Noak* 114 (K ; SRGH). **S. Rhodesia.** N : Urungwe, *Wild* 4086 (SRGH). C : Rusape, fl. & fr. vii.1952, *Dehn* in GHS 37565 (SRGH). E : Umtali, fl. & fr. vi.1919, *Eyles* 1692 (PRE ; SRGH). **Nyasaland.** S : Mt. Mlanje, 1830 m., fl. & fr. x.1891, *Whyte* 63 (BM). **Mozambique.** N : Mossuril, *Pedro & Pedrógão* 4761 (EA ; LMJ ; SRGH). Z : Milanje, fl. & fr. 25.viii.1942, *Hornby* 3789 (PRE ; SRGH). MS : Manica, between Dombe and Mavita, fl. & fr. 19.vi.1942, *Torre* 4372 (LISC ; SRGH). LM : Maputo, Quinta da Pedra, fl. & fr. 17.viii.1948, *Gomes e Sousa* 3797 (K ; LM ; PRE).
Widespread in tropical Africa and in the Transvaal, Swaziland, Zululand and Mauritius. Semi-ruderal and ± ubiquitous (roadsides to open woodland) and on a variety of soils ; from sea-level to c. 2000 m.

Forms with or without long patent hairs are found but all have the characteristic long-awned mericarps and are merely slight variations within the species.

10. **Abutilon lauraster** Hochr. in Ann. Conserv. Jard. Bot. Genève, **6** : 24 (1902) ; Fl. Madagasc., Malvac. : 142, t. 34 fig. 7–9 (1955). TAB. **93** fig. 11. Type from Madagascar.
     Abutilon zanzibaricum Boj. ex Mast. in Oliv., F.T.A. **1** : 186 (1868) pro parte quoad specim. Kirk. ex Senna.

Herb or soft-wooded shrub, 1–2 m. tall, annual or biennial ; stems subterete, tomentose, often with long patent hairs, glabrescent, sometimes somewhat gland-ular-viscid towards the apex. Leaf-lamina up to 20 × 18 cm. (smaller in the upper leaves), broadly ovate-cordate to cordate-acuminate, subentire, glabrescent above, tomentose or stellate-pubescent beneath ; petiole usually ± as long as the corresponding lamina (but shorter or almost absent in the upper leaves), subterete, finely pubescent. Flowers yellow, in axils of uppermost leaves forming narrow terminal pseudo-panicles ; pedicels usually less than 3 cm. long in flower, up to c. 6 cm. long in fruit, articulated in upper 10 mm., ferruginous-tomentose or pubescent. Calyx 10–12 mm. long, pubescent to tomentose and ciliate, divided about half-way down ; tube cupuliform ; lobes 5–8 × 2·5–5 mm., triangular or somewhat ovate-acute. Petals c. 20 mm. long, yellow. Staminal column pubescent on upper portion, glabrous towards the base. Fruit of 9–16 ultimately stellately spreading blackish mericarps ; mericarps 11 × 3 mm., glabrescent, gradually or abruptly attenuate at the apex into a subulate apical awn, 3-seeded. Seeds c. 1·5 × 1·5 mm., black, smooth, sometimes with a tuft of hairs near the hilum.

N. Rhodesia. E : Luangwa R., 520 m., fl. 25.iii.1955, *E.M. & W.* 1191 (BM ; LISC ; SRGH). **S. Rhodesia.** N : Urungwe Distr., Zambezi Valley, Rifa R., 520 m., fl. 24.ii.1953, *Wild* 4086 (BM ; PRE ; SRGH). **Nyasaland.** N : 19 km. W. of Karonga, fl. 14.iv.1954, *Williams* 250 (BM ; SRGH). **Mozambique.** N : Goa I., fl. 5.v.1947, *Gomes e Sousa* 3509 (K). Z : Chamo, fr. iii.1859, *Kirk* (K). T : opposite Sena, fr. 9.iv.1860, *Kirk* (K). MS : Vila Machado, fl. & fr. 14.iv.1948, *Mendonça* 3950 (LISC ; SRGH). SS : Vilanculos, *Barbosa & Balsinhas* (LM).
Also in the Transvaal, eastern tropical Africa and Madagascar. The paucity of specimens does not give enough information about the ecology of the species, but it seems to occur in habitats similar to those of *A. mauritianum*.

Superficially this species is very similar to *A. mauritianum* in habit, but the subentire, and, in the pseudo-panicles, subsessile leaves make the identification usually quite easy, even if fruits are lacking. It is much rarer than most of the other species of *Abutilon* and the cited specimens are all the records available.

11. **Abutilon guineense** (Schumach.) Bak. f. & Exell in Journ. of Bot. **74,** Suppl.
Polypet. Addend.: 22 (1936).—Exell & Mendonça, C.F.A. **1,** 1: 154 (1937).—
Keay, F.W.T.A. ed. 2, **1,** 2: 337 (1958). TAB. **93** fig. 12. Type from W. Africa.
Sida guineensis Schumach. in Kongel. Dansk. Vid. Selsk. Naturvid. Math. Afh.
4: 21 (1829). Type as above.
Abutilon asiaticum sensu Garcke in Peters, Reise Mossamb. Bot. **1**: 129 (1861).—
Mast. in Oliv., F.T.A. **1**: 184 (1868).—R.E.Fr., Wiss. Ergebn. Schwed. Rhod.-
Kongo-Exped. **1**: 143 (1914).—Brenan, T.T.C.L.: 298 (1949).
Abutilon hirsutissimum sensu Eyles in Trans. Roy. Soc. S. Afr. **5**: 413 (1916).

Herbaceous to suffruticose annual or biennial up to c. 1·5 m. tall, branched from the base and densely covered with a yellowish stellate indumentum; stems subterete, ultimately glabrescent and closely marked with short linear-rhombic shallow grooves. Leaf-lamina up to 8 (12) × 8 (12) cm., broadly ovate- to suborbicular-cordate, acute or obtuse or somewhat acuminate, upper surface deep yellowish-green; rugose-scabrid (rough to the touch) and thinly stellate-pubescent, lower surface paler and softly tomentose, margin finely and rather regularly (sometimes indistinctly) crenate or dentate; petiole about as long as the corresponding lamina. Flowers solitary in the axils of upper leaves of main stems; pedicels up to 8 (10) cm. long, terete, articulated in the upper 11 mm. Calyx cupuliform, 16–19 mm. long, lobed to about the middle; lobes triangular or ovate-triangular, acute, often mucronate, usually distinctly mid-veined, finely and densely ciliate, accrescent and ultimately equalling or slightly exceeding the ripe fruit. Petals c. 18 mm. long, yellow, obovate, glabrous except near the base. Staminal tube glabrous or with tufts of stellate hairs at the line of fusion with the petals. Fruit c. 15 × 20 mm., very densely pilose, depressed-globose, truncate to shallowly umbilicate at the apex; mericarps c. 20, their outer apical angle produced into a triangular-acuminate point or somewhat awned, densely pilose to bearded on the back and upper half of the flat sides, (2) 3-seeded. Seeds c. 3 × 2·5 mm., glabrous, often minutely rugulose-verrucose.

S. Rhodesia. W: Bulawayo, fl. & fr. xii.1897, *Rand* 7 (BM). S: Sabi-Lundi, Lower Sabi, W. bank, fl. & fr. 2.ii.1948, *Wild* 2480 (K; SRGH). **Nyasaland.** S: Shire R., N. of Chiromo, fl. & fr. 10.iii.1938, *Lawrence* 632 (K). **Mozambique.** Z: Alto Molocuè, Nhauela, fr. x.1945, *Pedro* 1333 (K; LMJ). T: Sena, fl. & fr. 7.iv.1860, *Kirk* (K). MS: between R. Zangui and Lacerdónia, *Pedro & Pedrógão* 8564 (LMJ; SRGH). LM: Sabiè, Moamba, fl. & fr. 27.v.1947, *Pedrógão* 203 (LMJ; PRE).
From Ghana to Angola and in the Transvaal, Swaziland, Natal and Madagascar. Semi-ruderal and in open places in bush, often on sandy soil; almost restricted to the drier warm lowland areas below 500 m.

12. **Abutilon grandiflorum** G. Don, Gen. Syst. **1**: 504 (1831).—Exell & Mendonça, C.F.A. **1,** 1: 154 (1937).—Exell, Cat. Vasc. Pl. S. Tomé: 116 (1944). TAB. **93** fig. 13, **94**. Type from S. Tomé.
Abutilon indicum sensu Mast. in Oliv., F.T.A. **1**: 186 (1868).—Ulbr. in Engl., Bot. Jahrb. **51**: 32 (1913).—R.E.Fr., Wiss. Ergebn. Schwed. Rhod.-Kongo-Exped. **1**: 143 (1914).—Arwidss. apud Norlindh & Weim. in Bot. Notis. **1934**: 94 (1934).—Burtt Davy & Hoyle, N.C.L.: 49 (1936).

Herb or soft-wooded shrub, 0·75–1·5 m. tall, annual or biennial, softly velutinous; stems terete, green or grey-green to olive, branched from the base, the older parts glabrescent and somewhat woody. Leaf-lamina up to 15 cm. long and 14 cm. broad but often much smaller, broadly ovate-cordate, green, lower surface slightly paler, usually long-acuminate with narrow acumen, margin rather coarsely and irregularly serrate to doubly serrate; petiole generally about as long as the corresponding lamina. Flowers axillary on main branches; pedicels up to 8 cm. long, articulated in the upper 11 mm. Calyx 15–17 mm. long, cupuliform, not much accrescent in fruit, lobed a little beyond the middle; lobes triangular to ovate-lanceolate, attenuate-acuminate into a sharp apex. Petals 15–20 mm. long, yellow. Staminal tube velutinous towards the base. Fruit c. 13 × 20 mm., depressed-globose, truncate to shallowly umbilicate, softly stellate-pilose and shortly stellate-pubescent. Mericarps c. 20, 2–3-seeded, the dorsal apical angle triangular-pointed to shortly awned, the back and a narrow zone along the upper

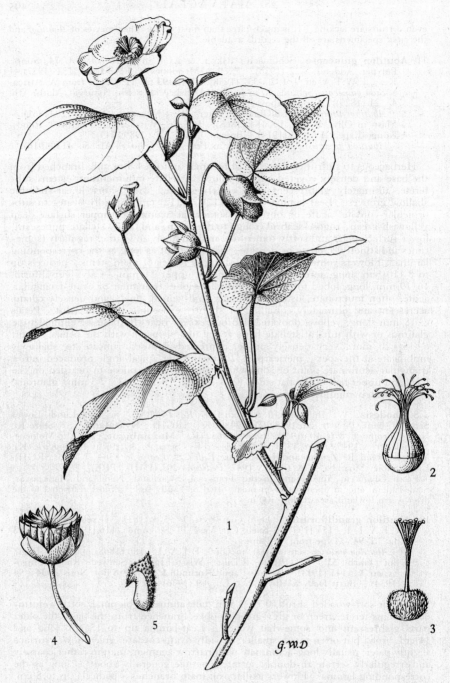

Tab. 94. ABUTILON GRANDIFLORUM. 1, flowering branch (×1) *Plowes* in GHS 40053;
2, androecium (×2) *Plowes* in GHS 40053; 3, gynoecium (×2) *Plowes* in GHS 40053;
4, fruit with persistent calyx (×1) *Codd* 4254; 5, mericarp (×1) *Codd* 4254.

edge with more or less pilose hairs, the lateral flat sides finely stellate-pubescent. Seeds c. 2×1.75 mm., punctate-lepidote, glabrous except for a tuft of hairs near the hilum.

S. Rhodesia. E: Lower Sabi, E. bank, Mtema, fr. 2.ii.1948, *Wild* 2409 (K; SRGH). S: West Nicholson, fl. & fr. 26.x.1952, *Plowes* 1516 (BM; SRGH). **Nyasaland.** S: 8 km. WNW. of Makanga, fl. & fr. 20.iii.1960, *Phipps* 2585 (PRE; SRGH). **Mozambique.** MS: Chemba, fr. 12.xi.1946, *Pedro & Pedrógão* 129 (LM; PRE). SS: Gaza, Chibuto, between Maniquenique and Licilo, fl. & fr. 6.viii.1958, *Barbosa & Lemos in Barbosa* 8311 (K; LMJ). LM: Sabiè, Moamba, fl. & fr. 27.xii.1946, *Pedrógão* 21 (LMJ; PRE).
Also in S. Tomé, Angola, Transvaal, Zululand and Madagascar. Mainly below 1200 m. in sunny to lightly shaded places on sandy, gravelly or rocky soil.

13. **Abutilon rehmannii** Bak. f. in Journ. of Bot. **31**: 217 (1893).—Ulbr. in Engl., Bot. Jahrb. **51**: 30 (1913).—Burtt Davy, F.P.F.T. **2**: 275 (1932). TAB. **93** fig. 14. Type from the Transvaal.
    Abutilon seineri Ulbr., op. cit. **48**: 369 (1912); op. cit. **51**: 16 (1913). Type from SW. Africa.

Soft-wooded shrub (probably annual) up to 1·5 m. tall, erect, usually not much branched, with a usually yellowish but occasionally dull-greyish-green dense short stellate-subvelutinous indumentum; stems terete or sometimes $\pm$ angular when very young, sulcate or ribbed, sometimes pilose towards the base. Leaf-lamina 2–10 (16) × 1–7 (9) cm., usually (at least those of the younger leaves) cordate-triangular, up to twice as long as broad (relatively broader and $\pm$ sub-orbicular-cordate in the older leaves), apex attenuate to long-acuminate or sometimes caudate (acute to shortly acuminate in the older leaves), margin irregularly or doubly serrate-dentate, sometimes shallowly crenate or serrate but rarely subentire (if so, youngest leaves distinctly serrate), upper surface rather dark green, minutely and scabridly stellate-pubescent, rarely somewhat velutinous, lower surface paler, somewhat glaucous, usually softly velutinous, rarely somewhat scabridly tomentose; petiole terete or sometimes sulcate, shorter than to longer than the lamina; stipules c. 5 mm. long, linear-lanceolate, velvety, usually very early caducous. Flowers yellow or pale yellow, solitary, mainly in upper axils of terminal and side-branches, often forming pseudo-racemes or pseudo-panicles because the upper leaves are usually small; pedicels usually under 3 cm. long (in fruit under 5 cm. long), articulated and usually more or less geniculate at or above the middle. Calyx 6–8 mm. long, cupuliform, lobed to about the middle; lobes ovate, acute to acuminate, often more or less apiculate or with a short narrow acumen, ciliate and distinctly mid-veined. Petals c. 13 mm. long, glabrous except for the ciliate narrow base. Staminal tube glabrous. Fruit c. 10×12–14 mm., subcylindrical-semiglobose, concave and umbilicate at the apex. Mericarps 10–20 (often 12–16), 7·5–9 × 4–5 mm., 3-seeded, rounded at the back and at the somewhat narrowed base, convex to obliquely truncate at the apex, toothed to shortly awned at the dorsal apical angle, 7·5–9 mm. long and 4–5 mm. broad measured across the large ventral tooth, the back and apical parts of lateral faces grey-velutinous. Seeds c. 2×1.5 mm., finely verruculose.

S. Rhodesia. S: Beitbridge, fl. & fr. 14.v.1959, *Drummond* 6145 (SRGH).
Also in SW. Africa, Griqualand-W. and northern Transvaal. On alkaline soils such as dolomite hillsides, limestone flats and salt pans at 700–1200 m.

This is the first record outside S. Africa.

14. **Abutilon engleranum** Ulbr. in Engl. Bot. Jahrb. **51**: 30 (1913). Syntypes from SW. Africa.
    Abutilon membranifolium Bak. f. in. Journ. of Bot. **77**: 17 (1939).—Exell & Mendonça, C.F.A. **1**, 2: 374 (1951). TAB. **93** fig. 15. Type from Angola.

Suffrutex up to c. 2 m. tall, sparsely branched, finely velutinous and in addition with usually rather dense patent long soft white hairs, the indumentum usually light-yellowish-grey but turning brown and gradually disappearing with age; stems rather stout, terete, ultimately light brown or purplish with numerous short fine longitudinal fissures. Leaf-lamina c. 3–8 (16) cm. in diam., thin but firm in texture, usually $\pm$ suborbicular, sometimes slightly 3-lobed, apex with a rather long acumen or occasionally rounded, margin irregularly and rather coarsely serrate, ± biserrate or somewhat crenate, base cordate (usually deeply so), upper surface finely velutinous, deep greyish-green to yellowish-green, lower surface paler, ash-grey or pale green with prominent veins ; petiole as long as or longer than the corresponding lamina, slender, terete or somewhat flattened at the base. Flowers yellow, axillary ; pedicels c. 7 cm. (up to 8 cm. in fruit) long, articulated towards the apex. Calyx c. 15 mm. long, campanulate, divided beyond the middle ; lobes 10 × 5 mm., ovate-lanceolate, long-acuminate to aristate, velutinous inside. Petals c. 16 × 10 mm., shortly pubescent outside and along the upper edge inside. Staminal tube c. 12 mm. long, conical, rather densely stellate-pubescent. Fruit about 15 mm. in diam. at the apex, about 12 mm. near the base, obconical-turbinate, shallowly umbilicate, stellate-hairy, the sides enclosed by the slightly accrescent calyx. Mericarps 16–20, c. 10 × 4 mm., somewhat oblique, usually 3-seeded ; the apical edge slanting, convex to nearly straight, the inner angle almost continuous with the large ventral tooth, the upper outer angle very acute or produced into a short awn up to 2 mm. long. Seeds 3 × 2·5 mm., minutely punctate-verruculose.

Caprivi Strip. Linyanti area, fl. & fr. 27.xii.1958, *Killick & Leistner* 3152 (BM ; K ; PRE ; SRGH). **Bechuanaland Prot.** N : Ngamiland, Mahlatlogo, fl. & fr. 10.v.1930, *van Son* in Herb. Transv. Mus. 28933 (PRE). **N. Rhodesia.** B : Sesheke, fl. 1860, *Kirk* (K). C : Chilanga, fl. & fr. 9.ix.1904, *Rogers* 8450 (BM ; GRA ; K). **S. Rhodesia.** S : Limpopo R., fl. & fr. x.1956, *Davies* 2172 (SRGH). **Mozambique.** SS : Guijà, fl. & fr. 6.vi.1947, *Pedrógão* 245 (LMJ ; PRE).
Also in Angola, SW. Africa and the Transvaal. Apparently on sandy soils in woodland or scrub, not on rocks.

Another species with W.–E. distribution (see note under *A. austro-africanum*).

15. **Abutilon ramosum** (Cav.) Guill. & Perr. in Guill., Perr. & Rich., Fl. Senegamb. Tent. **1** : 68 (1831).—Mast. in Oliv., F.T.A. **1** : 186 (1868).—Ulbr. in Engl., Bot. Jahrb. **51** : 27 (1913).—Burtt Davy, F.P.F.T. **2** : 275 (1932).—Exell & Mendonça, C.F.A. **1**, 1 : 152 (1937).—Keay, F.W.T.A. ed. 2, **1**, 2 : 337 (1958). TAB. **93** fig. 16. Type from Senegal.
  Sida ramosa Cav., Diss. **1** : 28, t. 6 fig. 1 (1785). Type as above.
  Abutilon harmsianum Ulbr. in Engl., Bot. Jahrb. **51** : 29 (1913). Type from SW. Africa.

Suffrutex up to c. 1·25 m. tall, erect or occasionally spreading (usually after having been grazed), usually branched from the base, with a short dense stellate usually somewhat rough pubescence and usually with additional ± sparse long patent hairs and often glandular-viscid when young ; stems terete, firm, green or yellowish-green, woody at the base, ultimately with a thin greyish bark. Leaf-lamina 4–10 (15) × 3–9 (13) cm., broadly ovate-cordate to suborbicular-cordate, sometimes shallowly 3-lobed, apex acuminate, margin usually shallowly but distinctly and regularly serrate or crenate with minutely apiculate serrations, upper surface dark green, finely and somewhat scabridly stellate-pubescent, lower surface slightly paler, similarly hairy ; petiole longitudinally sulcate, as long as or shorter than the corresponding blade. Flowers yellow to orange, in the axils of the upper leaves ; peduncle c. 4 cm. long, erect-patent, 2–4-flowered at the apex ; pedicels c. 2 cm. long, articulated in the upper 6 mm. (sometimes a solitary pedicel in the same axil as the peduncle but shorter than the latter). Calyx c. 8–12 mm. (accrescent to c. 9 × 12 mm. in fruit), shallowly cupular, lobes 4·5 mm. long, triangular or ovate-lanceolate, usually distinctly acuminate into a sulcate acumen. Petals 5–7 mm. long, ciliate at the base and sometimes also at the apex. Staminal tube densely stellate-hairy. Fruit of about 8 mericarps. Mericarps c. 9 × 3 × 2–2·5 mm., 2–3-seeded, ultimately light brown, ± radiate-spreading, pungently long-awned, with subulate curved outwardly spreading awns. Seeds c. 2 × 2 mm., dark brown, rugulose-papillose.

Bechuanaland Prot. N : Ngamiland, fl. & fr. 5.v.1930, *van Son* in Herb. Transv. Mus. 28934 (PRE ; SRGH). **N. Rhodesia.** E : between Minga and Petauke, fr. 4.vi.1958, *Fanshawe* 4523 (K ; PRE). S : Gwembe valley, 1·6 km. S. of Mambo's village, fr. 29.iii.1952, *White* 2366 (FHO ; K). **S. Rhodesia.** N : Chiswiti Reserve, fl. & fr. 27.i.1960, *Phipps* 2436 (BM ; EA ; LISC ; LMJ ; PRE ; SRGH). W : Wankie, fl. & fr. ii.1955, *Levy* 1080 (PRE ; SRGH). E : Birchenough Bridge, fl. & fr. i.1938, *Ober-meyer* in Herb. Transv. Mus. 37499 (BOL ; PRE ; SRGH). S : Gwanda, fl. v.1959, *Davies* 1245 (BM ; PRE ; SRGH). **Nyasaland.** S : Namgala R., SW. of Chiromo, fl. &

fr. 25.iii.1960, *Phipps* 2743 (BM ; LISC ; PRE ; SRGH). **Mozambique.** T : Muatize, *Mendonça* 4129 (LISC). SS : Guijà, fl. & fr. 6.v.1957, *Carvalho* 157 (LM ; PRE).
Widespread in tropical Africa, in SW. Africa and the Transvaal, and in NW. India. Found mainly in rocky places, often in semi-shade, sometimes gregarious.

The mericarps differ from those of all other African species of the genus, *A. ramosum* being the only representative of a group of otherwise exclusively American species characterised by this type of fruit.

12. WISSADULA Medic.

Wissadula Medic., Künstl. Geschlecht. Malv.-Fam. : 24 (1787).—R.E.Fr. in Kungl. Svensk Vet. Akad., Ny Földj. **43**, 4 : 1 (1908).

Small erect shrubs or suffrutices. Leaves usually cordate at the base, acuminate, petiolate, entire or denticulate ; stipules setaceous. Flowers orange, yellow or cream, in a lax panicle. Epicalyx absent. Calyx 5-cleft, stellate-hairy, shallowly saucer-shaped ; the lobes longer than the tube, deltoid-ovate. Petals obovate or obovate-spathulate. Staminal tube very short, ventricose ; free parts of filaments numerous, slender, elongate, linear. Ovary of 3–5, (1) 2–3-ovulate, free carpels ; style short, with 3–5 long terete branches ; stigmas capitate. Fruit of 3–5 free mericarps transversely divided by an internal false dissepiment formed by an oblique constriction of the lateral walls, beaked, pubescent, and ultimately septicidal, the anterior portion sometimes also loculicidal. Seeds 1–3 per mericarp, subglobose-reniform, pubescent ; cotyledons lying one within the other, forming a deeply concave structure.
A genus of about 40 species, predominantly American, but with a few (perhaps only one) in the Old World.

Wissadula rostrata (Schumach.) Hook. f. in Hook., Niger Fl. : 229 (1849).—Mast. in Oliv., F.T.A. **1** : 182 (1868).—Eyles in Trans. Roy. Soc. S. Afr. **5** : 413 (1916).— Exell in Bull. I.F.A.N. **21**, Sér. A : 452 (1959). TAB. **95**. Type from W. Africa.
   Sida rostrata Schumach. in Kongel. Dansk. Vid. Selsk. Naturvid. Math. Afh. **4** : 80 (1829). Type as above.
   Wissadula amplissima var. *rostrata* (Schumach.) R.E.Fr. in Kungl. Svensk Vetenskapakad. Handl., Ny Földj. **43**, 4 : 51, t. 6 fig. 13–14 (1908).—Exell & Mendonça, C.F.A. **1**, 1 : 150 (1937).—Brenan, T.T.C.L. : 308 (1949).—Keay, F.W.T.A. ed. 2, **1**, 1, 2 : 336 (1958). Type as above.
   Wissadula hernandioides var. *rostrata* (Schumach.) R.E.Fr. in Wiss. Ergebn. Schwed. Rhod.-Kongo-Exped. **1** : 143 (1914).—Robyns, Fl. Parc Nat. Alb. **1** : 584 (1948).
   Wissadula hernandioides sensu Eyles, loc. cit.

Suffrutex 0·75–1·25 m. tall, branched from the base ; stems stellate-pubescent, glabrescent and ultimately covered with a thin greyish bark. Leaf-lamina 4–13 ×1·5–8 cm., subentire, apex long-caudate, base truncate to cordate, upper surface minutely stellate-hairy and rather dark green, lower surface paler and more stellate-hairy to tomentose ; petiole up to 5 cm. long, floccose-tomentose ; pedicels 1–3 cm. long (elongating to c. 6 cm. in fruit), articulated towards the apex, puberulous to stellate-floccose. Calyx c. 3 mm. long, stellate-pubescent ; lobes acute. Petals c. 4 mm. long. Staminal tube glabrous. Fruit angularly obconic ; mericarps c. 7 ×3 mm., glabrescent. Seeds c. 3 ×3 mm., black.

Caprivi Strip. Fl. & fr. 10.iv.1946, *Kruger* F (K ; PRE). **Bechuanaland Prot.** N : Chobe-Zambezi confluence, fl. & fr. 11.iv.1955, *E.M. & W.* 1477 (BM ; LISC ; SRGH). **N. Rhodesia.** B : Gonye Falls, fr. 18.vii.1952, *Codd* 7120 (K ; PRE). W : Mufumbwe R., fl. & fr. 25.vi.1953, *Fanshawe* 124 (K). C : Chisamba, 1200 m., fl. & fr. 19.v.1957, *Best* 125 (K). S : Maala, 32 km. E. of Namwala, fl. & fr. 24.iv.1954, *Robinson* 733 (K ; SRGH). **S. Rhodesia.** W : Wankie, *Plowes* 1958 (PRE ; SRGH). **Nyasaland.** N : Kongwe and Karonga, 500–600 m., fl. & fr. vii.1896, *Whyte* 44 (K). **Mozambique.** N : Nampula, *Pedro & Pedrógão* 4726 (LMJ ; SRGH). MS : Moribane Forest, fl. & fr. 8.ii.1952, *Chase* 4455 (BM ; PRE ; SRGH). LM : Guijà, fl. & fr. 6.v.1957, *Carvalho* 161 (LM ; PRE).
Widespread in tropical Africa and in SW. Africa and the Transvaal. Usually in light shade on rocky and loamy soils and in riverine vegetation.

Some authors refer the African specimens to *W. periplocifolia* (L.) C. Presl, which they take to be pantropical, others refer it to *W. amplissima* (L.) R.E.Fr., or *W. hernandioides*

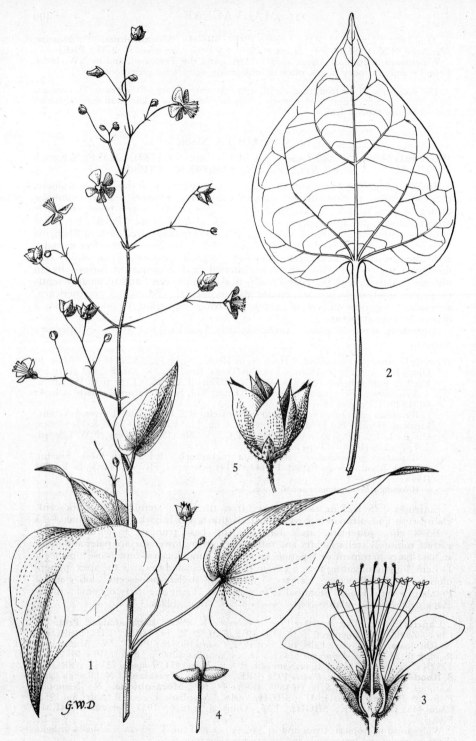

Tab. 95. WISSADULA ROSTRATA. 1, branch with flowers and fruits (× ½) *Chase* 4455; 2, leaf (× ½) *Eyles* 142; 3, vertical section of flower (× 5) *Chase* 4455; 4, anther (× 10) *Chase* 4455; 5, fruit (× 3) *Chase* 4455.

(L'Hérit.) Garcke and this is much a matter of personal opinion. Exell (l.c.) is of the opinion that the African form is best treated as a species and his treatment is followed here, although the differences between *W. amplissima*, *W. hernandioides* and *W. rostrata* indicated by him are small. It must be left to a future monographer of the genus to decide the issue.

13. MALVA

Malva L., Sp. Pl. **2**: 687 (1753); Gen. Pl. ed. 5: 308 (1754).

Annuals or perennials, herbaceous to somewhat suffruticose, erect to decumbent or prostrate, nearly glabrous to more or less densely pubescent. Leaves usually suborbicular-reniform in outline, crenate to serrate or shallowly to rather deeply palmatilobed, usually long-petioled. Flowers axillary, usually clustered or fasciculate; pedicels often short to 0. Epicalyx of 3 free or nearly free linear to ovate bracts. Calyx cupular, 5-lobed; lobes deltoid, acute. Petals usually obcordate-cuneate, emarginate at the apex, often with darker-coloured veins converging towards the glabrous or ciliate claw. Staminal tube antheriferous at the apex only. Ovary of 9–15 free 1-ovulate carpels arranged around a central torus; style-branches as many as there are carpels, obliquely clavate; stigmas linear, decurrent on the inside of the style-branches. Fruit discoid, separating at maturity into suborbicular-reniform free indehiscent muticous mericarps; mericarps smooth or variously sculptured on the back, glabrous or pubescent.

A genus of about 30 Old World species, mainly palaearctic, some occurring as weeds and now cosmopolitan. The species recorded from our region are almost certainly all introduced and two of them probably only quite recently. The genus *Malva* is, in Africa, probably only truly native in the Mediterranean zone north of the Sahara and in Ethiopia.

Flowers 2·5–5 cm. in diam., mauve-purple; epicalyx-lobes oblong, rounded at the apex; petal-claws bearded; mericarps reticulate on the back- - - 1. *sylvestris*
Flowers up to 2·5 cm. in diam., pale mauve, pinkish or blue-mauve to white; epicalyx-lobes linear or linear-lanceolate, acute to obtuse:
Mericarps with acutely angled reticular dorsal ridges, meeting laterally in a somewhat raised sharp edge; petals scarcely exceeding the calyx, with glabrous claws; plant usually decumbent - - - - - - - - 2. *parviflora*
Mericarps smooth to faintly reticulate on the dorsal surface, not meeting laterally in a sharp raised edge; petals distinctly longer than the calyx:
Plant usually erect; mericarps 10–12, usually glabrous; petal-claws glabrous or with a few weak hairs; cauline leaves shallowly but distinctly lobed, 4–11 cm. long, on 4–8 cm. long petioles - - - - - - 3. *verticillata*
Plant decumbent; mericarps 12–15, pilose; petal-claws bearded; leaves very slightly lobed, 1–6 cm. long, on 3–20 cm. long petioles - - 4. *neglecta*

1. **Malva sylvestris** L., Sp. Pl. **2**: 689 (1753).—Exell & Mendonça, C.F.A. **1**, 1: 147 (1937). Type from Europe.

Erect perennial up to c. 1 m. tall; stems sparsely hirsute with simple or stellate usually spreading hairs. Leaf-lamina of basal leaves 5–10 cm. in diam., shallowly crenate, of upper leaves ± deeply 5-lobed with crenulate-serrate lobes, glabrous except for a strigose pubescence on the main veins; petiole (2) 4–9 (16) cm. long, pubescent in an adaxial groove, otherwise glabrous; stipules 5–10 × 2–5 mm. obliquely ovate, serrate, acuminate. Flowers 2·5–5 cm. in diam., mauve-purple, or light magenta with darker purple veins, 2–5-fasciculate or in short racemes; pedicels 1–2 cm. long, glabrous or sparsely hirsute. Bracts of epicalyx ⅔ the length of the calyx, oblong-lanceolate. Calyx stellate-pilose, not enlarged in fruit; lobes ovate-deltoid. Petals 1·5–2 cm. long; claw barbate. Staminal tube setose. Mericarps 9–11, dorsally reticulate, usually glabrous.

S. Rhodesia. W: Bulawayo, *Miller* 2820 (SRGH).
Native of Europe. Ruderal. Sometimes an escape from cultivation.

2. **Malva parviflora** L., Demonstr. Pl.: 18 (1753).—Harv. in Harv. & Sond., F.C. **1**: 159 (1860).—Mast. in Oliv., F.T.A. **1**: 177 (1868).—Ulbr. in Engl., Bot. Jahrb. **51**: 36 (1913).—Burtt Davy, F.P.F.T. **2**: 273 (1932).—Exell & Mendonça, C.F.A. **1**, 1: 146 (1937). Type from Europe.

Prostrate or decumbent herb; stems usually under 40 cm. long, hairy, longi-tudinally sulcate. Leaf-lamina 1–5 × 1–5 cm., reniform or nearly suborbicular in outline, obtusely 3–5-angled, crenate or crenate-serrate, glabrous or sparsely hairy; petiole often longer than the lamina, usually with a few hairs near the apex. Flowers mauve or pale pink to white, subsessile, in few-flowered clusters in the leaf-axils; pedicels 2–4 mm. long. Bracts of epicalyx linear, deciduous. Calyx 2–5 mm. long, accrescent and enclosing the fruit; lobes ovate or roundish, mucronate, spreading. Petals scarcely exceeding the calyx; claw glabrous. Fruit glabrous or occasionally hairy; mericarps with acutely-angled reticular dorsal ridges and slightly raised and sharp lateral angles.

S. Rhodesia. C: Salisbury, fl. & fr. ix.1921, *Eyles* 3183 (K; PRE; SRGH).
Native of Europe and Asia, naturalised in many countries as a weed of cultivation and now almost cosmopolitan. Also in S. Africa and Madagascar. Found in waste places and in parks and gardens, but never a serious pest. The leaves are eaten by chickens and can be fed to cattle, pigs and various kinds of domestic birds.

3. **Malva verticillata** L., Sp. Pl. **2**: 689 (1753).—Mast. in Oliv., F.T.A. **1**: 177 (1868).—Burtt Davy, F.P.F.T. **2**: 274 (1932). Syntypes from China and Syria.

Biennial herb 0·5–1 m. tall; stems shortly stellate-pilose. Leaf-lamina 4–11 cm. in diam., of basal leaves reniform, of cauline leaves suborbicular and 5-lobed with rounded lobes on lower ones and those of upper ones triangular, margin crenulate-dentate; petiole 4–8 cm. long, longitudinally sulcate, glabrescent except that the grooves remain tomentose; stipules ovate-lanceolate. Flowers in fas-cicles; pedicels unequal, the longest up to about 3 cm., shortly stellate-pilose. Bracts of epicalyx 5–6 mm. long, linear, ciliate. Calyx somewhat inflated, sparsely stellate-hirsute; lobes triangular, acute. Petals about twice as long as the calyx; claw glabrous or with a few weak hairs. Staminal tube setose in upper portion. Mericarps 10–12, dorsally smooth or faintly rugose, rugose along the rounded angles, flabellately striate on the flat lateral sides.

S. Rhodesia. C: Rusape, Wick, 1370 m., fr. 21.ii.1940, *Hopkins* in GHS 7661 (BM; SRGH).
Native of Asia but introduced as a weed of cultivation into many parts of the world, e.g. Europe, S. Africa.

The leaves are relished by domestic birds and can also be fed to cattle and pigs. Ex-tensively cultivated in China as a pot green. It is said to be an excellent vegetable rich in vitamins and minerals.

4. **Malva neglecta** Wallr. in Syll. Ratisb. **1**: 140 (1824). Type from Europe.

Perennial herb with strong deep penetrating taproot up to 20 cm. in length; stems stout and woody in basal portion, those developing in the warmer season decumbent and forming flat rosettes up to about 1 m. in diam., those formed in colder seasons abbreviated with very short internodes, all sparsely and more or less appressed-stellate-pubescent. Leaf-lamina 1–6 × 1–6 cm., orbicular-reniform with deep triangular basal sinus, crenate-denticulate, that of leaves on elongated branches very shallowly 5–7-lobed, sparsely strigose on both sides, the hairs often simple on upper surface and stellate on lower surface; petiole 3–20 cm. long, softly and sparsely stellate-pilose; stipules ovate-acuminate or ovate-lanceolate, acute. Flowers in fascicles of up to 4 or (in lower portions of stems) solitary; pedicels 2–5 cm. long, sparsely stellate-pubescent. Epicalyx-lobes 3–4 mm. long, linear-lanceolate. Calyx 5–7 mm. long, stellate-pilose, lobed to or a little beyond the middle, later somewhat accrescent and enclosing the fruit; lobes triangular-ovate, acute, ± ciliate. Petals 10–13 mm. long, white, usually with mauve or purplish veins and mauve or pale purple towards the apex, the claws barbate. Staminal tube pubescent. Fruit 6–8 mm. in diam., of 12–15 smooth shortly stellate-pilose mericarps with rounded lateral angles.

S. Rhodesia. E: Chipinga, Sabi Valley Experimental Station, fl. & fr. x.1959, *Soane* 76 (BM; PRE; SRGH).

A ruderal or semi-ruderal species of Old World origin, naturalised as a weed of cultiva-tion in many parts of the world, e.g. in N. America and S. Africa. The cited gathering is the first record from southern tropical Africa, but specimens from S. Africa indicate that this species was introduced there about 50 years ago (before 1913).

14. MALVASTRUM A. Gray

Malvastrum A. Gray in Mem. Amer. Acad. Sci., N.S., **4** : 21 (1849).—Kearney in Leafl. West. Bot. **5** : 23 (1947) ; op. cit. **7** : 238 (1955) *nom. conserv.*

Annual to perennial low suffrutices, erect to somewhat spreading. Leaves ovate, dentate or ± palmatilobed, petiolate. Flowers yellow, axillary, solitary or in terminal racemes. Epicalyx of 3 linear or subulate free bracts. Calyx cupulate or saucer-shaped, foliaceous in fruit. Staminal tube antheriferous only at the apex ; filaments filiform. Ovary of 10 free 1-ovulate carpels ; style-branches 10, filiform ; stigmas capitate. Fruit discoid, of 10 reniform setose or strigose indehiscent 1-seeded mericarps ultimately separating from the columella. Seeds reniform ; cotyledons folded ; endosperm scanty.

Kearney has restricted *Malvastrum* A. Gray to 3 N. American species (2 of which are now pantropical weeds) and has excluded the other species formerly referred to the genus.

Malvastrum coromandelianum (L.) Garcke in Bonplandia, **5** : 295 (1857).—Keay, F.W.T.A. ed. 2, **1**, 2 : 350 (1958). Type from India.

    Malva coromandeliana L., Sp. Pl. **2** : 687 (1753). Type as above.
    Malva tricuspidata R.Br. in Ait., Hort. Kew. ed. 2, **4** : 210 (1812) *nom. illegit.* Type from the West Indies.
    Malvastrum tricuspidatum A. Gray, Pl. Wright. **1** : 16 (1852) *nom. illegit.*—Mast. in Oliv., F.T.A. **1** : 178 (1868). Type as above.
    Malvastrum spp.—Eyles in Trans. Roy. Soc. S. Afr. **5** : 413 (1916).

Annual or biennial suffrutex up to 1 m. tall (usually under 60 cm.) with the habit of a *Sida*, sparsely covered with simple and 4-rayed stellate-strigose hairs. Leaf-lamina 3–6 × 1–4 cm., coarsely serrate, apex usually obtuse ; petiole short ; stipules c. 5 mm. long, lanceolate. Flowers solitary in the upper leaf-axils ; pedicels c. 5 mm. long. Bracts of epicalyx 5 × 1 mm., linear. Calyx 8 mm. long, divided to c. ⅔ ; lobes ovate, acuminate, ciliate. Petals c. 8 × 4 mm., obovate. Staminal tube glabrous. Fruit c. 6 mm. in diam. ; mericarps 2 × 3 × 1 mm., with 2 short dorsal spines near the middle, crowned in the ventral area with the persistent base of the style, the inner upper portion produced into an acute slightly hooked beak with a narrow sinus below, dorsally and in part laterally sparsely stellate-hairy and also dorsally setose.

S. Rhodesia. C : Salisbury, fl. & fr. 11.vi.1956, *Drummond* 5127 (BM ; PRE ; SRGH). W : Bulawayo Municipal Park, fl. & fr. 30.iv.1958, *Drummond* 5517 (K ; PRE ; SRGH). **Nyasaland.** C : Kota Kota, *Benson* 78 (PRE). **Mozambique.** MS : Vila Machado, *Garcia* 986a (BM ; LISC). LM : Namaacha, *Carvalho* 235 (LM ; PRE ; SRGH).

Native of N. America but now pantropical and also in S. Africa. An introduced weed found along roadsides, on rubbish-heaps and in old cultivation.

15. URENA L.

Urena L., Sp. Pl. **2** : 692 (1753) ; Gen. Pl. ed. 5 : 764 (1754).

Suffrutices, usually stellate-pubescent. Leaves petiolate, subentire, 3–5-palmatilobed or sinuous along the margin, palmately 3–7-nerved, with a conspicuous gland near the base of the central nerve ; stipules small, linear to setaceous, usually caducous. Flowers axillary, solitary or fasciculate or crowded towards the end of the branches. Epicalyx campanulate, deeply 5-lobed ; bracts lanceolate, acute, striate towards the base. Calyx cupulate, deeply 5-fid ; lobes ovate or ovate-lanceolate, 1-nerved and keeled. Petals rose-pink or mauve, stellate-pilose outside. Staminal tube equalling the petals, the lower portion dilated and united with the petals at the base ; anthers subsessile. Ovary depressed-globose, hirtellous or glabrous, of 5 free 1-ovulate carpels ; styles 10, reflexed ; stigmas discoid, fimbriate at the apex. Fruit subglobose, of 5 mericarps ultimately separating from the torus ; mericarps trigonous, obovoid, coriaceous, indehiscent, the convex back stellately hirsute and with glochidiate spines, the flat lateral surfaces striate. Seeds obovoid-trigonous or reniform, glabrous.

A pantropical genus of c. 6 species. The bast fibres of some species, such as *U. lobata*, can be used as a substitute for jute and *Urena* has often been grown for this purpose (e.g. in Brazil and the Belgian Congo) but never with very satisfactory results.

Urena lobata L., Sp. Pl. **2**: 692 (1753).—Garcke in Peters, Reise Mossamb. Bot. **1**: 123 (1861).—Mast. in Oliv., F.T.A. **1**: 189 (1868).—Ficalho, Pl. Ut. Afr. Port.: 96 (1884).—Hochr. in Ann. Conserv. Jard. Bot. Genève, **5**: 131 (1901).—Sim, Forest Fl. Port. E. Afr.: 17 (1909).—Ulbr. in Engl., Bot. Jahrb. **51**: 52 (1913).—R.E.Fr., Wiss. Ergebn. Schwed. Rhod.-Kongo-Exped. **1**: 144 (1914).—Eyles in Trans. Roy. Soc. S. Afr. **5**: 414 (1916).—Burtt Davy & Hoyle, N.C.L.: 49 (1936).—Exell & Mendonça, C.F.A. **1**, 1: 155 (1937).—Brenan, T.T.C.L.: 308 (1949).—Keay, F.W.T.A. ed. 2, **1**, 2: 341 (1958). Type from China.

Suffrutex up to c. 1 m. tall; stems sparsely to densely stellate-tomentose. Leaves 3–5-palmatilobed with entire to pinnatisect lobes; lamina of lower leaves 4–6 × 5–7 cm., usually suborbicular, shallowly 3-lobed at the apex, rounded to subcordate at the base, of middle leaves 5–7 × 3–6·5 cm., often more ovate, of upper leaves ovate-elliptic; margin evenly or irregularly serrate; petiole 2–4 cm. long; stipules c. 3 mm. long, lanate. Flowers solitary; pedicels c. 3 mm. long, lanate. Epicalyx c. 6 mm. long, connate in the lower ⅓, stellate-pilose. Calyx shorter than the epicalyx, densely lanate at the base of the keels. Petals 10–15 (25) mm. long. Fruit c. 10 mm. in diam.

N. Rhodesia. B: near Senanga, fl. & fr. 30.vii.1952, *Codd* 7235 (BM; K; PRE; SRGH). N: Abercorn, Lucheche mouth, fl. & fr. 19.v.1936, *Burtt* 6368 (BM; K). W: Fort Rosebery Distr., Lake Bangweulu, Samfya, fl. & fr. 23.viii.1952, *Angus* 305 (BM; K; PRE). S: Livingstone, fl. & fr. xi.1956, *Gilges* 666 (SRGH). **S. Rhodesia.** W: Victoria Falls, Long I., fl. & fr. v.1904, *Eyles* 146 (BM; K; PRE; SRGH). C: Salisbury, cult., fl. & fr. 26.v.1949, *Wild* 2847 (BM; SRGH). E: Inyanga Distr., Honde Valley, fl. 17.iv.1958, *Phipps* 1072 (K; PRE; SRGH). S: Sabi R., Chitsa's Kraal, fr. 5.vi.1950, *Chase* 2330 (BM; SRGH). **Nyasaland.** N: Deep Bay, fl. & fr. 16.vii.1952, *Williamson* 25 (BM). C: Nchisi Mt., 1450 m., fl. & fr. 22.ii.1959, *Robson* 1710 (K; LISC; SRGH). S: Domasi Valley, fr. 25.iv.1949, *Wiehe* N/81 (K). **Mozambique.** N: Metangula, *Pedro & Pedrógão* 3854 (EA; LMJ; SRGH). Z: between Muobede and Tacuane, *Barbosa & Carvalho* 2844 (LM). T: Zobuè, Missão do Combate às Tripanosomas, *Mendonça* 4144 (LISC; SRGH). MS: Maronga Forest, *Simão* 444 (LM). SS: Gaza, Chongoene, fl. & fr. 2.v.1937, *Carvalho* 135 (LM; PRE; SRGH). LM: Polana, fr. 9.v.1920, *Borle* (PRE; SRGH).

Pantropical. A weed of cultivation usually in low hot regions and absent from most areas with a long dry winter and from those with a high rainfall and of higher altitude (such as the Eastern Border Mts. of S. Rhodesia).

Cultivated in Mozambique and S. Rhodesia on experimental farms as a potential fibre crop.

16. PAVONIA Cav.

Pavonia Cav., Diss. **2**, App. 2: [2] (1786); op. cit. **3**: 132, t. 45 (1787).— Ulbr. in Engl., Bot. Jahrb. **57**: 54 (1920–21).

Annual, biennial or perennial herbs or shrubs, usually erect. Leaves usually petiolate, varying in shape, undivided to palmatipartite, often cordate at the base. Flowers yellow to white, pink, mauve or reddish, pedicellate or subsessile, solitary or occasionally in terminal subcapitate racemes or spikes. Epicalyx of 5–16 ovate to linear or filiform bracts, free or ± connate, shorter or longer than the calyx. Calyx saucer-shaped to campanulate; lobes usually ovate. Petals longer than the calyx, sometimes clawed. Staminal tube dilated at the base, truncate and 5-toothed at the apex, bearing the free terete filaments on the upper portion (rarely nearly down to the base). Carpels 5, 1-ovulate; style-branches terete, usually 10 or occasionally fewer but always more than 5; stigmas capitate, often penicillate. Fruit of 5 indehiscent 1-seeded carpels separating from the torus, sometimes ribbed or winged, muticous or awned; pericarp thin or woody. Seeds reniform; cotyledons folded; endosperm absent.

Over 200 species in all warmer countries but mainly in S. America.

Bracts of epicalyx 10–15, filiform, longer than (often much exceeding) the calyx and forming a cage-like structure round the flower-bud (and later round the fruit):

Leaves palmatipartite; segments c. 5, lanceolate or oblong-lanceolate, usually serrate;
mericarps winged - - - - - - - - - 1. *clathrata*
Leaves not or only slightly lobed; mericarps not winged:
Epicalyx 9–12 mm. long, of 9–11 bracts; leaves ovate, often ± 3-lobed or sub-
hastate, coarsely and irregularly serrate or dentate, almost always greyish-tomentose
beneath and hence discolorous; mericarps not toothed on the lateral angles
2. *leptocalyx*
Epicalyx 15–21 mm. long, of 10–13 bracts; leaves elongate-triangular-hastate,
gradually attenuate at the subcaudate apex, rather regularly serrate-dentate, not
greyish-tomentose and only slightly paler beneath; mericarps with a tooth on
each lateral angle - - - - - - - - - 3. *rogersii*
Bracts of epicalyx 5–16, not filiform, shorter than to about as long as the calyx or only
slightly exceeding it:
Epicalyx rotate, of 5 (6) ± ovate-lanceolate or rhombic, spreading, basally connate
bracts persisting as a star-shaped structure under the fruit, up to c. 12 mm. long;
mericarps muticous, dorsally keeled and verrucose to muricate, c. 5 mm. long;
petals white to yellow or orange, 10–15 mm. long - - - 4. *patens*
Epicalyx of 5–16 bracts which are not stellately spreading in fruit; petals pink or
mauve to rose-coloured (or if yellow at least 20 mm. long):
Flowers pale yellow with deep-brown-purple or maroon centre; leaves suborbicular-
reniform, not or only shallowly lobed; lobes rounded; bracts of epicalyx 12–16
5. *hirsuta*
Flowers pink or mauve to white; leaves lobed to about the middle with ± triangular
lobes; bracts of epicalyx 5–12:
Mericarps muticous; calyx-lobes usually with 3 conspicuous longitudinal nerves
on a whitish background; epicalyx of 5 bracts - - - 6. *columella*
Mericarps with 3 apical finely and retrorsely aculeate protuberances or long awns;
calyx-lobes concolorous or nearly so; epicalyx of 6–12 bracts 7. *urens*

1. **Pavonia clathrata** Mast. in Oliv., F.T.A. **1**: 193 (1868).—Eyles in Trans. Roy. Soc.
S. Afr. **5**: 414 (1916).—Ulbr. in Engl., Bot. Jahrb. **57**: 178 (1921). Type: Bech-
uanaland Prot., Ngamiland, Norton Shaw Valley, *Baines* (K, holotype).
Pavonia schumanniana Gürke apud Schinz in Verh. Bot. Verein. Brand. **30**: 174
(1888).—Ulbr., op. cit. **51**: 63 (1913); op. cit. **57**: 177 (1921). Type from SW.
Africa.
Pavonia kraussiana sensu R.E.Fr., Wiss. Ergebn. Schwed. Rhod.-Kongo-Exped. **1**:
144 (1914).—Suesseng. in Proc. & Trans. Rhod. Sci. Ass. **43**: 104 (1951).

Erect probably perennial herb or soft-wooded shrub, 0·75–1·25 m. tall, usually
branched mainly at the base, with short often glandular hairs and longer stiff
simple or bifurcate hairs. Leaf-lamina up to c. 8×6 cm., palmatipartite; segments usually 5, 3–7 cm. long, lanceolate, acute, sharply serrate or dentate-lobulate,
sparsely stellate-pubescent on both surfaces and glandular-ciliate on the margin;
petiole 2–7 cm. long, terete; stipules 7–10 mm. long, filiform, pubescent. Flowers
pale yellow, solitary in the upper axils, often forming a leafy terminal pseudo-
raceme; pedicels 2–5 cm. long, terete. Epicalyx of 10–15 bracts; bracts 25–35
mm. long, linear-filiform, acute, nearly free, shortly glandular-pubescent and
with sparse bulbous-based long stiff spreading hairs. Calyx c. 10 mm. long, pubescent and glandular, divided nearly to the base; lobes ovate-lanceolate, acuminate,
3-nerved. Petals 2–3 cm. long, with fine soft stellate hairs on the outside. Staminal
tube glabrous. Mericarps 9–12 mm. long, broadly elliptic-suborbicular, pale
brown, emarginate at the apex, with an inflated portion containing the seed, 2-
winged, 1·5–2·5 mm. broad, with an unveined marginal zone and many fine cross-
veins on the inner zone extending from transverse ridges on the back of the central
portion. Seeds c. 5 mm. long, ovoid, brown-pubescent.

Caprivi Strip. Singalamwe area, fl. 3.i.1959, *Killick & Leistner* 3263 (PRE; SRGH).
Bechuanaland Prot. N: Chobe to Norton Shaw Valley, fl. i–iii.1863, *Baines* (K).
SE: Mochudi, fl. & fr. v.1898, *Rand* 291 (BM). **S. Rhodesia.** W: Wankie Game
Reserve, fl. & fr. 22.ii.1956, *Wild* 4786 (BM; PRE; SRGH). C: Marandellas, *Dehn*
711 (M).
Also in SW. Africa and the Transvaal.

2. **Pavonia leptocalyx** (Sond.) Ulbr. in Engl., Bot. Jahrb. **57**: 151 (1920). Type from
Natal.
Hibiscus leptocalyx Sond. in Linnaea, **23**: 7 (1850).—Harv. in Harv. & Sond.,
F.C. **1**: 175 (1860). Type as above.
Pavonia odorata sensu Garcke in Peters, Reise Mossamb. Bot. **1**: 123 (1861).

Pavonia discolor U!br., tom. cit.: 148 (1920). Syntypes from Kenya, Tanganyika and Mozambique, Cabaceira Grande, *Prelado Barroso* 25 (COI), Querimba, *Peters* 57 (K), 66 (B†), 68 (B†).

Pavonia fruticulosa Ulbr., tom. cit.: 152 (1920). Syntypes from Tanganyika, (Tanga) and Mozambique, Lourenço Marques, *Schlechter* 11708 (B, holotype† ; BOL ; GRA ; K ; L ; PRE).

Erect or low-spreading suffrutex, 20–90 cm. tall, often branched from the base ; stems slender, terete, yellowish- or brownish-hirtellous-stellate-pubescent to stellate-tomentose when young, sometimes with long stellate or simple patent hairs and sometimes also ± glandular-viscid, glabrescent, ultimately woody and covered with a greyish to pale-reddish-brown finely longitudinally marked bark. Leaf-lamina 1–6 × 0·5–4 cm., usually ± ovate, sometimes hastate-triangular, somewhat angular or subtrilobed, apex usually acute, margin irregularly and coarsely serrate or crenate to lobulate or occasionally almost entire, base rounded to subcordate, upper surface usually dark green drying brown and usually sparsely and somewhat harshly stellate-pubescent to almost glabrous, lower surface usually more densely stellate-pubescent to tomentose with a whitish or greyish indumentum and hence usually discolorous ; petiole up to c. 6 cm. long, hairy like the stems ; stipules 2–4 mm. long, setaceous, caducous. Flowers axillary, solitary, mostly towards the tips of branches and twigs ; pedicels up to 4 cm. long, articulated in the upper 7 mm. Epicalyx of 9–11 bracts ; bracts 9–12 mm. long, nearly free, filamentous-setaceous or narrowly linear, hispid-setose. Calyx 5–7 mm. long, subcampanulate, membranous, scarious-papyraceous in fruit, stellate-setulose to almost glabrous, divided beyond the middle ; lobes 1·5–2·5 mm. broad near the base, acute or acuminate. Petals c. 16–18 mm. long, stellate-pubescent on the outside. Staminal tube glabrous or with small tufts of hairs at the line of fusion with the petals. Mericarps 3–4 mm. long, triquetrous, ± keeled dorsally and with 3 weak transverse ridges, not winged, glabrous or finely stellate-pubescent mainly near the apex. Seeds 2–3 mm. long, ovoid-triquetrous, brown, sparsely to densely pubescent with very small yellowish or brownish hairs.

S. Rhodesia. W : Wankie, fl. iv.1932, *Levy* 29 (PRE). S : Ndanga, fl. & fr. 29.i.1957, *Phipps* 202 (PRE ; SRGH). **Mozambique.** N : Cabo Delgado, between Porto Amélia and Missão de S. Paulo, fl. 27.x.1942, *Mendonça* 1085 (BM ; LISC). SS : Mabote, *Barbosa & Balsinhas in Barbosa* 5065 (LM ; SRGH). LM : Maputo, fl. & fr. 15.iv.1947, *Hornby* 2654 (K ; L ; LMJ ; PRE ; SRGH).

Also in the Transvaal and Natal and in the coastal areas of E. Africa. In semi-shade in forests, scrub, riverine bush and gallery forests, on sandy and loamy alluvial soils, usually in the coastal regions in areas with a fairly high to high rainfall but occasionally further inland ; also on gravelly soils and in woodland (Rhodesia and the Transvaal) up to c. 1000 m. ; sometimes gregarious.

This species is rather variable in habit, leaf-shape and degree of pubescence, so that it has been described several times. The larger-leaved plants from E. Africa, referred by Ulbrich (tom. cit.: 150 (1920)) to *P. mollissima* (Garcke) Ulbr., do not seem to be more than luxuriant forms of *P. leptocalyx*.

3. **Pavonia rogersii** N.E.Br. in Kew Bull. **1932**: 95 (1932). Type : S. Rhodesia, Wankie, *Rogers* 13242 (BM ; BOL ; K, holotype).

Erect suffrutex or soft-stemmed shrub, probably perennial, c. 0·75 m. tall, branched mainly at the base with subvirgate branches, with spreading partly glandular hairs ; stems terete, ultimately glabrescent, with a faintly wrinkled bark. Leaves deflexed from the apex of the petiole ; lamina 1–6 × 0·7–3 cm., elongate-ovate-hastate to triangular-hastate, gradually attenuate to slightly acuminate into the subacute minutely mucronate apex, margin coarsely and remotely and rather regularly serrate-dentate, base cordate-hastate, thinly to rather densely pubescent on both surfaces with simple and few-rayed stellate hairs, less densely so above ; petiole up to c. 6 cm. long, slender, terete ; stipules up to c. 6 mm. long, filiform, pubescent. Flowers yellow, solitary, mainly in the upper leaf-axils, pedicels up to 4 cm. long, elongating to 6 cm. in fruit. Epicalyx of 10–13 (often 11) bracts ; bracts 15–21 mm. long, nearly free, green, filiform-setaceous with sparse setaceous bulbous-based hairs and shorter glandular and eglandular hairs. Calyx free to 1–2 mm. from the base, pubescent outside ; lobes 5–6 mm. long, ovate-lanceolate to lanceolate, acute, ciliate, indistinctly 3-nerved. Petals 20–25 mm. long, glabrous or stellate-puberulous. Staminal tube glabrous. Mericarps c. 6 mm. long, triquetrous, the sharp somewhat raised lateral angles with a small subulate tooth in the middle, the back faintly longitudinally carinate and transversely ridge-veined. Seeds very minutely puberulous.

S. Rhodesia. W: Wankie, *Levy* 1179 (PRE; SRGH).

Apparently a local endemic of the Wankie region. Probably on deep sandy soil in mixed woodland in semi-shade; but ecological notes are scanty.

4. **Pavonia patens** (Andr.) Chiov. in Ann. di Bot. **13**: 409 (1915).—Exell & Mendonça, C.F.A. **1**, 1: 156 (1937).—Martineau, Rhod. Wild Fl.: 48, t. 17 fig. 3 (1954). Type a cultivated plant originating from Ethiopia.
  Sida patens Andr., Bot. Rep. **9**: t. 571 (1809). Type as above.
  Althaea burchellii DC., Prodr. **1**: 438 (1824).—Harv. in Harv. & Sond., F.C. **1**: 159 (1860). Type from Cape Prov.
  Pavonia kraussiana Hochst. in Flora, **27**: 293 (1844).—Ulbr. in Engl., Bot. Jahrb. **51**: 57 (1913); op. cit. **57**: 125 (1920). Type from S. Africa.
  Lebretonia glechomifolia A. Rich., Tent. Fl. Abyss. **1**: 54 (1847). Type from Ethiopia.
  Pavonia macrophylla E. Mey. ex Harv. in Harv. & Sond., F.C. **1**: 169 (1860) *nom. illegit.*—Mast. in Oliv., F.T.A. **1**: 190 (1868).—Eyles in Trans. Roy. Soc. S. Afr. **5**: 414 (1916). Type from S. Africa.
  Pavonia glechomifolia (A. Rich.) Garcke in Schweinf., Beitr. Fl. Aethiop. **1**: 54 (1867).—Mast., loc. cit.—Ulbr., op. cit. **57**: 119 (1920). Type as for *Lebretonia glechomifolia*.
  Pavonia procumbens sensu Garcke in Peters, Reise Mossamb. Bot. **1**: 123 (1861).
  Pavonia leptoclada Ulbr., op. cit. **51**: 60, fig. 3 (1913); op. cit. **57**: 123 (1920). Type from SW. Africa.
  Pavonia spp.—Eyles, loc. cit. pro parte quoad specim. *Chubb* 338 et *Rand* 94.
  Pavonia glechomifolia var. *tomentosa* Ulbr. op. cit. **57**: 121 (1920).—Burtt Davy, F.P.F.T. **2**: 277 (1932). Syntypes from Somaliland, Kenya, Transvaal and Mozambique, Sena, *Peters* 47 (B†).
  Pavonia burchellii (DC.) R. A. Dyer in Kew Bull. **1932**: 152 (1932).—Burtt Davy, tom. cit.: 278 (1932).—Arwidss. apud Norlindh & Weim. in Bot. Notis. **1934**: 96 (1934).—Martineau, Rhod. Wild Fl.: 49 (1954). Type as for *Althaea burchellii*.

Erect to spreading, occasionally decumbent to subscandent suffrutex or shrub up to c. 1 m. tall, biennial or perennial, varying from sparsely and sometimes minutely stellate- or glandular-pubescent to stellate-tomentose, sometimes also with longer patent hairs or fairly densely pilose; stems herbaceous or wiry, terete, slender, reddish or yellowish or greenish, ultimately glabrous and covered with a thin dark-grey to brown dark purple or black thin smooth or finely fissured bark. Leaf-lamina 2–8 × 1–7 cm., usually ovate-pentagonal in outline, often shallowly 3–5-lobed, with a usually deep and narrow basal sinus (the basal lobes sometimes overlapping) and subacute to acute lobes, margin as a rule coarsely and somewhat irregularly crenate to crenate-serrate, both surfaces usually stellate-pubescent but the paler lower surface more densely so or occasionally tomentose; petiole up to c. 8 cm. long, usually slender and hairy like the stems; stipules up to c. 5 mm. long, setaceous, pubescent, usually caducous. Flowers white, pale yellow, cream or orange, with or without a reddish centre, solitary in the upper leaf-axils; pedicels up to c. 5 cm. long, rarely longer, slender, terete, pubescent, articulated in the upper 10–15 mm. Epicalyx rotate; bracts 5 (6), 7–11 mm. long, oblong to ovate-lanceolate or subrhomboid, acuminate to acuminate-caudate, usually obtuse, 5-nerved, with or without long simple hairs in addition to a stellate pubescence, usually ciliate, green and herbaceous but accrescent in fruit and ultimately light brown. Calyx campanulate-cupulate, paler and thinner in texture than the epicalyx, variously pubescent, usually divided a little beyond the middle; lobes c. 5 mm. long, broadly ovate to somewhat rhombic, acute or shortly acuminate, 5-nerved. Petals 10–15 mm. long, often sparsely and finely stellate-pubescent outside towards the apex. Staminal tube usually glabrous. Mericarps c. 5 × 4 × 3 mm., yellow or light brown when ripe, obliquely triquetrous-obovoid, muticous, dorsally keeled and with warty protuberances or muricate, laterally with 4–5 ridges, the latter and the protuberances often crowned with sparse stellate hairs (at least when immature). Seeds c. 3·5 × 2 mm., obliquely obovoid-pyriform, somewhat angular, glabrous.

Caprivi Strip. Okovango R., Bagani Pontoon, fr. 19.i.1956, *de Winter* 4335 (K).
Bechuanaland Prot. N: Francistown, fl. & fr. i.1926, *Rand* 28 (BM). SE: Kanye, fl.
14.xi.1948, *Hillary & Robertson* 508 (PRE). **N. Rhodesia.** N: Lumi R., fl. 9.ii.1955,
Richards 4394 (K). C: 11 km. E. of Lusaka, 1250 m., fl. 9.ii.1958, *King* 418 (K). S:
Choma Distr., between Sinazongwe and Zongwe R. mouth, 550 m., fl. 29.xii.1958, *Robson*
1008 (BM; K; LISC; SRGH). **S. Rhodesia.** N: Mazoe, fl. xii.1905, *Eyles* 205 (BM;
SRGH). W: Matobo Distr., Lucydale, fl. & fr. v.1945, *West* 2146 (BM; PRE; SRGH).
C: Marandellas, fl. & fr. 6.ii.1954, *Corby* 7911 (PRE; SRGH). E: Melsetter Distr.,
Hotsprings, fl. & fr. 24.ii.1952, *Chase* 4382 (BM; PRE; SRGH). S: Ndanga, Triangle
Sugar Estate, fl. 25.i.1949, *Wild* 2745 (BM; SRGH). **Nyasaland.** S: near Liwonde
Ferry, fl. 13.iii.1955, *E.M. & W.* 849 (BM; LISC; SRGH). **Mozambique.** T:
Lupata, fr. iv.1860, *Kirk* (K). MS: Chimba-Molima, *Pedro & Pedrógão* 8535 (LMJ;
SRGH). SS: Gaza, between Guijà and Mabalane, fl. 3.vi.1959, *Barbosa & Lemos* in
Barbosa 8581 (K; LMJ). LM: Sabiè, Bundoio, fl. 8.ii.1945, *Esteves de Sousa* 8 (PRE;
SRGH).
From Ethiopia, Eritrea and Somaliland southwards in the drier regions of eastern
tropical Africa to the Transvaal and Natal, and in Angola, SW. Africa and Griqualand-W.
Mainly in light shade on a variety of soils, often on sandy and loamy types, also rocks,
sometimes on limestone, from about sea-level to c. 1500 m., generally avoiding moist
climatic conditions and areas with severe ground-frosts. Under trees, in dry river-beds,
on river banks, among rocks and boulders, rarely in the open.

The extreme variability of this species accounts for the fairly numerous synonyms and
varietal names. All the forms seem to intergrade and a sharp distinction is not possible.

5. **Pavonia hirsuta** Guill. & Perr. in Guill., Perr. & Rich., Fl. Senegamb. Tent. **1**: 51
(1831).—Mast. in Oliv., F.T.A. **1**: 191 (1868).—Ulbr. in Engl., Bot. Jahrb. **51**:
56 (1913); op. cit. **57**: 116 (1920).—R.E.Fr., Wiss. Ergebn. Schwed. Rhod.-
Kongo-Exped. **1**: 144 (1914).—Exell & Mendonça, C.F.A. **1,** 1: 156 (1937).—Keay,
F.W.T.A. ed. 2, **1,** 2: 341 (1958). Type from Senegambia.
   Hibiscus baumii Gürke in Warb., Kunene-Samb.-Exped. Baum: 299 (1903).
Type from Angola (Bié).
   Pavonia zawadae Ulbr., op. cit. **48**: 371 (1912); op. cit. **51**: 57 (1913); op. cit.
57: 118 (1920). Type from SW. Africa.
   Pavonia sp.—Eyles in Trans. Roy. Soc. S. Afr. **5**: 414 (1916) pro parte quoad
specim. *Rogers* 5168 et 5710.

A spreading soft-stemmed shrub, probably biennial, with long somewhat decumbent nearly prostrate or ascending branches, forming dense semi-globose
bushes up to 1·25 m. tall and 2 m. in diam.; stems terete or somewhat angular,
usually somewhat harshly stellate-tomentose, ultimately with a thin greyish or
brownish bark. Leaf-lamina 3–8 × 3–10 cm., suborbicular- or reniform-cordate,
sometimes angular to shallowly 5–7–9-palmatilobed, apex rounded rarely acute,
margin irregularly coarsely crenate to lobulate or serrate or subentire, base cordate
to truncate, both surfaces stellate-hairy but the upper one glabrescent, the lower
one paler less harshly pubescent and often subtomentose; petiole up to 11 cm.
long, stellate-pubescent to tomentose; stipules c. 10 mm. long, filiform. Flowers
up to 8 cm. in diam., pale- or sulphur-yellow with a port-wine to maroon centre,
solitary, axillary, mainly in the upper leaf-axils, often forming pseudo-racemes;
pedicels usually short, under 4 cm. long, stellate-pubescent to ± densely tomentose.
Epicalyx of 12–16 bracts; bracts narrowly linear, tomentose-pubescent, shorter
than the calyx. Calyx 10–16 mm. long, campanulate, stellate-tomentose, lobed to
beyond the middle; lobes ovate-acute or ovate-triangular, 3-nerved with raised
veins which are occasionally darker and appear as grey-green stripes. Petals
25–45 mm. long, distinctly veined, fimbriate-ciliate at the narrow base. Staminal
tube usually glabrous. Mericarps 8–10 mm. long, dorsally foveolate-rugose and
shortly retrorsely aculeate-hispid, with 3 short protuberances or short blunt spines
at the apex, the lateral edges narrowly winged and the lateral faces flat, smooth and
glabrous. Seeds c. 6 mm. long, glabrous except for the area of the hilum.

Caprivi Strip. Katima Mulilo, fl. & fr. 24.xii.1958, *Killick & Leistner* 3052 (BM; K;
PRE; SRGH). **Bechuanaland Prot.** N: Chobe-Zambezi confluence, fl. & fr.
11.iv.1955, *E.M. & W.* 1476 (BM; LISC; SRGH). SE? "Guive Pits", *van der
Merwe* (BOL). **N. Rhodesia.** B: Sesheke, fl. & fr. i.1922, *Borle* 343 (PRE; SRGH).
C: Lusaka, fl. & fr. 21.iv.1957, *Noak* 217 (PRE; SRGH). S: Namwala, fl. & fr.
1934–35, *Read* 5 (BM; K; PRE). **S. Rhodesia.** N: Sebungwe Distr., fl. ix.1955,
Davies 1511 (BM; SRGH). W: Wankie, fl. & fr. 17.ii.1956, *Wild* 4756 (BM; PRE;

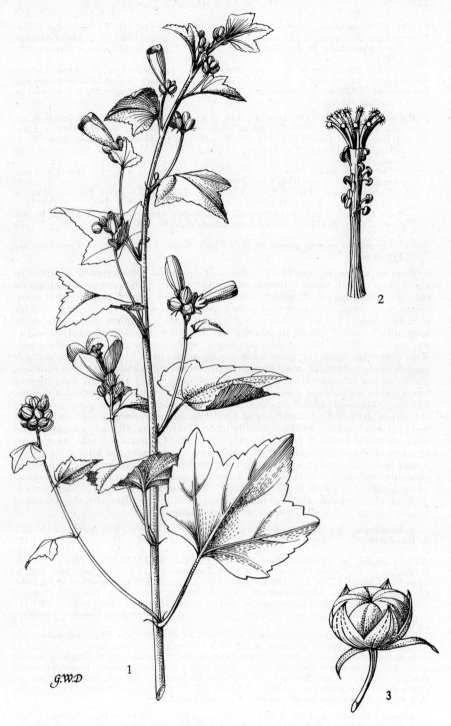

Tab. 96. PAVONIA COLUMELLA. 1, stem with flowers (× 1); 2, staminal tube with projecting style-branches (× 3); 3, fruit with persistent calyx (× 3). All from *Chase* 3608.

SRGH) C: Chilimanzi, Umvuma, Mtao, *Eyles* 4753 (K; SRGH). S: Beitbridge Distr.,
Umzingwane R. bridge, fr. 25.iii.1959, *Drummond* 6036 (K; SRGH). **Mozambique.**
SS: Gaza, Massingire, R. dos Elefantes, fl. 19.xi.1957, *Barbosa & Lemos* in *Barbosa* 8182
(K; LMJ).
Also in the drier areas of W. Africa from Senegal southwards to SW. Africa and the
Transvaal and in the Sudan. Mainly on sandy or loamy soils in woodland in association
with species of *Grewia*, *Terminalia* etc., or along rivers and in seasonally dry river-beds,
usually in light shade.

6. **Pavonia columella** Cav., Diss. **3**: 138, t. 48 fig. 3 (1787).—Bak. f. in Journ. Linn.
Soc., Bot. **40**: 27 (1911).—Eyles in Trans. Roy. Soc. S. Afr. **5**: 414 (1916).—Ulbr.
in Engl., Bot. Jahrb. **57**: 135 (1920).—Brenan in Mem. N.Y. Bot. Gard. **8**, 3:
224 (1953).—Hochr., Fl. Madag., Malvac.: 155 (1955). TAB. **96**. Type from
Réunion.
   Pavonia mollis E. Mey. ex Harv. in Harv. & Sond., F.C. **1**: 169 (1860) non
Kunth (1822). Type from Cape Prov.
   Pavonia meyeri Mast. in Oliv., F.T.A. **1**: 191 (1868).—Eyles, loc. cit.—Phillips
in Fl. Pl. S. Afr. **9**: t. 334 (1929).—R. A. Dyer in Kew Bull. **1932**: 153 (1932).—
Burtt Davy, F.P.F.T. **2**: 278 (1932).—Arwidss. apud Norlindh & Weim. in Bot.
Notis. **1934**: 96 (1934).—Martineau, Rhod. Wild Fl.: 49, t. 17 fig. 1 (1954). Type
as for *P. mollis*.
   Pavonia galpiniana Schinz in Bull. Herb. Boiss. **4**: 43 (1896). Type from the
Transvaal.

Erect or somewhat spreading biennial or perennial soft-stemmed suffrutex or
shrub, 1–2 m. tall, usually densely and rather shortly pubescent to subvelutinous
(but both very hairy and sparsely pubescent extremes occur); stems rather stout,
terete, with greyish or brownish pubescence, glabrescent. Leaf-lamina variable
in dimensions, 2·5–12 × 2–15 cm., but usually of fairly uniform size on one specimen, usually broadly ovate to suborbicular in outline, usually 3–5-lobed almost to
the middle, central lobe always the largest with usually gradually acute somewhat acuminate apex, the other lobes usually acute, base ± cordate, both surfaces
usually finely pubescent or subhispid mainly on the veins and on the lower surface,
the latter not infrequently subtomentose; petiole usually 1·8–6 cm. long (that of
lower leaves occasionally much longer), pubescent to velutinous; stipules minute,
filiform, pubescent, usually caducous. Flowers mauve, pink or rose, solitary in
the upper axils and on short lateral branches, often forming a terminal narrow
leafy pseudo-panicle; pedicels usually under 9 mm. long, slender, tomentose and
with additional long soft hairs. Epicalyx of 5 bracts; bracts 5–8 mm. long, linear
or narrowly lanceolate, pubescent, nearly free. Calyx about as long as or a little
longer than the epicalyx, cupuliform-campanulate, pubescent and ciliate, lobed to
about the middle or a little beyond; lobes triangular, acute to apiculate, conspicuously 3-nerved. Petals 2–2·5 cm. long. Staminal tube glabrous. Mericarps c.
3 mm. long, muticous, glabrous or nearly so, dorsally reticulate with slightly raised
ridges. Seeds 3–3·5 × 2–2·5 mm., reniform, glabrous.

   N. Rhodesia. C: Chilanga Distr., Quien Sabe, fl. & fr. 9.xi.1929, *Sandwith* 157 (K).
S. Rhodesia. N: Mazoe, fl. & fr. iv.1906, *Eyles* 355 (BM; BOL; PRE; SRGH).
C: Makoni, fl. & fr. vi.1917, *Eyles* 799 (K). E: Inyanga, fl. & fr. i.1958, *Miller* 4944a
(PRE; SRGH). S: Belingwe East, fl. & fr. 7.vii.1953, *Wild* 4129 (BM; PRE; SRGH).
Nyasaland. N: Nyika Plateau, fl. 1.ix.1935, *Lawrence* 179 (K). S: Cholo Mt., fl. & fr.
22.ix.1946, *Brass* 17740 (K; PRE; SRGH); Limbe, 1220 m., fl. & fr. 1.viii.1948,
Goodwin 143 (BM). **Mozambique.** N: Maniamba, Serra Geci, *Pedro & Pedrógão*
4075 (EA; LMJ). Z: highest part of Serra Guruè, *Mendonça* 2103 (LISC; SRGH).
T: Angónia, Monte Dómuè, *Torre* 6056 (LISC). MS: Gorongosa, Gogogo, *Mendonça*
2404 (LISC).
   Also from the Transvaal, Swaziland and Natal to Eastern Cape Prov., extending into
Tanganyika and Uganda and also in Réunion; introduced in Madagascar. Typical of
moister and generally frost-free habitats, hence in the Transvaal and S. Rhodesia usually
in the mountains and in sheltered river-valleys with a dense vegetation; in moister cooler
climates (Natal, E. Cape) also at lower altitudes, but almost invariably in forest edges and
clearings, riverine bush etc.; usually in soil or soil pockets rich in humus. The distribution clearly reflects this plant's ecological requirements.

7. **Pavonia urens** Cav., Diss. **3**: 137, t. 49 fig. 1 (1787).—Harv. in Harv. & Sond., F.C.
2: 586 (1862).—Ulbr. in Engl., Bot. Jahrb. **57**: 104 (1920).—Brenan, T.T.C.L.:
306 (1949); in Mem. N.Y. Bot. Gard. **8**, 3: 223 (1953).—Exell & Mendonça,
C.F.A. **1**, 2: 374 (1951).—Hochr., Fl. Madag., Malvac.: 156 (1955). Type from
the Mascarene Is.

Pavonia schimperana Hochst. ex A. Rich., Tent. Fl. Abyss. **1**: 52 (1847).—Mast. in Oliv., F.T.A. **1**: 192 (1868).—Bak. f. in Journ. Linn. Soc., Bot. **40**: 27 (1911).— R.E.Fr.,Wiss.Ergebn.Schwed.Rhod.-Kongo-Exped.**1**: 145 (1914).—Eyles in Trans. Roy. Soc. S. Afr. **5**: 414 (1916).—Ulbr., tom. cit.: 107 (1920).—Burtt Davy & Hoyle, N.C.L.: 49 (1936).—Exell & Mendonça, tom. cit., **1**: 156 (1937).— Brenan, T.T.C.L.: 306 (1949).—Keay, F.W.T.A. ed. 2, **1**, 2: 341 (1958). Type from Ethiopia.

Poavnia stolzii Ulbr., tom. cit.: 115 (1920). Type from southern Tanganyika.

Erect suffrutex or shrub, 1–3 m. tall, usually densely pubescent on stems and leaves, sometimes also with longer soft hairs and/or with coarse harsh and slightly pungent hairs (the latter causing a slight irritation of the skin) but indumentum varying from pubescent to densely to sparsely tomentose-velutinous. Leaf-lamina 3–20 cm. in diam., suborbicular-cordate in outline, the lower surface usually more densely pubescent than the usually ± scabrid upper surface, usually 3–5-lobed; lobes triangular, acute and coarsely serrate-dentate; petiole 0·4–7 cm. long, usually fairly uniform in length in one specimen; stipules c. 5 mm. long (occasionally longer), filiform, usually caducous. Flowers pale pink to mauve or rather deep mauve-red, in subsessile clusters or in pedunculate few-flowered inflorescences, occasionally solitary, in the upper leaf-axils of the main branches and also sometimes on axillary short shoots; peduncle up to 5 cm. long; pedicels usually very short but up to 7 cm. long in some flowers. Epicalyx of 6–8 bracts, nearly as long as the calyx; bracts free, linear, pubescent and ciliate. Calyx 8–10 mm. long, cupuliform, densely pubescent and ciliate, glabrescent in fruit, lobed to about the middle; lobes usually narrowly triangular, acute to acuminate. Petals up to 2 cm. long. Staminal tube glabrous. Mericarps c. 5 mm. long (excluding the 3 apical protuberances or awns which are retrorsely spinose-hispid and (1)5(7) mm. long), dorsally reticulate with slightly raised ridges, somewhat pubescent. Seeds 4 × 2 mm., reniform, faintly longitudinally striate.

N. Rhodesia. N: Mpongwe, Lake Kashiba, fr. 5.vi.1957, *Fanshawe* 3323 (K). **S. Rhodesia.** E: Imbesa Forest Estate, fl. & fr. 15.v.1949, *Chase* 1644 (BM; SRGH). **Nyasaland.** N: Masuku Plateau, *Whyte* (K). C: Kota Kota, Nchisi Mt., fl. 28.vii.1946, *Brass* 16997 (K; PRE; SRGH). S: Zomba Mt., *Lawrence* 21 (K). **Mozambique.** N: Maniamba, Serra Geci, *Pedro & Pedrógão* 4055 (EA; LMJ; SRGH). MS: Manica, Vumba Mts., *Pedro & Pedrógão* 6875 (LMJ; SRGH).

Widespread in tropical Africa and in Natal, Madagascar and Réunion. Mainly in light woodland and scrub but also (as a remnant of the original flora) in places where the original vegetation has been destroyed by man; in frost-free areas at c. 600–2000 m. where the rainfall is not too low.

P. urens is a polymorphic species or a species-complex with a bewildering number of forms varying in pubescence, leaf-shape, inflorescence, flower-colour etc. I agree with Hochreutiner (loc. cit.) that these should all be merged. Brenan (loc. cit.) maintains all the varieties distinguished by Ulbrich (loc. cit.) under *P. urens* and *P. schimperana* as varieties of *P. urens* but I cannot separate them clearly in the fairly rich range of material studied. Plants agreeing with the type of *P. stolzii* are distinct in the much shorter awns of the mericarps but there is no other difference from typical *P. urens* and intermediates occur between long- and short-awned forms, such as *Buchanan* 145 (BM; K) from Nyasaland (Mt. Chiradzulu).

29. BOMBACACEAE

By H. Wild

Trees, sometimes very large; bark often smooth. Leaves alternate, petiolate, simple or digitate; stipules deciduous. Flowers bisexual, actinomorphic, large and showy. Calyx tubular, truncate, shortly toothed or lobed (lobes valvate in bud), or rarely deeply 5-lobed with slightly imbricate lobes, often subtended by an epicalyx. Petals free, often elongated, sometimes absent. Stamens 15–∞, free or filaments united into a tube below or in phalanges of 2–3; anthers 1-thecous, reniform or linear; pollen smooth. Ovary superior, 2–10-locular, with axile placentas; ovules 2 or more per loculus; style simple; stigma capitate or lobed. Fruit a woody capsule or indehiscent. Seeds embedded in a powdery matrix or often in hairs on the wall of the fruit; endosperm scanty or absent; cotyledons flat or contorted or plicate.

Stamens numerous; flowers large, 4·8–15 cm. long:
    Calyx deeply 5-lobed; fruit indehiscent with a floury acid pulp   -   1. **Adansonia**
    Calyx truncate or with short rounded lobes; fruit dehiscent, full of silky hairs 2. **Bombax**
Stamens 15, united in 5 phalanges each bearing 2–3 coiled, 1-thecous anthers resembling
    single stamens; flowers c. 3 cm. long, usually fasciculate; calyx shortly 5-lobed;
    fruit dehiscent, full of silky hairs - - - - - - 3. **Ceiba**

1. ADANSONIA L.

Adansonia L., Sp. Pl. 2: 1190 (1753); Gen. Pl. ed. 5: 497 (1754).

Trees, trunk often of great girth. Leaves digitately 3–9-foliolate, leaflets entire. Flowers large, solitary in the leaf-axils; pedicels with 2 bracteoles. Calyx deeply 5-lobed, villous inside. Petals 5. Stamens ∞, united into a tube below; anthers reniform and often somewhat coiled. Ovary 5–10-locular, loculi multiovulate; stigma 5–10-lobed. Fruit indehiscent, woody, oblong-cylindric to globose. Seeds many, embedded in a pulpy or floury matrix; endosperm in a very thin layer; cotyledons contorted.

Adansonia digitata L., Sp. Pl. 2: 1190 (1753).—Peters, Reise Mossamb. Bot., 1: 129 (1861).—Mast. in Oliv., F.T.A. 1: 212 (1868).—Ficalho, Pl. Ut. Afr. Port.: 100 (1884).—Sim, For. Fl. Port. E. Afr.: 16, t. 11 (1909).—Eyles in Trans. Roy. Soc. S. Afr. 5: 417 (1916).—Burtt Davy, F.P.F.T. 1: 268 (1926).—Steedman, Trees etc. S. Rhod.: 48 (1933).—Verdoorn in S. Afr. Journ. Sci.: 30: 255 (1933).—Gomes e Sousa in Bol. Soc. Estud. Col. Moçamb. 32: 81 (1936); Dendrol. Moçamb. 4: 39, 41, 43 cum tab. (1958).—Exell & Mendonça, C.F.A. 1, 1: 143 (1937).—Garcia in Bol. Soc. Brot., Sér. 2, 20: 35 (1946).—Hutch., Botanist in S. Afr.: 314 cum tab. (1949).—Brenan, T.T.C.L.: 74 (1949).—Pardy in Rhod. Agr. Journ. 50: 5 cum tab. (1953).—O.B.Mill. in Journ. S. Afr. Bot. 18: 56 (1956). Type from Senegal.

Massive deciduous tree not usually more than 20 m. tall but with a trunk up to 10 m. in diam.; bark smooth, reddish-brown or greyish-brown; primary branches stout but gradually tapering; young branches often tomentose, rarely quite glabrous. Leaves collected at the ends of the branches, c. 20 cm. in diam.; petiole up to 12 cm. long, densely pubescent, rarely glabrous; leaflets 5–7, sessile or very shortly petiolulate; leaflet-lamina 5–15 × 3–7 cm., oblong-elliptic to ob-ovate-elliptic, apex acuminate, base cuneate, decurrent, margin entire, covered below when young with stellate hairs but soon glabrescent, rarely quite glabrous; stipules very caducous, c. 2 mm. long, subulate. Flowers large and showy, usually drooping, on axillary tomentose peduncles up to 20 cm. long; bracteoles small and very caducous, borne near the apex of the peduncle; bud ovoid to globose with a conical or apiculate apex. Calyx 5–8 × 4–7 cm., divided to $\frac{1}{2}$–$\frac{3}{4}$ of the way, shortly tomentose outside, velvety inside; lobes 3–7 × 1·8–2·5 cm., oblong, apex obtuse or subacute. Petals white, becoming brown when dry, 6 (10) × 7 (12) cm., from very broadly obovate to oblate, apex rounded, base shortly clawed, very sparsely hairy, but densely so towards the claw inside. Stamens very numerous; staminal column 1·5–3 (4·5) cm. long; free portion of filaments slender, equalling the column in length. Ovary ovoid-globose, silky-tomentose; style exserted c. 1·5 cm. beyond the anthers, hairy near the base; stigma with 5–10 fimbriate-papillose lobes up to 8 mm. long. Fruit up to c. 25 × 12 cm., ovoid to oblong-cylindric, variable and sometimes rather irregular in shape, pointed or obtuse at the apex, covered with a velvety tomentum of pale brownish hairs, ± filled with a mealy pulp which is quite dry in ripe fruits. Seeds many, c. 1·3 × 0·9 cm., reniform; testa smooth, dark brown.

Bechuanaland Prot. N: Sigara Pan, 48 km. W. of Nata R. mouth, fr. 26.iv.1957, *Drummond & Seagrief* 5244 (SRGH). **N. Rhodesia.** W: Ndola, cult. seedling, 3.iii.1954, *Fanshawe* 925 (K). S: Gwembe valley, 4 km. N. of Sinazezi, fr., *White* 2627 (FHO). **S. Rhodesia.** N: well known in the Zambezi valley but no herbarium material recorded. W: Victoria Falls, fl. 11.xi.1942, *Pardy* in GHS 9341 (SRGH). E: Sabi valley, Hot

Springs, fl. 22.x.1948, *Chase* 3728 (BM; SRGH). S: Triangle Sugar Estate, fl. 6.x.1946, *Bates* in GHS 15721 (K; SRGH). **Nyasaland.** Herbarium material not seen but well known at lower altitudes and particularly in the Shire valley. **Mozambique.** N: Lake Nyasa, Metangula, fl. & fr. 10.x.1942, *Mendonça* 740 (LISC; SRGH). MS: Chupanga, fl. xii.1862, *Kirk* (K). SS: between Mopai and Mabelane, fl. 4.xi.1944, *Mendonça* 2751 (LISC). LM: Magude, Mapulanguene, fl. 30.xi.1944, *Mendonça* 3179 (LISC).

Throughout the hotter and drier parts of tropical and subtropical Africa and in Madagascar.

The well-known Baobab.

Often found in *Colophospermum mopane* woodland and in *Acacia* woodland or thicket, rarely found above 1000 m. Natives use the bark for rope making. The white pulp of the fruits contains tartaric acid and makes a refreshing drink with water. The leaves are used as a native vegetable. The seedlings have simple leaves and a swollen turnip-like tap root and so are not often recognised in the field.

2. BOMBAX L.

Bombax L., Sp. Pl. **1**: 511 (1753); Gen. Pl. ed. 5: 227 (1754).

Trees. Leaves digitately (3) 5–7-foliolate. Flowers moderately large, solitary or cymose in the leaf axils. Calyx truncate or subtruncate. Petals 5. Stamens ∞, united in a tube below; anthers straight or somewhat curved. Ovary 5-locular, loculi multiovulate; stigma 5-lobed or style-apex ± entire. Fruit a 5-valved woody capsule with many seeds embedded in a more or less abundant wool.

Staminal tube 1–1·5 cm. long; style apex entire; petals up to 11 cm. long, straight in
full flower; capsule wool very copious:
  Style quite glabrous; calyx glabrous or appressed-stellate-puberulous:
    Branchlets and leaves glabrous - - - 1. *rhodognaphalon* var. *rhodognaphalon*
    Branchlets and leaves, at least when young, tomentose
                                                    1. *rhodognaphalon* var. *tomentosum*
  Style hairy in the upper ⅓; calyx densely stellate-pubescent to tomentellous; petals
    up to c. 5 cm. long - - - - - - - 2. *mossambicense*
Staminal tube 3·5–5 cm. long; style apex 5-lobed; petals 14–17 cm. long, recurved in
  full flower; capsule wool not very copious - - - - - 3. *oleagineum*

1. **Bombax rhodognaphalon** K. Schum. [ex Engl. in Abh. Königl. Akad. Wiss.
   Berl.: **1894**: 33 (1894) *nom. nud.*] ex Engl., Pflanzenw. Ost-Afr. **C**: 269 (1895).—
   Engl., Pflanzenw. Afr. **1, 1**: 393, fig. 332 (1910).—Ulbr. in Notizbl. Bot. Gart.
   Berl. **6**: 27, fig. 4 (1913); in Engl., Bot. Jahrb. **49**: 539, fig. 3 (1913).—Bakh. in
   Bull. Jard. Bot. Buitenz., Sér. 3, **6**: 184 (1924) pro parte excl. syn.—Gomes e Sousa,
   Dendrol. Mozamb.: 113 cum tab. (1951).—A. Robyns in Bull. Jard. Bot. Brux.
   27: 665 (1957). Type from Tanganyika.

Tree to 35 m. tall, with smooth greenish-yellow bark. Leaves collected towards the ends of the branches, (3) 5 (7)-foliolate; petiole up to 7 cm. long; leaflet-lamina oblong-elliptic, apex bluntly acuminate or occasionally mucronate, base cuneate, with c. 12 pairs of slightly raised nerves looping within the margin; petiolule up to 1 cm. long. Flowers single or in fascicles of up to 5 in the leaf-axils; pedicel rather stout, up to 2·5 cm. long. Calyx c. 1 × 0·8 cm., campanulate, ± truncate, glabrous or stellate-puberulous. Petals apricot or pale yellow, up to 9 (11) cm. long, straight in full flower, narrowly lanceolate-linear, apex blunt or sub-acute, tomentellous on both sides. Staminal tube up to 1·5 cm. long; filaments red, 5–6 cm. long, attached just above the anther-base; anthers c. 1·5 mm. long. Ovary ovoid, tomentellous; style c. 9 cm. long, glabrous; stigma ellipsoid, entire. Fruit c. 6 × 3 cm., ellipsoid to obovoid; valves glabrous or minutely tomentellous outside; wool very copious, reddish-brown. Seeds numerous, c. 1 cm. in diam., irregularly globose; testa smooth, brown.

Also in Kenya and Tanganyika. Dry woodland, often with *Acacia* spp. In Tanganyika it is also recorded in evergreen forest.

Var. **rhodognaphalon**. TAB. **97** fig. A.

Branches and leaves (even when young) quite glabrous.

Mozambique. N: Chiure, Lúrio, fl. 20.vii.1948, *Andrada* 1281 (COI; LISC), Z: Maganja da Costa, fl. 5.ix.1949, *Andrada* 1919 (COI). MS: Inhaminga, fl. 12.vii.1941, *Torre* 3084 (LISC).

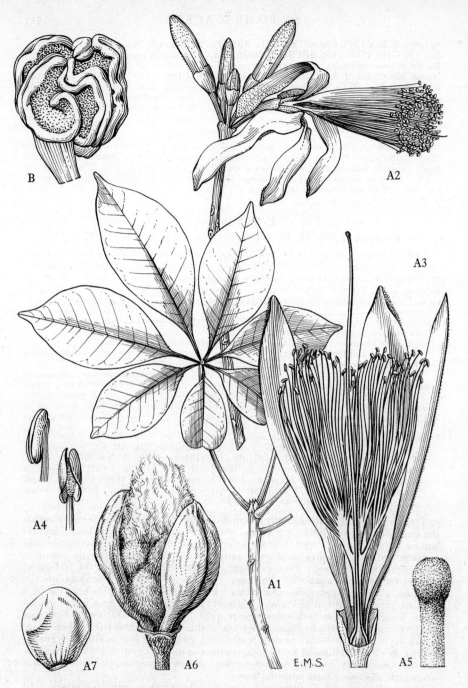

Tab. 97. A.—BOMBAX RHODOGNAPHALON VAR. RHODOGNAPHALON. A1, leaf and portion of stem (×⅔); A2, inflorescence (×⅔); A3, longitudinal sect. of flower (×1); A4, anthers (×6); A5, tip of style (×12); A6, fruit (×⅔); A7, seed (×2). 1 and 2 from *Hubbert* 2032, 3–5 *Battiscombe* 474, 6 and 7 *Bally* 4687. B.—CEIBA PENTANDRA, apex of anther triad (×6).

Also in Kenya and Tanganyika. Ecology as for the species. The whole flower shown in the illustration has its petals apparently somewhat recurved but this flower was probably wilted when pressed.

Var. **tomentosum** A. Robyns in Bull. Jard. Bot. Brux. **27**: 666 (1957). Type: Mozambique, Cabo Delgado, *Andrada* 1368 (COI, holotype; LISC).

Young branches tomentose; young leaves often pubescent.

Mozambique. N: Cabo Delgado, *Andrada* 1368 (COI; LISC). MS: Inhaminga, fl. 25.v.1948, *Mendonça* 4387 (LISC; SRGH).
Not known outside Mozambique. Open woodland.

2. **Bombax mossambicense** A. Robyns in Bull. Jard. Bot. Brux. **27**: 664 (1957).
   Type: Mozambique, Malema, *Gomes e Sousa* 868 (COI, holotype; K).
   Bombax buonopozense sensu Sim, For. Fl. Port. E. Afr.: 16, t. 12 (1909).—Bakh.
   in Bull. Jard. Bot. Buitenz., Sér. 3, **6**: 184 (1924) pro parte quoad specim. Sim.

Tree up to 15 m. tall; branchlets densely pubescent. Leaves 5–7-foliolate; petiole up to 10 cm. long, densely pubescent; leaflet-lamina 5–9·5 × 2–3·5 cm., obovate-elliptic to obovate-oblong, apex bluntly acuminate, sometimes mucronate, base cuneate, sparsely stellate-pubescent above, rather more densely so below, with 12–14 pairs of nerves slightly raised above and below and looping within the margin; petiolules up to 1 cm. long. Flowers usually in 2-flowered cymes with a tendency to form terminal fascicles; peduncle c. 0·3 cm. long, stellate-puberulous; pedicels similar, c. 0·8 cm. long. Calyx 6–8 × 7–8 mm., campanulate, very shortly lobed, stellate-tomentellous outside but later tomentum becoming sparser, silvery tomentose with simple hairs within. Petals whitish, 4·8–5·2 × 0·8 cm., oblong-lanceolate, apex blunt, straight in full flower, tomentellous outside, tomentellous inside except towards the base. Staminal tube c. 1·2 cm. long; filaments c. 3 cm. long; attached just above the anther-base, anthers c. 1·2 mm. long, oblong. Ovary pyriform, glabrous when young but soon becoming tomentose; style c. 5 cm. long, densely hirsute in the upper third; stigma ellipsoid, entire. Mature fruit not known.

Mozambique. N: Malema R., fl. 22.viii.1931, *Gomes e Sousa* 868 (COI; K). Z: Quelimane, fide Sim.
Not yet known outside Mozambique. According to Sim this species is cultivated around Quelimane and the trunks are used for dugout canoes.

This species is very near *B. stolzii* from Tanganyika but differs principally from this latter species by its hairy style. A much wider range of material is needed before the real importance of this apparently rather minor difference can be properly assessed.

3. **Bombax oleagineum** (Decne.) A. Robyns in Bull. Jard. Bot. Brux. **29**: 26 (1959).
   Type a specimen cultivated in the Jardin du Hamma, Algiers, origin unknown (P, holotype).
   Pachira oleaginea Decne. in Fl. des Serres **23**: 49 (1880) ("oleagina").—Bakh.
   in Bull. Jard. Bot. Buitenz., Sér. 3, **6**: 173 (1924). Type as above.
   Bombax kimuenzae De Wild. & Dur. in Bull. Herb. Boiss., Sér. 2, **1**: 740 (1901).—
   Ulbr. in Engl., Bot. Jahrb. **49**: 545 (1913).—Exell & Mendonça, C.F.A. **1**, 1: 145
   (1937).—A. Robyns in Bull. Jard. Bot. Brux. **27**: 666 (1957). Type a cultivated or introduced plant from the Belgian Congo.
   Pachira affinis sensu Bakh., tom. cit.: 171 (1924) pro parte quoad syn. *P. kimuenzae.*
   Bombax affine sensu Ducke in Arch. Jard. Bot. Rio de Jan. **5**: 162 (1930) pro parte quoad descr.
   Bombax sessile sensu Keay, F.W.T.A. ed. 2, **1**, 2: 335 (1958).

Small tree c. 3–5 m. tall; branches sparsely stellate-puberulous when young, soon glabrescent. Leaves 5–7-foliolate; petiole up to c. 15 cm. long, stellate-puberulous or glabrescent; leaflet-lamina up to c. 19 × 8 cm., elliptic-oblong or slightly obovate-elliptic-oblong, apex acute, often mucronate, base cuneate, minutely and sparsely stellate-puberulous especially below or glabrescent, nerves in 14–20 pairs, slightly raised above, strongly raised below, looping within the margin; petiolules 0–5 mm. long. Flowers usually solitary in the axils; peduncle up to c. 2 cm. long, stellate-puberulous. Calyx c. 1·7 × 1·3 cm., campanulate, very shallowly lobed, sparsely stellate-puberulous. Petals pale green or greenish-white, 14–17 × c. 0·7 cm., linear, apex subacute, recurved in full flower, very finely tomentellous on both sides. Staminal tube 3·5–4·7 cm. long; filaments many, c. 11 cm. long, attached just above the anther-base; thecae c. 2·5 mm. long, linear. Ovary globose, tomentellous; style c. 15 cm. long, pubescent at the base; stigma 5-lobed. Capsule c. 9 × 8 cm., ovoid, smooth; wool whitish, not very copious. Seeds many, c. 2·3 × 2 cm.; testa greyish or greyish-brown with c. 8 pale striations.

N. Rhodesia. N: Abercorn, Chilongowelo, cult., fr. vi.1953, *Gamwell* (K) **S. Rhodesia.** C: Salisbury, cult., fl. 16.x.1953, *McGregor* in GHS 44154 (BM; SRGH). Cultivated widely in tropical America, throughout Africa and in Asia. Widely recorded as an escape in parts of tropical Africa but not so far in our area.
The nuts are very oily and are edible.

The nomenclature of this species has caused considerable trouble in the past, due largely to the fact that, although K. Schumann in Mart., Fl. Bras. **12,** 3: 232 (1886) reduced *Carolinea affinis* Mart. to synonymy with *Pachira insignis* (Sw.) Sav., this was not taken up by Bakhuizen (tom. cit.: 171), who retained the former taxon as a distinct species, *Pachira affinis* (Mart.) Decne. Re-examination of the type of *Carolinea affinis* shows that Schumann was right and also that it has nothing to do with our plant. Moreover, it is evident from Bakhuizen's key to the genus *Pachira* (tom. cit.: 170) in which he describes *P. affinis* as " calyx . . . pilis stellatis sparse obductus " that he is not referring to *P. affinis* in the true sense but to our species, as the type of *P. affinis* has a tomentellous calyx. The type of *P. oleaginea* has been examined and it is evident that here we have the earliest known name for our material although it is unfortunate that, like the type of *Bombax kimuenzae*, it is a cultivated or introduced specimen whose country of origin is unknown. In the absence of reliable evidence it is impossible to say more than that this species may be of tropical American origin.

3. CEIBA Mill.

Ceiba Mill., Gard. Dict. Abridg. ed. 4, **1** (1754).

Tall trees, often spiny. Leaves digitately 3–9 (or very rarely more)-foliolate. Flowers moderately large, axillary or subterminal, solitary in the axils or often fasciculate. Calyx 5-lobed. Petals 5. Stamens (10) 15, united in 5 phalanges, each bearing 2–3 coiled anther-thecae and united below in a tube. Ovary 5-locular, loculi multiovulate; stigma club-shaped, pentagonal. Capsule a woody or coriaceous 5-valved capsule. Seeds many, embedded in a copious wool.

Ceiba pentandra (L.) Gaertn., Fruct. **2**: 244, t. 133 (1791).—Exell & Mendonça, C.F.A. **1**, 1: 145 (1937).—Brenan, T.T.C.L.: 75 (1949).—Keay, F.W.T.A. ed. 2, **1**, 2: 335 (1958). TAB. **97** fig. B. Type from " Indiis ".
    Bombax pentandrum L., Sp. Pl. **1**: 511 (1753). Type as above.
    Eriodendron anfractuosum DC., Prodr. **1**: 479 (1824) *nom. illegit.*—Mast. in Oliv., F.T.A. **1**: 214 (1868).—Sim, For. Fl. Port. E. Afr.: 17, t. 13 (1909). Type as above.

Tree up to 18 m. tall (often much taller outside our area) with horizontally branched crown and very straight trunk; bark pale grey, smooth, with scattered conical spines 1–1·5 cm. long, at least when young. Leaves 5–9-foliolate (usually 7, very rarely more than 9); petiole up to c. 15 cm. long, glabrous; leaflet-lamina articulated to a ± suborbicular disk at the apex of the petiole, up to 16 × 6·5 cm., narrowly elliptic or oblanceolate-elliptic, acuminate at the apex, cuneate at the base, margin undulate or rarely obscurely toothed near the apex, slightly paler below, glabrous on both sides, with 7–10 pairs of nerves slightly raised above, scarcely raised below, looping within the margin; petiolules up to 0·8 cm. long. Flowers in axillary fascicles of 1–3; peduncle 2–3 cm. long, glabrous. Calyx c. 1·5 × 1·2 cm., campanulate, lobed ⅕ of the way down with rounded lobes, glabrous outside, pubescent within. Petals rose-coloured or white, c. 3·5 × 0·7 cm., oblong-spathulate, rounded at the apex, tomentose outside, glabrous within. Staminal tube c. 5 mm. long; phalanges of 2–3 filaments c. 2 cm. long. Ovary glabrous or nearly so; style c. 3·3 cm. long. Capsule c. 8–20 × 4–11 cm., oblong-ellipsoid, valves smooth, brown, the whole filled with a copious whitish silky wool. Seeds many, c. 6 mm. in diam., dark brown, subglobose.

N. Rhodesia. B: Mongu-Lealui, cult., fr. 18.ii.1952, *White* 2093, (FHO; K) N: 27 km. NW. of Abercorn, fl. 19.vii.1930, *Hutchinson & Gillett* 3944 (K). **Nyasaland.** N: 9 km. N. of Rukuru R., fl. 10.vi.1938, *Pole Evans & Erens* 688 (K; PRE). **Mozam-**

bique. Z: Quelimane, fr. 1908, *Sim* (PRE). MS: Chupanga, fl. viii.1858, *Kirk* (K). SS: Mambone, fl. 6.viii.1907, *Johnson* 267 (K).

Native in tropical America, tropical Africa and India but also widely cultivated throughout the tropics. Most specimens in our area are cultivated but the species has been recorded as spontaneous at times (e.g. *Hutchinson & Gillett* 3944).

This is the Silk Cotton or Kapok Tree which produces the Kapok fibre of commerce.

30. STERCULIACEAE
By H. Wild

Herbs, shrubs or trees, almost invariably with stellate hairs, sometimes with simple hairs intermingled. Leaves usually alternate, simple and pinnately or palmately nerved, entire to toothed or lobed, or digitately compound; stipules usually present. Flowers actinomorphic, bisexual, unisexual or polygamous. Calyx often persistent, more or less deeply divided into 5 or rarely 3 or 4 or 6 valvate lobes, rarely splitting irregularly into 2 valves, or sepals quite free. Petals usually 5, hypogynous, free or adhering to the staminal column, contorted in bud, or small and scale-like or absent. Stamens 5–∞, with filaments united into a tube or free or confluent at the apex of an androphore, sometimes with staminodes alternating with the stamens or with fascicles of stamens or with an inner ring of staminodes; anthers 2-thecous and opening outward by slits or occasionally by apical pores. Ovary superior, sessile or stipitate, 2–5-locular or with 2–5 ± coherent carpels which separate in fruit, rarely 10–12-locular or reduced to a single carpel; styles simple or equalling the number of loculi. Fruit various but often a dehiscent capsule or of woody or membranous follicles or of indehiscent fruiting carpels, occasionally winged. Seeds 1–∞, sometimes winged; endosperm present or absent; cotyledons flat or folded, thin or fleshy.

Guazuma ulmifolia Lam., a native of tropical America, with small yellow flowers, a muricate globose fruit and discolorous obliquely oblong leaves reminiscent of *Grewia*, is cultivated in Mozambique as a decorative plant. It is not closely related to any of our indigenous genera.

Flowers all bisexual; petals present:
  Staminodes present and conspicuous:
    Fruit an ovoid capsule or of winged carpels; seeds not winged; pedicels not articulated:
      Fruit an ovoid or globose capsule; mostly perennial herbs or shrubs or, if large trees, then petals less than 2 cm. long:
        Bracts caducous; petals persistent and becoming papery or scarious in fruit, white, pink or purplish - - - - - - - 1. **Dombeya**
        Bracts persistent, forming a 3-partite epicalyx; petals yellow, becoming detached in fruit but remaining twisted above the capsule - 2. **Melhania**
      Fruit of 5 (or fewer by abortion) winged carpels; tall tree; petals c. 3·5 × 2·5 cm. 6. **Triplochiton**
    Fruit an obconic winged capsule excavated at the apex; seeds winged; pedicels articulated in the upper part - - - - - 7. **Nesogordonia**
  Staminodes absent or very minute and inconspicuous:
    Ovary 5-locular:
      Loculi 2-ovulate; filaments more or less connate into a tube; anthers oblong-elliptic, glabrous - - - - - - - - 3. **Melochia**
      Loculi 3–∞-ovulate; filaments free or slightly connate at the base, more or less obovate or tuberculate-cruciform; anthers tapering to an acuminate apex, pubescent or pilose or ciliolate - - - - - - 5. **Hermannia**
    Ovary 1-locular - - - - - - - - - 4. **Waltheria**
Flowers unisexual (monoecious, dioecious or polygamous); petals absent:
  Calyx campanulate; fruiting carpels woody and fibrous:
    Fruiting carpels not strongly keeled; seeds 1 to many:
      Anthers irregularly disposed in a capitate head at the apex of the androphore 8. **Sterculia**
      Anthers arranged in one whorl (in our species) which may be straight or markedly zigzag:
        Seeds not winged, without endosperm - - - - 9. **Cola**

Seeds with a large wing, with endosperm - - - - 10. **Pterygota**
Fruiting carpels strongly keeled ; seed single ; a mangrove species   12. **Heritiera**
Calyx tubular ; fruiting carpels papery - - - - - 11. **Hildegardia**

1. DOMBEYA Cav.

Dombeya Cav., Diss. Bot. 2 App. : 1 (1786) ; op. cit. 3 : 121, t. 38 fig. 1 (1787)
nom. conserv.

Shrubs or trees with petiolate, usually cordate, palmately nerved, unlobed or palmately lobed leaves; indumentum of stellate hairs often mixed with simple or glandular hairs ; stipules present. Flowers bisexual, white, pinkish or reddish, in axillary or terminal cymes or umbellate cymes. Bracts usually 3, free, caducous, foliaceous and moderately large or minute, usually arising just below the calyx, rarely scattered on the pedicels. Calyx of 5 lobes, reflexed in mature flowers ; tube very short. Petals 5, usually obliquely asymmetrical, persistent and becoming papery or scarious with age. Stamens united below in a tube ; filaments in groups of 2–3 (4), alternating with 5 narrow staminodes ; anthers oblong, opening by slits. Ovary sessile, 3–5-locular ; ovules 2–∞ per loculus ; loculi sometimes hairy inside ; style simple, with 3–5 stigmas. Fruit an ovoid or globose, dehiscent capsule. Seeds 1–several in each loculus ; endosperm present.

The bark of practically all the African shrubby species is used for rope-making by Africans.

Ovary 5-locular; stigmas 5 ; bracteoles never linear ; petals 1–3·5 cm. long :
  Inflorescence cincinnate, i.e. the peduncle bearing at its apex 1–2 axes curled backwards
    in bud and with the buds arranged racemosely along one side of each axis
                                                          1. *cincinnata*
  Inflorescence not cincinnate : (except occasionally in bud):
    Leaves with a short dense indumentum completely concealing the lower surface of
      the lamina and nerves:
      Leaf-lamina ovate to orbicular, not lobed, apex rounded - - 3. *brachystemma*
      Leaf-lamina broadly ovate-acuminate, not lobed or 3–5-lobed :
        Leaf-lamina not lobed, margin very regularly dentate-crenate, nerves and veins
          only slightly raised below - - - - - - - 4. *wittei*
        Leaf-lamina 3–5 lobed (rarely not lobed), margin irregularly serrate or dentate,
          nerves and veins very prominently raised below :
          Stipules from broadly lanceolate to ovate-lanceolate ; peduncles stout,
            1–2 mm. in diam. - - - - - - - 7. *nyasica*
          Stipules narrowly lanceolate ; peduncles slender, up to 1 mm. in diam.
                                      5. *greenwayi*
    Leaves glabrous or with a sparse to ± dense indumentum not concealing the lower
      surface of the lamina and nerves :
      Stems persistently tomentose or hispid with mainly simple hairs :
        Stipules lanceolate-acuminate to ovate-acuminate :
          Leaf-apex rounded or emarginate; lobes shallowly rounded   2. *claessensii*
          Leaf-apex acute or acuminate; lobes acute or acuminate :
            Bracts linear to narrowly lanceolate - - - - 19. *sp. A.*
            Bracts lanceolate to broadly ovate :
               Indumentum of the stems of 1–3-branched stellate hairs   6. *burgessiae*
              Indumentum of the stems of many-branched densely crowded
                stellate hairs :
                Leaves densely tomentose below - - - 9. *johnstonii*
                Leaves densely pilose below (epidermis just visible between the
                  hairs) - - - - - - - - - 10. *lasiostylis*
        Stipules linear to linear-lanceolate - - - - - 11. *calantha*
      Stems very soon glabrescent :
        Ovules 4 or more per loculus ; young stems with a few short glandular hairs
          or softly pilose ; shrubs up to c. 4 m. tall :
          Stems glabrous when young or with a few scattered glandular hairs ; petals
            pink or white, c. 1·8 cm. long ; ovules 4 per loculus   8. *tanganyikensis*
          Stems softly pilose when young ; petals vermilion, up to 3·5 cm. long ;
            ovules 5–6 per loculus in a single row - - - - 12. *lastii*
        Ovules 2 per loculus ; young stems lepidote-stellate with some simple hairs ;
          tree up to 14 m. tall - - - - - - 13. *erythroleuca*
Ovary 3 (4)-locular; stigmas 3 (5) ; bracts filiform to linear-lanceolate ; petals c. 1 cm.
  long :

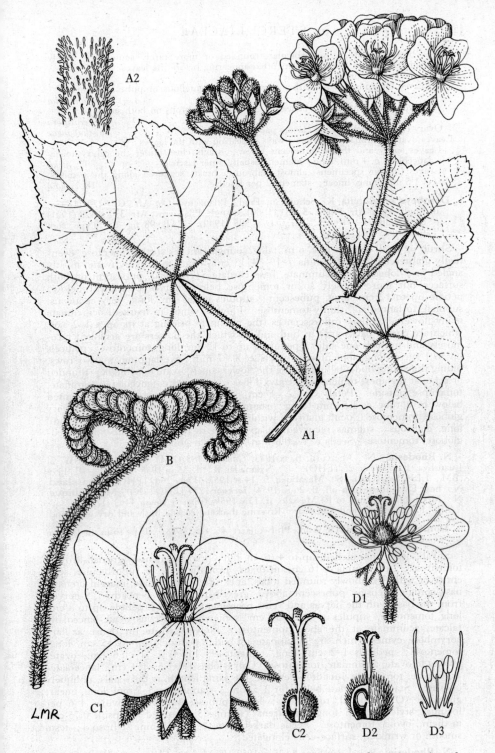

Tab. 98. A.—DOMBEYA TANGANYIKENSIS. A1, flowering branch (×⅔); A2, portion of flowering branch (×12), both from *Richards* 5573. B.—DOMBEYA CINCINNATA, young inflorescence (×2) *Riddelsdell* 95. C.—DOMBEYA SHUPANGAE. C1, flower with part of stamen tube removed (×4); C2, gynoecium with one quarter of ovary removed (×6), both from *Brass* 17577. D.—DOMBEYA ROTUNDIFOLIA. D1, flower with one petal removed (×3); D2, gynoecium, with one quarter of ovary removed (×6); D3, 3 stamens and 2 staminodes (×3); all from *Plowes* 1631.

Leaves broadly ovate to suborbicular, rounded or more rarely acute (very rarely
   acuminate) at the apex; flowers often appearing before the leaves:
  Ovary tomentose or stellate-pilose:
    Leaves tomentellous beneath; calyx-lobes tomentellous or pubescent outside
                                                               14. *rotundifolia*
    Leaves stellate-pubescent beneath; calyx-lobes glabrous on both sides
                                                            16. *cerasiflora*
  Ovary covered with minute clavate papillae   -   -   -   -   **15.** *shupangae*
Leaves ovate-acuminate; flowers always appearing with the leaves:
  Leaves softly and thinly pubescent below; bracts close under the calyx, linear-
   lanceolate, c. 4 mm. long; stamens usually 3 per fascicle - **17.** *kirkii*
  Leaves (in our specimens) almost glabrous; bracts scattered along the pedicels,
   c. 1·5 mm. long, linear; stamens 2 per fascicle - - - **18.** *cymosa*

1. **Dombeya cincinnata** K. Schum. in Engl., Pflanzenw. Ost-Afr. **C**: 270 (1895);
   in Engl., Mon. Afr. Pflanz. **5**: 31 (1900).—Engl., Pflanzenw. Afr. **3**, 2: 428 (1921)
   (" consinuata ").—Brenan, T.T.C.L.: 595 (1949). TAB. **98** fig. B. Type from
   Tanganyika.

Shrub or small tree up to c. 6 m. tall; young branches softly tomentose, eventually glabrescent. Leaf-lamina 7–19 ×4–13 cm., ovate to broadly ovate, mostly
angled or 3-lobed, apex acuminate, base cordate 5–7-nerved, pubescent on both
surfaces but more densely so or tomentose below, nerves prominent below;
petiole up to 12 cm. long, pubescent; stipules c. 1·5 cm. long, very caducous,
subulate to lanceolate, thinly tomentose. Flowers white, in markedly cincinnate
many-flowered axillary inflorescences (the peduncle bearing at its apex 1–2 axes
which are curled backwards in bud and on which the flowers are arranged race-
mosely and towards one side); peduncle up to 12 cm. long, tomentose; pedicels
up to 1·5 cm. long, tomentose; bracts c. 8 ×7 mm., overlapping, in two rows,
hiding the young buds, falling before the flowers open, ovate to orbicular, rounded
or acute at the apex, tomentose. Calyx-lobes c. 1 ×0·3 cm., lanceolate-acuminate,
tomentose outside. Petals c. 1·2 ×1 cm., obliquely obovate. Stamens united
below in a tube c. 1·5 mm. long; staminodes c. 7 mm. long, linear. Ovary
globose, tomentose; loculi glabrous within; ovules 2 per loculus; style c. 7 mm.
long, glabrous; stigmas recurved. Capsule up to 6 mm. in diam., depressed-
globose, tomentose. Seeds c. 1 ×0·6 mm., brown, trigonous.

N. Rhodesia. N: Abercorn, fl. ix.1933, *Gamwell* 183 (BM). W: 40 km. SW. of
Luanshya, fl., *Holmes* 25 (FHO). E: Nyamadzi R., fl. 25.iii.1955, *E.M. & W.* 1173
(BM; LISC; SRGH). S: Mazabuka, fr. 14.iv.1952, *White* 2669 (FHO). **Nyasaland.**
N: near Ekwendeni turn-off, fl. 26.iii.1954, *Jackson* 1279 (FHO; K). **Mozambique.**
N: Mocimboa da Praia to R. Messalo, fl. 13.v.1959, *Gomes e Sousa* 4463 (K).
   Also in Tanganyika and Zanzibar. Riverine thickets, coastal bush and dry woodland.

2. **Dombeya claessensii** De Wild., Pl. Bequaert. **4**: 415 (1928). Type from the Belgian
   Congo.

Shrub, often unbranched, up to 4 (6) m. tall; branches densely greyish-brown-
tomentose. Leaf-lamina 5–16 cm. in diam., broadly ovate to suborbicular, subentire or with 3 shallowly rounded lobes, apex often emarginate, margin crenate,
base cordate, densely pubescent above, greyish-tomentose beneath, 7–9-nerved
from the base with the nerves and veins prominent beneath; petiole up to 8 cm.
long, tomentose; stipules up to 2 ×0·8 cm., somewhat foliaceous, ovate-lanceolate,
falcate, acuminate at the apex, tomentose. Flowers white, in dense axillary
corymbose cymes which are cincinnate in bud; peduncle up to 15 cm. long,
tomentose; pedicels 1–2 cm. long, tomentose; bracts c. 1 ×0·6 cm., ovate-
oblong to ovate, acuminate, tomentose. Calyx-lobes 1·0–1·5 ×0·4 cm., lanceolate-
acuminate, tomentose outside. Petals c. the same length as the sepals, obliquely
obovate. Stamens connate below for 2–5 mm.; staminodes 5–7 mm. long, linear,
glabrous. Ovary ovoid, tomentose; loculi pubescent within; ovules 4–6 per
loculus; style c. 4 mm. long, pubescent; stigmas 5, recurved. Capsule 0·75 cm.
in diam., ovoid, tomentose. Seeds dark brown, c. 3 ×1·5 mm., ellipsoid; testa
smooth or wrinkled, surface-cells elongated.

N. Rhodesia. W: Luanshya, fl. 27.iv.1952, *Holmes* 24 (FHO). C: Walamba, fl.
22.v.1954, *Fanshawe* 1233 (K).
   Also in the Belgian Congo. In the more open patches of *Brachystegia* woodland or in
riverine thickets.

3. **Dombeya brachystemma** Milne-Redh. in Hook., Ic. Pl. **32**: t. 3195 (1933). Type: N. Rhodesia, Solwezi Distr., *Milne-Redhead* 657 (K, holotype).

Weak or subscandent shrub up to 6 m. tall; young branches minutely puberulous, older branches quite glabrous. Leaf-lamina 8–11 cm. in diam., ovate to suborbicular, apex rounded, margin crenate, base deeply cordate, 5–7-nerved, densely puberulous above, glaucous-tomentellous beneath, nerves somewhat raised below; petiole up to 11 cm. long, glandular-puberulous or glabrous; stipules c. 1 cm. long, caducous, subulate. Flowers pink in axillary, umbelliform, 3–5-flowered cymes; peduncle 2·5–3 cm. long, pubescent with some glandular papilliform hairs; pedicels 1·2–1·5 cm. long, stellate-tomentellous with some simple hairs; bracts up to 9 mm. long, caducous, lanceolate-subulate, tomentellous. Calyx-lobes 1·9 × 0·3 cm., lanceolate-acute, tomentellous outside. Petals 2·2 × 1·7 cm., obliquely obovate. Stamens connate below for 2–3·5 mm.; staminodes 1·2–1·5 cm. long, linear, glabrous. Ovary ovoid, tomentose; loculi densely pubescent within; ovules 6 per loculus; style c. 1 cm. long, densely hirsute below, glabrous above; stigmas recurved, sparsely hairy. Ripe capsule not yet known.

N. Rhodesia. W: Solwezi Distr., Mbulungu Stream, 3.vii.1930, *Milne-Redhead* 657 (K).
Endemic in N. Rhodesia. Evergreen fringing forest.

4. **Dombeya wittei** De Wild. & Staner, Contr. Fl. Katanga, Suppl. **4**: 63 (1932). Type from the Belgian Congo.

Shrub up to 5 m. tall; branches pubescent at first but soon glabrous. Leaf-lamina 2–10 × 1·5–7·5 cm., broadly ovate, apex acuminate, margin dentate or crenate, base cordate, 7–9-nerved, tomentellous on both sides, dark green above, pale below, veins prominent below; petiole up to 6 cm. long, slender, with short stellate hairs and longer simple hairs or glabrescent; stipules 8–9 × 0·5–1 mm., linear, tomentellous, ciliate at the margins and apex. Flowers pink, in axillary usually 3–5-flowered subumbellate cymes; peduncle and pedicels with an indumentum similar to that of the young branches, peduncles up to 7 cm. long; bracts 6–8 × 2 mm., lanceolate, acute, tomentellous. Calyx-lobes 1·6–1·9 × 0·4 cm., lanceolate-acuminate, tomentellous outside. Petals ± equal in length to the sepals, obliquely obovate. Stamens connate below for c. 2 mm.; staminodes c. 1 cm. long, linear, glabrous. Ovary ovoid, tomentose or densely pilose; loculi pubescent inside; ovules 4 per loculus; style 0·9–1·2 cm. long, pilose, at least below; stigmas recurved. Capsule c. 1·2 × 0·9 cm., oblong-ovoid. Seeds c. 3 × 3·5 mm., dark brown, angular; testa corrugated.

N. Rhodesia. W: Ndola, fl. 3.ix.1949, *Fanshawe* 1421 (K). C: Kafulafuta R., fl. vii.1909, *Rogers* 8313 (K). S: Mumbwa, fl., *Macaulay* 934 (K).
Also in the Belgian Congo. River-sides and stream banks.

There is a rather fragmentary specimen of this species in the Kew Herbarium (*Rogers* 6037) which, according to its label, was collected in August 1911 at Francistown, Bechuanaland. I feel some doubt about the locality of this specimen and think it may in fact come from N. Rhodesia and be wrongly labelled.

5. **Dombeya greenwayi** Wild in Bol. Soc. Brot., Sér. 2, **33**: 35 (1959). Type: N. Rhodesia, Shiwa Ngandu, *Greenway* 5527 (EA; K, holotype).

Lax shrub c. 2·6 m. tall; young branches, petioles, peduncles, pedicels, stipules, bracteoles and outside of sepals with a mixture of stellate and simple glandular hairs; older branches glabrescent. Leaf-lamina 5·5–8·5 × 4–6·5 cm., very broadly ovate, often shallowly 3-lobed, apex acuminate, margin irregularly dentate, base deeply cordate, greenish-grey-tomentellous above, whitish-tomentellous below, veins reticulate and impressed above, nerves and veins prominent below; petiole up to 6·5 cm. long; stipules up to 1 cm. long, narrowly lanceolate-acuminate. Flowers in 2–3-flowered axillary umbelliform cymes; peduncle up to 7·5 cm. long; pedicels up to 2·5 cm. long; bracts up to 1 cm. long, linear-lanceolate, apex acuminate; buds c. 1·4 × 0·7 cm., narrowly ovoid, acuminate at the apex. Calyx-lobes up to 2 × 0·4 cm., narrowly lanceolate, apex subulate-acuminate, tomentellous outside, glabrous within. Petals c. 2·2 × 2·2 cm., deep pink, obliquely triangular-obovate. Stamens connate below for c. 2·5 mm.; staminodes c. 1·1 cm. long, linear-spathulate, glabrous. Ovary ovoid, tomentose; loculi

pubescent within ; ovules normally 4 per loculus ; style c. 0·9 cm. long, densely
stellate-pubescent in the lower half. Ripe capsule not seen.

N. Rhodesia. N : 48 km. S. of Shiwa Ngandu, fl., *Angus* 874a (FHO)
Not so far recorded outside the Northern Province of N. Rhodesia. Stream-banks or
forest-edges.

6. **Dombeya burgessiae** Gerr. ex Harv. in Harv. & Sond., F.C. **2** : 590 (1862).—Harv.,
Thes. Cap. **2** : 24, t. 137–8 (1863).—Mast. in Oliv., F.T.A. **1** : 228 (1868).—
K. Schum. in Engl., Mon. Afr. Pflanz. **5** : 28 (1900).—Sim, For. Fl. Port. E. Afr. :
20, t. 8 fig. B (1909).—Engl., Pflanzenw. Afr. **3, 2** : 428 (1921).—Burtt Davy,
F.P.F.T. **1** : 260 (1926). Type from Natal.
    Dombeya rosea Bak. f. in Journ. Linn. Soc., Bot. **40** : 29 (1911).—Eyles in Trans.
Roy. Soc. S. Afr. **5** : 418 (1916).—Engl., Pflanzenw. Afr. **3, 2** : 428 (1921).—
Steedman, Trees, etc. S. Rhod. ; 49 (1933). Type : S. Rhodesia, Chirinda, *Swyn-
nerton* 196 (BM, holotype ; K ; SRGH).
    Dombeya spectabilis sensu Mast. in Oliv., F.T.A. **1** : 227 (1868) quoad specim.
Meller.
    Dombeya parvifolia K. Schum. in Engl., Mon. Afr. Pflanz. **5** : 30 (1900) pro parte
quoad specim. Kirk.
    Dombeya sp. nr. *dawei* Sprague et *burgessiae* Gerr. ex Harv. & Sond.—Brenan in
Mem. N.Y. Bot. Gard. **8**, 3 : 227 (1953).

Bush or shrub 2–4 m. tall, branching low down ; stems with a brown bark,
densely villous when young with 1–3-fid stellate hairs or hispid with mainly simple
hairs, longer hairs mixed with shorter glandular hairs ; indumentum of petioles,
peduncles and pedicels similar but denser than that of the stems. Leaf-lamina up
to 18 × 18 cm., very broadly ovate-cordate, entire or crenate, unlobed or more
usually 3–5-lobed with blunt or acuminate lobes, villous to sparsely pubescent on
both surfaces, sometimes rather discolorous, 5–9-nerved at the base ; petiole up
to c. 12 cm. long ; stipules 1–1·5 cm. long, lanceolate-acuminate to ovate-acuminate, pubescent or glabrescent. Flowers in many-flowered (rarely few-flowered)
axillary corymbose cymes or more rarely in subumbellate cymes ; peduncle ±
equalling the petiole ; pedicels up to 3 cm. long ; bracts up to 1·2 cm. long,
lanceolate-acuminate to broadly ovate, densely pubescent or villous. Calyx-lobes
1–2 × 0·3–0·5 cm., narrowly lanceolate-acuminate, densely pubescent to villous
outside. Petals 1·5–2·7 × 1·5–2·4 cm., white with rose-pink guide-lines or rose-
pink, asymmetrically obovate. Stamens connate below for c. 2 mm. ; staminodes
up to c. 1·4 cm. long, linear, glabrous. Ovary ovoid, villous ; loculi pubescent
within ; normally 4 ovules per loculus ; style 0·5–1·2 cm. long, pubescent, at
least in the lower half ; stigmas 5, recurved. Capsule c. 1 cm. in diam., ovoid,
villous. Seeds 2·3 × 1·7 mm., dark brown, ellipsoid, up to 4 per loculus ; testa
minutely tessellated, with the surface-cells elongated.

N. Rhodesia. B : Balovale, fl. 24.v.1954, *Gilges* 370 (K ; SRGH). W : Mwinilunga,
fl. 15.ii.1938, *Milne-Redhead* 4582 (K). **S. Rhodesia.** N : Lomagundi, fl. v.1926, *Jack*
(SRGH). W : Bulawayo, fl. iv.1915, *Rogers* 5555a (K). C : Salisbury, Cleveland Dam,
fl. 2.ix.1946, *Wild* 1207 (K ; SRGH). E : Hondi Gorge, fl. 22.vi.1948, *Chase* 792 (BM ;
K ; SRGH). S : Belingwe, fl. 18.viii.1948, *West* 2771 (SRGH). **Nyasaland.** N : N.
of Lake Nyasa, *Thomson* (K). S : Zomba, fl. 26.vi.1946, *Shortridge* in *Brass* 16026 (K ;
SRGH). **Mozambique.** N : Nampula, fl. 17.vii.1936, *Torre* 615 (COI ; LISC). Z :
Namagoa, fl. vii.1943, *Faulkner* 92 (BM ; K ; PRE). T : Angónia, Vila Mousinho, fl.
13.v.1948, *Mendonça* 4211 (LISC). MS : Chimoio, Bandula, fl. 28.iii.1948, *Garcia* 785
(BM ; LISC). LM : Libombo Mts., Mt. Mponduim, fl. 22.ii.1955, *E.M. & W.* 526
(BM ; LISC ; SRGH).
Also in Uganda, Kenya, Tanganyika, Transvaal, Natal and Swaziland. Forest-margins
or in woodland where there is above average humidity, and some degree of shade.

Dombeya burgessiae is here treated in a very broad sense and includes a number of forms
which have in the past been recognised as distinct species, for instance, *D. mastersii* Hook.
f. originally described from cultivated material but probably originating in E. Africa, and
D. nairobensis Engl., from Kenya. Our material falls mainly into two groups : a form in
the Libombo Mts. with villous stems and acute villous bracts matching the type of *D.
burgessiae* and another with hispid stems and abruptly acuminate bracts matching the
type of *D. rosea* Bak. f. The villosity of the stems of *D. burgessiae* sens. strict. is produced however, merely by a greater density of the hispid hairs of the *D. rosea* form
and a greater tendency for the hairs to be 2–3-branched from the base. In particular
there appears to be little geographical segregation of the various forms, and specimens
with the acute bracts and villous stems of *D. burgessiae* sens. strict. have been found in

material from Uganda and Kenya, although the abruptly acuminate bracteoles of the *D. rosea* form are otherwise predominant to the north of the Union of S. Africa. There is also a relatively rare form with umbellate inflorescences (e.g. *Gilges* 370 from Balovale) which appears again on the Mozambique coast (*Faulkner* 92). At first sight this seems very distinct but there are intermediates. In addition hybrids of *D. burgessiae* and related species are known in cultivation and may well occur in the wild state, thus further complicating the situation. In the present state of our knowledge of this species therefore it is not practicable to consider *D. burgessiae* as more than one very variable species.

7. **Dombeya nyasica** Exell in Journ. of Bot. **77**: 166 (1939). Type: Nyasaland, between Dowa and Kota Kota, *Burtt* 6085 (BM, holotype; K).
    Dombeya sp. nr. *nyasica* Exell.—Brenan in Mem. N.Y. Bot. Gard. **8**, 3: 227 (1953).

Pink-flowered shrub c. 2 m. tall, similar in general appearance to *D. burgessiae* and *D. tanganyikensis*, but differing from both in that the leaves are pale-grey- or whitish-tomentellous below (less densely greenish-grey-tomentellous above). Stems soon quite glabrous, as in *D. tanganyikensis*, but the pedicels much stouter than in that species (1–2 mm. in diam. rather than 0·5–0·75 mm. in diam.) and the inflorescences less strictly umbellate.

Nyasaland. N: Vipya, Mzimba, fl. 12.vi.1947, *Benson* 1283 (BM). C: Nchisi Mt., fl. 25.vii.1946, *Brass* 16929 (K; SRGH). **Mozambique.** N: Massangulo, fl. v.1933, *Gomes e Sousa* 1455 pro parte (COI). T: Fíngoè, fl. 27.vi.1949, *Andrada* 1677 (COI). Endemic in our area. Mountain slopes at about 1500 m., at stream-sides or in grassland. The glabrous mature stems and tomentellous leaves distinguish this species from all forms of *D. burgessiae*.

8. **Dombeya tanganyikensis** Bak. in Kew Bull. **1897**: 244 (1897).—K. Schum. in Engl., Mon. Afr. Pflanz. **5**: 40 (1900). TAB. **98** fig. A. Type: Nyasaland, Fort Hill, *Whyte* (K, holotype).
    Dombeya platypoda K. Schum. in Engl., tom. cit.: 29 (1900).—Brenan, T.T.C.L.: 596 (1949). Type: N. Rhodesia, Abercorn Distr., Fwambo, *Carson* (K, holotype).
    Dombeya gamwelliae Exell in Journ. of Bot. **70**: 104 (1932). Type: N. Rhodesia, Abercorn, *Gamwell* 50 (BM, holotype).

Spreading bush 2·5–4 m. tall, very similar to *D. burgessiae* but the stems and petioles quite glabrous except for a few short, scattered glandular hairs when young. Stipules ovate-acuminate and glabrous. Flowers in dense umbelliform cymes. Peduncle and pedicels sparsely stellate-pubescent with glandular hairs intermingled. Bracts c. 0·8 ×0·3 cm., lanceolate-acuminate. Petals pink or white, c. 1·8 × 1·5 cm.

N. Rhodesia. N: Abercorn Distr., Chilongowelo, *Richards* 5384 (K; SRGH). **Nyasaland.** N: Fort Hill, fl., *Whyte* (K). S: Zomba, fl. 1937, *Townsend* 184 (FHO). Probably also in Tanganyika. Forest margins or in localities protected from fire.

9. **Dombeya johnstonii** Bak. in Kew Bull. **1898**: 301 (1898).—K. Schum. in Engl., Mon. Afr. Pflanz. **5**: 33 (1900). Syntypes: Nyasaland, between Mpata and Tanganyika Plateau, *Whyte* (K); Nyika Plateau, *Whyte* (K).

Bush up to 3 m. tall with pale brown, densely tomentose stems. Leaf-lamina up to 18 ×18 cm., very broadly cordate-ovate or orbicular, entire or shallowly 3–5-lobed, lobes acute, margin irregularly crenate, softly and shortly tomentose on both surfaces, somewhat paler or whitish and more densely so below, 5–7-nerved at the base; petiole up to 20 cm. long, densely tomentose; stipules up to 1·5 cm. long, densely pubescent, caudate-ovate. Flowers pink, in dense subumbellate cymes; peduncle up to 10 cm. long, brown-tomentose; pedicels c. 2 cm. long, densely pubescent; bracts c. 1 cm. long, oblong to ovate-oblong, acute, tomentellous. Calyx-lobes and petals as in *D. burgessiae*. Stamens united at the base for 2–3 mm.; staminodes c. 1·5 cm. long, linear, papillose-puberulous. Ovary globose, tomentose; loculi pubescent within; normally 4 ovules per loculus; style c. 1 cm. long, pubescent; stigmas 5, recurved, ± pubescent. Capsule c. 1 ×1·2 cm., tomentose, ovoid. Seeds c. 3·5 ×1·5 mm., ellipsoid; testa shining brown, surface cells elongated.

Nyasaland. N: North Vipya, opposite Rumpi Gorge, fl. v.1953, *Chapman* 169 (BM). C: Dedza, fl. 10.ix.1929, *Burtt Davy* 21581 (FHO). **Mozambique.** N: Vila Cabral,

Metónia, fl. & fr. 12.x.1942, *Mendonça* 780A (K; LISC; SRGH). T: Marávia, Fíngoè, fl. 11.viii.1941, *Torre* 3238 (LISC; SRGH).
Endemic in our area. Forest margins between 1000 and 2500 m.

10. **Dombeya lasiostylis** K. Schum. in Engl., Mon. Afr. Pflanz. **5**: 24 (1900). Type: Nyasaland, Muata Manga, between Zomba and Mt. Chiradzulu, *Kirk* (K, holotype).
　　Dombeya burgessiae sensu Mast. in Oliv., F.T.A. **1**: 228 (1868) quoad specim. Kirk.

Shrub 2–4 m. tall, very similar to *D. johnstonii* but indumentum on the under surface of the leaves not so pale and whitish, being less dense and coarser. Epidermis visible between the hairs on all but the young leaves (i.e. indumentum densely pilose rather than tomentose). Bracteoles lanceolate-acuminate rather than oblong to ovate-oblong and acute.

N. Rhodesia. E: Nyika, fl. 3.v.1952, *White* 2568 (FHO). **Nyasaland.** C: Dedza, fl. 9.vii.1954, *Adlard* 117 (FHO). S: Zomba Distr., Mulunguzi R., fl. 15.vi.1954, *Banda* 5 (BM). **Mozambique.** N: Vila Cabral, Lichinga, fl. x.1953, *Torre* 4 (COI; LISC).
Endemic in our area. Open woodland, or sometimes a constituent of secondary vegetation following the destruction of forest. Appears to occur in more open conditions than *D. johnstonii*.

D. johnstonii and *D. lasiostylis* are very close, and further material and accurate field observations are needed to decide whether they are in fact distinct.

11. **Dombeya calantha** K. Schum. in Engl., Mon. Afr. Pflanz. **5**: 28 (1900).—Sprague in Bot. Mag.: t. 8424 (1912).—Engl., Pflanzenw. Afr. **3**, 2: 428 (1921). Type: Nyasaland, Zomba, *Whyte & McClounie* (K, holotype).

Shrub c. 4 m. tall with brown tomentose stems. Leaf-lamina up to 30 × 30 cm., very broadly ovate-cordate, 3–5-lobed with acuminate lobes, margin coarsely crenate, densely pubescent especially beneath, 7-nerved at the base; petiole up to 15 cm. long, tomentose; stipules up to 2 cm. long, linear to linear-lanceolate, pubescent. Flowers pink, in 10–20-flowered corymbose cymes; peduncle up to 20 cm. long, brown-tomentose; pedicels up to 3 cm. long, softly tomentose; bracts up to 2 × 0·8 cm., caducous, ovate-lanceolate, caudate-acuminate, densely pubescent. Calyx-lobes up to 2 × 0·6 cm., lanceolate, acute, densely pubescent outside. Petals c. 2 × 2 cm., asymmetrically obovate. Stamens united below for c. 4 mm.; staminodes c. 2 cm. long, linear-lanceolate. Ovary subglobose, tomentose; loculi densely hairy inside; 6–8 ovules per loculus; style c. 1 cm. long, densely pubescent below, less so above; stigmas revolute. Capsule c. 1·2 × 1 cm. oblong-ovoid, tomentose, with up to 8 seeds per loculus. Seeds c. 3·3 × 1·4 mm., dark brown, ellipsoid; testa as in *D. burgessiae*.

Nyasaland. S: Blantyre Distr., fl. 1891, *Buchanan* 539 (BM; K).
Endemic in Nyasaland. Nothing is known so far of the ecology of this species and it is represented in herbaria only by the type gathering, the other specimen quoted and a cultivated specimen.

12. **Dombeya lastii** K. Schum. in Engl., Pflanzenw. Ost-Afr. **C**: 270 (1895); in Engl., Mon. Afr. Pflanz. **5**: 25, t. 2 fig. C (1900).—Engl., Pflanzenw. Afr. **3**, 2: 428 (1921). Type: Mozambique, Zambezia, Namuli, *Last* (B, holotype †; K).

Shrub c. 2 m. tall; branches softly pilose with simple or little-branched hairs, soon glabrescent. Leaf-lamina up to 24 cm. in diam., very broadly ovate to orbicular, usually shallowly 3-lobed with acuminate lobes, deeply cordate and 7-nerved at the base, pilose above, more densely so or subtomentose and paler below, nerves and veins prominent below; petiole up to 15 cm. long, pubescent or glabrescent; stipules up to 1·5 × 0·7 cm., submembranous, lanceolate to ovate-lanceolate, acuminate, densely pilose. Flowers vermilion, in axillary subumbellate 3–9-flowered cymes; peduncle up to 9 cm. long, softly pilose; pedicels up to 4 cm. long, densely pilose; bracts c. 1·6 × 0·7 cm., lanceolate-acuminate, pilose. Calyx-lobes up to 2·5 × 0·7 cm., narrowly lanceolate-acuminate, densely pubescent outside. Petals up to 3·5 × 2·5 cm., obliquely rhomboid-obovate. Stamens connate below for 5 mm.; staminodes c. 1 cm. long, linear, minutely papillose. Ovary oblong-ovoid, tomentose; loculi glabrous within; ovules 5–6 per loculus, in a single row; style c. 2 cm. long, glabrous except for a few mm. at the base; stigmas recurved, c. 2 mm. long. Ripe capsule not so far seen.

Mozambique. Z: Gúruè Mts., fl. 26.vi.1943, *Torre* 5618 (K; LISC; SRGH). Endemic in Mozambique. Woodlands on mountain-sides.

This is certainly our most handsome *Dombeya* and well worth cultivation.

13. **Dombeya erythroleuca** K. Schum. in Engl., Bot. Jahrb. **30**: 353 (1901).—Brenan, T.T.C.L.: 596 (1949). Type from Tanganyika.

Tree up to 14 m. tall with smooth dark brown or blackish bark; young branches at first lepidote-stellate and with some long simple hairs, soon glabrescent. Leaf-lamina 4–13 × 2·5–7 cm., ovate or ovate-oblong, apex acuminate or obtuse, margin crenate or subentire, base deeply cordate and 7-nerved, sparsely stellate-puberulous above, more densely so below, nerves prominent below, venation minutely reticulate below; petiole up to 6 cm. long, pubescence similar to that of the young stems; stipules caducous, oblong-ovate, apex acuminate, tomentellous. Flowers white, pink-veined, in axillary corymbose cymes; peduncle up to 6 cm. long; pedicels c. 1 cm. long; indumentum of inflorescence-branches similar to that of the young stems but becoming denser towards the flowers; bracts caducous, not seen. Calyx 1–1·3 cm. long, tomentellous outside; lobes lanceolate-acute. Petals 1·4–1·7 cm. long, obliquely obovate. Stamens connate below for c. 3 mm.; staminodes 6–9 mm. long, linear, minutely papillose. Ovary ovoid, tomentose; loculi only slightly pubescent within; ovules 2 per loculus; style c. 8 mm. long, pubescent; stigmas twisted, 2–3 mm. long. Ripe capsules not yet known.

N. Rhodesia. E: Nyika Plateau, c. 3 km. N. of rest house, fl. 27.xi.1955, *Lees* 97 (K).
Nyasaland. N: Nyika Plateau, fl. vi.1953, *Chapman* 116 (FHO; K).
Also in Tanganyika. Evergreen forests at about 2500 m.

14. **Dombeya rotundifolia** (Hochst.) Planch. in Fl. Serres, **6**: 225 (1850–51).—Harv. in Harv. & Sond., F.C. **1**: 221 (1860).—K. Schum. in Engl., Mon. Afr. Pflanz. **5**· 35 (1900).—R.E.Fr., Wiss. Ergebn. Schwed. Rhod.-Kongo-Exped. **1**: 148 (1914). —Eyles in Trans. Roy. Soc. S. Afr. **5**: 418 (1916).—Engl., Pflanzenw. Afr. **3**, 2: 430 (1921).—Burtt Davy, F.P.F.T. **1**: 259 (1926).—Steedman, Trees etc. S. Rhod.: 49, t. 48 (1933).—O. B. Mill. in Journ. S. Afr. Bot. **18**: 57 (1952).— Brenan in Mem. N.Y. Bot. Gard. **8**, 3: 227 (1953). TAB. **98** fig. D. Type from Natal.
   Xeropetalum rotundifolium Hochst. in Flora, **27**: 295 (1844). Type as above.
   Dombeya densiflora Planch. ex Harv. in Harv. & Sond., F.C. **2**: 589 (1862).— Eyles, loc. cit. Type from the Transvaal.
   Dombeya spectabilis sensu Mast. in Oliv., F.T.A. **1**: 227 (1868) pro parte quoad specim. Meller.
   Dombeya multiflora sensu Mast. loc. cit. pro parte quoad specim. Meller.— Sim, For. Fl. Port. E. Afr.: 20, t. 8 fig. C (1909).
   Dombeya multiflora var. *vestita* K. Schum., tom. cit.: 34 (1900), pro parte quoad specim. Nyas., verisim. etiam specim. Mossamb.
   Dombeya reticulata sensu K. Schum., tom. cit.: 36 (1900) pro parte quoad specim. Nyas.—Bak. f. in Journ. Linn. Soc., Bot. **40**: 29 (1911).—Eyles, loc. cit.

Shrub or small tree up to c. 8 m. tall; bark dark and rough; buds ferruginously tomentose; young branches thinly tomentose but soon glabrous. Leaf-lamina 3–15 cm. in diam., leathery, broadly ovate to suborbicular, apex rounded or more rarely acute, margin irregularly dentate or subentire, base cordate and 5–7-nerved, scaberulous with minute appressed stellate hairs above, pale-tomentellous below, nerves prominent below, veins prominent and reticulate; petiole up to 8 cm. long but usually less, tomentellous; stipules c. 3 mm. long, very caducous, narrowly lanceolate-acuminate, pubescent. Flowers white or more rarely pink, in many-flowered axillary panicles, often appearing before the leaves; peduncle often very short, up to c. 3 cm. long, slender, floccose-tomentellous; pedicels up to c. 1·5 cm. long, with a similar indumentum; bracts 2–4 mm. long, very caducous, linear-lanceolate, tomentose. Calyx-lobes c. 7 × 2·5 mm., lanceolate-acuminate, tomentellous or densely pubescent outside. Petals c. 1 × 0·7 cm., very obliquely tri-angular-obovate. Stamens joined below in a very short tube 0·5–1 mm. long; staminodes 6–9 mm. long, filamentous to linear-lanceolate, glabrous. Ovary depressed-globose, tomentose; loculi glabrous within; ovules 2 or rarely 3 per loculus; style c. 2 mm. long, sparsely pilose, at least near the base; stigmas 3 or rarely 5, c. 1·5 mm. long, spreading or recurved. Capsule c. 6 mm. in diam., globose, tomentose. Seeds brown, c. 3 × 2.5 mm., trigonous; testa slightly wrinkled.

Bechuanaland Prot. N: Francistown, fl. vii.1911, *Rogers* 6037 (BM; PRE). SE: Mochudi, fl. i.1915, *Harbor* (K; PRE; SRGH). **N. Rhodesia.** N: Abercorn, fl. 22.vii.1930, *Hutchinson & Gillett* 4007 (K). W: Solwezi, fl. 16.vii.1930, *Milne-Redhead* 716 (K). C: Chilanga, fl. 10.ix.1909, *Rogers* 8476 (K; SRGH). E: Fort Jameson, fl. 25.viii.1950, *Gilges* 46 (SRGH). S: Livingstone, fl. 31.viii.1947, *Brenan & Greenway* 7788 (FHO; K). **S. Rhodesia.** N: Sinoia, fl. viii.1926, *Rand* 243 (BM). W: Nya-mandhlovu, fl. 13.ix.1953, *Plowes* 1631 (K; SRGH). C: Rusape, fl. 2.x.1947, *Munch* 39 (K; SRGH). E: Umtali, fl. 23.vii.1949, *Chase* 1699 (BM; K; SRGH). S: Zimbabwe, fl. 10.viii.1941, *Hopkins* in GHS 8119 (K; SRGH). **Nyasaland.** N: Vipya, fl. 28.ix.1950, *Jackson* 183 (K). C: Dedza, fl. 24.ix.1931, *Galpin* 15052 (K; PRE). S: Limbe, fl. 1.x.1946, *Brass* 17885 (BM; K; SRGH). **Mozambique.** N: Vila Cabral, Litunde, fl. 9.x.1942, *Mendonça* 703 (LISC; SRGH). Z: between Ile and Gúruè, fl. 26.ix.1941, *Torre* 3518 (LISC). T: Zobuè, fl. 3.x.1942, *Mendonça* 559 (LISC). MS: Beira, fl. viii.1954, *Gomes e Sousa* 4253 (K; PRE; SRGH). LM: Catuane, fl. viii.1930, *Gomes e Sousa* 38 (COI).

Also in Kenya, Uganda, Tanganyika, Belgian Congo, SW. Africa, Swaziland, Transvaal and Natal. Most types of open woodland, often found on termite mounds.

Like other members of Subgen. *Xeropetalum* Planch., this species normally has 3 stigmas. Specimens with 3–5-stigmas do occur, however, but they invariably have ovaries with but 3 loculi, never 5 as in Subgen. *Dombeya*.

15. **Dombeya shupangae** K. Schum. in Engl., Mon. Afr. Pflanz. **5**: 39 (1900) "mu-pangae ".—Engl., Pflanzenw. Afr. **3**, 2: 432 (1921).—Sprague in Journ. of Bot. **59**: 349 (1921).—Brenan in Mem. N.Y. Bot. Gard. **8**, 3: 227 (1953). TAB. **98** fig. C. Type: Mozambique, Chupanga, *Kirk* (K, holotype).
    Dombeya spectabilis sensu Sim, For. Fl. Port. E. Afr.: 20, t. 8 fig. A (1909).

Small tree sometimes reaching 10 m. in height, with white or pink flowers. Closely resembles *D. rotundifolia* in herbarium material but leaves usually rather larger, often c. 20 cm. in diam. and pubescent instead of tomentellous below with a minutely reticulate pattern of veinlets visible between the hairs at a magnification of × 10. Ovary not tomentose but covered with minute brown clavate papillae (the most important difference between the two species).

Nyasaland. N: Mugesse Forest Reserve, fl. ix.1953, *Chapman* 134 (FHO; K). C: Kota Kota, Nchisi Mt., fl. 9.ix.1946, *Brass* 17577 (BM; K; SRGH). **Mozambique.** N: Mutuali, fl. 2.ix.1953, *Gomes e Sousa* 4102 (COI; K; PRE; SRGH). Z: Mocuba, Mulange, fl. 3.viii.1943, *Torre* 5747 (LISC). MS: Gorongosa, fl. viii.1911, *Dawe* 383 (K).

Also in Tanganyika. Appears to occur in higher-rainfall areas than *D. rotundifolia* and although itself deciduous is often found at the margins of evergreen forest as well as in more open woodland.

16. **Dombeya cerasiflora** Exell in Journ. of Bot. **65**, Suppl. Polypet.: 40 (1927).— Exell & Mendonça, C.F.A. **1**, 2: 187 (1951). Type from Angola (Malange).

Small tree up to 4 (10) m. tall; branches stiff and rather thick, pubescent or glabrescent (buds ferruginously tomentose), often furrowed. Leaf-lamina 12 × 10–11 cm., broadly ovate to suborbicular, entire or very shallowly 3-lobed, apex acute, rarely acuminate, margin remotely denticulate, base cordate and c. 7-nerved, thinly stellate-pubescent above, more densely so below; petiole 7–8 cm. long, thinly stellate-pubescent; stipules caducous (not seen). Flowers white, in sub-umbellate panicles collected towards the ends of the stems, appearing before the leaves; peduncle up to 10 cm. long, sparsely puberulous or glabrous; pedicels up to 1·2 cm. long, slender, sparsely puberulous; bracts very caducous, c. 2 mm. long, linear, stellately tomentose. Calyx-lobes reddish tinged, 6 mm. long, lanceolate, apex acute to acuminate, glabrous on both sides. Petals 7–9 × 4–5 mm., very obliquely cuneate. Stamens usually 10, 4–5 mm. long, united below for c. 1 mm., arranged in pairs between the linear 8 mm. long staminodes. Ovary globose, stellately pilose or tomentose; ovules 2–3 per loculus; style c. 2·5 mm. long, glabrous; stigmas 3, recurved. Mature capsule not seen.

N. Rhodesia. N: Lake Mweru, Puta, fl. 17.viii.1958, *Fanshawe* 4701 (K). Also in Angola and the Belgian Congo. Open woodland.

Angola specimens are reported to have dark corky bark and to flower profusely giving the appearance of cherry blossom.

17. **Dombeya kirkii** Mast. in Oliv., F.T.A. **1**: 227 (1868).—K. Schum. in Engl., Mon. Afr. Pflanz. **5**: 39 (1900). Syntypes: Nyasaland, approx. 16° S. 35° E., *Meller* (K); Mozambique, R. Zambezi, Lupata Gorge, *Kirk* (K).

Dombeya gilgiana K. Schum. in Engl., Pflanzenw. Ost-Afr. **C**: 270, t. 30 (1895); in Engl., Mon. Afr. Pflanz. **5**: 38 (1905).—Brenan, T.T.C.L.: 598 (1949). Type from Tanganyika (Usambara).

Dombeya gilgiana var. *scaberula* K. Schum. in Engl., Mon. Afr. Pflanz. **5**: 39 (1900). Type from Tanganyika (Usambara).

Dombeya umbraculifera K. Schum. in Engl., Mon. Afr. Pflanz. **5**: 38 (1900). Type from Kenya.

Dombeya laxiflora K. Schum. in Engl., Mon. Afr. Pflanz. **5**: 37 (1900). Syntypes from Kenya and Nyasaland, Southern Province, *Buchanan* 345 (B†; K).

Much-branched shrub or small tree up to c. 9 m. tall; young branches tomentellous or densely pubescent, dark brown and glabrescent later; bark smooth and light grey. Leaf-lamina up to 13 × 10 cm., ovate, usually unlobed, rarely shallowly 3-lobed, apex acuminate, cordate at the base, margin irregularly crenate-serrate, base cordate and 5 (7)-nerved, thinly scaberulous-pubescent above, softly and rather thinly pubescent below, nerves fairly prominent below; petiole up to 5·5 cm. long, densely pubescent or glabrescent; stipules c. 5 mm. long, very caducous, subulate, tomentose. Flowers white, in many-flowered axillary panicles; peduncles slender, densely pubescent with both long and short stellate hairs; pedicels up to c. 2 cm. long, with similar indumentum; bracts c. 4 mm. long, very caducous, linear-lanceolate, densely pubescent. Calyx-lobes c. 7 × 1·5 mm., narrowly lanceolate-acuminate, tomentellous outside. Petals c. 1 × 0·6 cm., very obliquely cuneate. Stamens united below for c. 1 mm.; staminodes c. 6 mm. long, narrowly linear, glabrous. Ovary depressed-globose, tomentose; loculi glabrous inside; ovules 2 (3) per loculus; style 2 mm. long, glabrous; stigmas 3, recurved. Capsule c. 5 mm. in diam., depressed-globose, somewhat 3-lobed, tomentose. Seeds c. 3 × 2 mm., dark brown, usually 1 per loculus, obovoid; testa almost smooth or very minutely reticulate.

N. Rhodesia. N: Lake Mweru, Chiengi, fl. 18.viii.1958, *Fanshawe* 4727 (K). C: Chisamba, fl. 8.iv.1933, *Michelmore* 666 (K). S: Choma, Mapanza, fl. 27.ii.1958, *Robinson* 2771 (K; SRGH). **S. Rhodesia.** N: Urungwe, Msuku R., fl. v.1951, *Lovemore* 53 (K; SRGH). W: Wankie Distr., Sebungwe R., fl. 10.v.1955, *Plowes* 1806 (K; SRGH). E: Umvumvumvu R., fl. 14.iv.1948, *Chase* 639 (BM; K; PRE; SRGH). S: Gwanda, Bubye R., fl. v.1955, *Davies* 1334 (K; SRGH). **Nyasaland.** S: Chikwawa, fl. 8.vii.1955, *Banda* 134 (BM). **Mozambique.** Z: Massingire, fl. 21.v.1943, *Torre* 5345 (K; LISC; SRGH). T: Muatize, fl. 7.v.1948, *Mendonça* 4121 (LISC). MS: R. Sabi, Maringa, fl. 30.vi.1950, *Chase* 2477 (BM; SRGH). SS: between Mapai and Caniçado, fl. 7.v.1944, *Torre* 6590 (K; LISC; SRGH).

Also in the Transvaal, Tanganyika and Kenya. Riverine thickets at altitudes of about 1000 m. and lower.

A very variable species. Typical *D. kirkii* has narrow, very acute buds, a profusion of patent grey hairs on the inflorescence branches and grey indumentum on the young branches. Material from Kenya, Tanganyika and the Northern Province of N. Rhodesia which would previously have been separated as *D. gilgiana* or *D. umbraculifera* has more ovoid, blunter buds and more ferruginous hairs on the inflorescence branches and young parts. These differences, however, break down too often to give a reliable segregation and the whole is best considered as one rather variable species.

18. **Dombeya cymosa** Harv. in Harv. & Sond., F.C. **2**: 589 (1862).—K. Schum. in Engl., Mon. Afr. Pflanz. **5**: 33 (1900). Type from Cape Prov.

Shrub or small tree up to 8 m. tall; bark whitish; young branches thinly pubescent or quite glabrous. Leaf-lamina up to 10 × 7·5 cm., ovate, apex acuminate, margin crenate, base cordate and 5–7-nerved, minutely and sparsely pubescent on both surfaces or almost entirely glabrous with a few hairs near the base of the midrib; petiole up to 4 cm. long, slender, pilose or glabrous; stipules 2–3 mm. long, very caducous, filiform, pubescent or tomentose. Flowers white, in axillary panicles; peduncle up to 2·7 cm. long, slender, sparsely and densely flocculose-pubescent; pedicels up to 1 cm. long, with a similar indumentum; bracts c. 1·5 mm. long, linear, dispersed along the pedicels, pubescent. Calyx-lobes up to

7·5 × 2 mm., narrowly lanceolate-acuminate, pubescent outside or quite glabrous except at the base. Petals c. 1·7 cm. long, very obliquely cuneate. Stamens united below in a tube c. 1 mm. long; staminodes c. 5 mm. long, linear. Ovary depressed-globose, densely pubescent to shortly tomentose; loculi glabrous within; ovules 2 per loculus; style c. 3 mm. long, glabrous; stigmas 3, 1–1·5 mm. long, spreading or recurved. Capsule c. 3·5 mm. in diam., globose, pubescent, often single-seeded. Seed c. 3 × 2·5 mm., brown, ovoid; testa very minutely roughened.

Mozambique. LM: Goba, fl. 23.viii.1944, *Mendonça* 1823 (LISC; SRGH).
Also in Natal, Transvaal and the Eastern Cape Prov. River or stream banks and coastal bushland.

The Mozambique form has less hairy branchlets, leaves, sepals, and ovaries than the type but intermediates occur in the Transvaal giving a fairly continuous range of variation in the species.

Species insufficiently known

19. **Dombeya** sp. A.
     Dombeya sp. nr. *D. platypoda.*—Brenan in Mem. N.Y. Bot. Gard. **8**, 3 : 227 (1953).

Shrub c. 2 m. tall, closely related to *D. burgessiae* but with the indumentum of young branches and inflorescence of much shorter hairs; leaves simple or shallowly 3-lobed, with venation more strongly reticulate below; bracts narrowly lanceolate (c. 8 × 1·5–3 mm.).

Nyasaland. C: Kota Kota Distr., Nchisi Mt., fl. 26.vii.1946, *Brass* 16973 (K; SRGH).
Montane forest margins.

This may be merely one of the many variants of *D. burgessiae* but it may be an unde-scribed species as the indumentum of the young branches is very characteristic. The more shallowly lobed leaves, the more reticulate venation and narrower bracteoles, however, are not in themselves sufficient to distinguish it from *D. burgessiae* and it is desirable that more material be examined before a final decision is reached. Another specimen from the Nyika Plateau in the Northern Province of Nyasaland (*Brass* 17183 (K; SRGH)) has a denser indumentum and younger leaves but probably belongs here. No other specimens exactly matching these are known as yet.

2. MELHANIA Forsk.

Melhania Forsk., Fl. Aegypt.-Arab.: CVII, 64 (1775).

Perennial herbs or shrubs, herbaceous parts usually densely stellate-tomentose or tomentellous. Leaves serrate-crenate or entire; stipules linear or subulate. Flowers bisexual, yellow, solitary or geminate or several on a common peduncle in the upper leaf-axils. Bracts of the epicalyx 3, from linear-lanceolate to broadly ovate, persistent. Sepals 5, free almost to the base. Petals 5, usually obovate, unequal-sided, convolute in aestivation. Stamens 5, opposite the petals, connate at the base and alternating with 5 ligulate staminodes; anthers sagittate, dehiscing longitudinally. Ovary sessile, often tomentose, 5-locular; ovules 1–∞ per loculus; style short or long; stigmas 5, linear. Fruit a loculicidally dehiscent capsule splitting from above; seeds 1–several per loculus; testa smooth or variously rugose or tuberculate; cotyledons plicate, 2-partite.

Leaves lorate or narrowly lanceolate :
     Leaves glabrous above or with sparse long silky hairs; epicalyx-bracts narrowly
         ovate-acuminate, rounded or slightly cordate at the base -   -   1. *prostrata*
     Leaves tomentellous above; epicalyx-bracts lanceolate-acuminate :
         Sepals 1 cm. long or less; peduncles exceeding the petioles; leaves entire   2. *randii*
         Sepals c. 1·6 cm. long; peduncles ± equalling the petioles; leaves dentate, especially
            towards the apex - - - - - - - - - 3. *burchellii*
Leaves broadly oblong-elliptic, oblong or ovate :
     Epicalyx-bracts not exceeding 7 × 6 mm.; herbaceous parts thinly greyish-tomentose;
         leaves 0·8–3 × 0·6–2·3 cm. :
         Epicalyx-bracts broadly ovate - - - - - - 4. *rehmannii*
         Epicalyx-bracts linear-subulate - - - - - - 5. *griquensis*

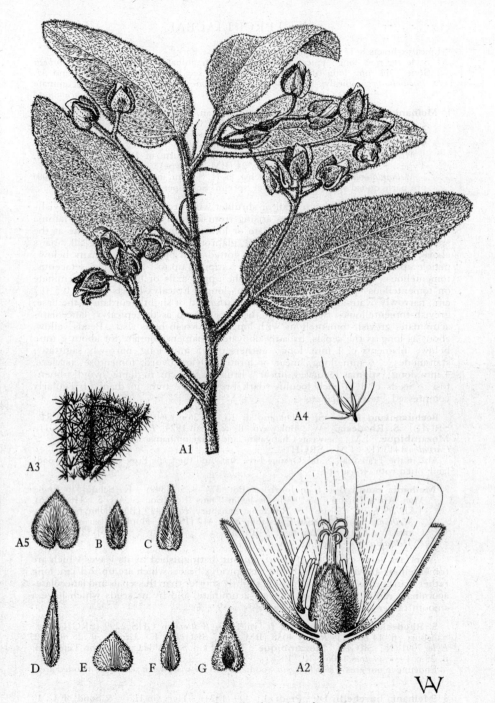

Tab. 99. A.—MELHANIA FORBESII. A1, flowering branch (×⅖); A2, longitudinal sect. of flower with part of staminal tube removed (×4); A3, portion of leaf surface (×8); A4, hair from leaf surface (×14); A5, epicalyx-bract (×⅖), all from *Johnson* 111. B.—MELHANIA BURCHELLII, epicalyx-bract (×2) *Holub* s.n. C.—MELHANIA ACUMINATA VAR. ACUMINATA, epicalyx-bract (×2) *Goldsmith* 35/55. D.—MELHANIA GRIQUENSIS, epicalyx-bract (×2) *Dinter* 8300. E.—MELHANIA REHMANNII, epicalyx-bract (×2) *Davies* 2312. F.—MELHANIA RANDII, epicalyx-bract (×2) *Cecil* 237. G.—MELHANIA PROSTRATA, epicalyx-bract (×2) *Schlechter* 11946.

Epicalyx-bracts 8–15 mm. long:
  Style 1–3 mm. long; epicalyx-bracts acute or acuminate   -   -   6. *forbesii*
  Style 7–11 mm. long (if style c. 5 mm. long see hybrids of *M. forbesii* and *M. acuminata*); epicalyx-bracts abruptly acuminate or sometimes caudate-acuminate
7. *acuminata*

1. **Melhania prostrata** DC., Prodr. **1**: 499 (mid-Jan. 1824).—Burch., Trav. Int. S. Afr. **2**: 263 (1824).—Harv. in Harv. & Sond., F.C. **1**: 222 (1860).—K. Schum. in Engl., Mon. Afr. Pflanz. **5**: 9 (1900).—Burtt Davy, F.P.F.T. **1**: 260 (1926). TAB. **99** fig. G. Type from Cape Prov. (Griqualand-W.).
    Melhania prostrata forma *latifolia* Bak. f. in Journ. of Bot. **37**: 425 (1899). Type: S. Rhodesia, Bulawayo, *Rand* 24 (BM, holotype).
    Melhania ovata var. *oblongata* Hochst. ex K. Schum. in Engl., tom. cit.: 9 (1900) pro parte quoad specim. Junod. excl. specim. Schimper.

Perennial herb up to 0·6 m. tall or shrublet with prostrate or suberect greyish-brown thinly tomentose branches arising from a woody rootstock. Leaf-lamina up to 11 × 1·7 cm., narrowly lanceolate or lorate, rounded and mucronate at the apex, rounded and 3-nerved at the base, glabrous or with sparse long silky hairs above, densely greyish-tomentellous, ± dotted with brown stellate hairs below; petiole up to 1 cm. long, thinly tomentose; stipules up to 1·2 cm. long, setaceous, tomentellous. Flowers borne singly in the upper axils or occasionally geminate on tomentellous peduncles up to 4 cm. long. Epicalyx-bracts 1–1·5 × 0·5–0·7 cm., narrowly ovate-acuminate or caudate, rounded or slightly cordate at the base, greyish-tomentellous. Sepals about the same length as the epicalyx, lanceolate-acuminate, greyish-tomentellous with longer brownish hairs also. Petals yellow, about as long as the sepals, broadly obovate. Stamens connate for about 2 mm. below; filaments c. 1 mm. long; anthers c. 2·5 mm. long, narrowly sagittate; staminodes c. 6 mm. long, linear or lorate. Ovary ovoid, tomentose; style c. 4 mm. long; stigmas recurved, linear. Fruit up to 1 cm. in diam., ovoid, tomentose. Seeds c. 3 in each loculus, dark brown, c. 2 mm. in diam., irregularly compressed; testa reticulate.

Bechuanaland Prot. SE: Dikomo di Ki, fl. 26.ii.1960, *Wild* 5179 (K; PRE; SRGH). **S. Rhodesia.** W: Bulawayo, fl. & fr. iii.1924, *Eyles* 6955 (K; SRGH). **Mozambique.** LM: between Changalane and Mazeminhama, fl. 20.xii.1952, *Myre & Carvalho* 1413 (K; LM; SRGH).
Also in the Transvaal, Natal, Orange Free State and the Cape Prov. Rather dry woodland, often with *Acacia* spp.

2. **Melhania randii** Bak. f. in Journ. of Bot. **37**: 425 (1899).—K. Schum. in Engl., Mon. Afr. Pflanz. **5**: 6 (1900).—Eyles in Trans. Roy. Soc. S. Afr. **5**: 418 (1916). TAB. **99** fig. F. Type: S. Rhodesia, Salisbury, *Rand* 439 (BM, holotype).
    Melhania prostrata sensu Eyles tom. cit.: 417 (1916).—Hopkins, Bacon & Gyde, Comm. Veld Fl.: 73 cum fig. (1940).
    Melhania sp.—Eyles, loc. cit.

A species very similar to *M. prostrata* but distinguished by its leaves which are tomentellous above as well as below, its epicalyx bracts which are up to 1 cm. long rather than approaching 1·5 cm. long, slightly shorter than the sepals and lanceolate-acuminate rather than narrowly ovate-acuminate, and by its seeds which have a smooth rather than reticulate testa.

S. Rhodesia. N: Mazoe, fl. & fr. viii.1917, *Walters* in GHS 2299 (SRGH). C: Salisbury, fl. 14.ix.1911, *Rogers* 4058 (BM; K; SRGH). E: Umtali, fl. 26.iii.1932, *Eyles* 7091 (K; SRGH). **Mozambique.** T: 112 km. S. of Vila Coutinho on Tete road, fl. 25.ix.1935, *Galpin* 15064 (K; PRE).
Endemic in our area. *Brachystegia* woodland and the margins of seasonally flooded areas (vleis).

3. **Melhania burchellii** DC., Prodr. **1**: 499 (1824).—Harv. in Harv. & Sond., F.C. **1**: 222 (1860).—Burtt Davy, F.P.F.T. **1**: 260 (1926). TAB. **99** fig. B. Type from Cape Prov.
    Melhania linearifolia sensu Eyles in Trans. Roy. Soc. S. Afr. **5**: 417 (1916).

Very like the two preceding species and resembling *M. randii* in having its leaves tomentellous on both sides. Flowers almost twice the size of those of *M. randii*, the sepals being c. 1·6 cm. long. Epicalyx bracts lanceolate-acuminate, differing from those of both *M. randii* and *M. prostrata* in being about half the length of the sepals. Peduncles not elongated, as in both the other species, at maturity but shorter than, or only slightly exceeding, the petioles in length. Leaf-margins usually dentate, especially towards the apex, those of *M. randii* and *M. prostrata* being usually quite entire. Seeds slightly roughened or tuberculate but scarcely reticulate.

Bechuanaland Prot. SW: 80 km. N. of Kang, fl. & fr. 18.ii.1960, *Wild* 5052 (PRE; SRGH). SE: Eastern Bamangwato Territory, fl. & fr., *Holub* (K). **S. Rhodesia.** W: 128 km. N. of Bulawayo, fl. & fr. xii.1902, *Eyles* 1127 (SRGH).
Also in the Transvaal and Cape Prov. Drier types of woodland in the SW. of our area.

4. **Melhania rehmannii** Szyszyl., Polypet. Thalam. Rehm.: 138 (1887).—K. Schum. in Engl., Mon. Afr. Pflanz. **5**: 10 (1900).—Burtt Davy, F.P.F.T. **1**: 260 (1926). —O. B. Mill. in Journ. S. Afr. Bot. **18**: 57 (1952). TAB. **99** fig. E. Type from the Transvaal.
Melhania griquensis Bolus in Journ. Linn. Soc., Bot. **24**: 172 (1887) pro parte quoad specim. Holub.

Small woody shrublet 0·3 m. tall with all parts greyish-tomentose. Leaf-lamina 0·8–3 × 0·6–2·3 cm., broadly oblong-elliptic or broadly elliptic, truncate and mucronulate at the apex, rounded or broadly cuneate at the base, margin coarsely crenate-dentate; petiole up to 1·3 cm. long; stipules c. 5 mm. long, filamentous. Flowers solitary or geminate in the upper axils on peduncles c. 1 cm. long; pedicels c. 1 cm. long. Epicalyx-bracts 5–7 × 4–6 mm. broadly ovate, acute at the apex, subcordate at the base. Sepals equalling or somewhat exceeding (particularly in fruit) the epicalyx, lanceolate, acute. Petals yellow, c. 6 × 4 mm., broadly obovate. Stamens connate for 1 mm. below; filaments 2 mm. long; anthers 1 mm. long, oblong; staminodes c. 3·5 mm. long, linear. Ovary ovoid, tomentose; style 1–2 mm. long; stigmas linear. Fruit c. 0·8 × 0·6 cm., ovoid, tomentose. Seeds c. 1·8 mm. in diam., c. 3 per loculus, brown, irregularly compressed, testa vermiform-tuberculate.

Bechuanaland Prot. N: Francistown, fl. i.1926, *Rand* 26 (BM). SE: Bakwena Territory, *Holub* (K). **S. Rhodesia.** S: Beitbridge, fl. 15.ii.1955, *E.M. & W.* 426 (BM; LISC; SRGH).
Also in the Transvaal and Cape Prov. In drier types of woodland, often associated with *Colophospermum mopane*.

5. **Melhania griquensis** Bolus in Journ. Linn. Soc., Bot. **24**: 172 (1887) excl. specim. Burchell. Holub. Rehmann. et Orpen. pro parte—K. Schum. in Engl., Bot. Jahrb. **10**: 41 (1888); in Verh. Bot. Verein. Brand. **30**: 229 (1888); in Engl., Mon. Afr. Pflanz. **5**: 5 (1900). TAB. **99** fig. D. Type from Cape Prov. (Griqualand).
Melhania bolusii Burtt Davy, F.P.F.T. **1**: 261 (1926). Type as for *M. griquensis*.

Branching perennial up to 0·3 m. tall and very like *M. rehmannii*, but the epicalyx bracts are always linear-subulate, never ovate. The grey indumentum also tends to be thinner on older leaves and they become glabrescent.

Bechuanaland Prot. N: Ngamiland, *Curson* 170 (PRE). SW: Takatshwane Pan, fl. & fr. 20.ii.1960, *Wild* 5089 (K; PRE; SRGH).
Also in Cape Prov. and SW. Africa. In our driest type of woodland or bush.

The name of this species has in the past been the cause of considerable confusion. It is now evident, however, that Bolus intended to base his new species on four specimens on the *Orpen* sheet in the Bolus Herbarium. So much is plain from his description of the epicalyx-bracts as linear-subulate. Unfortunately, he failed to notice in the first place that the fifth branchlet on the *Orpen* sheet was a specimen of *Melhania rehmannii* with ovate epicalyx-bracts. Secondly, on a visit to Kew in 1881 he made a note in his diary that *Burchell* 2050 in Herb. Kew was the same as the *Orpen* plant. Not having the *Orpen* sheet with him, his judgment proved at fault, since *Burchell* 2050 is also a good specimen of *M. rehmannii*. On his return to Cape Town he copied this note from his diary on to the *Orpen* sheet and then presumably described the species. When the description was finally published N. E. Brown added another *Burchell* specimen, and the *Holub* and *Rehmann* specimens to the citations since these matched *Burchell* 2050. They are also specimens of *M. rehmannii*. Because of this confusion *M. griquensis* has often in herbaria been considered to be a synonym of *M. rehmannii* but it is, in fact, a perfectly distinct species.

6. **Melhania forbesii** Planch. ex Mast. in Oliv., F.T.A. **1**: 231 (1868).—K. Schum. in Engl., Mon. Afr. Pflanz. **5**: 12 (1900).—Eyles in Trans. Roy. Soc. S. Afr. **5**:

417 (1916).—Burtt Davy, F.P.F.T. **1**: 261 (1926).—Exell & Mendonça, C.F.A. **1**, 2: 190 (1951). TAB. **99** fig. A. Syntypes: Mozambique, Chupanga, *Kirk* (K); without precise locality, *Hutton* (K).

Melhania serrulata R.E.Fr., Wiss. Ergebn. Schwed. Rhod.-Kongo-Exped. **1**: 157 (1914). Type: S. Rhodesia, Victoria Falls, *Fries* 74 (UPS).

Melhania didyma sensu O. B. Mill. in Journ. S. Afr. Bot. **18**: 57 (1952) quoad specim. Lugard.

Branching shrublet c. 0·6 m. tall, densely covered in all the herbaceous parts with a greyish-tomentellous indumentum intermingled, particularly on the stems, petioles and nerves on the under surface of the leaves, with longer stellate-ferruginous hairs. Leaf-lamina up to 11 × 5 cm., oblong, narrowly ovate-oblong or ovate, somewhat discolorous, apex obtuse or rarely acute, margin serrate or crenate, base rounded; petiole up to 2·5 cm. long. Flowers yellow, 1–3 together on axillary peduncles up to 9 cm. long; pedicels 0·5–1·5 cm. long, fasciculate. Epicalyx-bracts 1–1·5 × 0·8–1 cm., very broadly ovate, acute or slightly acuminate at the apex, subcordate at the base, densely greyish-white-tomentose. Sepals of ⅔ or equalling the length of the epicalyx, lanceolate-acuminate, tomentose. Petals about the same length as the sepals, very broadly obovate. Stamens connate for 1–2 mm. below; filaments flattened, 1–2 mm. long; anthers lanceolate, 2–5 mm. long; staminodes c. 7 mm. long, linear or lorate. Ovary ovoid, tomentose; style 1–3 mm. long; stigmas linear, recurved. Fruit slightly shorter than the sepals, ovoid, tomentose. Seeds c. 6 per loculus, c. 2 mm. in diam., dark brown, irregularly compressed; testa smooth or occasionally slightly rugose.

Caprivi Strip. E. of Kwando R., fr. x.1945, *Curson* 1098 (PRE). **Bechuanaland Prot.** N: Kwebe Hills, fl. 22.i.1898, *Lugard* 123 (K). **N. Rhodesia.** S: Livingstone, fl. 5.xii.1931, *Trapnell* 557 (K). **S. Rhodesia.** N: Darwin Distr., near Bopoma R. and Mazoe R., fl. & fr. 10.v.1957, *Crehan* 191 (SRGH). W: Shangani Reserve, fl. iii.1949, *Davies* 6 (K; SRGH). E: Melsetter North, fl. 16.viii.1932, *Brain* 9592 (K; SRGH). S: Nuanetsi, Tswiza, fl. & fr. 1.xi.1955, *Wild* 4698 (K; SRGH). **Mozambique.** MS: between Gombalançae & Maringuè, fl. & fr. 29.ix.1949, *Pedro & Pedrógão* 8420 (LMJ; SRGH). SS: Alto de Maxixe, fl. 25.ii.1955, *E. M. & W.* 567 (BM; LISC; SRGH). LM: Delagoa Bay, fl. 4.i.1898, *Schlechter* 11982 (BM; COI; K; PRE).

Also in the Transvaal, SW. Africa and Angola. Low-altitude or drier types of woodland.

Each flower in this species opens for one day only and often they are open in the mornings only. *M. acuminata* and probably *M. velutina* behave similarly. The photograph in Hopkins, Bacon and Gyde, Common Veld Fl.: 74 (1940) is either of this species or *M. acuminata*. Some material of *M. forbesii*, viz. *van Son* in Herb. Transv. Mus. 29014 (BM; K; PRE; SRGH) and *Brain* 9232 (K; SRGH) from the Victoria Falls, is near *M. velutina* Forsk. (described from Arabia but reaching Tanganyika and Angola to the south) in having rather acuminate bracts. They are not good specimens, however, and were probably collected in dry conditions.

7. **Melhania acuminata** Mast. in Oliv., F.T.A. **1**: 231 (1868).—K. Schum. in Engl. Mon. Afr. Pflanz. **5**: 13 (1900).—Eyles in Trans. Roy. Soc. S. Afr. **5**: 417 (1916).—Burtt Davy, F.P.F.T. **1**: 261 (1926).—Gomes e Sousa in Bol. Soc. Estud. Col. Moçamb. **32**: 81 (1936). Type: Mozambique, Sena, *Kirk* (K, holotype).

Melhania velutina sensu Exell & Mendonça, C.F.A. **1**, 2: 190 (1951) pro parte.

Shrublet with the habit and general appearance of *M. forbesii* but differing as follows: herbaceous parts greyish-tomentose and not so markedly ferruginous-hairy on the stems and nerves; bracts of the epicalyx abruptly acuminate or even caudate at the apex and often noticeably cordate at the base; style from 0·7–1·1 cm. long rather than 0·1–0·3 cm. long; and seeds slightly rugose.

Var. **acuminata**. TAB. **99** fig. C.

Leaves ovate-oblong or ovate. Inflorescence-branches and epicalyx lacking the scattered dark-brown stellate hairs of var. *agnosta*. Sepals and petals from 1–1·5 cm. long.

Bechuanaland Prot. N: Francistown, fl. i.1926, *Rand* 29 (BM). SE: Kwena, fl. 6.v.1955, *Reyneke* 329 (K; PRE). **N. Rhodesia.** B: Sesheke, fl. xii.1875, *Holub* 855 (K). S: Livingstone, fl. & fr. iv.1909, *Rogers* 7051 (K). **S. Rhodesia.** N: Mtoko, Mazoe R., fl. & fr. i.1953, *Phelps* 18 (K; SRGH). W: Bulawayo, fl., *Rogers* 13538 (K; PRE; SRGH). C: Salisbury, Prince Edward Dam, fl. 28.ii.1937, *Eyles* 8951 (K; SRGH). E: Lower Sabi, Mtema, fl. 28.i.1948, *Wild* 2401 (K; SRGH). S: West

Nicholson, fl. & fr. 23.iii.1953, *Plowes* 1578 (K; SRGH). **Mozambique.** T: Changara, fl. 13.v.1949, *Gerstner* 7047 (K; PRE; SRGH). MS: Chimoio, Belas Mt., fl. 2.iv.1948, *Garcia* 847 (BM; LISC; SRGH). LM: Magude, Mapulanguene, fl. 19.ix.1942, *Mendonça* 3184 (LISC).

Also in Angola and the Transvaal. Open woodland, often occurring in the same areas as *M. forbesii* but also occurring at higher altitudes in *Brachystegia* woodland.

Var. **agnosta** (K. Schum.) Wild in Bol. Soc. Brot., Sér. 2, **33**: 36 (1959). Type from the Transvaal.

    Melhania agnosta K. Schum. in Engl., Mon. Afr. Pflanz. **5**: 11 (1900).—Burtt Davy, F.P.F.T. **1**: 261 (1926). Type as above.

    Melhania obtusa N.E.Br. in Kew Bull. **1906**: 99 (1906). Type: S. Rhodesia, near Bulawayo, *Cecil* 94 (K, holotype).

    Melhania sp. cf. *M. acuminata* Mast.—O. B. Mill. in Journ. Bot. S. Afr. **18**: 57 (1952).

Differs from var. *acuminata* in having oblong or narrowly oblong leaves and scattered dark brown stellate hairs on the inflorescence, branches and epicalyx. Sepals and petals rarely exceeding 1 cm. long. Seeds 4 per loculus instead of c. 6 in var. *acuminata*.

Bechuanaland Prot. SE: Kanye, fl. v.1958, *Miller* 586 (PRE). **N. Rhodesia.** S: Livingstone, fl. 6.iv.1956, *Robinson* 1440 (K). **S. Rhodesia.** N: Mrewa, fl. & fr. v.1956, *Davies* 1936 (K; SRGH). W: Matopos, fl. iii.1918, *Eyles* 980 (BM; K; SRGH). C: Rusape, fl. v.1931, *Eyles* 7533 (SRGH). E: Umtali, Mpembi Mt., fl. 2.iv.1950, *Chase* 2130 (BM; K; SRGH). S: S. of Lundi R., on Beitbridge road, fl. 15.xi.1955, *E.M. & W.* 373 (BM; LISC; SRGH). **Mozambique.** MS: Beira, fl. ii.1912, *Rogers* 5952 (K; SRGH).

Also in the Transvaal. Ecology similar to that of var. *acuminata*.

The range of distribution of this variety is approximately that of the type variety. Var. *agnosta* is particularly common around Bulawayo and the Matopos in S. Rhodesia.

? Melhania acuminata Mast. × **Melhania forbesii** Planch. ex Mast.

1. Putative hybrids with the facies and epicalyx-bracts of *M. forbesii* but with styles 4–6 mm. long (intermediate between the two putative parents).

N. Rhodesia. C: Lusaka, fl. 26.i.1956, *Noak* 70 (SRGH). S: Mumbwa, fl. 1911, *Macaulay* 778 (K). **S. Rhodesia.** N: Karoi Exp. Farm, fl. & fr. 7.iii.1947, *Wild* 1834 (K; SRGH). C: Salisbury, Twentydales, fl. 1.iii.1947, *Wild* 1871 (K; SRGH). **Nyasaland.** S: Blantyre, fl., *Buchanan* 918 (K).

Melhania forbesii has two main centres of distribution, one in the southern part of Mozambique and the other in the Cuanza and Luanda provinces of Angola. In between these areas the species is much rarer but occasional specimens are recorded, principally from the Zambezi valley. *Melhania acuminata*, on the other hand, is mainly concentrated on the escarpment country and the plateaux to the south of the Zambezi in S. Rhodesia and Mozambique as well as in the Transvaal and Bechuanaland. The former species is characterised by a short style of 1–3 mm. in length and the latter by a long style of from 0·7–1·1 cm. in length. Along the Zambezi valley particularly, but also as far south as the Transvaal, specimens occur, however, which look superficially like one or other of these species but have styles of about 5 mm. in length. Although intermediate in style-length these specimens are not as a rule intermediate in other characters and the difference in styles in the two species is so extreme that it is considered most unlikely that we are dealing with a single variable species here but rather two distinct species hybridising freely where the ranges of distribution overlap. One group of hybrids superficially resembles *M. forbesii* as above. (? back-crosses with *M. forbesii*). Others superficially resemble *M. acuminata* (see below). It is very desirable that breeding experiments and cytological investigations should be carried out to test the validity of these suggestions. *Chase* 3703 (BM; K; SRGH) from Dora Ranch, Umtali, presents useful evidence for the above hypothesis since the British Museum sheet has a mixture of *M. acuminata* var. *agnosta* and *M. forbesii* × *M. acuminata* (? var. *agnosta*) i.e. the hybrid and one parent were collected in the same spot in one gathering. The remaining sheets bearing this number consist only of *M. acuminata* var. *agnosta*.

2. Putative hybrids with the facies and epicalyx-bracts of *M. acuminata* but with styles 4–6 mm. long.

Bechuanaland Prot. SW: near Chukude Pan, fl. & fr. 24.vi.1955, *Story* 4973 (K;

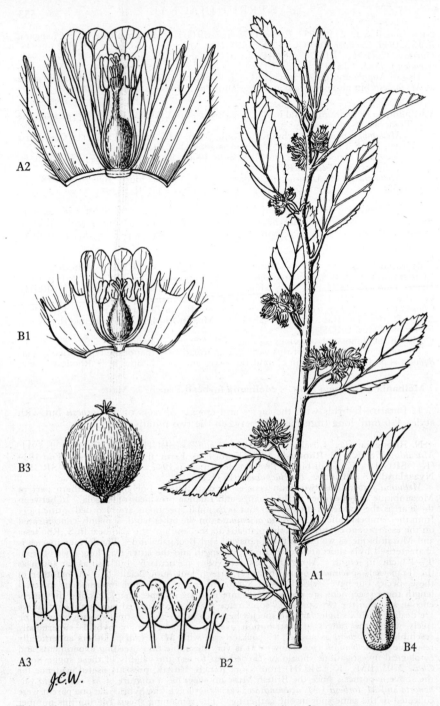

Tab. 100. A.—WALTHERIA INDICA. A1, flowering branch (×⅔); A2, flower with calyx opened out and one petal and one stamen removed (×6); A3, arrangement of anthers and upper margin of staminal tube (×12), all from *Brass* 18010. B.—MELOCHIA CORCHORIFOLIA. B1, flower with calyx opened out and one petal and one stamen removed (×6); B2, arrangement of anthers and upper margin of staminal tube (×12); B3, fruit (×4); B4, seed (×6), all from *Faulkner* 306.

PRE). **S. Rhodesia.** W: Bulalima-Mangwe, Embakwe, fl. & fr. 1.iv.1942, *Feiertag* in GHS 45373 (SRGH).

This putative hybrid (? the result of back-crosses with *M. acuminata*) also occurs in the Transvaal (Soutpansberg, *Codd* 4137 (K; PRE), etc.).

3. MELOCHIA L.

Melochia L., Sp. Pl. **2**: 674 (1753); Gen. Pl. ed. 5: 304 (1754).

Herbs or shrubs. Leaves simple, petiolate, stipulate. Flowers bisexual, in terminal or axillary cymose clusters; pedicels bracteate. Calyx campanulate or inflated, shortly 5-lobed. Petals 5, usually obovate or oblong-spathulate, marcescent. Stamens 5, opposite the petals and sometimes coherent with them at the base; filaments more or less connate; anthers 2-thecous, thecae parallel. Ovary sessile or stipitate, 5-locular; ovules 2 per loculus; styles 5, free or connate at the base. Fruit a loculicidal 5-valved capsule; loculi 1-seeded. Seeds ascending; embryo straight, with some endosperm; cotyledons flat.

Flowers in spreading or subglobose, terminal or axillary cymes; stems with a line of
    stellate hairs decurrent from the base of the stipules - - - 1. *corchorifolia*
Flowers in axillary subsessile cymes; stems pubescent or hispidulous all round
                                        2. *melissifolia*

1. **Melochia corchorifolia** L., Sp. Pl. **2**: 675 (1753).—Mast. in Oliv., F.T.A. **1**: 236 (1868).—K. Schum. in Engl., Mon. Afr. Pflanz. **5**: 41, t. 3 fig. G (1900).—Eyles in Trans. Roy. Soc. S. Afr. **5**: 419 (1916).—Exell & Mendonça, C.F.A. **1**, 2: 191 (1951). TAB. **100** fig. B. Type from India.

Erect annual herb up to c. 2 m. tall but often less; stems glabrescent but with a line of stellate hairs decurrent from the stipule bases. Leaf-lamina up to c. 7 × 4·5 cm., oblong-lanceolate, narrowly ovate or ovate, apex acute, margin acutely serrate, base broadly cuneate or truncate and 5-nerved, thinly hispidulous at least on the nerves below; petiole up to 2·3 cm. long, pubescent at least on the upper surface; stipules c. 1 cm. long, subulate-lanceolate, margins ciliolate. Flowers numerous, in dense terminal or axillary cymes; peduncle up to 5 cm. long; pedicels 1–3 mm. long, hispidulous; bracts 4–10 mm. long, numerous, subulate-lanceolate, ciliolate at the margins. Calyx 3 mm. long, campanulate, with very short abruptly acuminate teeth, ciliolate. Petals white, yellow at the base inside, c. 8 mm., obovate. Stamens adherent to the petals below; filaments united in a tube almost to the apex, about ½ the length of the petals, free portion of filaments c. 1 mm. long; pairs of anther-thecae connate in the lower half, free above. Ovary oblong-ovoid, densely pilose; styles 5, slightly connate below, c. 2 mm. long, hirsute. Capsule 5 mm. in diam., subglobose, hispidulous, 5-valved, surrounded by the persistent calyx and bracts. Seeds c. 3 × 2·5 mm., 1 per loculus, dark brown or greyish, 3-sided; testa minutely striate.

N. Rhodesia. B: Mongu-Lealui, fl. & fr. 11.ii.1952, *White* 2054 (FHO). **S. Rhodesia.** W: Victoria Falls, fl. & fr. xii.1906, *Allen* 397 (K; SRGH). **Nyasaland.** N: Nyika Plateau, Mwanemba, ii–iii.1903, *McClounie* 135 (K). C: Senga Bay, fl. & fr. 17.ii.1959, *Robson* 1644 (K; SRGH). S: Fort Johnston, fr. xii.1893, *Scott Elliot* 8425 (K). **Mozambique.** N: Nampula, fl. & fr. 20.iii.1937, *Torre* 1286 (COI; LISC). Z: Mocuba, Namagoa, fl. & fr. iv.1943, *Faulkner* 306 (K; PRE; SRGH). T: Lower Shire, fr. xi.1861, *Kirk* (K). MS: Vila Machado, fl. & fr. 26.ii.1948, *Mendonça* 3811 (LISC). LM: Umbeluzi, cult., fl. 10.v.1949, *Myre & Balsinhas* 708 (LM).

Widely distributed through the tropical regions of the Old World. Often a weed of cultivation.

The leaves are eaten as a spinach and also used for smoking (Allen).

2. **Melochia melissifolia** Benth. in Hook., Journ. Bot. **4**: 124 (1841).—Mast. in Oliv., F.T.A. **1**: 236 (1868).—Exell & Mendonça, C.F.A. **1**, 2: 191 (1951). Syntypes from British and French Guiana.
    Melochia welwitschii Hiern, Cat. Afr. Pl. Welw. **1**: 91 (1896). Type from Angola.
    Melochia melissifolia var. *welwitschii* (Hiern) K. Schum. in Engl., Mon. Afr. Pflanz. **5**: 43 (1900). Type as for *M. welwitschii*.
    Melochia melissifolia var. *mollis* K. Schum. in Engl., loc. cit.—Keay, F.W.T.A. ed. 2, **1**, 2: 318 (1958). Syntypes from Cameroons, Congo, Sudan and Uganda.
    Melochia mollis (K. Schum.) Hutch. & Dalz., F.W.T.A. **1**: 250 (1928). Syntypes as for *M. melissifolia* var. *mollis*.

Annual herb up to c. 1·6 m. tall; branches pubescent or hispidulous. Leaf-lamina up to 6·5 ×4·8 cm., ovate, apex acute, margin serrate-crenate, base sub-cordate or truncate and 5-nerved, appressed-hispidulous on both surfaces, some-times densely so, or glabrescent; petiole up to 1·5 cm. long, pubescent or his-pidulous; stipules up to 1·3 cm. long, narrowly lanceolate-acuminate, margins ciliolate. Flowers in subsessile axillary cymes; pedicels up to c. 5 mm. long, hispidulous; bracts c. 1 cm. long, subulate, hispidulous. Calyx c. 5 mm. long, campanulate, ciliolate, with short abruptly acuminate lobes. Petals white, c. 8 mm long, obovate. Stamens adherent to the petals below; filaments c. 5 mm. long, united in a tube almost to the apex or to about ½ way up; anther-thecae as in *M. corchorifolia*. Ovary oblong-ovoid, silky-hairy; styles c. 5 mm. long. Capsule 3·5 mm. in diam., depressed-globose, 5-valved, hispidulous or glabrescent, sur-rounded by the persistent calyx and bracts. Seeds c. 2·5 ×1·3 mm., 1 per loculus, subellipsoid, 3-sided; testa very minutely rugulose.

N. Rhodesia. N: Niomkolo, fl. vii.1890, *Carson* (K). **Mozambique.** Z: between Ile and Alto Molocuè, fl. 21.vi.1943, *Torre* 5541 (LISC; SRGH). MS: Inhaminga, fl. 22.v.1948, *Mendonça* 4357 (LISC).
Widely distributed in tropical America and tropical Africa. Swampy grasslands or waste ground.

A polymorphic species divided by Schumann (loc. cit.) into 6 varieties. If these varietal distinctions were retained here the *Torre* and *Mendonça* specimens would be var. *mollis* K. Schum. and the *Carson* one would be var. *welwitschii* (Hiern) K. Schum. respec-tively. However, I consider var. *welwitschii* to be an intermediate between var. *melissifolia* from tropical America and var. *mollis*, and so there does not seem much point in dis-tinguishing these varieties.

4. WALTHERIA L.

Waltheria L., Sp. Pl. 2: 673 (1753); Gen. Pl. ed. 5: 304 (1754).

Herbs or small shrubs with a stellate indumentum. Leaves petiolate, simple, crenate or serrate, 3–5-nerved from the base; stipules present. Flowers small, bisexual, in globose clusters, scorpioid spikes, or branching cymes or corymbs; peduncle short or elongated; bracts and bracteoles present. Calyx 5-lobed. Petals 5, ± obovate or spathulate, marcescent. Stamens 5, opposite the petals and with the filaments united in a tube; anthers 2-thecous, thecae parallel. Ovary sessile, 1-locular; ovules 2; style somewhat excentric, clavate or penicillate-fimbriate above. Capsule 2-valved, 1 (2)-seeded. Seed ascending, with endo-sperm; embryo straight; cotyledons flat.

Waltheria indica L., Sp. Pl. 2: 673 (1753) emend. excl. syn. Hort. Cliff. pro parte quoad descr. et specim.—Harv. in Harv. & Sond., F.C. 1: 180 (1860).—Exell & Mendonça, C.F.A. 1, 2: 192 (1951).—Brenan in Mem. N.Y. Bot. Gard. 8, 3: 228 (1953). TAB. 100 fig.A. Type from India.
    Waltheria americana L., loc. cit.—Mast. in Oliv., F.T.A. 1: 235 (1868).—K. Schum. in Engl., Mon. Afr. Pflanz. 5: 45, t. 3 fig. J (1900) pro parte.—Bak. f. in Journ. Linn. Soc., Bot. 40: 30 (1911).—Eyles in Trans. Roy. Soc. S. Afr. 5: 419 (1916).—O. B. Mill. in Journ. S. Afr. Bot. 18: 57 (1952). Type from the West Indies or perhaps Dutch Guiana.
    Waltheria americana var. *indica* (L.) K. Schum. in Engl., tom. cit.: 47 (1900).—Burtt Davy, F.P.F.T. 1: 268 (1926).—O. B. Mill., tom. cit.: 58 (1952). Type as for *W. indica*.
    Waltheria americana var. *subspicata* K. Schum., loc. cit.—Gomes e Sousa, Bol. Soc. Estud. Col. Moçamb. 26: 43 (1935). Syntypes from Sudan, Kenya, Tan-ganyika, SW. Africa, Angola and Mozambique, Sena, *Peters* 34 (B†).
    Waltheria wildii Suesseng. in Proc. & Trans. Rhod. Sci. Ass. 43: 107 (1951). Type: S. Rhodesia, Marandellas, *Dehn* 81 (M, holotype).

Bushy herb usually 0·6–1·3 m. tall, rather sparsely stellate-pubescent to stellate-tomentose in its vegetative parts. Leaf-lamina up to 9 ×4 cm., ovate-oblong to oblong, apex obtuse or subacute, margin serrate-crenate, base rounded and 5-nerved, nervation somewhat impressed above, prominent below; petiole up to 3·3 cm. long. Flowers yellow, in dense subsessile or long-pedunculate, axillary, globose heads, or the upper inflorescences forming a short interrupted or continuous spike, or irregularly collected into dense, leafy or leafless cymes or corymbs;

bracts and bracteoles about as long as the flowers, linear or linear-lanceolate, tomentose outside. Calyx 2–3 mm. long, campanulate, villous, divided somewhat less than ½ way into 5 triangular lobes. Petals as long as or slightly longer than the calyx, obovate-oblong, shortly clawed, glabrous or the apex stellate-ciliolate. Stamens with the filaments united into a tube shorter than the petals ; anther-thecae parallel, bluntly mucronate at base and apex. Ovary oblong-ovoid, hirsute in the upper half ; style pubescent, somewhat excentric, c. 1·5 mm. long ; stigma penicillate. Capsule c. 3 × 2 mm., obovoid, villous above, 1- or rarely 2-seeded. Seeds c. 2 × 1·4 mm., dark brown, obovoid ; testa smooth.

Bechuanaland Prot. N : Tsau, fl. 4.viii.1955, *Story* 5101 (K ; PRE). SW : 410 km. NW. of Molepolole, fl. 24.vi.1955, *Story* 4974 (K ; PRE). SE : Mochudi, fl. v.1915, *Harbor* (SR). **N. Rhodesia.** B : Sesheke, fl. iv., *Macaulay* 451 (K). N : Mpika, fl. 17.vii.1930, *Hutchinson & Gillett* 3785 (K). W : Mwinilunga, fl. 29.xii.1937, *Milne-Redhead* 3866 (BM ; K). C : Lusaka, fl. & fr. 2.vi.1956, *Angus* 1312 (K ; SRGH). E : Fort Jameson, fl. 30.viii.1929, *Burtt Davy* 21020 (FHO). S : Victoria Falls, fl. *Rogers* 7159 (K ; SRGH). **S. Rhodesia.** N : Mazoe, fl. 28.v.1931, *Brain* 4558 (SRGH). W : Nyamandhlovu, fl. 29.i.1951, *West* 3206 (K ; SRGH). C : Hartley, fl. 11.v.1950, *Hornby* 3177 (K ; SRGH). E : Umtali, fl. 6.ii.1956, *Chase* 6005 (BM ; K ; SRGH). S : Sabi-Lundi Junction, fl. 8.v.1950, *Wild* 3417 (K ; SRGH). **Nyasaland.** N : Mzimba, fl. 6.iv.1955, *Jackson* 1602 (K ; SRGH). C : Lilongwe, Chitedze, fl. 22.iii.1955, *E.M. & W.* 1107 (BM ; LISC ; SRGH). S : Chikwawa Distr., Lower Mwanza R., fl. 6.x.1946, *Brass* 18010 (K). **Mozambique.** N : Nampula, fl. 13.iv.1936, *Torre* 690 (COI ; LISC ; K). Z : Maganja da Costa, fl. vi.1946, *Pedro* 1495 (K ; LMJ). T : Tete, 21.viii.1931, *Guerra* 48 (COI). MS : Mossurize, Maconi-Madanda, fl. 5.ii.1907, *Johnson* 97 (K). SS : Massinga, fl. iv.1936, *Gomes e Sousa* 1720 (COI ; K ; LISC). LM : Marracuene, fl. 23.iv.1947, *Barbosa* 160 (COI ; LM).

Widely distributed through the tropics and subtropics. Common species of grassland or woodland, common in waste places and as a weed of cultivation.

5. HERMANNIA L.

Hermannia L., Sp. Pl. **2** : 673 (1753) ; Gen. Pl. ed. 5 : 304 (1754).

Herbs or small shrubs, usually stellately hairy, sometimes also with simple or glandular hairs. Leaves subentire, dentate or incised ; stipules foliaceous or minute or rarely absent, entire or variously divided. Flowers bisexual, yellow, reddish, purplish or white, 1–several on axillary peduncles ; peduncle usually bracteate. Calyx campanulate or globose, 5-lobed. Petals 5, narrowed to the base with margins often inrolled in the lower half, glabrous or variously pubescent. Stamens 5, opposite the petals ; filaments free or connate at the base, subulate, obovate and with membranous wings, or cruciform-tuberculate at the middle ; anthers linear with attenuate-acuminate apices, usually pubescent or ciliolate. Ovary sessile or shortly stipitate, 5-locular, loculi 3–∞ ovulate ; styles 5, slender, more or less united. Capsule dehiscent, with 5 valves, apex rounded or with short horn-like appendages. Seeds fairly numerous, with endosperm, subreniform ; testa brown or black, smooth, granulate or minutely tubercular, sometimes with a few widely spaced wrinkles.

Filaments always with membranous wings or cruciform-tuberculate in the middle :
  Filaments suddenly dilated and cruciform-tuberculate in the middle :
    Calyx globose, inflated in fruit ; bracts palmatifid with linear segments
                                           1. *grandistipula*
  Calyx campanulate, not inflated in fruit ; bracts entire :
    Leaves pubescent or glabrescent :
      Petals c. 8 mm. long ; leaves oblong to ovate :
        Leaves green, oblong to oblong-lanceolate, base broadly cuneate, margins serrate or crenate - - - - - - - 2. *quartiniana*
        Leaves often bronze-green, ovate to ovate-oblong, base rounded to sub-cordate, margins irregularly crenate - - - - 3. *depressa*
      Petals c. 4 mm. long ; leaves linear or oblong-linear - - 4. *parvula*
    Leaves densely greyish- or yellowish-tomentose - - - 5. *staurostemon*
Filaments slightly connate at the base, oblong or obovate, wings membranous :
  Calyx inflated, c. 1 cm. in diam. - - - - - - 6. *comosa*
  Calyx not inflated, or if slightly inflated not more than 5 mm. in diam. :
    Flowers (1) 2 or more together on axillary peduncles 1–several together in the upper axils forming a terminal panicle ; leaves ovate ; capsule not horned
                                            7. *floribunda*

Flowers solitary or rarely paired on peduncles solitary in the upper axils ; leaves
   linear to oblong, obovate-oblong, elliptic or obovate ; capsule often horned :
  Leaves greyish-tomentose or greyish-tomentellous :
    Leaves narrowly elliptic-oblong or narrowly obovate-oblong ; petals much
     shorter than the calyx ; seeds smooth :
      Petals spoon-shaped with margins of both claw and blade inrolled ; pros-
       trate plant - - - - - - - - 8. *tomentosa*
      Petals obovate-cuneate with apex truncate and margin of blade not inrolled ;
       erect plant - - - - - - - - 9. *guerkeana*
    Leaves obovate to broadly obovate ; petals longer than the calyx ; seeds
     wrinkled - - - - - - - - - 12. *torrei*
  Leaves glabrescent to densely pubescent but never tomentose or tomentellous :
   Ovary and capsule stellate-tomentose, not glandular :
    Branchlets not glandular :
     Petals spoon-shaped, margin of blade inrolled ; branches prostrate
                                           10. *angolensis*
     Petals obovate-cuneate, subauriculate, margin of blade not inrolled ;
      branches erect - - - - - - - 11. *micropetala*
    Branchlets glandular with some longer stellate or simple hairs
                                          12. *boraginiflora*
   Ovary and capsule with glandular hairs and often some simple or stellate hairs
    also but never tomentose :
    Petals shorter than the calyx, with the margins of the lower half strongly
     inrolled and with a pubescent line across the middle - 13. *glanduligera*
    Petals as long as or longer than the calyx, with margins only slightly inrolled
     towards the base, glabrous within :
     Leaves narrowly lanceolate to ovate-lanceolate ; testa granular :
      Petals yellow, narrowly oblanceolate, 3–5 × 0·75–1 mm. ; capsule ovoid,
       c. 3 × 4 mm. - - - - - - - - 15. *tigreensis*
      Petals pink or purplish, narrowly obovate, 5·5–10 × 2·5–6·5 mm. ;
       capsule truncate-ovoid to oblong-ovoid, 5–8 × 5–7 mm. 16. *kirkii*
     Leaves strictly linear ; testa of seeds minutely tuberculate   17. *modesta*
Filaments ± subulate - - - - - - - - - 18. *oliveri*

1. **Hermannia grandistipula** (Buching. ex Harv.) K. Schum. in Engl., Mon. Afr.
  Pflanz. **5**: 63 (1900).—Burtt Davy, F.P.F.T. **1**: 264 (1926). TAB. **101** fig. A.
  Type from Natal.
    Mahernia grandistipula Buching. ex Harv. in Harv. & Sond., F.C. **1**: 209 (1860).
  Type as above.

Stellately pilose shrublet with annual stems up to c. 30 cm. tall arising from a
woody rootstock. Leaf-lamina up to 4 × 1·8 cm., oblong to obovate-oblong, apex
obtuse, margin dentate or crenate or subentire, base broadly cuneate, sparsely and
softly pilose on both surfaces; petiole up to 3 mm. long; stipules c. 1 × 0·3 cm.,
leafy, palmately 2–4-partite with lanceolate segments or entire and lanceolate.
Flowers usually in pairs, on a common peduncle in the upper axils; peduncle up
to 1·5 cm. long, densely pilose; pedicels similar, up to 0·7 cm. long; bracts
palmatifid with linear segments, densely pilose. Calyx up to 1·2 cm. in diam.,
depressed-globose, villous, inflated, with deltoid teeth c. 3 mm. long. Petals c.
1·0 × 0·3 cm., yellow, just exserted, narrowly oblong-spathulate, glandular-pubescent, margins inrolled in the lower ⅓. Stamens with cruciform-tuberculate
pilose filaments c. 4 mm. long; anthers c. 3·5 mm. long, pilose. Ovary subglobose,
densely pubescent; styles 4 mm. long, puberulous at the base. Calyx inflated
and persistent in fruit, enclosing the ripe capsule. Capsule 6–7 mm. in diam.,
globose, pubescent. Seeds almost black, c. 2 × 1·5 mm., compressed, hemispherical-
reniform.

Mozambique. LM: Namaacha, fl. 22.xii.1944, *Torre* 6942a (LISC ; SRGH).
Also in Natal, Cape Prov., Transvaal and Swaziland. Confined in our area to the open
grasslands of the Libombo Mts.

2. **Hermannia quartiniana** A. Rich., Tent. Fl. Abyss. **1**: 75 (1847).—Eyles in Trans.
  Roy. Soc. S. Afr. **5**: 418 (1916).—Burtt Davy, F.P.F.T. **1**: 264 (1926). TAB. **101**
  fig. B. Type from Ethiopia.
    Mahernia abyssinica Hochst. ex Harv. in Harv. & Sond., F.C. **1**: 216 (1860).—
  Mast. in Oliv., F.T.A. **1**: 234 (1868). Type from Ethiopia.
    Hermannia abyssinica (Hochst. ex Harv.) K. Schum. in Engl. & Prantl, Nat.
  Pflanzenfam. **3**, 6: 80 (1895); in Engl., Mon. Afr. Pflanz. **5**: 68, t. 3 fig. A (1900).—

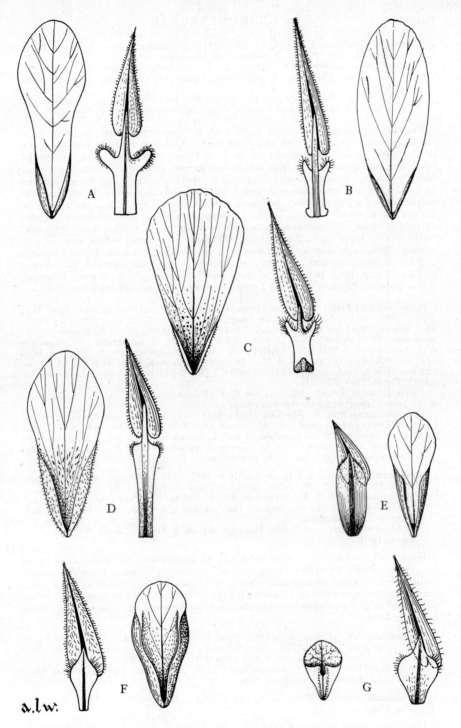

a.l.w.

Tab. 101. PETALS AND STAMENS OF HERMANNIA SPP. (all ×6). A.—H. GRANDISTIPULA, *Gomes e Sousa* 383. B.—H. QUARTINIANA, *Wild* 4732. C.—H. DEPRESSA, *Drummond* 4890. D.—H. STAUROSTEMON, *Wild* 4670. E.—H. FLORIBUNDA, *Eyles* 3878. F.—H. TOMENTOSA, *Rogers* 6043. G.—H. GUERKEANA, *Erens* 330.

Exell & Mendonça, C.F.A. **1, 2**: 196 (1951).—Suesseng. in Proc. & Trans. Rhod. Sci. Ass. **43**: 106 (1951). Type as for *M. abyssinica*.
  Hermannia adenotricha K. Schum. in Notizbl. Bot. Gart. Berl. **2**: 306 (1899); in Engl. Mon. Afr. Pflanz. **5**: 68 (1900).—Burtt Davy, loc. cit. Type from the Transvaal.
  ? Hermannia rhodesiaca Engl., Bot. Jahrb. **39**: 587 (1907). Type: Bechuanaland Prot., between Machoda and Palapye, *Engler* 2909a (B, holotype †).

Prostrate spreading perennial herb with its vegetative parts stellately pubescent or sometimes also with glandular hairs; branches sparsely pubescent, up to c. 50 cm. long. Leaf-lamina up to 7 × 2·4 cm., narrowly oblong to oblong-lanceolate, apex obtuse, margin serrate or crenate sometimes coarsely so, base broadly cuneate, pubescent on both surfaces; petiole up to 7 mm. long; stipules up to 1·1 × 0·5 cm., lanceolate-acuminate, variously incised or entire. Flowers usually paired on axillary peduncles; peduncle slender, up to 2 cm. long; pedicels up to 1·3 cm. long, slender; bracts c. 3 × 1·5 mm., lanceolate, connate below. Calyx up to 7 × 7 mm., campanulate, pubescent with simple or stellate hairs, with narrowly deltoid lobes up to 5 mm. long. Petals c. 8 × 3 mm., yellow, pink or red, narrowly elliptic to elliptic. Stamens with cruciform-tuberculate pilose filaments c. 2·5 mm. long; anthers c. 6 mm. long, pubescent. Ovary ellipsoid, pubescent; styles c. 6 mm. long, pubescent in the lower ⅔. Capsule c. 8 × 6 mm., ellipsoid, thinly pubescent. Seeds c. 1·5 × 1 mm., dark grey-brown, subreniform.

Bechuanaland Prot. N: 98 km. from Nata on Maun road, fl. 9.ix.1954, *Story* 4631 (PRE). SW: 98 km. from Ghanzi on Maun road, fl. 26.vii.1955, *Story* 5048 (K; PRE). SE: between Machoda and Palapye, fl. 7.ix.1905, *Engler* 2909a (B†). **N. Rhodesia.** B: Nangweshi, fl. 28.vii.1952, *Codd* 7216 (K; PRE; SRGH). **S. Rhodesia.** N: Shawanoe R., fl. 14.i.1937, *Eyles* 8933 (K; SRGH). W: Wankie Game Reserve, fl. 15.ii.1956, *Wild* 4732 (K; SRGH). C: Hartley, fl. 20.iii.1956, *Hornby* 3158 (K; SRGH). E: Moosgwe, 128 km. S. of Umtali, fl. 19.v.1935, *Eyles* 8423 (K; SRGH). S: Gwanda, Sezane, fl. v.1955, *Davies* 1289 (SRGH).
  Also in Ethiopia, Eritrea, Angola, Cape Prov., Transvaal and SW. Africa. Commonest in the grassy areas bordering seasonal swamps or vleis.
  Two specimens from S. Rhodesia (*Eyles* 8423 (K; SRGH) from Moosgwe in the Chipinga Distr. and *Godman* 169 (BM) from Zimbabwe) match the type of *H. adenotricha* K. Schum. which has glandular as well as stellate hairs. A whole range of intermediates exists, however, between the types of *H. quartiniana* and *H. adenotricha* so this glandular form is included here in *H. quartiniana*.

3. **Hermannia depressa** N.E.Br. in Kew Bull. **1897**: 245 (1897).—K. Schum. in Engl., Mon. Afr. Pflanz. **5**: 70 (1900).—Eyles in Trans. Roy. Soc. S. Afr. **5**: 418 (1916).—Burtt Davy, F.P.F.T. **1**: 265 (1926). TAB. **101** fig. C. Type from the Transvaal.
  Mahernia erodioides var. *latifolia* Harv. in Harv. & Sond., F.C. **1**: 214 (1860). Type as for *H. depressa*.
  Mahernia quartiniana sensu Hopkins, Bacon & Gyde, Comm. Veld Fl.: 74, fig. 4 (1940).

Habit and general appearance similar to *H. quartiniana* but leaf-lamina up to 4 × 2·5 cm., ovate to ovate-oblong, margin irregularly crenate, rounded to subcordate, subglabrous and often purplish or bronzy above. Petals c. 8 × 4 mm., orange-yellow or reddish and somewhat broader than in *H. quartiniana*, and calyx, ovary and petal-bases glandular (as also in some forms of *H. quartiniana*). Style c. 3 mm. long.

S. Rhodesia. N: Mazoe, fl. ix.1906, *Eyles* 423 (BM; SRGH). W: Bulawayo, fl. ix.1945, *Martineau* 791 (SRGH). C: Salisbury, fl. 6.x.1955, *Drummond* 4890 (K; SRGH). E: Umtali, fl. 11.x.1948, *Chase* 1619 (BM; K; SRGH).
  Also in Cape Prov., Natal, Transvaal and Orange Free State. Open grassland, often at the margins of seasonal swamps or vleis.

4. **Hermannia parvula** Burtt Davy, F.P.F.T. **1**: 42, 266 (1926). Type from the Transvaal.

Dwarf perennial herb c. 8 cm. tall with a woody rootstock; stems sparsely stellate-pubescent and with a few minute glands. Leaf-lamina up to 23 × 4 mm., linear to oblong-linear, apex acute, base cuneate, margin serrate, sparsely stellate-pubescent on both sides and with a few minute glands; petiole 1–3 mm. long, sparsely stellate-pubescent and glandular; stipules 3–5 mm. long, lanceolate, obliquely falcate, sparsely stellate-pubescent and glandular. Flowers in pairs in the upper axils; peduncles 8–10 mm. long, slender, stellate-pubescent; bracts 1–3 mm. long, linear-lanceolate, stellate-pubescent and glandular; pedicels up to 10 mm. long, stellate-pubescent. Calyx 4 mm. long, campanulate, stellate-pubescent and glandular; lobes 1·5 mm. long, triangular-acuminate. Petals orange, c. 4 × 2 mm., obovate, margins scarcely inrolled, puberulous near the base. Stamens with cruciform-tuberculate pilose filaments c. 1 mm. long; anthers c. 2 mm. long, pubescent. Ovary obovoid, glandular-puberulous; styles c. 1·7 mm. long, glabrous. Ripe capsules not seen.

Bechuanaland Prot. SE: 3 km. S. of Lobatsi, fl. 17.i.1960, *Leach & Noel* 153 (SRGH).
Also in the Transvaal. Bush steppe or bush savanna species.

5. **Hermannia staurostemon** K. Schum. in Notizbl. Bot. Gart. Berl. **2**: 305 (1899); in Engl., Mon. Afr. Pflanz. **5**: 76, t. 3 fig. E (1900).—Burtt Davy, F.P.F.T. **1**: 266 (1926). TAB. **101** fig. D. Type from the Transvaal.

Much-branched shrublet up to 40 cm. tall, greyish-stellate-tomentellous with some glandular hairs on all its vegetative parts. Leaf-lamina up to 8 × 2·2 cm., narrowly oblong-elliptic to lanceolate, apex acute, base cuneate, margin entire or serrate, tomentellous on both surfaces; petiole up to c. 6 mm. long; stipules up to 2 × 0·8 cm. but usually smaller, oblong-lanceolate, acute at the apex, with 3 nerves from the base. Flowers (1) 3–5 in axillary pseudo-umbels; peduncle c. 4 mm. long, tomentellous; pedicels similar, c. 4 mm. long; bracts up to 1 × 0·5 cm., solitary, tomentellous, lanceolate, entire, sparsely toothed or 2–3-partite. Calyx 7–10 × 6 mm., campanulate, lobed about half-way with acute narrowly triangular lobes, greyish-stellate-tomentellous but densely glandular also. Petals 0·8–1·2 × 0·4–0·6 mm., yellow, oblong-spathulate, densely pubescent in the lower half, the margins inrolled in the lower half. Stamens with cruciform-tuberculate pilose filaments 5–6 mm. long; anthers c. 4 mm. long, pubescent. Ovary shortly stipitate, oblong-obovoid, densely glandular; styles c. 4 mm. long, pubescent near the base. Ripe capsule not known.

S. Rhodesia. E: Inyanga, fl. 28.v.1954, *Chase* 5259 (BM; K; SRGH).
Also in the Transvaal and Natal. Submontane ericoid scrub or by mountain streams.
The S. Rhodesian examples of this species have more noticeably serrate leaves than those from the Transvaal and Natal and usually have rather smaller flowers. In floral structure, however, they match very well.

6. **Hermannia comosa** Burch. ex DC., Prodr. **1**: 493 (1824).—Harv. in Harv. & Sond., F.C. **1**: 164 (1860).—K. Schum. in Engl., Mon. Afr. Pflanz. **5**: 60 (1900).—Burtt Davy, F.P.F.T. **1**: 267 (1926). Type from Cape Province.

Perennial herb c. 20 cm. tall; branches densely stellate-tomentose. Leaf-lamina c. 3 × 2 cm., ovate-oblong, apex acute, base broadly cuneate, margin irregularly serrate, crenate or undulate, tomentose on both sides, ± discolorous, appearing ± plicate due to nerves being ± deeply immersed above and raised below; petiole up to c. 3 cm. long, tomentose; stipules up to 5 mm. long, subulate, tomentose. Flowers in pairs in the axils of the upper leaves; peduncles c. 1–1·5 cm. long, tomentose; pedicels up to 5 mm. long, tomentose; bracts and bracteoles 2–5 mm. long, subulate, tomentose. Calyx up to c. 1 cm. in diam., inflated, globose-campanulate, toothed about $\frac{1}{5}$–$\frac{1}{3}$ of the way down, tomentose, often with stalked stellate hairs or appendages on the angles. Petals yellow, slightly exceeding the calyx, spathulate with the margins infolded in the lower half, stellate-pubescent on both sides. Stamens with narrowly obovate filaments c. 4 mm. long and pubescent on the shoulders; anthers c. 4 mm. long, pubescent. Ovary subglobose, shortly stipitate, apex truncate with 5 blunt processes, tomentose; styles 5 mm. long, puberulous near the base. Ripe capsules not seen.

Bechuanaland Prot. SW: Tsabong, 25.ii.1960, *Wild* 5145 (BOL; SRGH).
Widely distributed also in the Northern Cape, Transvaal, Orange Free State and SW. Africa. Confined to low-rainfall areas.
The Bechuanaland form is densely tomentose but outside our area much less tomentose forms are to be found.

7. **Hermannia floribunda** Harv. in Harv. & Sond., F.C. **1**: 201 (1860).—K. Schum. in Engl., Mon. Afr. Pflanz. **5**: 56 (1900).—Burtt Davy, F.P.F.T. **1**: 267, fig. 409 (1926).—Suesseng. in Proc. & Trans. Rhod. Sci. Ass. **43**: 107 (1951). TAB. **101** fig. E. Type from the Transvaal.

Small branching shrub c. 60 cm. tall; young branches densely stellate-pubescent or tomentose. Leaf-lamina up to c. 6 × 5 cm., ovate, apex obtuse, margin irregularly crenate, base truncate or subcordate, stellate-pubescent above, tomentose and paler beneath; petiole up to 2·5 cm. long, densely pubescent; stipules up to c. 1 cm. long, subulate to lanceolate-subulate, stellate-pubescent. Flowers 1–2 or more together on axillary peduncles in the upper axils (particularly near the apices of the stems the peduncles may be several per axil and a terminal paniculate inflorescence produced); peduncle 1–3 cm. long, stellate-pubescent; pedicels similar but slightly longer; bracts 2–3 mm. long, subulate, pubescent. Calyx c. 6 × 6 mm., broadly campanulate, densely stellate-pubescent, lobed about ⅓ of the way down; lobes broadly triangular-acuminate. Petals yellow turning orange at the tips, slightly exceeding the calyx, oblong-spathulate. Stamens with spathulate-oblong filaments, c. 3 × 0·5 mm., ciliate on the shoulders; anthers c. 2·5 mm. long, pubescent. Ovary oblong-ovoid, densely tomentose, shortly stipitate; styles c. 3 mm. long, sparsely pubescent in the lower half. Calyx somewhat inflated and membranous in fruit. Capsule c. 6 mm. in diam., depressed-globose, pubescent. Seeds almost black, c. 1·5 × 1 mm., subreniform.

S. Rhodesia. W: Matobo, fl. v.1955, *Miller* 2875 (SRGH). C: Marandellas, Cave Farm, fl. 5.iv.1950, *Wild* 3251 (K; SRGH). E: Umtali, fl. & fr. 21.v.1950, *Chase* 2882 (BM; SRGH).
Also in the Transvaal and Cape Prov. Often in habitats protected from fire and grazing, e.g. among boulders on rocky outcrops in *Brachystegia* woodland.

8. **Hermannia tomentosa** (Turcz.) Schinz ex Engl., Bot. Jahrb. **55**: 371 (1919).— Burtt Davy, F.P.F.T. **1**: 267 (1926).—O. B. Mill. in Journ. S. Afr. Bot. **18**: 57 (1952). TAB. **101** fig. F. Type from the Transvaal.
    Mahernia tomentosa Turcz. in Bull. Soc. Imp. Nat. Mosc. **31**: 218 (1858).— Harv. in Harv. & Sond., F.C. **1**: 219 (1860). Type as above.
    Hermannia brachypetala Harv. in Harv. & Sond., tom. cit.: 202 (1860).—K. Schum. in Engl., Mon. Afr. Pflanz. **5**: 87, t. 3 fig. C (1900).—Eyles in Trans. Roy. Soc. S. Afr. **5**: 418 (1916). Syntypes from the Transvaal.

Spreading or prostrate perennial herb or shrublet, greyish-stellate-tomentellous on all its vegetative parts; branches up to c. 40 cm. long. Leaf-lamina up to 5 × 1·5 cm., narrowly elliptic-oblong or narrowly obovate-oblong, apex obtuse or subacute, base cuneate, margins finely serrate, nerves somewhat impressed above and prominent below, tomentellous on both surfaces; petioles up to 1 cm. long; stipules up to 3 mm. long, subulate. Flowers reflexed, 1 or rarely 2 together on slender axillary peduncles up to 3 cm. long; pedicels 2–3 mm. long; bracts c. 2 mm. long, subulate. Calyx 6–7 × 7 mm., with spreading or recurved narrowly triangular lobes up to 5 mm. long, campanulate, tomentellous or densely pubescent outside. Petals c. 4 × 1·75 mm., creamy-yellow, spoon-shaped with inrolled margins to both the claw and blade. Stamens with narrowly obovate filaments c. 2 mm. long, pubescent on the shoulders; anthers c. 6·5 mm. long, pubescent. Ovary oblong-ovoid, tomentose; styles c. 6 mm. long, puberulous except in the upper ⅓. Capsule c. 6 × 6 mm., obovoid, rounded or crowned with 5 short blunt truncate processes, tomentose or densely pubescent. Seeds brown, 1·5 × 1 mm., reniform, with a few radiating wrinkles.

Bechuanaland Prot. N: Tsao, fl. 4.viii.1955, *Story* 5104 (K; PRE). SE: Gaberones, fl. 10.iii.1930, *van Son* in Herb. Transv. Mus. 29010 (BM; K; PRE; SRGH). **S. Rhodesia.** W: Matopos, fl. iv.1934, *Brain* 10591 (K; SRGH).
Also in Cape Prov., Orange Free State, Natal, Transvaal and SW. Africa. Drier types of woodland, bushland or at the margins of seasonal swamps to the west of our area.

9. **Hermannia guerkeana** K. Schum. in Verh. Bot. Verein. Brand. **30**: 231 (1888); in Engl., Mon. Afr. Pflanz. **5**: 57 (1900). TAB. **101** fig. G. Type from SW. Africa (Amboland).

Small shrublet very near *H. tomentosa* but having an erect habit, a more yellowish indumentum and, most important diagnostically, petals 2·5–4 × 1·4–2 mm., obovate-cuneate, truncate and often retuse at the apex and with the margins not inrolled in the upper part but distinctly so in the lower half. Capsule and seeds not seen in a perfectly mature state but apparently similar to those of *H. tomentosa*.

Bechuanaland Prot. N: Maun, fl. 30.vi.1937, *Erens* 330 (K; PRE; SRGH). Also in SW. Africa. River banks and the dunes of dry river beds.

10. **Hermannia angolensis** K. Schum. in Warb., Kunene-Samb.-Exped. Baum: 73, 302 (1903).—Exell & Mendonça, C.F.A. **1**, 2: 195 (1951). TAB. **102** fig. A. Type from Angola.

Prostrate perennial herb similar to *H. tomentosa* but with its leaf-lamina up to 4 ×2 cm., oblong-ovate, margin often coarsely serrate or crenate and the indumentum appressed-yellowish-stellate-pubescent. Young branchlets densely yellowish-hispidulous like the ovary, and capsule with much longer hairs than in *H. tomentosa*. Petals similar in shape and size to those of *H. tomentosa* but pink rather than yellow. Seeds as in *H. tomentosa*.

Caprivi Strip. Kwando R., fl. x.1945, *Curson* 1203 (PRE). **Bechuanaland Prot.** N: 6 km. from Nata R. on Francistown Rd., fl. & fr. 24.iv.1957, *Seagrief* 2454 (K; SRGH). SW: Ghanzi, fl. 25.vii.1955, *Story* 5036 (K; PRE). **N. Rhodesia.** B: near Senanga, fl. 2.viii.1952, *Codd* 7355 (K; PRE; SRGH). **S. Rhodesia.** W: Wankie Game Reserve, fl. & fr. 15.ii.1956, *Wild* 4733 (K; SRGH).
Also in Angola and SW. Africa. Dry woodland or in grassland at the edge of seasonal swamps.

Story describes the leaves of this species as " crystalline pale green " obviously a very good description of the appearance created by the discrete yellowish stellate hairs each with a solid centre seen against the green background of the leaf in *H. angolensis* and very different from the continuous greyish tomentum of the leaves of *H. tomentosa*.

11. **Hermannia micropetala** Harv. in Harv. & Sond., F.C. **1**: 201 (1860).—K. Schum. in Engl., Mon. Afr. Pflanz. **5**: 58 (1900). TAB. **102** fig. B. Type: Mozambique, Delagoa Bay, *Forbes* 22 (K).
  Hermannia phaulochroa K. Schum. in Notizbl. Bot. Gart. Berl. **2**: 303 (1899). Syntypes: Mozambique, Lourenço Marques, *Schlechter* 11576 (B†; BM; COI; K); Delagoa Bay, *Junod* 29 (Z).

Erect branching perennial herb or shrublet, c. 1 m. tall; young branches tomentose, later glabrescent. Leaf-lamina up to 4·2 ×1·8 cm., narrowly oblong to narrowly oblong-obovate, apex obtuse, margins entire or coarsely serrate especially towards the apex, base broadly cuneate, thinly stellate-pubescent above, densely so beneath; petiole up to 8 mm. long, densely pubescent; stipules c. 2 mm. long, narrowly deltoid, stellate-pubescent. Flowers borne singly in the upper axils; peduncle slender, up to 1·7 cm. long, densely stellate-pubescent; pedicel similar, up to 6 mm. long; bracts c. 0·5 mm. long, deltoid, stellate-pubescent. Calyx c. 7 ×6 mm., campanulate, densely stellate-pubescent; lobes c. 5 mm. long, narrowly triangular-acuminate. Petals pink, c. 2·5 ×1·5 mm., obovate-cuneate, apex truncate, crenate, margins inrolled in the lower half and subauriculate at the apex of the inrolled portion. Stamens with obovate filaments c. 3·5 mm. long, pubescent on the shoulders; anthers c. 6 mm. long, pubescent. Ovary oblong-ovoid, truncate above, stellate-tomentose; styles c. 3·5 mm. long, puberulous near the base. Capsule c. 7 ×7 mm., oblong-ovoid, apex truncate with 5 short blunt processes, stellate-tomentose. Seeds dark-brown, c. 1·5 ×1 mm., subreniform, minutely granulate and wrinkled.

Mozambique. MS: Chiloane, fl. 12.viii.1887, *Scott* (K). SS: Gaza, Manjacaze, fl. 7.xii.1944, *Mendonça* 3310 (K; LISC; SRGH). LM: Lourenço Marques, fl. viii.1886, *Bolus* 1109 (BM; BOL; K).
Apparently endemic in southern Mozambique. Sandy soils in coastal areas.

12. **Hermannia boraginiflora** Hook., Ic. Pl. **6**: t. 597 (1843).—Harv. in Harv. & Sond., F.C. **1**: 201 (1860).—K. Schum. in Engl., Mon. Afr. Pflanz. **5**: 81 (1900).— Eyles in Trans. Roy. Soc. S. Afr. **5**: 418 (1916) pro parte quoad spec. Kolbe.— Burtt Davy, F.P.F.T. **1**: 267 (1926). TAB. **102** fig. C. Type from the Transvaal.

Perennial herb or shrublet, up to 0·6 m. tall; branches pubescent with glandular hairs often mixed with longer stellate hairs. Leaf-lamina up to 3 ×1 cm., narrowly oblong to obovate-oblong or elliptic-oblong, apex obtuse or acute, margin

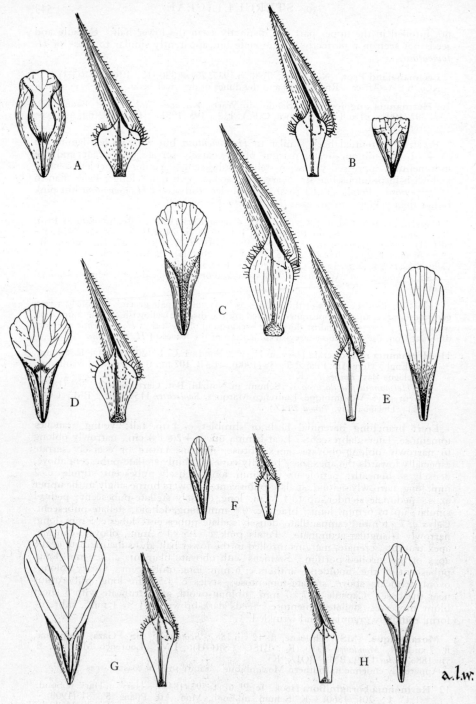

a.l.w.

Tab. 102. PETALS AND STAMENS OF HERMANNIA SPP. A.—H. ANGOLENSIS (×6) *Kolbe*
3161. B.—H. MICROPETALA (×6) *Forbes* 22. C.—H. BORAGINIFLORA (×6) *Kolbe*
3179. D.—H. GLANDULIGERA (×6) *Pedrógão* 326. E.—H. TORREI, petal (×4), stamen
(×6). F.—H. TIGREENSIS (×6) *Chase* 4375. G.—H. KIRKII (×4) *Lugard* 145.
H.—H. MODESTA (×6) *Davies* 67.

serrate, at least towards the apex, base cuneate or rounded, thinly stellate-pubescent and glandular above, more densely so beneath; petiole up to 7 mm. long, glandular and stellate-pubescent; stipules c. 1 mm. long, deltoid-acuminate, glandular and stellate-pubescent. Flowers on 1-flowered thinly stellate-pubescent and glandular axillary peduncles up to 1·5 cm. long; pedicel similar, 2–4 mm. long; bracts like the stipules. Calyx c. 7 × 5 mm., campanulate, stellately pubescent with some glandular hairs; lobes c. 5 mm. long, narrowly triangular. Petals c. 5 × 2 mm., pink, oblong-obovate, margins inrolled in the lower third, pubescent within the inrolled portion. Stamens with narrowly obovate filaments, c. 3·5 mm. long, pilose on the shoulders; anthers c. 5·5 mm. long, purple, connivent, exserted, pubescent. Ovary obovoid-oblong, tomentose; styles c. 4 mm. long, pubescent except near the apex. Capsule c. 8 × 5 mm., oblong-ovoid, truncate at the apex with 5 spreading horns 2 mm. long, tomentose. Seeds dark brown, c. 1·5 × 1 mm., reniform with radiating wrinkles; testa minutely granulated.

Bechuanaland Prot. SE: Mochudi, fl. v.1914, *Rogers* 6485 (BM). **S. Rhodesia.** W: Shangani, Gwampa Forest Reserve, fl. vii.1955, *Goldsmith* 162/55 (K; SRGH). E: Chipinga, Giriwayo, fl. & fr. 13.iii.1957, *Phipps* 601 pro parte (SRGH). S: Nuanetsi, fl. x.1955, *Davies* 1624 (K; SRGH). **Mozambique.** LM: Goba-Fronteira, fl. & fr. 13.xii.1947, *Barbosa* 717 (BM; K; LISC; SRGH).
Also in the Transvaal, Natal and the northern Cape Prov. Drier types of woodland, often accompanying *Colophospermum mopane* or *Acacia* spp.

One specimen, *Wild* 3426 (K; SRGH) from the Sabi-Lundi Junction in S. Rhodesia differs from typical *H. boraginiflora* in having smaller narrower leaves and some glands on the ovary. I do not consider, however, that these minor differences make it more than a form of this species.

13. **Hermannia glanduligera** K. Schum. in Verh. Bot. Verein. Brand. **30**: 232 (1888); in Engl., Mon. Afr. Pflanz. **5**: 57 (1900). TAB. **102** fig. D. Type from SW. Africa.
  Hermannia inamoena K. Schum. in Engl., Pflanzenw. Ost-Afr. **C**: 270 (1895); in Engl., Mon. Afr. Pflanz. **5**: 56 (1900). Type: Nyasaland, without precise locality, *Buchanan* 1251 (B, holotype †).
  Hermannia nyasica Bak. in Kew Bull. **1897**: 245 (1897).—K. Schum. in Engl., tom. cit.: 82 (1900).—N.E.Br. in Kew Bull. **1909**: 94 (1909). Type: Nyasaland, Monkey Bay, *Whyte* (B†; K, holotype).
  Hermannia cyclophylla K. Schum. in Notizbl. Bot. Gart. Berl. **2**: 303 (1899); in Engl., tom. cit.: 82 (1900) e descript. Type: Mozambique, Quelimane, *Peters* (B, holotype, †).
  Hermannia viscosa sensu Bak. f. in Journ. of Bot. **37**: 426 (1899).—Eyles, loc. cit.—Burtt Davy, F.P.F.T. **1**: 267 (1926).—Martineau, Rhod. Wild Fl.: 54 (1954).
  Hermannia viscosa var. *randii* Bak. f. in Journ. of Bot. **39**: 128 (1901).—Eyles in Trans. Roy. Soc. S. Afr. **5**: 419 (1916) (" viscida "). Type: S. Rhodesia, Bulawayo, *Rand* 295 (BM, holotype).
  Hermannia kirkii sensu Bak. f. in Journ. Linn. Soc., Bot. **40**: 29 (1911).
  Hermannia boraginiflora sensu Eyles, tom. cit.: 418 (1916) pro parte quoad specim. Chubb.

Perennial herb or shrublet, up to c. 1 m. tall, viscid-pubescent with glandular hairs on all parts except the corolla but also with long simple hairs and stellate hairs. Leaf-lamina up to 5·5 × 2·3 cm. but usually less, lanceolate to narrowly oblong-lanceolate, apex acute or subacute, margin serrate, base cuneate or rounded, pilose on both surfaces with appressed stellate hairs and some glandular hairs; petiole up to 1·4 cm. long; stipules 1–2 mm. long, lanceolate-subulate. Flowers solitary in the upper axils; peduncle slender, up to 1·3 cm. long; pedicel c. 3 mm. long; bracts c. 0·5 mm. long, deltoid-subulate. Calyx 5–6 × 5–6 mm., campanulate, lobed ⅔ of the way; lobes narrowly triangular with long simple hairs and shorter glandular hairs. Petals c. 4–5 × 3 mm., pink, red or rarely whitish, obovate, apex denticulate, retuse or truncate, margin subauriculate and inrolled in the lower half, pilose across the opening created by this inrolling. Stamens with obovate filaments c. 2·5 mm. long, pilose on the shoulders; anthers 5–6 mm. long, pubescent. Ovary shortly stipitate, narrowly obovoid, densely glandular with a few hispidulous hairs at the apex and on the ribs; styles c. 5 mm. long, pubescent and glandular except near the apex. Capsule c. 8 × 7 mm., oblong-obovoid, densely glandular but hispidulous on the ribs, with short spreading horns c. 1·5 mm. long at the apex. Seeds c. 1·5 × 1 mm., subreniform, wrinkled and granulated.

Caprivi Strip. Kwando R., fl. & fr. x.1945, *Curson* 1239 (PRE). **Bechuanaland Prot.** N: Okovango valley, fl. vi.1898, *Lugard* 236 (K). **N. Rhodesia.** B: Sioma Mission, Zambezi R., fl. 29.iv.1925, *Pocock* 1736 (PRE). C: Rufunsa, fl. & fr. 19.ix.1953, *Fanshawe* 289 (K; SRGH). S: Makoli, fl. vi.1909, *Rogers* 8282 (K; SRGH). **S. Rhodesia.** N: Sebungwe, Zambezi R., fl. & fr. ix.1955, *Davies* 1471 (K; SRGH). W: Bula-wayo, Khami, fl. & fr. 25.iv.1946, *Wild* 1064 (K; SRGH). C: Gwelo, fl. & fr. iv., *Eyles* 6963 (K; SRGH). E: Umtali, fl. iv.1947, *Chase* 345 (BM; K; SRGH). S: Sabi-Lundi Junction, Chitsa's Kraal, fl. & fr. 4.vi.1950, *Chase* 2292 (SRGH). **Nyasaland.** N: Nyika Plateau, Mwanemba, fl. ii.1903, *McClounie* 73 (K). C: Salima Bay, fl. & fr. 23.ix.1935, *Galpin* 15043 (K; PRE). S: W. of Fort Johnston, fl. & fr. 14.iii.1955, *E.M. & W.* 862 (BM; LISC; SRGH). **Mozambique.** Z: Massingire, Megaza, fl. 2.x.1944, *Mendonça* 2324 (K; LISC; SRGH). T: Boroma, fl. & fr. 9.vii.1950, *Chase* 2657 (K; SRGH). MS: Espungabera, fl. 22.vii.1949, *Pedro & Pedrógão* 7623 (LMJ; SRGH). SS: Massengena, fl. & fr. 22.vii.1932, *Smuts* 351 (K; PRE). LM: Goba, fl. 22.iv.1944, *Torre* 6510 (LISC; SRGH).
Also in Angola, SW. Africa and the Transvaal. River banks and open woodland in the drier parts of our area, and also as a weed of cultivation.

This common species has been identified in various publications and herbaria with *H. viscosa* Hiern. The latter is, however, a rather rare plant confined to the Mossâmedes desert of Angola with differently shaped, glabrous petals. *Pedrógão* 326 (K; LMJ; PRE) and *Pedrógão* 376 (LMJ; PRE) from Sul do Save, belong to *H. glanduligera* but are peculiar in that they have short blunt sterile anthers. This may be due to disease or per-haps to a mutation.
Although the type of *H. inamoena* is destroyed and no duplicate can be traced at Kew, the British Museum or Edinburgh, there is no doubt that it belongs here.

14. **Hermannia torrei** Wild in Bol. Soc. Brot., Sér. 2, **33**: 38 (1959). TAB. **102** fig. E.
    Type: Mozambique, Guijà, Posto do Alto Changano, *Torre* 8071 (LISC, holotype).

Perennial herb or shrublet up to 50 cm. tall; branchlets at first stellate-tomen-tose with some glandular hairs, soon becoming pubescent, and purplish-brown. Leaf-lamina 0·7–1·5 × 0·4–1 cm., obovate to broadly obovate, apex truncate or rounded, margin coarsely serrate but entire towards the base, base broadly cuneate, shortly greyish-stellate-tomentose on both surfaces and with a few glands, nerves in 2–4 pairs somewhat impressed above and prominent beneath; petiole up to 8 mm. long, stellate-tomentose and glandular; stipules c. 2 mm. long, tomentose, ovate, acute. Flowers solitary in the upper axils, nodding; peduncle c. 1·2 cm. long, slender, stellate-tomentose, with 2 minute bracts near the apex. Calyx 6–7 × 6–7 mm., campanulate, densely stellate-pubescent and glandular; lobes c. 4·5 mm. long, narrowly triangular-acuminate. Petals c. 1 × 0·4 cm., red or purplish, nar-rowly obovate, margin inrolled in the lower ⅓, pubescent in the lower ⅓. Stamens with narrowly oblanceolate filaments c. 3·5 mm. long, pubescent on the shoulders; anthers c. 5 mm. long, pubescent. Ovary shortly stipitate, oblong-ovoid, minutely horned at the apex, stellate-tomentose and glandular; style c. 4 mm. long, puberulous in the lower ½. Capsule c. 8 × 6 mm., oblong-ovoid, with 5 spreading apical horns 0·5–1 mm. long, shortly stellate-tomentose and glandular. Seeds brown, c. 1·5 × 1 mm., subreniform, wrinkled.

Mozambique. SS: Guijà, Posto do Alto Changano, fl. & fr 14.vii.1948, *Torre* 8071 (LISC).
Endemic, as far as is known, to our area. Sandy brackish flats at river margins.

15. **Hermannia tigreensis** Hochst. ex A. Rich., Tent. Fl. Abyss. **1**: 74, t. 17 (1847).—Mast. in Oliv., F.T.A. **1**: 233 (1868) (" tigrensis ").—K. Schum. in Engl., Mon. Afr. Pflanz. **5**: 85, t. 4 fig. B (1900).—Exell & Mendonça, C.F.A. **1**, 2: 194 (1951).—Suesseng. in Proc. & Trans. Rhod. Sci. Ass. **43**: 107 (1951). TAB. **102** fig. F. Type from Ethiopia.
    Hermannia stenopetala K. Schum. in Notizbl. Bot. Gart. Berl. **2**: 304 (1899); in Engl., Mon. Afr. Pflanz. **5**: 86 (1900). Type: Nyasaland, without precise locality, *Buchanan* (B, holotype, †; K).

Branching annual herb up to c. 30 cm. tall, stellate-pubescent in all its vegetative parts with sometimes a few glandular hairs scattered on the young stems. Leaf-lamina up to 4 × 1·5 cm., lanceolate to ovate-lanceolate, apex acute, margin serrate, base rounded, thinly pubescent on both surfaces; petiole up to 5 mm. long; stipules 1·5–5 mm. long, subulate. Flowers solitary in the upper axils; peduncle slender, up to 2·5 cm. long; pedicel c. 2 mm. long; bract c. 0·5 mm. long, solitary, subulate. Calyx c. 4 × 3 mm., campanulate, with scattered simple and glandular hairs; lobes c. 2 mm. long, deltoid-acuminate. Petals 3–5 × 0·75–1 mm., yellow, narrowly oblanceolate, margin slightly inrolled in the lower half. Stamens with obovate filaments 1·5–2 mm. long, often entirely without hairs on the shoulders; anthers 1–1·5 mm. long, pubescent. Ovary globose, very shortly stipitate, glandular; styles 0·75–1·5 mm. long, puberulous at the base. Capsule 3 × 4 mm., ovoid-truncate, with short horns c. 0·5 mm. long, stellate-pubescent and with some short glandular hairs. Seeds c. 1·3 × 0·75 mm., brown, subreni-form, wrinkled, very minutely granular.

S. Rhodesia. C: Salisbury, Rumani, fl. & fr. 10.v.1948, *Wild* 2487 (K; SRGH). E: Umtali, fl. 23.ii.1952, *Chase* 4375 (BM; K; SRGH). **Nyasaland.** N: Fort Hill, fl. & fr. vii.1896, *Whyte* (K). C: Dedza Distr., Mua-Livulezi Forest Reserve, fl. 9.iii.1955, *E.M. & W.* 1048 (BM; LISC; SRGH). S: Blantyre, fl. & fr., *Buchanan* 547 (K). **Mozambique.** Z: R. Zambezi, fl. & fr. 1884–5, *Carvalho* (COI).

Also in Ethiopia, Sudan, Tanganyika, West Africa and Angola. Riverine fringes and in the soil pockets of rocky outcrops.

16. **Hermannia kirkii** Mast. in Oliv., F.T.A. **1**: 233 (1868).—K. Schum. in Engl., Mon. Afr. Pflanz. **5**: 84, t. 4 fig. E (1900). TAB. **102** fig. G. Syntypes: Mozam-bique, Sena, *Kirk* (K); R. Zambezi, Lupata, *Kirk* (K); Tete, *Peters* 6 (B†; K); Bechuanaland Prot., near Ghanzi, *Baines* (K).

 Hermannia tigreensis sensu Klotzsch in Peters, Reise Mossamb. Bot. **1**: 132 (1861).

 Hermannia filipes var. *elatior* K. Schum. in Verh. Bot. Verein. Brand. **30**: 235 (1888). Type from SW. Africa.

 Hermannia modesta var. *elatior* (K. Schum.) K. Schum. in Engl., Mon. Afr. Pflanz. **5**: 84 (1900).—Engl., Bot. Jahrb. **55**: 367 (1919). Type as for *H. filipes* var. *elatior*.

 Hermannia lugardii N.E.Br. in Kew Bull. **1909**: 94 (1909).—Burtt Davy, F.P.F.T. **1**: 268 (1926). Syntypes: Bechuanaland Prot., Kwebe Hills, *Lugard* 125 (K) and 145 (K).

 Hermannia holubii Burtt Davy, F.P.F.T. **1**: 42, 268 (1926). Type from the Transvaal.

Slender or much-branched annual herb up to 60 cm. tall; branches slender, pubescent with short glandular hairs and longer simple hairs. Leaf-lamina 1·5–4 × 0·3–1·6 cm., narrowly lanceolate to narrowly oblong-lanceolate, apex obtuse or acute, margin serrate, base rounded or cuneate, thinly appressed-stellate-pub-escent on both surfaces and also with glandular hairs in some specimens; petiole up to 1·5 cm. long, with glandular and simple hairs; stipules c. 2 mm. long, subulate, pubescent. Flowers solitary in the upper axils, on filamentous puber-ulous peduncles up to 3·5 cm. long; pedicel up to 8 mm. long, articulated by a slightly swollen joint to the peduncle, puberulous; bracts absent. Calyx 3–6·5 × 3–7 mm., campanulate, pubescent with simple and glandular hairs, lobed ⅔ of the way down; lobes ovate-acuminate. Petals pink, reddish-orange or purplish, 5·5–10 × 2·5–6·5 mm., narrowly obovate, margins slightly inrolled in the lower half, as long as or considerably longer than the calyx. Stamens with obovate filaments 3·5–4 mm. long, with hairs on the shoulders; anthers 3·5–6 mm. long, pubescent. Ovary oblong-ovoid, stipitate, glandular; styles 3–4 mm. long, puberulous in the lower half. Capsule 5–8 × 5–7 mm., truncate-ovoid to oblong-ovoid, thinly stellate-pubescent with some glands or glabrescent, with spreading horns at the apex up to 1·5 mm. long. Seeds c. 1·75 × 0·75 mm., brown, subreni-form, wrinkled and granulate.

Bechuanaland Prot. N: NE. tip of Lake Ngami, fl. 18.ix.1954, *Story* 4717 (PRE). SW: Ghanzi, fl. & fr. ii.1952, *de Beer* 48 (PRE; SRGH). SW: Takatshwane Pan, fl. 21.ii.1960, *Wild* 5110 (K; PRE; SRGH). SE: Kuke Pan, fl. & fr. 24.iii.1930, *van Son* in Herb. Transv. Mus. 29012 (BM; K; PRE; SRGH). **S. Rhodesia.** N: Mtoko,

Nyangombe R., fl. & fr. 11.viii.1950, *Whellan* 465 (K; SRGH). W: Wankie Game Reserve, fl. 15.ii.1956, *Wild* 4734 (K; SRGH). E: Sabi valley, Hot Springs, fl. 26.xii.1947, *Chase* 477 (BM; K; SRGH). S: Gwanda, fl. 15.xii.1956, *Davies* 2311 (K; SRGH). **Mozambique.** T: Ulandi, fl., *Guerra* 78 (COI). MS: Chemba, fl. 12.iv.1960, *Lemos & Macuácua* 67 (BM; LMJ).

Also in Kenya, Tanganyika, Angola, SW. Africa and the Transvaal. In our drier and hotter areas, occurring in woodland, the grassy margins of seasonal swamps and also as a weed of cultivation.

Engler, Schumann (loc. cit.) and others have considered this plant to be a variety of *H. modesta* but in my opinion the differences of leaf-shape, petiole-length, seed-colour and testa-pattern render such a concept untenable. Within *H. kirkii* itself there is considerable variation. The Lupata syntype of *H. kirkii* represents one form with smaller leaves, a more branching habit, petals scarcely exceeding the calyx and a more ovoid capsule. The syntypes of *H. lugardii* represent the other forms with larger leaves, less branching habit, petals much exceeding the calyx and a more elongate capsule. There are, however, plenty of intermediates and no geographical segregation of the two forms.

17. **Hermannia modesta** (Ehrenb.) Mast. in Oliv., F.T.A. **1**: 232 (1868).—K. Schum. in Engl., Mon. Afr. Pflanz. **5**: 83, t. 4 fig. C (1900).—Burtt Davy, F.P.F.T. **1**: 268 (1926). TAB. **102** fig. H. Type from Saudi Arabia.
 Trichanthera modesta Ehrenb. in Linnaea, **4**: 402 (1829). Type as above
 Hermannia filipes Harv. in Harv. & Sond. F.C. **1**: 206 (1860).—Eyles in Trans. Roy. Soc. S. Afr. **5**: 418 (1916). Type from S. Africa.

Slender or much-branched annual herb up to c. 60 cm. tall; branches slender, pubescent with short glandular hairs and longer simple hairs. Leaf-lamina 1–4 × 0·1–0·25 cm., strictly linear, apex obtuse, margin entire, thinly appressed-stellate-pubescent on both surfaces and also with glandular hairs in some specimens; petiole c. 2 mm. long, with glandular and simple hairs; stipules c. 2 mm. long, subulate, pubescent. Flowers solitary in the upper axils on filamentous puberulous peduncles up to 3·5 cm. long; pedicel up to 8 mm. long, articulated by a slightly swollen joint to the peduncle, puberulous; bracts absent. Calyx campanulate, 3–5 × 3–6 mm., pubescent with simple and glandular hairs, lobed ⅔ of the way down; lobes from lanceolate to ovate-acuminate. Petals c. 6 × 2·5 mm., pink, red or purplish, narrowly obovate, margins scarcely inrolled in the lower half. Stamens with obovate filaments c. 2·5 mm. long, without hairs on the shoulders; anthers c. 3 mm. long, pubescent. Ovary oblong-ovoid, stipitate, glandular; styles c. 2 mm. long, puberulous in the lower half. Capsule c. 8 × 6 mm., oblong-ovoid, thinly stellate-pubescent with some glands or glabrescent, with spreading horns at the apex c. 1 mm. long. Seeds c. 1·5 × 0·75 mm., almost black, corniform (arcuate and thicker at one end than the other), wrinkled and minutely tuberculate.

Bechuanaland Prot. SW: 464 km. NW. of Molepolole, fl. & fr. 5.vii.1955, *Story* 5002 (PRE). **S. Rhodesia.** W: Bulalima-Mangwe, Embakwe, fl. & fr. 28.iii.1942, *Feiertag* in GHS 45508 (K; SRGH). S: Gwanda, Sezane Reservoirs, fl. & fr. v.1955, *Davies* 1291 (K; SRGH).

Also in Arabia, Sudan, Egypt, Eritrea, Somaliland, Angola, SW. Africa, Cape Prov., Orange Free State and the Transvaal. In the driest types of woodland and bushland in our area.

18. **Hermannia oliveri** K. Schum. in Engl., Bot. Jahrb. **15**: 134 (1893); Engl., Mon. Afr. Pflanz. **5**: 52 (1900).—Brenan, T.T.C.L.: 600 (1949). Type from Kenya (Taveta).
 Mahernia exappendiculata var. *tomentosa* Oliv. in Trans. Linn. Soc., Ser. 2, Bot. **2**: 329 (1887) *nom. nud.*

Shrublet c. 1 m. tall; branches stellate-tomentose, eventually glabrescent. Leaf-lamina up to 4 (5·8) × 2 (4) cm., ovate-oblong to ovate, apex obtuse, margin crenate-dentate, stellate-pubescent above, stellate-tomentose below, nerves prominent beneath; petiole up to c. 1·7 cm. long, stellate-tomentose; stipules up to 2 (5) mm. long, stellate-tomentose. Flowers in ± diffuse panicles; branches of inflorescence stellate-tomentose; pedicels c. 5 (9) mm. long; bracts and bracteoles c. 2 mm. long, subulate, stellate-tomentose. Calyx c. 6 mm. long, campanulate, stellate-tomentose, lobed ⅗ of the way down; lobes narrowly lanceolate. Petals yellow, c. 6 mm. long, narrowly obovate, margins not inrolled, glabrous. Stamens with ± subulate, glabrous filaments c. 1·5 mm. long; anthers rather long-ciliate. Ovary oblong-ovoid, subsessile, stellate-tomentose, with 5 thickened longitudinal ribs; styles c. 3·5 mm. long, glabrous. Capsule c. 8 × 5 mm., oblong-ovoid, with thickened ribs, stellate-tomentellous, not horned at the apex. Mature seeds not seen.

S. Rhodesia. N: Darwin Distr., Chimanda Reserve, fl. 10.ix.1958, *Phipps* 1330 (SRGH).
Also in Kenya and NE. Tanganyika. Occurs at about 600 m. A rare plant in our area; little is known of its ecology.

Our material agrees very well with the type, differing only in that the calyx-lobes are somewhat longer and the indumentum of the inflorescence is less coarse. The apparent gap in the distribution of the species is remarkable.

6. TRIPLOCHITON K. Schum.

Triplochiton K. Schum. in Engl., Bot. Jahrb. **28**: 330 (1900) *nom. conserv.*

Trees with petiolate, palmately lobed leaves; stipules present. Flowers bisexual, in reduced axillary cymes or many-flowered panicles; bracts present and also 3 bracteoles forming a deciduous involucre beneath the calyx. Calyx campanulate with 5 valvate lobes. Petals 5, with a contorted aestivation, tomentose, with a basal claw. Androgynophore conspicuous. Stamens many; staminodes 5, ovate, with a contorted aestivation. Ovary of 5 coherent carpels and with 5 coherent styles; ovules c. 12, 2-seriate. Fruit of 5 samara-like, single-seeded carpels each with a unilaterally produced wing.

Triplochiton zambesiacus Milne-Redh. in Kew Bull. **1935**: 271 (1935). TAB. **103** fig. A. Type: S. Rhodesia, Wankie, *Eyles* 8295 (K, holotype; SRGH).
Sterculiacea—Mast. in Oliv., F.T.A. **1**: 239 (1868).

Large tree up to 18 m. tall with a straight bole and light-grey smooth flaking bark; branchlets slender, glabrous. Leaf-lamina up to 12 × 14 cm., palmately 5–9-lobed, cordate, lobes ovate to oblong, shortly acuminate at the apex, sparsely stellate-pubescent when young, later glabrous except sometimes for a few hairs at the base of the nerves on the lower surface; petiole up to 5 cm. long, sparsely stellate-pubescent or glabrous; stipules c. 7 mm. long, very caducous, subulate, stellate-pubescent. Flowers in reduced 1–4-flowered axillary or terminal cymes; pedicels c. 1·5 cm. long, stellate-pubescent; bracts up to 8 mm. long, caducous, ovate, obtuse, concave, sparsely stellate-pubescent; bracteoles c. 7 mm. long, broadly ovate, densely pubescent, forming an involucre round the base of the calyx. Calyx c. 2 cm. long, with deltoid acute lobes c. 8 × 7 mm., broadly funnel-shaped, stellate-tomentose on both sides. Petals 5, c. 3·5 × 2·5 cm., whitish or yellowish but deep red towards the base, broadly obovate, apex rounded or emarginate, clawed at the base, with an inconspicuous appendage c. 1 mm. long at the base of the lamina, pubescent on both surfaces of the lamina, densely hirsute inside just above the claw. Disk inconspicuous, annular. Androgynophore 9 mm. long, 3 mm. in diam., sulcate or angular, densely appressed-hirsute above. Stamens very numerous; filaments 6–8 mm. long, joined below in pairs or triads; staminodes 5, c. 6 × 4 mm., contorted, ovate to ovate-oblong, very concave, scarious, glabrous. Ovary ovoid, 5-angled and 5-locular, densely hirsute; style 2 mm. long, pubescent; ovules 10–12 per loculus. Fruit of 5 tomentose carpels, each with a unilateral, obovoid-oblong, stiffish, pubescent wing up to 7 × 2·8 cm.

N. Rhodesia. S: Gwembe valley, fl. 1932, *Macrae* 5 (K). **S. Rhodesia.** N: Urungwe, western end of Kariba Gorge, fr. 25.xi.1953, *Wild* 4259 (K; SRGH). W: Wankie, Deka R., fl. 25.xii.1952, *Lovemore* 345 (K; SRGH).
Endemic in our area. Often on termitaria but also on flood plains of silty sand and river banks.

A very fine shade-giving tree with dense foliage and large handsome flowers. The wood is hard and is used for yokes.

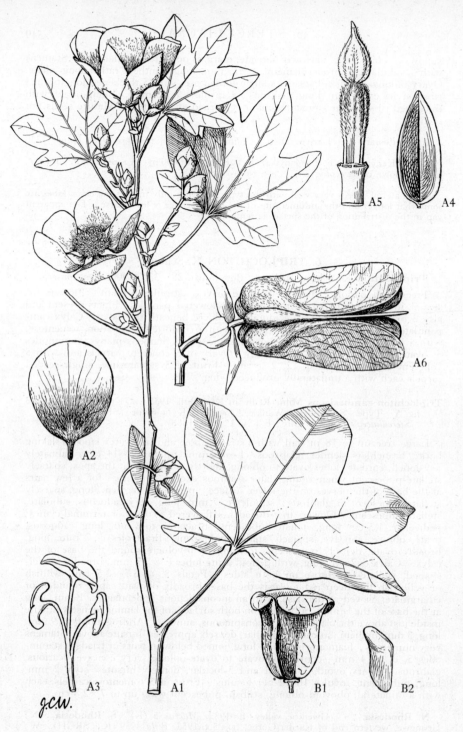

Tab. 103. A.—TRIPLOCHITON ZAMBESIACUS. A1, flowering branch (×⅔); A2, petal (×1);
A3, stamen triad (×4); A4, staminode (×6); A5, androphore and ovary (×2), all
from *Eyles* 8295; A6, fruit (×⅔) *Wild* 4259. B.—NESOGORDONIA PARVIFOLIA. B1,
fruit (×⅘); B2, seed (×⅘), both from *Drummond & Hemsley* 3602.

7. NESOGORDONIA Baill.

Nesogordonia Baill. in Bull. Soc. Linn. Par. **1**: 555 (1886).

Small or large trees, more or less stellate-pubescent at least on the young parts. Leaves petiolate, simple; stipules caducous, small. Flowers bisexual, solitary or axillary, in pedunculate, few-flowered cymes; pedicels articulated. Sepals 5, valvate. Petals 5, often somewhat twisted. Stamens in an uninterrupted ring or in fascicles of 2–4 opposite the sepals; filaments free or more or less united within the fascicles; and with an inner ring of 5 free fertile stamens or 5 free stami-nodes opposite the petals. Ovary obconic or subglobose, pubescent or tomentose, 5-locular; loculi 2-ovulate; styles 5, free or united into a single style; stigmas fleshy, papillose. Capsule obconic, truncate and excavated at the apex, with 5 rather thick wings, dehiscent into 5 valves, with 1–2 seeds per loculus. Seeds with a unilateral wing.

Nesogordonia parvifolia (M.B.Moss) Capuron in Notul. Syst. **14**, 4: 259 (1953). TAB. **103** fig. B. Type from Kenya (K, holotype).
  Cistanthera parvifolia M. B. Moss in Kew Bull. **1937**: 411 (1937). Type as above.

Tree up to 12 m. tall in our area but reaching 25 m. farther north; bark dark and rough; young branches stellate-pilose. Leaf-lamina up to 6·5 × 3·5 cm., oblong to oblong-lanceolate or narrowly ovate, apex subacute or narrowly obtuse, shortly mucronate, margin entire or undulate or crenate, base rounded, glabrous above except for the puberulous midrib, minutely stellate-pubescent below or glabrescent, with tufts of hairs in the nerve-axils, midrib prominent below, nerves in 6–7 pairs, scarcely prominent below; petiole up to 1·6 (2) cm., stellate-pubes-cent; stipules c. 5 mm. long, very caducous, subulate, pubescent. Flowers (description based on Kenya and Tanganyika material) solitary or in short reduced axillary few-flowered cymes; peduncle 6–8 mm. long, stellate-pubescent; pedicels somewhat shorter, articulated in the upper half. Sepals 8 × 2·5 mm., free, oblong-lanceolate, patent or reflexed, shortly stellate-pubescent outside, puberulous inside. Petals cream, 8 × 2 mm., oblong, glabrous. Stamens 10, in unequal pairs opposite the petals; filaments scarcely 1 mm. long, coherent in pairs; anthers 3–4 mm. long; staminodes 5, 6–7 mm. long, alternating with the stamen-pairs, linear-lanceolate. Ovary subglobose, minutely pubescent; style c. 1 mm. long; stigma fleshy, 5-lobed. Capsule 1·5 × 1 cm., obconic, widened and depressed in the centre at the apex. Seeds 3 × 2·5 mm., brown, smooth, ellipsoid, but flattened and somewhat excavated on one side, with a unilateral membranous obliquely oblong wing 5 × 3·5 mm.

Mozambique. Z: Maganja da Costa, fr. 27.ix.1949, *Barbosa & Carvalho* 4220 (LM; PRE).
  Also in Kenya and Tanganyika. In coastal evergreen forest.

A useful timber tree with pink heartwood. This species is very near *N. holtzii* (Engl.) Capuron from Tanganyika but differs in that the anther-filaments are connate in pairs. This character shows the affinity of our species with the Madagascar species of the genus.

8. STERCULIA L.

Sterculia L., Sp. Pl. **2**: 1007 (1753): Gen. Pl. ed. 5: 438 (1754).

Trees with entire, lobed or occasionally digitate leaves. Inflorescences pani-culate or rarely racemose, terminal or axillary, appearing before or with the leaves. Flowers bisexual* or unisexual, monoecious or dioecious. Calyx 4–5 (6)-lobed. Petals 0. Male flower: anthers c. 10–20 in a capitate or capitate-globose cluster on a slender androphore. Female flower: ovary of 4–5 coherent carpels with 2–∞ ovules per carpel; often borne on a short gynophore with vestigial anthers at its base; styles coherent; stigma peltate or 4–5-lobed. Carpels separating at maturity and becoming follicular, usually rather woody or leathery. Seeds 1–∞.
  The bast fibre of almost all African species is commonly used by natives for rope making.

* Flowers apparently bisexual have often only vestigial anthers and are thus functionally female (all our species fall into this category).

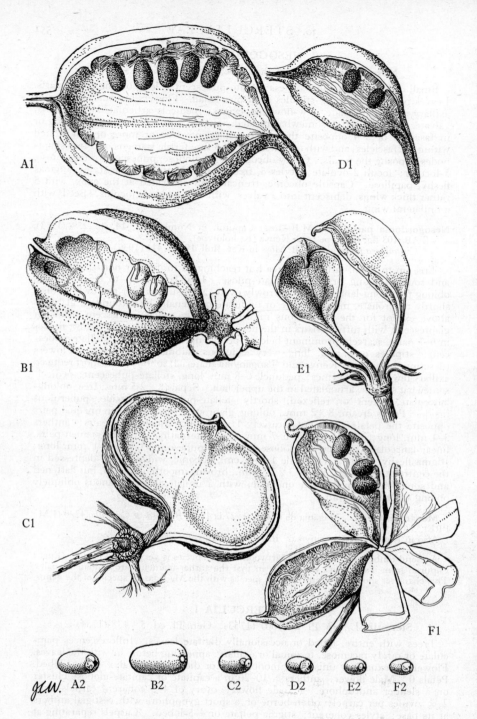

Tab. 104. FRUITS (A1, B1, etc.) AND SEEDS (A2, B2, etc.) OF STERCULIA SPP. (all natural size). A.—S. AFRICANA. B.—S. APPENDICULATA. C.—S. TRAGACANTHA. D.—S. ROGERSII. E.—S. QUINQUELOBA. F.—S. MHOSYA.

Leaves simple, entire or lobed:
  Leaves lobed or if scarcely lobed then orbicular or at least broadly ovate:
    Flowers c. 1 cm. long; follicles spreading, sessile or subsessile:
      Follicles broadly ovoid or ovoid-globose; flowers greenish, yellowish or yellowish
        with red guide-lines; branches of inflorescence ± palely tomentose:
        Follicles with a curled or spreading apical horn; seeds not embedded in a
          yellowish, dry pulp; leaf-lobes acuminate or obtuse or leaves subentire:
          Tree 10–25 m. tall with a stout trunk; leaves 5–15 × 4–13 cm., ± concolorous;
            follicles c. 10 cm. long - - - - - - 1. *africana*
          Tree up to 5 m. tall, often branching low down; leaves up to 6 × 5 cm.,
            whitish-tomentellous below; follicles up to 7·5 cm. long 2. *rogersii*
        Follicles scarcely apiculate; seeds embedded in a dry yellowish pulp; leaf-
          lobes often caudate-acuminate, basal pair subsagittate (leaves rarely sub-
          entire) - - - - - - - - - - 3. *appendiculata*
      Follicles oblong-cylindric; flowers dull red; branches of inflorescence with dark
        purplish glandular pubescence - - - - - - 4. *mhosya*
    Flowers up to 5 mm. long; follicles held suberect on stipes c. 1 cm. long
                          5. *quinqueloba*
  Leaves not lobed, broadly oblong to oblong-obovate:
    Leaves ferruginously tomentose below - - - - - 6. *tragacantha*
    Leaves not ferruginously tomentose below:
      Leaves closely appressed-silvery-tomentellous below; calyx-lobes somewhat
        connivent at their apices or erect - - - - - 7. *subviolacea*
      Leaves rather sparsely appressed-pubescent below; calyx-lobes reflexed
                          8. *schliebenii*
Leaves digitately compound (cultivated species) - - - - 9. *foetida*

1. **Sterculia africana** (Lour.) Fiori in Agric. Colon. Ital. **5**, Suppl.: 37 (1912).—Merr., Comm. Lour. Fl. Cochinch.: 263 (1935).—Brenan in Mem. N.Y. Bot. Gard. **8**, 3: 226 (1953).—Gomes e Sousa, Dendrol. Moçamb. **4**: 132, 139 cum tab. (1958). TAB. **104** fig. A. Type: Mozambique, Mossuril, opposite Moçambique I., *Loureiro* (P, holotype).

    Triphaca africana Lour., Fl. Cochinch.: 577 (1790). Type as above.
    Sterculia triphaca R. Br. in Benn., Pl. Jav. Rar.: 228 (1844).—Mast. in Oliv., F.T.A. **1**: 216 (1868).—K. Schum. in Engl., Mon. Afr. Pflanz. **5**: 105 (1900) pro parte.—Engl., Pflanzenw. Afr. **3**, 2: 452 (1921). Type as for *Sterculia africana*.
    Sterculia ipomoeifolia Garcke in Peters, Reise Mossamb. Bot. **1**, 130 (1861). Type: Mozambique, Sena, *Peters* (B†).
    Sterculia triphaca var. *rivaei* K. Schum. in Engl., Mon. Afr. Pflanz. **5**: 106 (1900) pro parte quoad specim. Buchanan.
    Sterculia tomentosa sensu Sim, For. Fl. Port. E. Afr.: 18, t. 7B (1909).—O. B. Mill. in Journ. S. Afr. Bot. **18**: 57 (1952).
    Cola cordifolia sensu Sim, tom. cit.: 19 (1909).

Tree 10–25 m. tall, with a stout trunk up to 1 m. in diam. somewhat resembling that of the Baobab; bark peeling in papery flakes, brownish, yellowish, whitish or liver-coloured, inner layers green and with the sapwood pink; primary branches stiff and very stout. Leaves collected at the ends of the branches. Leaf-lamina 5–15 × 4–13 cm., very broadly ovate-cordate, apex acuminate, almost entire or 3–5-lobed with somewhat acuminate lobes, c. 7 nerved at the base, from thinly to densely and harshly pubescent or tomentose, sometimes glabrescent; petiole up to 10 cm. long, coarsely pubescent. Flowers in clustered, usually terminal panicles appearing before the leaves and up to 9 cm. long; branches of inflorescence densely pubescent; bracteoles c. 1·5 mm. long, linear-oblong, acute, pubescent; pedicels up to 1 cm. long, articulated near the middle, pubescent. Calyx up to 1·2 cm. long, yellowish with reddish guide-lines within, campanulate, divided rather more than half way into 5–6 acute lobes, tomentellous outside, glabrous within except near the apex. Male flower: stamens c. 10, in a capitate discoid cluster; andro-phore 5–6 mm. long, slender, glabrous. Female flower: ovary ovoid, tomentose, with a few vestigial stamens at its base, on a puberulous gynophore c. 4 mm. long; style c. 3 mm. long, pubescent, often reflexed. Follicles 3–5, c. 10 cm. long, spreading, subsessile, oblong-ovoid, with a horn-like often curled apiculus at the apex, golden-tomentellous and finely longitudinally ridged outside, opening widely after dehiscence; placentas with very dense acicular hairs. Seeds numerous, c. 2 × 1 cm., oblong-ellipsoid, with a whitish aril at one end; testa smooth, dull-blackish.

Caprivi Strip. E. of Kwando R., fl. x.1945, *Curson* (PRE). **Bechuanaland Prot.**

N: between Tutumi and Sebena, fr. 14.xii.1929, *Pole Evans* 2607 (K; PRE; SRGH)
N. Rhodesia. E: Petauke, fl. & fr. 20.iv.1952, *White* 2422 (FHO). S: Victoria Falls,
fl. 22.xi.1949, *Wild* 3153 (K; SRGH). **S. Rhodesia.** N. Sebungwe, Muzaza Hill, fl.
& fr. 1.xi.1947, *Whellan* 311 (K; SRGH). W: Bulalima-Mangwe, Embakwe, fl.
28.ix.1941, *Feiertag* in GHS 45459a (SRGH). C: Hartley, near Halfway Hotel, fr.
1.ii.1950, *Birkett* 44 (K; SRGH). **Nyasaland.** N: Mlali to Chisiombe, fr., *Lewis* 85
(FHO). C: Dedza Distr., Mua, fl. 1932, *Townsend* 14 (FHO). S: Zomba, fl. 1936,
Clements 621 (FHO; K). **Mozambique.** N: Mutuali, fl. 21.ix.1953, *Gomes e Sousa*
4122 (COI; K; PRE; SRGH); Mossuril, opposite Moçambique I., fl., *Loureiro* (P).
T: Mutarara, Inhangoma I., fl. 3.x.1944, *Mendonça* 2340 (LISC). MS: Cheringoma,
Inhaminga, fr. 26.v.1948, *Mendonça* 4389 (LISC; SRGH).

Also in SW. Africa and Tanganyika. In the hotter and drier parts of our area mainly
in the coastal plain of Mozambique and the Zambezi valley.

This species varies between forms with glabrescent leaves and forms with roughly
tomentose leaves and more acute lobes to the blade. These latter forms have sometimes
been named in various herbaria *S. setigera* Del. (*S. tomentosa* Guill. & Perr.) but this
is quite a different species occurring to the north of our area with much smaller fruits not
exceeding 7·5 cm. long with a short beak, smaller seeds, and flowers which lack prominent
reddish guide-lines within. *Buchanan* 1025 (B†; K) belongs with these hairy forms and
was placed in his var. *rivaei* by K. Schumann. It is not, however, a good match for the
type of this variety from Somaliland. Intermediate forms are common in our area
between the very hairy and glabrescent forms so no attempt is made here to distinguish
any varieties.

2. **Sterculia rogersii** N.E.Br. in Kew Bull. **1921**: 290 (1921).—Burtt Davy, F.P.F.T.
   1: 259 (1926).—O. B. Mill. in Journ. S. Afr. Bot. **18**: 57 (1952). TAB. **104** fig. D
   Type from the Transvaal.

Small tree up to 5 m. tall with rather slender branches and often branching
low down; bark brown and peeling in papery flakes. Leaves borne singly or
clustered on the ends of short side-shoots; lamina up to 6 × 5 cm., very broadly
cordate-ovate, entire or very shallowly and sinuately 3–5-lobed, apex obtuse or
subacute, minutely puberulous above, whitish and tomentellous below, 5–7-nerved
at the base; petiole up to 2 cm. long, tomentellous. Flowers in clusters on the
short side-shoots, appearing before the leaves; pedicels up to 5 mm. long, tomentellous, articulated near the middle; bracteoles c. 1·5 mm. long, linear, pubescent,
caducous. Flowers yellowish-green with reddish guide-lines within, very similar
in size and structure to those of *S. africana*. Follicles 3–5, up to 7·5 cm. long,
oblong-ovoid, spreading, subsessile, apical horn up to 1·3 mm. long, golden-
tomentellous outside, pale and tomentellous inside; placentas with very dense
acicular hairs. Seeds numerous, like those of *S. africana* but only 1–1·3 × 0·7 cm.

Bechuanaland Prot. SE: Topsi Siding, fr. xii. *Miller* B/803 (FHO; PRE). **S.
Rhodesia.** W: Bulalima-Mangwe, fl. & fr. 28.ix.1941, *Feiertag* in GHS 45459 (SRGH).
E: Hot Springs, fl. 1.i.1949, *Chase* 1508 (BM; K; SRGH). S: Lower Sabi, west bank,
fr. 29.i.1948, *Wild* 2434 (K; SRGH). **Mozambique.** SS: Guijà, Caniçado, fl.
4.vii.1947, *Pedro & Pedrógão* 1287 (COI; K; LMJ; SRGH). LM: Goba, fr. 15.xi.1940,
Torre 1998 (K; LISC; SRGH).

Also in the Transvaal. Very characteristic species of dry woodland and bush in the
Limpopo valley and adjacent areas.

3. **Sterculia appendiculata** K. Schum. in Engl., Pflanzenw. Ost-Afr. **C**: 272, t. 24
   (1895); in Engl., Mon. Afr. Pflanz. **5**: 105, t. 9 fig. C (1900).—Engl., Pflanzenw.
   Afr. **3**, 2: 452 (1921).—Brenan, T.T.C.L.: 602 (1949). TAB. **104** fig. B. Type
   from Tanganyika.

Tall tree up to 40 m. tall or even more; bark pale yellow and smooth. Leaves
towards the ends of the branches, on glabrous or glabrescent petioles c. 12 cm.
long; lamina 8–20 × 7–15 cm., broadly cordate-ovate with 3–7 usually caudate-
acuminate lobes or angles, glabrous at maturity, c. 7-nerved at the base; petiole
up to c. 12 cm. long, glabrous or nearly so. Flowers in terminal yellowish-hairy
racemes or panicles 10–12 cm. long; bracteoles c. 5 mm. long, caducous, oblong,
tomentellous on both sides. Calyx 1·2–1·4 cm. long, yellowish-brown or greenish,
campanulate, 5–6-lobed to about half-way, tomentose outside and inside except
towards the base. Male flower: anthers numerous, in a capitate-globose cluster
on a glabrous androphore c. 4 mm. long with a tomentose disk at its base. Female
flower: ovary ovoid, tomentose, with a ring of vestigial subsessile anthers at its

base, on a disk-like gynophore c. 1 mm. long; style c. 3 mm. long, tomentose Follicles 2–3, 7–9 × 5–6 cm., ovoid-globose, scarcely apiculate, brown-tomentellous outside. Seeds c. 2 × 1·3 cm., dark brown, numerous, cylindric or radially compressed, surrounded by a yellow aril or dry pulp similar to that of baobab seeds.

Nyasaland. S : Lower Shire R., fl. vii.1936, *Topham* 989 (FHO ; K). **Mozambique.** N : Montepuez, Metuge, fr. 9.ix.1948, *Barbosa* 2026 (K ; LISC ; SRGH). T : Chioco, fr. 26.ix.1942, *Mendonça* 448 (LISC). MS : Vila Machado, fl. 29.ii.1938, *Mendonça* 3833 (BM ; LISC).
Also in Tanganyika. In coastal and riverine forest.

4. **Sterculia mhosya** Engl., Pflanzenw. Afr. **3**, 2 : 455, fig. 208 A–H (1921).—Brenan, T.T.C.L.: 602 (1949). TAB. **104** fig. F. Type from Tanganyika.
Sterculia sp. nr. *mhosya*. Brenan, loc. cit.

Shrub or small tree up to c. 8 m. tall; bark dark grey and peeling. Leaves collected at the ends of the branches; lamina up to 15 cm. in diam., very broadly cordate-ovate to orbicular, more or less divided into 3–5 acuminate lobes, pubescent on both sides, c. 7-nerved at the base; petiole up to 12 cm. long, glandular-pubescent. Flowers dull red or purplish, appearing with the leaves, in terminal panicles up to 12 cm. long; branches of inflorescence sticky with dark purplish, glandular-viscid hairs; bracteoles c. 5 mm. long, linear, glandular-pubescent. Calyx up to 1·2 cm. long, campanulate, 5–6-lobed rather more than ½-way, densely pubescent and somewhat glandular outside, pubescent towards the tips of the lobes inside. Male flower: androphore c. 7 mm. long, slender, glabrous; anthers numerous, in a capitate cluster. Female flower: ovary ovoid, densely pubescent, with a ring of vestigial stamens at the base, on a glabrous gynophore c. 5 mm. long ; style 3–4 mm. long, reflexed, pubescent. Follicles c. 5, 5–7 × 1–1·5 cm., subsessile, spreading, oblong-cylindric with a very short apiculus, brown-tomentellous outside. Seeds 10–20 per follicle, 1–1·5 × 0·7 cm., oblong-ellipsoid; testa black ; aril white or orange.

N. Rhodesia. N : Lake Tanganyika shore near Kisanza, fl. 18.v.1936, *Burtt* 6307 (K). Also in Tanganyika. Scattered along the rocky areas of the Lake Tanganyika shore.

Engler in his description of this species describes the stipes of the follicles as being 1 cm. long, but illustrates a follicle without a stipe. The species is well known in Tanganyika and the follicles are invariably sessile, so although the type was not designated and no specimens were cited, we can take it that the description is in error in this respect. As the material on which the species is based is probably destroyed the illustration accompanying the description must be selected as the lectotype. Fortunately it is an excellent illustration and compares very well with material from the type locality and elsewhere.

5. **Sterculia quinqueloba** (Garcke) K. Schum. in Engl., Bot. Jahrb. **15** : 135 (1892) ; in Engl., Mon. Afr. Pflanz. **5** : 104, t. 9 fig. D (1900).—Engl., Pflanzenw. Afr., **3**, 2 : 450 (1921).—Exell & Mendonça, C.F.A. **1**, 2 : 200 (1951).—Brenan in Mem. N.Y. Bot. Gard. **8**, 3 : 226 (1953).—Gomes e Sousa, Dendrol. Moçamb. **4** : 141 cum tab. (1958). TAB. **104** fig. E. Syntypes: Mozambique, Sena, *Peters* (B†) ; Macanga, *Peters* (B†).
Cola quinqueloba Garcke in Peters, Reise Mossamb. Bot. **1** : 130 (1861). Syntypes as above.
? *Sterculia livingstoneana* Engl., Bot. Jahrb. **39** : 592 (1907). Type: N. Rhodesia, Victoria Falls, *Engler* 2936 (B†, holotype).
Sterculia quinqueloba Sim, For. Fl. Port. E. Afr.: 18, t. 6 (1909) *nom. illegit.* Type: Mozambique, Maganja da Costa, *Sim* 998 (PRE, holotype).

Small tree, or occasionally reaching 25 m. tall; bark smooth and peeling off in flakes, silvery or pale in the inner layers; branches thick and stiff. Leaves at the ends of the branches; lamina up to 40 × 40 cm., very broadly cordate-ovate, with 3–5 (7) usually acuminate lobes, greyish-tomentose below, less so above; petiole up to 27 cm. long, harshly tomentose. Flowers appearing with the young leaves in terminal ample many-flowered panicles 9–30 cm. long; branches of inflorescence tomentose or tomentellous and glandular; bracteoles 5–6 mm. long, caducous, lanceolate, tomentellous, glandular. Calyx c. 4 × 3·5 mm., greenish, campanulate, 5-lobed about ⅓ of the way down, tomentellous outside and inside. Male flower: stamens many, in a capitate-globose cluster on a slender glabrous

H

androphore c. 2 mm. long. Female flower: ovary ovoid, tomentellous; style c. 1 mm. long, glabrous; vestigial stamens in three clusters of about 3 on short filaments ⅓ the length of the ovary; gynophore c. 0·5 mm. long, glabrous. Follicles 3–5, up to 6 × 3 cm., held rather erect, on tomentellous stipes c. 1 cm. long, ovoid, shortly apiculate or acute, brown-tomentellous outside. Seeds several per follicle, c. 8 × 6 mm., oblong-ellipsoid; testa blackish; aril small.

N. Rhodesia. N: Abercorn, Mbete, fl. 22.v.1936, *Burtt* 6308 (BM; K). W: Mwinilunga, fl. 12.ii.1938, *Milne-Redhead* 4539 (BM; K). C: Lunsemfwa R., fr. 24.viii.1929, *Burtt Davy* 20885 (FHO; K). E: Mvuvye R., fr. 16.viii.1955, *Lees* 29 (K). S: 30 km. N. of Livingstone, fl. 18.iii.1952, *White* 2282 (FHO). **S. Rhodesia.** N: Concession, fl. 6.i.1939, *McGregor* 2/39 (FHO; K; SRGH). W: Mafungabusi Plateau, st., *Goldsmith* 23/47 (FHO; SRGH). C: Salisbury, Enterprise, fl. & fr. 5.ii.1952, *Wild* 3759 (K; SRGH). **Nyasaland.** N: Nyungwe R., fl. 19.ix.1930, *Migeod* 941 (BM). C: Dedza, fl. & fr. viii.1954, *Adlard* 191 (FHO). S: Chikwawa, fr. 5.x.1946, *Brass* 17993 (K; SRGH). **Mozambique.** N: Cuamba, fl. 12.vii.1935, *Torre* 860 (COI; LISC). Z: Mbobo-Mopeia, fl. & fr., 18.v.1943, *Torre* 5332 (K; LISC; SRGH) T: above Tete, fr. xi.1860, *Kirk* (K). MS: Chimoio, Belas Mt., fl. 1.iv.1948, *Garcia* 825 (BM; LISC; SRGH).

Also in Angola, Belgian Congo and Tanganyika. In dry woodland at low altitudes including the coastal plain of Mozambique, but also common on rocky outcrops and hills in *Brachystegia* woodland.

6. **Sterculia tragacantha** Lindl., Bot. Reg. **16**: t. 1353 (1830).—Mast. in Oliv., F.T.A. **1**: 216 (1868).—Ficalho, Pl. Ut. Afr. Port.: 105 (1884).—K. Schum. in Engl., Mon. Afr. Pflanz. **5**: 102, t. 9 fig. F (1900).—R.E.Fr., Wiss. Ergebn. Schwed. Rhod.-Kongo-Exped. **1**: 148 (1914).—Engl., Pflanzenw. Afr., **3**, 2: 450 (1921).—Exell & Mendonça, C.F.A., **1**, 2: 202 (1951). TAB. **104** fig. C. Type a cultivated specimen grown from seed collected in Sierra Leone.

Large tree 15 m. tall or more, with smooth greyish bark and stiff rugose branchlets. Leaves collected towards the ends of the branches; lamina up to 20 × 12 cm., leathery, entire, broadly oblong, obovate-oblong, or ovate-oblong, apex obtuse or abruptly acuminate, base rounded or slightly cordate, penninerved, thinly pubescent or glabrescent above, more or less ferruginously tomentose beneath; petiole up to 5 cm. long. Flowers reddish, in harshly pubescent panicles c. 12 cm. long, clustered at the ends of the branches. Calyx campanulate, divided to about half-way into 4–6 narrowly oblong lobes which are coherent at their tips, pubescent outside and on the lobes within. Male flower: stamens numerous, in a capitate globose cluster on a glabrous androphore c. 2 mm. long surrounded by a ring of hairs at its base. Female flower: ovary ovoid, tomentose, with a ring of vestigial anthers at its base, on a glabrous gynophore up to 1 mm. long; style c. 1 mm. long, tomentose. Follicles c. 5 cm. long, ellipsoid with an acute apiculus, ferruginously tomentose outside and inside. Seeds c. 1·6 × 1·2 cm., oblong-ellipsoid; testa black; aril small and yellow.

N. Rhodesia. N: Lake Mweru, Nchelengi, fr. 23.iv.1951, *Bullock* 3830 (K; SRGH). Also in Angola, Belgian Congo, Tanganyika and tropical West Africa. In swamp forest or fringing forest or occurring as an isolated tree in lake-shore thickets.

7. **Sterculia subviolacea** K. Schum. in Engl., Pflanzenw. Ost-Afr., **C**: 271 (1895); in Engl., Mon. Afr. Pflanz. **5**: 102, t. 9 fig. H (1900).—Engl., Pflanzenw. Afr., **3**, 2: 449 (1921).—Brenan, T.T.C.L.: 601 (1947). Type from Tanganyika.
　Sterculia ambacensis Welw. ex Hiern, Cat. Afr. Pl. Welw. **1**: 83 (1896).—K. Schum. in Engl., Mon. Afr. Pflanz. **5**: 104 (1900).—Engl., tom. cit.: 450 (1921).—Exell & Mendonça, C.F.A. **1**, 2: 201 (1951). Type from Angola.

Massive tree up to 30 m. tall, with a buttressed bole. Leaves similar in shape and size to those of *S. tragacantha* but deep green and glabrous above and very densely and closely appressed-whitish-stellate-tomentellous below. Flowers in axillary or terminal panicles c. 14 cm. long; branches of inflorescence violet-tomentose. Calyx up to 8 mm. long, campanulate, divided to half-way or a little more into 5 oblong lobes somewhat coherent at their tips, stellate-tomentose outside with some larger purple hairs among the paler ones, pubescent inside with some longer purple simple hairs and some subsessile simple glandular hairs. Male flower: stamens 8–10, in a capitate globose cluster on a glabrous androphore c. 1·5 mm. long. Female flower: ovary ovoid, tomentose, with a ring of rudimentary stamens at its base, on a glabrous gynophore c. 0·5 mm. long; style c. 1 mm. long, tomentose; stigmas recurved. Follicles similar in size and shape to those of *S. tragacantha*. Seeds c. 1·4 cm. long, oblong-ellipsoid; testa black.

N. Rhodesia. N: Mkupa, fr. 7.x.1949, *Bullock* 1173 (K; SRGH). W: Mwinilunga, fr. 14.viii.1955, *Holmes* 1183 (K).
Also in Angola and Tanganyika. Swamp forest or riverine forest.

Unfortunately the type of *S. subviolacea* (*Stuhlmann* 3549 (B) from Ost-Usindja (=Mwanza Distr., Tanganyika Territory)) is destroyed. However, excellent material is available from the type locality which fits the description very well and it is proposed that one sheet of this, *Carmichael* 473 (EA; K, neotype) from Tanganyika, Mwanza Distr., Geita, Iliagamba be selected as the neotype.

8. **Sterculia schliebenii** Mildbr. in Notizbl. Bot. Gart. Berl. **12**: 519 (1935).—Brenan, T.T.C.L.: 604 (1949). Type from Tanganyika.

Tree 10–16 m. tall; young branches glabrous, with greyish bark. Leaves collected at the ends of the stems; lamina up to 14 × 9·5 cm., entire, obovate-oblong, apex rounded or acuminate, base slightly cordate and 5–7-nerved, sparsely appressed-stellate-pubescent above, a little more densely so beneath, not discolorous ; nerves and midrib prominent below, venation prominent and laxly reticulate ; petiole c. 5 cm. long, thinly pubescent. Flowers appearing before the leaves in narrowly pyramidal panicles 4–8 cm. long at the ends of the branches; branches of inflorescence densely purple-tomentose. Calyx 5–6 mm. long, campanulate, lobed to half-way or a little more into 5 deltoid reflexed lobes, with both pale and deep purple stellate hairs outside, inside densely pubescent with short simple somewhat stellate hairs mixed with subsessile glandular hairs. Male flower: stamens 10, in a capitate-globose cluster on a glabrous androphore c. 2 mm. long. Female flower: ovary ovoid, tomentose, with a ring of vestigial anthers at its base, on a glabrous gynophore c. 0·5 mm. long; style 1 mm. long, tomentose; stigmas recurved. Follicles c. 5, 2·5 × 2 cm., ovoid, tomentellous outside, scarcely apiculate, on stipes 2–3 mm. long. Seeds not seen.

Mozambique. N: Cabo Delgado, Tungue, fl. & fr. 18.ix.1948, *Andrada* 1369 (COI ; K; LISC; SRGH).
Also in Tanganyika. In open woodland.

This species was described from a leafless specimen and was considered by Mildbraed (loc. cit.) to be related to *S. quinqueloba* but, now we have more complete material, it is obviously more closely related to *S. subviolacea*.

Cultivated Species

9. **Sterculia foetida** L., Sp. Pl. **2**: 1008 (1753). Type from Ceylon.

Tree up to 30 m. tall, with digitately compound leaves; leaflets 8–11, subsessile, lanceolate, glabrous or glabrescent. Flowers in terminal panicles, red or scarlet, smelling of carrion; calyx c. 1·25 cm. long, tomentose. Follicles ovoid, seeds large.

Mozambique. N: Naguema, Mossuril, fl. 13.viii.1948, *Pedro & Pedrógão* 4758 (LMJ; SRGH). Z: Quelimane, fl. 20.vii.1943, *Torre* 5708 (LISC; SRGH).
A native of tropical Asia cultivated as a decorative species in Mozambique for many years.

Cultivated Species of Brachychiton

Three other decorative trees cultivated in our area belong to an Australian genus closely related to *Sterculia*. One is *Brachychiton acerifolium* (A. Cunn.) F. Muell. (*Sterculia acerifolia* A. Cunn.), sometimes called the Flame Tree, from New South Wales. It is a glabrous tree up to 20 m. or more in height with deeply 5–7-lobed leaves, the lobes oblong-lanceolate to rhomboid, and with brilliant scarlet flowers. It is grown in the Municipal Gardens, Lourenço Marques, and in S. Rhodesia. A second is *Brachychiton populneum* (Schott) R. Br. (*Sterculia diversifolia* G. Don), the Kurrajong of Queensland and Victoria. It also reaches 20 m. in height, is glabrous except for the flowers and has ovate-lanceolate leaves which are entire or 3–5-lobed with slender petioles. The flowers are yellowish-white and often dark-spotted and may be reddish inside. This is grown as a street tree in the Rhodesias and Mozambique. Finally, there is *Brachychiton discolor*

F. Muell. from New South Wales and Queensland, which is also cultivated in Lourenço Marques and Salisbury. It has 5–7-lobed broadly cordate-ovate discolorous leaves and carmine flowers, and reaches 15 m. in height.

9. COLA Schott & Endl.

Cola Schott & Endl., Melet. Bot.: 33 (1832) *nom. conserv.*

Trees with entire, palmatilobed or digitately divided, petiolate leaves; petioles often with a swollen apical pulvinus, sometimes swollen at the base also. Flowers unisexual or polygamous, usually dioecious, in axillary racemes, panicles or fascicles; pedicels usually articulated. Calyx 4–5 (6)-lobed. Petals absent. Male flower: anthers 5–12 in 1 or apparently 2 superposed rings on an androphore; vestigial carpels often sunk in the top of the androphore. Female flower: carpels (3) 4–5 (10) coherent; ovules 2–∞ per carpel; styles as many as the loculi but often united into a single column; stigmas equalling the number of carpels. Fruit of 3–several (or fewer by abortion) leathery or woody carpels finally splitting lengthwise. Seeds 1–∞, without endosperm; cotyledons thick, 2 (in our species) or more.

Leaves ± cuneately narrowed to a narrowly obtuse or subacute base, rarely somewhat
   rounded, oblanceolate or obovate-elliptic, with 6–12 pairs of lateral nerves :
   Fruiting carpels club-shaped, narrowing to a distinct stipe 5–8 mm. long   1. *clavata*
   Fruiting carpels globose or obovoid-ellipsoid, sessile or very shortly stipitate :
      Indumentum on outside of flowers and pedicels dark brown; rays of stellate hairs
        short, less than the diameter of the pedicel; pedicels articulated above the base;
        young vegetative parts with a short appressed indumentum   -   2. *greenwayi*
      Indumentum on outside of flowers and pedicels light ferruginous-brown; rays of
        stellate hairs equalling or exceeding the diameter of the pedicels; pedicels not
        articulated; young vegetative parts with a floccose pale brown indumentum
                                    3. *mossambicensis*
Leaves broadly rounded and cordate or subcordate at the base, ovate or ovate-elliptic,
   lateral nerves in 4–6 pairs with 1–2 pairs of basal nerves - 4. *discoglypremnophylla*

1. **Cola clavata** Mast. in Oliv., F.T.A. **1**: 222 (1868).—Sim, For. Fl. Port. E. Afr.: 19 (1909). Syntypes: Mozambique, " Zambezia ", Chamo, *Kirk* (K); *Meller* (K).

Large tree with ash-coloured bark; young branchlets stellate-puberulous but soon glabrous. Leaf-lamina up to 12·5 × 5·6 cm., coriaceous, narrowly obovate to elliptic, apex obtuse, margin entire or undulate, base cuneate, midrib prominent below, nerves in c. 10 pairs, venation reticulate but not very prominent on both sides; petiole up to 6·5 cm. long, slightly swollen and minutely stellate-puberulous at base and apex; stipules c. 3 mm. long, caducous, subulate, stellate-puberulous. Flowers in axillary fascicles on 1–3-year-old branches; pedicels c. 1 cm. long, 3-nate, pubescent. Calyx stellate-pubescent. Follicles 1–2 by abortion, c. 2 × 1·1 cm., clavate, narrowing to a slightly curved stipe 0·5–0·8 cm. long, minutely and sparsely stellate-puberulous, 1-locular, 1-seeded. Seed c. 1·75 × 1·5 cm., oblong-ellipsoid; testa shining brown; cotyledons flat, thick.

Mozambique. Z: Chamo, fr. i.1862, *Kirk* (K).
Perhaps confined to Mozambique. Probably a riverine species.

Brenan (in Kew Bull. **1956** : 148 (1956)) considered that a number of specimens from Kenya and Tanganyika represented flowering material of this species and one of his reasons for doing so was that their flowers, which are unusually small for this genus, match the dried-up calyces on the two syntypes of *C. clavata*, especially in size. A careful re-examination of these reveals, however, that none of these calyx-segments on the synltypes is complete to the apex and therefore their original length is uncertain. It is at least as likely that flowering material is represented by a specimen which has since come to light collected only 60 km. from the type locality at Inhamitanga, in the Manica e Sofala Province of Mozambique (*Simão* 1202 (K; LM; SRGH), 27.xi.1946). This has bisexual flowers (corresponding male-flowered specimens still remain to be collected). If this be the case the Kenya and Tanganyika material represents a different and probably undescribed species. After re-examination of the situation both Brenan and myself consider that there is room for doubt as to which is the true flowering material of *C. clavata* and it is very desirable that more good material should be collected in the type locality. A short diagnosis of the flowers of *Simão* 1202 follows: calyx rotate-campanulate, deeply lobed; tube 1·5–2 mm. long, stellate-puberulous outside, glabrous within; lobes 5 × 2·5 mm., oblong-ovate, apex acute, densely stellate-pubescent outside, with globose papillae within, with c. 7 parallel nerves. Carpels 4–5, densely stellate-tomentose,

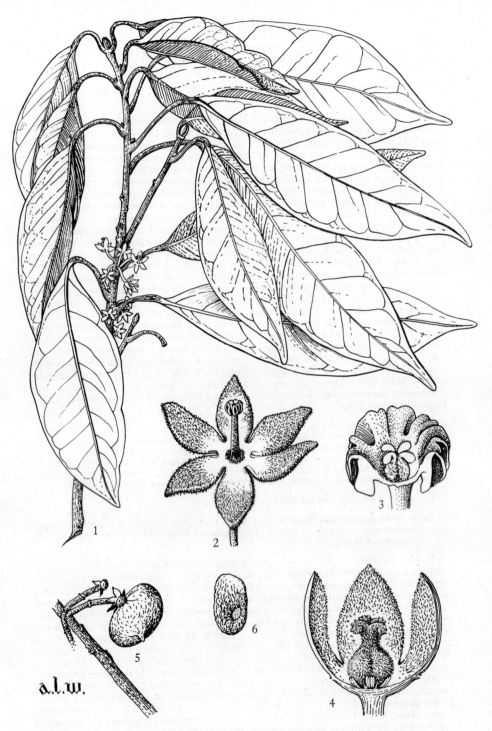

Tab. 105. COLA GREENWAYI. 1, male flowering branch (×⅔); 2, male flower (×4); 3, stamens with front portion of top of the androphore removed to show vestigial ovary (×14) all from *Grant 39/39*; 4, longitudinal sect. of female flower (×4) *Drummond & Hemsley* 2965; 5, fruit and portion of stem (×1) *Greenway* 7931; 6, seed (×1) *Greenway* 7931.

2-ovulate; styles c. 1 mm. long, tomentose; stigmas 1 ×1 mm., spreading, papillose; surrounded at the base by 8 sessile anthers. A specimen in bud from Nyasaland (S: Chikwawa, Likabula R., *Jackson* 1707 (FHO)) also probably belongs here.

2. **Cola greenwayi** Brenan in Kew Bull. **1956**: 144 (1956). TAB. **105**. Type from Tanganyika.

Tree 6–17 (24) m. tall, evergreen, dioecious; young branches densely dark-brown-tomentellous; bark rough. Leaf-lamina up to 15 ×7 cm., subcoriaceous, purplish-red when young, elliptic to oblanceolate or obovate-elliptic, apex obtusely acuminate, margin entire or sinuate, base broadly cuneate or narrowly rounded, glabrous on both surfaces, midrib prominent on both surfaces, nerves in 8–12 pairs, prominent on both surfaces but particularly beneath, venation reticulate; petiole up to 5·5 cm. long, with a dark-brown-stellate-tomentellous pulvinus just below the lamina; stipules 5 mm. long, caducous, subulate-lanceolate, brown-tomentose. Flowers 1–8-nate in fascicles on 1–3-year-old wood. Male flowers (not collected in our area as yet, description based on Tanganyika material) on ferruginous-stellate-pubescent pedicels 4–7 mm. long articulated above the base; calyx deeply 4–6-lobed, densely ferruginous-stellate-tomentose outside, tube up to 1·5 mm. long, glabrous within; lobes 5–10 ×2·5–5 mm., elliptic to obovate-elliptic with acute apices, with 3–5 longitudinal nerves, stellate-pubescent within the upper half, minutely papillose below; androphore 2·5–4·5 mm. long, glabrous except for a circle of stellate hairs at the base or densely pubescent, with 5–6 anthers in a single series at the apex and 4 vestigial pubescent or subglabrous carpels immersed in the top. Female flowers as in the male but with the pedicels up to 1 (2) cm. long and the part above the articulation markedly sulcate; carpels 3–4, densely stellate-tomentose, biovulate; styles up to 3 mm. long, tomentose; stigmas 1 ×0·5–1·25 mm., patent or recurved, papillose; anthers 5, sessile, surrounding the base of the carpels. Follicles 1–1·8 ×1·4–2 cm., sessile or sub-sessile, obliquely ellipsoid to obovoid, rounded or mucronate at the apex, ferruginously tomentellous, 1–2-seeded. Seeds 13–16 ×8–11 mm., ellipsoid, with a brown rugulose testa.

N. Rhodesia. W: Ndola Distr., Hippo Pools, fr., *Holmes* 478 (FHO). **S. Rhodesia.** E: Vumba, fl. 1.xi.1946, *Wild* 1588 (K; SRGH). **Nyasaland.** N: Nyika Plateau, Chowo Rock, fr., *Chapman* 294 (FHO; K). S: Mangoche Mt., *Chapman* 456 (FHO). **Mozambique.** T: Monte Zobuè, st. 3.x.1942, *Mendonça* 628 (K; LISC; SRGH). MS: Chimoio, Monte Garuzo, fl. 29.xi.1943, *Torre* 6247 (K; LISC; PRE; SRGH). Also in Tanganyika and Kenya. In evergreen forest between 1500 and 2200 m.

The wood is used for making bows in S. Rhodesia.

3. **Cola mossambicensis** Wild in Bol. Soc. Brot., Sér. 2, **33**: 39 (1959). Type: Mozambique, Manica e Sofala, Espungabera, Gogoi Mt., *Torre* 4308 (LISC, holotype; SRGH).

Evergreen tree up to 27 m. tall; young branches softly and ferruginously stellate-tomentose, soon glabrescent; bark ashy-grey. Leaf-lamina up to 22 ×9·5 cm., coriaceous or subcoriaceous, obovate-elliptic to obovate, apex abruptly acuminate, margin entire or sinuate, base broadly cuneate or narrowly rounded, glabrous on both surfaces except when very young, midrib prominent on both surfaces, nerves in 12–15 pairs, slightly prominent above, very prominent beneath, venation reticulate, prominent beneath; petiole up to 6 cm. long, softly tomentellous but soon glabrescent, with a persistently puberulous pulvinus; stipules up to 7 mm. long, caducous, linear-lanceolate, ferruginously tomentose. Flowers in few- to many-flowered fascicles on the 1–3-year-old wood. Male flowers ferruginously tomentose; pedicels c. 1 cm. long, not articulated; calyx campanulate-rotate, deeply 5–6-lobed, densely light-brown-pubescent or -tomentose outside; tube glabrous within, 1·5 mm. long; lobes 3·5–6 ×2·5–3 mm., narrowly ovate, apex acute, with 3 longitudinal nerves, densely stellate-pubescent in the upper half within, with minute globose papillae below; androphore c. 3 mm. long, densely stellate-pubescent to near the apex, crowned by a uniseriate ring of 5–6 anthers; vestigial carpels 4, immersed in the top. Female flowers as in the male but with 4 stellate-tomentose, 2-ovulate carpels; style 1 mm. long, stellate-tomentose; stigmas 1 ×0·75 mm., patent, papillose; anthers 6–7, sessile, surrounding the base of the carpels. Follicles 1–2, 1·3–1·4 × 1·3–1·7 cm., sessile or subsessile, fubglobose, apex rounded or bluntly mucronate, slightly narrowed at the base, serruginously stellate-tomentose, 1–2-seeded. Seeds ellipsoid, c. 1·2 × 0·9 cm.; testa brown, rugulose.

Nyasaland. S: Nama Kokwi R. to Ncheu, fr. 16.ix.1929, *Burtt Davy* 21732 (FHO). **Mozambique.** Z: Milange, 16.ix.1942, *Hornby* 2794 (K; LM; PRE; SRGH). MS: Moribane, fr. 5.x.1955, *Pedro* 4219 (K; LMJ; PRE).

Endemic in our area. In evergreen forest up to about 600 m. (above this altitude apparently replaced by *C. greenwayi*).

In *Torre* 4308 the flowers are monoecious, but this is presumably unusual. The sterile specimens *Mendonça* 2303 (K; LISC; SRGH) from between Mutuali and Milange on the borders of the Zambezia and Niassa Provinces and *Clements* 67 (FHO; K) from Zomba Distr. in Nyasaland should perhaps be referred here.

Doubtfully recorded species

4. ? **Cola discoglypremnophylla** Brenan & Jones in Bull. Jard. Bot. Brux. **18**: 2 (1946).—Brenan in Kew Bull. **1956**: 150 (1956). Type from Tanganyika (Lindi Distr.).

Tree 6–10 m. tall, evergreen; very young branchlets minutely stellate-pilose, soon glabrous; bark whitish and longitudinally sulcate. Leaf-lamina up to 10 × 6–7 cm., coriaceous, broadly ovate to ovate-elliptic, apex obtuse or subacuminate, margin entire or undulate, base broadly rounded or slightly cordate, glabrous on both surfaces except sometimes for a few hairs near the base of the midrib, midrib prominent especially beneath, nerves in 4–6 pairs with 1–2 pairs of basal nerves, prominent especially beneath, venation prominently reticulate beneath; petiole up to 6·6 cm. long, basal and apical pulvini minutely stellate-puberulous, rest glabrous; stipules c. 8 mm. long, caducous, linear, subglabrous. Flowers 1–7-nate in axillary fascicles on 1–3-year-old wood. Male flowers on unarticulated stellate-puberulous pedicels up to 7 mm. long: calyx cyathiform, deeply 4–5-lobed, densely greyish-puberulous outside; tube up to 1·5 mm. long, glabrous at the base within; lobes 3·5–5 × 2·5–3·5 mm., ovate-elliptic, with acute apex, longitudinally 5–8-nerved, with a few stellate hairs inside towards the apex, otherwise with subglobose papillae; androphore up to 2·4 mm. long, glabrous, with 8–9 uniseriate anthers; vestigial carpels 4, glabrous, immersed among the anthers. Female flowers and fruit not so far known.

Mozambique. N: Mossuril, Naguema-Monapo, 5.v.1948, *Pedro & Pedrógão* 3153 (LMJ; PRE; SRGH).

Also in Tanganyika. Nothing is known of the ecology of this species.

The specimen cited is in young bud and it cannot be confirmed that it is this species without doubt. It was collected, however, not very far (c. 550 km.) from the type locality.

10. PTERYGOTA Schott & Endl.

Pterygota Schott & Endl., Melet. Bot.: 32 (1832).

Trees with entire or palmatilobed, petiolate leaves. Flowers unisexual, in axillary panicles. Calyx campanulate, 5-lobed. Petals absent. Male flowers: androphore with an apical ring of anthers in one or more rows. Female flowers with a perianth as in the male; ovary of 5 sessile or shortly stipitate coherent carpels; styles connate, stigmas 5, recurved; ovules ∞. Fruit of 1–5 large woody follicles. Seeds many, with a large wing; endosperm present.

Leaves broadly ovate, subentire or shallowly 3-lobed, 7–9-nerved at the base
1. ? *mildbraedii*
Leaves oblong-ovate, entire, 5-nerved at the base - - - - 2. ? *alata*

1. ? **Pterygota mildbraedii** Engl., in Mildbr., Wiss. Ergebn. Deutsch. Zentr.-Afr. Exped. 1907–8, **2**: 506 (1910). Type from Ruanda-Urundi.

Tall tree up to 60 m. tall with a smooth trunk and pale grey bark, branching high up; young branches minutely stellate-puberulous. Leaf-lamina up to 40 × 30 cm., broadly ovate, apex subacute, deeply cordate at the base, subentire or shallowly 3-lobed, glabrous on both sides, 7–9-nerved from the base, nervation prominent on both sides, venation reticulate, prominent on both sides; petiole 15 cm. or more long, glabrous or stellate-puberulous near the apex. Flowers and fruits not seen from our area.

N. Rhodesia. N: Abercorn Distr., Tasker's Deviation, st. 28.ii.1952, *Richards* 926 (K).

Pterygota mildbraedii is recorded from the Belgian Congo, Tanganyika and Uganda. The inflorescence is paniculate, few-flowered, ferruginously tomentose with very concave bracts subtending the flowers. Calyx 5-lobed; anthers arranged in a zig-zag sequence at the apex of the androphore. Fruit an ellipsoid dehiscent capsule or follicle c. 17 × 14 cm. with many seeds arranged horizontally. Seeds c. 8 × 2·7 cm., with an oblong unilateral wing.

Mrs. Richards's specimen is not matched by any other material from our area and was growing near a waterfall. Flowers and fruits are needed to check its identity.

2. **? Pterygota alata** (Roxb.) R. Br. in Benn., Plant. Jav. Rar.: 234 (1844).—K. Schum., Mon. Afr. Pflanz. **5**: 135 (1900). Type from India.

    Sterculia alata Roxb. [Hort. Beng.: 50 (1814) *nom. nud.*]; Pl. Coromand. **3**: 84, t. 287 (1820). Type as above.

    Pterygota schumanniana Engl., Pflanzenw. Afr. **3**, 2: 469 (1921) *nom. nud.*

Tall tree with straight trunk buttressed at the base; bark grey, rather smooth; young branches minutely stellate-puberulous. Leaf-lamina 8–22 × 6–17 cm., ovate to oblong-ovate, apex acuminate, base glabrous, entire, glabrous on both surfaces, 5-nerved from the base, veins prominent below. Flowers and fruits not seen from our area.

Mozambique. N: Muatua, Nametil, st. 21.vii.1948, *Pedro & Pedrógão* 4611 (EA; LMJ).

Pterygota alata is a widespread tropical Asian species sometimes cultivated in Africa. Our specimen, however, is indigenous and resembles (as far as one can tell from poor sterile material) indigenous material from coastal Tanganyika. When fertile material is found in our area it will be possible to decide whether our plant is indeed *P. alata* or a distinct African species, as Engler (loc. cit.) evidently thought with regard to the Tanganyika plant.

11. HILDEGARDIA Schott & Endl.

Hildegardia Schott & Endl., Melet. Bot.: 33 (1832).

Shrubs or trees with entire cordate-ovate petiolate leaves. Flowers unisexual, in axillary racemes. Calyx elongate-campanulate or fusiform-clavate, with short lobes. Petals absent. Anthers c. 10, linear, in 2 rows in a globose head borne on a slender androphore. Ovary of 5 coherent carpels; ovules 1–2 per carpel; gynophore present; styles coherent, sometimes very short; stigma-lobes 5, reflexed. Fruiting carpels stipitate, membranaceous, sometimes winged. Seeds 1–2, pubescent.

Hildegardia migeodii (Exell) Kosterm. in Comm. For. Res. Inst. Indon. **54**: 32 (1956). Type from Tanganyika.

    Firmiana migeodii Exell in Journ. of Bot. **68**: 83 (1930); op. cit. **69**: 100 (1931). Type as above.

    Erythropsis migeodii (Exell) Ridl. in Kew Bull. **1934**: 216 (1934).—Brenan, T.T.C.L.: 599 (1949). Type as above.

Shrub c. 2 m. tall or sometimes a tree up to 12 m. tall, glabrous in all parts (except for the corolla and seeds); bark greyish. Leaf-lamina up to 16 × 18 cm., chartaceous, very broadly cordate-ovate, obtuse at the apex, margin undulate or entire, 7-nerved from the base; petiole up to 30 cm. long. Flowers salmon-coloured, in axillary racemes c. 12 cm. long borne near the ends of the branches and appearing before the leaves; pedicels 3–4 mm. long, articulated near the apex. Calyx up to 2 × 0·5–0·8 cm., elongate-campanulate, slightly curved, slightly swollen and truncate at the base, glabrous outside, villous at the base within, pubescent above; lobes 2–3 mm. long, triangular, acute, erect or reflexed. Male flowers: androphore up to 2·6 cm. long, pubescent except on the exserted portion. Female flowers: ovary ovoid, glabrous; gynophore c. 7 mm. long, densely pubescent except near the apex; stigma-lobes recurved, subsessile. Fruiting carpels 5·5 × 2 cm., on an elongated gynophore up to 3 cm. long, elliptic, purplish, membranaceous, glabrous, with basal stipes up to 1 cm. long. Seeds 1–2, c. 0·7 cm. in diam., globose, pubescent.

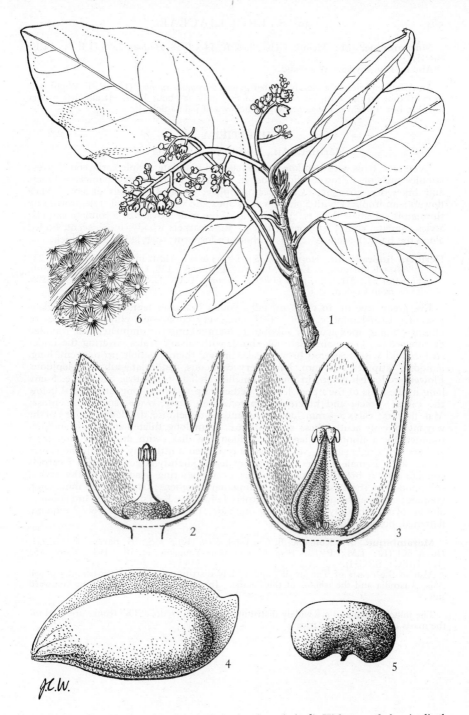

Tab. 106. HERITIERA LITTORALIS. 1, flowering branch (× $\frac{2}{3}$) *Kirk* s.n.; 2, longitudinal sect. of male flower (× 10) *Kirk* s.n.; 3, longitudinal sect. of female flower (× 10) *Tanner* 2459; 4, fruit (× $\frac{2}{3}$) *Faulkner* 750; 5, seed (× $\frac{2}{3}$) *Faulkner* 750; 6, leaf surface (× 24) *Kirk* s.n.

Mozambique. M: Memba-Geba, fl. & fr. 14.x.1948, *Barbosa* 2396 (LISC; PRE; SRGH).
Also in Tanganyika. Woodland.

An Asiatic species in a closely related genus, *Firmiana simplex* (L.) W. F. Wight, was in cultivation in the Fort Victoria area of S. Rhodesia about 1909. It also has papery follicles, but the leaves are tomentellous below and palmatilobed.

12. HERITIERA Ait.

Heritiera Ait., Hort. Kew. **3**: 546 (1789).

Evergreen trees with lepidote branchlets and petiolate simple coriaceous leaves; stipules present. Flowers dioecious, in axillary many-flowered panicles; bracts and bracteoles present. Calyx campanulate, 5-lobed. Petals absent. Male flowers: androphore with 5 anthers at its apex and with a vestigial style protruding through it. Female flowers: perianth as in the male; ovary of 5 connate carpels; styles connate, short; stigmas 5, thick. Ripe carpels woody, indehiscent, keeled along the back, 1-seeded. Seeds without endosperm; cotyledons very thick.

Heritiera littoralis Ait., Hort. Kew. **3**: 546 (1789).—Mast. in Oliv., F.T.A. **1**: 225 (1868).—K. Schum. in Engl., Mon. Afr. Pflanz. **5**: 136, t. 10 fig. C (1900).—Sim, For. Port. E. Afr.: 19, t. 14 (1909).—Brenan, T.T.C.L.: 599 (1949). TAB. **106**. Type from Ceylon.

Evergreen tree up to c. 16 m. tall; young branches lepidote, but soon glabrescent. Leaf-lamina c. 9–30 × 4–15 cm., very coriaceous, elliptic to ovate- or elliptic-oblong, apex acute or subobtuse, margin entire or undulate, base rounded or subcordate, green and glabrous above, with silvery scales covering the under surface and with scattered brown scales among them; petiole up to 2 cm. long, lepidote; stipules 5–6 mm. long, very caducous, lanceolate-subulate, lepidote. Flowers yellowish-green, in much-branched stellate-tomentose panicles c. 5 cm. long in the axils of the upper leaves; pedicels up to 5 mm. long, articulated below the calyx; bracts and bracteoles up to 2 mm. long, ovate, acute, tomentose. Male flowers: calyx 5–6 mm. long, campanulate, tomentose, divided to about ⅓ of the way into 5 ovate acute lobes; androphore c. 1 mm. long, dilated at the base and surrounded by a minutely glandular cushion-like disk, with an apical ring of 5 anthers in a single series; vestigial style produced a little way through the centre of the ring. Female flowers: as in the male but slightly larger and with 5 carpels connate into a broadly ovoid glabrous ovary tapering gradually to the styles, surrounded at the base by 5 free rudimentary stamens; styles 0·5–1 mm. long, connate; stigmas recurved. Ripe carpels 1–4, 6–8 × 3–4 cm., brown and shining, oblong-ovoid, strongly keeled along one side, 1-seeded. Seeds c. 3 × 2 × 1 cm., flattened, oblong-ellipsoid, brown.

Mozambique. Z: Kongone, fl. 6.vi.1868, *Kirk* (K). MS: R. Savane, fr. ix.1931, *Honey* 661 (K; LM; PRE). SS: Morrumbene-Yangamo, st. 18.xi.1941, *Torre* 3853 (K; LISC; SRGH).
Also on the coasts of Tanganyika and widely distributed along the coasts of tropical Asia, Australia and the islands of the Pacific. In mangrove swamps on their landward side.

The follicles float and are widely distributed by sea currents. The trunks are used for the masts of dhows in E. Africa.

ADDITIONS AND CORRECTIONS TO PART 1

p. 83. Under 5. **Encephalartos villosus** (Gaertn.) Lem. insert:

Encephalartos villosus var. **umbeluziensis** (R. A. Dyer) J. Lewis, stat. nov.
  Encephalartos umbeluziensis R. A. Dyer in Fl. Pl. Afr. 28: t. 1100 (1951).
  Type: Mozambique, Lourenço Marques, R. Umbeluzi, *Key* in Nat. Herb.
  Pret. 28429 (PRE, holotype).

A variant described as less robust than *E. villosus*, with leaves that lack
spines towards the base and with smaller cones. In the northern part of
the area of *E. villosus*, i.e. near Goba and Namaacha in our area and just
over the borders in Swaziland and the Transvaal. Under trees and along
water-courses.

p. 85. The record of *Podocarpus milanjanus* from Gwaai (*Eyles* 7445) must have
been a cultivated specimen.

p. 86. After 2. **Podocarpus milanjianus** Rendle insert:

3. **Podocarpus ensiculus** Melville in Kew Bull. **1954**: 566 (1955); F.T.E.A.
   Gymnosp.: 12 (1958). Type from Tanganyika (W. Usambaras).

A long-leaved species of *Podocarpus*, like *P. milanjianus* but with pendulous relatively more slender leaves and even larger mature seeds (c. 3 cm.
long). The testa is not woody and the receptacles are narrower, hard and
non-fleshy.

  Nyasaland. N: Matipa forest, fr. 20.xii.1955, *Chapman* 260 (BM; FHO;
  SRGH).
  This species, most closely allied to the S. African *P. henkelii* Stapf, reaches its
  most southerly station in this new record for our area.
  Dr. Melville has kindly confirmed the identification.

p. 122, line 12. For " Stamens 8 " read " Stamens ∞ ".

p. 177. Under **Nymphaea capensis** Thunb. insert:

  S. Rhodesia. N: Urungwe, fl. 23.vi.1952, *Lovemore* 262 (SRGH).

p. 193. Under **Coronopus integrifolius** (DC.) Spreng. insert:

  N. Rhodesia. S: Lusitu, fl. & fr. 26.ix.1959, *Fanshawe* 5231 (BM; LISC;
  SRGH). **S. Rhodesia.** N: near Binga, fl. & fr. 6.xi.1958, *Phipps* 1362 (SRGH).
  E: Chimanimani Mts., weed on gravel near Mountain hut, fl. & fr. 26.x.1959,
  Goodier & Phipps 300 (SRGH).

p. 198. 4. **Cleome ciliata** Schumach. should be replaced by 4. **Cleome rutido-
sperma** DC., Prodr. 1: 241 (1824). Type from Sierra Leone.

p. 208. Under **Cadaba aphylla** (Thunb.) Wild insert:

  S. Rhodesia. S: Shashi-Limpopo confluence, st. 22.iii.1959, *Drummond*
  5945 (SRGH).

p. 212. Tab. 33. A.—CLADOSTEMON KIRKII. " A3, fruit (×5) " should read
" A3, fruit (×⅔) ".

p. 276. Replace **Scolopia zeyheri** (Nees) Szyszyl. by **Scolopia zeyheri** (Nees)
Harv. in Harv. & Sond., F.C. **2**: 584 (1862).

p. 308. In the synopsis of *Polygala* species (line 7 *et seq.*) the numbering of the
species in the groups should read: (a) *Species* 1–5; (b) *Species* 6–8;
(c) *Species* 9–12; (d) *Species* 13–15; (e) *Species* 16–19; (f) *Species*
20–24; (g) *Species* 25–39; (h) *Species* 40–50; (i) *Species* 51–55.

p. 331. Under 42. **Polygala schinziana** Chod. insert:

> **N. Rhodesia.** B : near Kalabo, fl. 17.xi.1959, *Drummond & Cookson* 6568 (SRGH).

p. 332. The two top lines should be transferred to the bottom of the page.

p. 335. After 50. **Polygala westii** Exell insert:

> 50a. **Polygala robsonii** Exell in Bol. Soc. Brot., Sér. 2, **34** : 93 (1960).

> Annual herb up to 30 cm. high, with white or pale violet flowers, with wing sepals 4 × 3·5 mm., differing from *P. westii* in having broader leaves, wing sepals 5-veined instead of 3-veined and a larger capsule with a broader scalloped margin.

> **Nyasaland.** C : N. of Chitala, 700 m., fl. & fr. 12.ii.1959, *Robson* 1578 (BM ; K, holotype ; SRGH).

INDEX TO BOTANICAL NAMES

STERCULIA (*contd.*)
  subviolacea, 553, **556**
  tomentosa, 553, 554
  tragacantha, 553, **556**, tab. **104** fig. C
  triphaca, 553
    var. *rivaei*, 553, 554
Sterculiacea, 549
STERCULIACEAE, 517
Streblocarpus pubescens, 222
  scandens, 219
Strombosia, 14, 18
SYMPHONIA, 379, **394**
  gabonensis, 394
  globulifera, **394**, tab. **77**
    var. *gabonensis*, 394
  oligantha, 396
  urophylla, 396
Symphyostemon strictus, 198
Synclisia delagoensis, 151
  zambesiaca, 151, 153

TALINUM, 363, **369**
  arnotii, 370, **372**, tab. **71** fig. B
  caffrum, **370**, tab. **71** fig. C
  crispatulatum, **370**, tab. **71** fig. A
  cuneifolium, 372
  portulacifolium, 370, **372**, tab. **71** fig. D
  transvaalense, 372
Taxus falcata, 85
Ternstroemia polypetala, 405
Ternstroemiaceae, 405
Ternstroemieae, 405
TETRACERA, **103**
  alnifolia, 103, **104**, tab. **7** fig. C
  boiviniana, **103**, 104, tab. **7** fig. B
  masuiana, 103, **104**, tab. **7** fig. A
  strigillosa, 104
Tetrastemma, 119
Tetratelia, 203, 205
  maculata, 202
  nationae, 205
  tenuifolia, 203, 205
    var. *maculatiflora*, 205
THALICTRUM, 89, **96**
  innitens, 96
  rhynchocarpum, **96**
  zernyi, **96**
THEACEAE, 405
THESPESIA, 420, **421**, 423
  acutiloba, **421**, tab. **82**
  garckeana, 432
  lampas, 432
  populnea, **421**
    var. *acutiloba*, 421
  rehmannii, 428
  rogersii, 432
  sp., 434
  trilobata, 432
THESPESIOPSIS, 420, **423**
  mossambicensis, **423**, tab. **83**
Thlaspi africanum, 190
  bursa-pastoris, 192
THYLACIUM, 195, **213**
  africanum, **213**, tab. **34** fig. B
  ovalifolium, 213
  querimbense, 213
  verrucosum, 213
TILIACORA, 150, 151, **153**

funifera, **155**
glycosmantha, 155
johannis, 155
pynaertii, 155
warneckei, 155
TINOSPORA, 150, 151, **159**
  caffra, 159, **161**, 163, tab. **21** fig. C
  mossambicensis, **159**
  tenera, 159, **161**, tab. **21** fig. D
Trichanthera modesta, 548
TRICLISIA, 150, 151, **155**
  sacleuxii, **156**
    var. ovalifolia, **156**
    var. *sacleuxii*, 156
  sp., 156
TRIMERIA, 262, **296**
  grandifolia, 296, 298
  rotundifolia, **296**, tab. **53** fig. A
Triphaca africana, 553
TRIPLOCHITON, 517, **549**
  zambesiacus, **549**, tab. **103** fig. A

Uapaca kirkiana, 401
Unona acutiflora, 137
  aethiopica, 136
  buchananii, 128
  ferruginea forma, 126
  obovata, 123
  oxypetala, 137
  parvifolia, 125
  sp., 123, 115
URENA, 421, **503**
  lobata, **504**
UVARIA, **106**, 114, 116
  acuminata, 107, **112**
  aethiopica, 136
  angolensis, **107**, 108, 110
    subsp. angolensis, **107**
    subsp. *guineensis*, 108
  bukobensis, 107, 108
  caffra, 107, **111**, tab. **8** fig. B
  divaricata, 115
  edulis, 107, **108**
  fruticosa, 110
  gazensis, 108
  gracilipes, 107, **111**, tab. **8** fig. C
  hexaloboides, 114
  holstii, 112
  huillensis, 116
  kirkii, 107, **110**
  leptocladon, 112
    var. *holstii*, 112
  lucida, 110
  monopetala, 116
  nyassensis, 107, 108
  pecoensis, 110
  pulchra, 108
  scheffleri, 111
  schelei, 110
  smithii, 110
  sp. nova ?, 110
  stuhlmannii, 110
  sect. *Uvariodendron*, 114
  valvata, 111
  variabilis, 107
  versicolor, 110
  virens, 107, **108**, 111, tab. **8** fig. A
  welwitschii, 107, **110**